Modes

Modes

Anna B. Romanowska
Faculty of Mathematics and Information Sciences
Warsaw University of Technology, Poland

Jonathan D. H. Smith
Department of Mathematics
Iowa State University, USA

World Scientific
New Jersey • London • Singapore • Hong Kong

Published by

World Scientific Publishing Co. Pte. Ltd.
P O Box 128, Farrer Road, Singapore 912805
USA office: Suite 1B, 1060 Main Street, River Edge, NJ 07661
UK office: 57 Shelton Street, Covent Garden, London WC2H 9HE

Library of Congress Cataloging-in-Publication Data
Romanowska, A. B. (Anna B.)
  Modes / Anna B. Romanowska, Jonathan D.H. Smith
  p. cm.
  Includes bibliographical references and indexes.
  ISBN 981024942X (alk. paper)
  1. Geometry, Algebraic. 2. Moduli theory. I. Smith, Jonathan D. H., 1949–   II. Title.

QA564 .R584 2002
516.3'5--dc21   2002024968

British Library Cataloguing-in-Publication Data
A catalogue record for this book is available from the British Library.

Copyright © 2002 by World Scientific Publishing Co. Pte. Ltd.

All rights reserved. This book, or parts thereof, may not be reproduced in any form or by any means, electronic or mechanical, including photocopying, recording or any information storage and retrieval system now known or to be invented, without written permission from the Publisher.

For photocopying of material in this volume, please pay a copying fee through the Copyright Clearance Center, Inc., 222 Rosewood Drive, Danvers, MA 01923, USA. In this case permission to photocopy is not required from the publisher.

This book is printed on acid-free paper.

Printed in Singapore by Uto-Print

PREFACE

This book is an introduction to the theory and application of modes. Modes are algebras that are idempotent (i.e. each element forms a subalgebra) and entropic (i.e. each operation is a homomorphism). Modes appear in many different branches of mathematics, for instance in affine geometry, convex analysis, linear programming, differential geometry, combinatorics, and program semantics. They find application in computer science, economics, physics, and biology. Examples of modes include semilattices and more general normal bands, many interesting classes of groupoids appearing in combinatorics and geometry, convex sets and their generalization to barycentric algebras, affine spaces under affine combinations, and various reducts of such algebras.

Modes were studied under several guises by many authors since the late nineteen-thirties, and have attracted much attention over recent years. For a long time, many results were scattered throughout the literature, and were often obtained independently of the work of others in the field. The first systematic approach to modes (and the closely related algebras known as modals) was presented in our 1985 monograph "Modal Theory, an Algebraic Approach to Order, Geometry and Convexity," published by Heldermann, Berlin. We will refer to this book as [MT] throughout the present volume. The subsequent development of modal theory has proceeded in many directions, essentially enlarging almost all the areas covered in [MT], and bringing many new applications. Surveys have appeared in [Romanowska 1992] and [Smith 1999]. This new book is intended as an introduction to the current state of the theory of modes. Bearing in mind that [MT] was formulated in the condensed style of a research monograph, we decided that the present volume should begin at a gentler pace, more readily accessible to non-specialists and beginning graduate students. There is a gradual progression to the more demanding and "sophisticated" latter chapters. Some of the material from [MT] is reprised here, albeit in different form, but the emphasis is on topics that were not covered in [MT].

Readers are assumed to have a solid grounding in undergraduate mathematics, especially in the classical algebraic structures: groups, rings, fields, and vector spaces. Following a brief summary of some preliminaries, we begin with a chapter discussing the basic concepts of general algebra, illustrated by descriptions of some of the non-classical algebraic structures to be studied in more detail later. The next chapter reviews the basic tools of category theory that will be required. Chapter 3 covers the fundamentals of the theory of varieties, prevarieties, and quasivarieties. Taken together, the first three chapters should provide the basis for a full understanding of the subsequent material, but we will occasionally refer the reader to other sources, especially to our monograph "Post-Modern Algebra" (Wiley, 1999), cited throughout as [PMA].

Chapter 4 provides a general theory for the construction of new algebras as "sums" of component algebras, usually better known algebras of the same type, and indexed by elements of another algebra. These sums provide an excellent method for representing modes in certain quasivarieties. The true introduction to modes starts in Chapter 5, where we discuss the basic properties and basic examples. Later chapters proceed to more advanced topics. After a brief introduction to general Mal'cev varieties, Chapter 6 studies Mal'cev modes. The main theorem of this chapter demonstrates the equivalence between Mal'cev modes and affine spaces. Subsequent sections are devoted to certain classes of groupoids and quasigroups equivalent to affine spaces. Chapter 7 discusses subreducts of affine spaces, and their sums. The general results are applied to describe the structure of barycentric algebras and commutative groupoid modes, and to classify the quasivarieties of barycentric algebras. The main theorems show that each cancellative mode embeds as a subreduct into an affine space, and more generally that certain sums of cancellative modes embed as subreducts into functorial sums of affine spaces. Chapter 8 studies the algebraic properties of binary or groupoid modes. It describes the free binary modes, characterizes the simple objects, and discusses the classification of varieties. Binary modes provide a prototype for more general modes. Until one has reached a full understanding of the intricacies of binary modes, it will be hard to proceed to an effective analysis of the general case.

The ninth chapter is devoted to one sample application of the theory of modes, namely the development of hierarchical statistical mechanics. The hierarchy here is understood in the truly qualitative sense, not just in the quantitative sense where the levels of the hierarchy are differentiated only by orders of magnitude of comparable quantities. Traditional statistical mechanics is ultimately founded on convex sets. The hierarchical statistical mechanics discussed in Chapter 9 is founded on barycentric algebras. Modal theory shows how barycentric algebras decompose as sums of convex sets indexed by a semilattice, the semilattice replica of the barycentric algebra. This decomposition is the key to the whole application. The semilattice replica of a barycentric algebra describes the structure of the underlying hierarchy of a complex system.

The book concludes with Chapter 10, discussing further topics, recent developments, and open problems that are the subject of current research. As mentioned earlier, much of the work on modes is scattered throughout the literature. It is not always easy to find, and not always readily accessible. At the end of each chapter from the fourth to the eighth, we have tried to collate at least the most essential references in a series of "historical" notes. Nevertheless, these notes are by no means intended to be definitive or comprehensive. The general bibliography appears at the end of the book, but again we make no claim to completeness. References to the bibliography are by name of author(s) and year of publication (with additional letters where necessary to distinguish multiple citations of the same author(s) in the same year). An internal reference such as "Theorem 6.2.3" denotes Theorem 2.3 of Chapter 6. Occasionally, references such as "Proposition 6.4.4.2" are necessary, denoting part of the discussion of Example 6.4.4 apppearing in the fourth section of Chapter 6.

A wide range of exercises is offered throughout most chapters of the book, usually placed at the end of each section and indexed alphabetically. Some exercises are simply designed to familiarize readers with the notation and concepts appearing. Others are more difficult, extending the content of the sections in which they appear, and intended to give a foretaste of further research. Some of the material in the book has been used in a one-semester graduate-level course at Warsaw University of Technology. We are grateful to our colleague, Barbara Roszkowska, and to our students, particularly Katarzyna Matczak, Krzystof Pszczoła, Michał Stronkowski, and Anna Zamojska, for correcting various mistakes in earlier versions of the text. The mathematics departments of Warsaw University of Technology and Iowa State University contributed substantially towards the timely completion of this book by granting the authors the leave and facilities necessary for its production. In particular, Anna Romanowska was partially supported by a series of Warsaw University of Technology Statutory Grants.[1] It is also a great pleasure to acknowledge the contributions of Ruth DeBoer, Krzysztof Pszczoła and Agnieszka Świątkiewicz in the preparation of the typescript and bibliography, and to thank E.H. Chionh of World Scientific for her fine editorial work that brought the project to fruition.

---

[1]Numbers 504/052/0178, 504G/1172/1120/0004/000, 504G/1120/0015/000, and 504G/1120/0027/000.

CONTENTS

PREFACE v

0   PRELIMINARIES 1

I   ALGEBRAS 9
1.1   Algebras and subalgebras 9
1.2   Homomorphisms and congruences 20
1.3   Direct products 30
1.4   Word algebras 35
1.5   Identities and equationally definable classes 44
1.6   Further examples of algebras 54
1.7   Quasi-identities and equational implications 69

II CATEGORIES OF ALGEBRAS 77
2.1   Categories and subcategories 78
2.2   Functors 83
2.3   Some categorical constructions 93

III VARIETIES, PREVARIETIES AND QUASIVARIETIES 105
3.1   Subdirect product representation 106
3.2   Relative subdirect product representation 112
3.3   Replicas and free algebras 121
3.4   Prevarieties 135
3.5   Varieties 141
3.6   Quasivarieties and directed colimits 153
3.7   Quasivarieties and reduced products 164

IV CONSTRUCTING ALGEBRAS AND QUASIVARIETIES 181
4.1   Algebraic quasi-ordering of an algebra 182
4.2   Functorial sums 187
4.3   Płonka sums and regularizations 192

4.4	Płonka sums and quasiregularizations	203
4.5	Non-functorial sums	213

V INTRODUCTION TO MODES ... 235
- **5.1** Modes of submodes ... 241
- **5.2** Semigroup and diagonal modes ... 244
- **5.3** Affine spaces ... 247
- **5.4** Quasigroup modes ... 252
- **5.5** Numbers and binary modes ... 256
- **5.6** Differential groups and groupoids ... 268
- **5.7** Multiply differential groups and groupoids ... 281
- **5.8** Convex sets and barycentric algebras ... 289
- **5.9** Free modes on two generators ... 298

VI MAL'CEV MODES AND AFFINE SPACES ... 311
- **6.1** Mal'cev varieties ... 311
- **6.2** Mal'cev algebras and entropic laws ... 317
- **6.3** Mal'cev modes and affine spaces ... 325
- **6.4** Affine spaces equivalent to groupoids ... 336
- **6.5** Affine spaces equivalent to quasigroups ... 352
- **6.6** Commutative quasigroup modes ... 355

VII SUBREDUCTS OF AFFINE SPACES ... 365
- **7.1** Tensor products of varieties and affinization ... 366
- **7.2** Subreducts of Mal'cev modes ... 372
- **7.3** Subreducts of Płonka sums of Mal'cev modes ... 381
- **7.4** Subreducts of functorial sums of cancellative modes ... 385
- **7.5** Barycentric algebras and commutative binary modes ... 391
- **7.6** Varieties and quasivarieties of barycentric algebras ... 403
- **7.7** Embedding cancellative modes into affine spaces ... 415
- **7.8** Embedding sums of cancellative modes ... 427

VIII BINARY MODES ... 439
- **8.1** Standard form for binary mode words ... 440
- **8.2** Generators of the variety of binary modes ... 451
- **8.3** Simple binary modes and minimal varieties ... 457
- **8.4** Reductive modes ... 466
- **8.5** Reductive and affine binary mode varieties ... 480
- **8.6** Beyond reductive and affine binary modes ... 488

IX HIERARCHICAL STATISTICAL MECHANICS ... 505
- **9.1** Plan of the chapter ... 506
- **9.2** Barycentric operations in sum notation ... 509
- **9.3** The predual of a barycentric algebra ... 509

9.4	Concavity and logarithmic convexity	511
9.5	The Structure Theorem for barycentric algebras	513
9.6	Canonical distributions and entropy	517
9.7	Properties of the partition function	522
9.8	Canonical separability and the Legendre transform	524
9.9	Elementary examples	528
9.10	Competition between species	531
9.11	Competition with mutability	533

X RECENT DEVELOPMENTS AND OPEN PROBLEMS 539

10.1	The structure of binary and general modes	539
10.2	Semi-affine spaces	542
10.3	Simple and subdirectly irreducible modes	545
10.4	More about varieties and quasivarieties of binary modes	547
10.5	Modes of submodes	550
10.6	Modals	553
10.7	Dualities and equivalences	557

BIBLIOGRAPHY 563

INDEX OF SYMBOLS 607

INDEX 613

CHAPTER 0

PRELIMINARIES

This chapter presents basic notation, terminology and background information.

0.1 Sets. Our approach to set theory is informal. We use both classes and sets, and the naive theory of sets and classes is sufficient for our purposes. Standard set-theoretical notation is used to denote the *membership* $a \in A$ of an element a in a set A, the empty set $\varnothing$, the *inclusion* $A \subseteq B$ of a set A in a set B, set construction notation $\{- \mid -\}$, the *union* or *sum* $A_1 \cup \cdots \cup A_n$ or $\bigcup_{i=1}^{n} A_i$ of sets A_i, the *intersection* $A_1 \cap \cdots \cap A_n$ or $\bigcap_{i=1}^{n} A_i$ of sets A_i, the *difference* $A - B$, and the *cartesian product* $A_1 \times \cdots \times A_n$ of sets A_i. In particular, $A^n := A \times \cdots \times A = \{(a_1, \ldots, a_n) \mid a_i \in A\}$ is the set of n-tuples of elements of A. The cartesian product A^0 is defined to be $\{\varnothing\}$. The *power set* $\mathcal{P}(A)$ of a set A is the set of all subsets of A. We also consider the *disjoint union* $A_1 \sqcup \cdots \sqcup A_n$ or $\bigsqcup_{i \in I} A_i$ of sets A_i, which may be realized as $\bigcup_{i \in I} (A_i \times \{i\})$. The symbols $\mathbb{N}, \mathbb{Z}, \mathbb{Z}^+, \mathbb{Q}, \mathbb{R}$ and $\mathbb{C}$ respectively denote the sets of natural numbers (including 0), integers, positive integers, rationals, reals and complex numbers.

0.2 Relations. A subset of a cartesian product $A_1 \times \cdots \times A_n$ is called a *relation* between the sets A_i. An *n-ary relation* on a set A is a subset of $A^n = A \times \cdots \times A$. A 2-ary relation is called *binary*. We sometimes write $a \rho b$ to denote $(a, b) \in \rho$ for a binary relation ρ. The *converse* ρ^{-1} of a binary relation ρ on A is given by $(a, b) \in \rho^{-1}$ iff (i.e. if and only if) $(b, a) \in \rho$. A binary relation ρ is:

(a) *reflexive* if $(a,a) \in \rho$ for each $a \in A$;
(b) *symmetric* if $(a,b) \in \rho$ implies $(b,a) \in \rho$;
(c) *transitive* if $(a,b) \in \rho$ and $(b,c) \in \rho$ imply $(a,c) \in \rho$;
(d) *antisymmetric* if $(a,b) \in \rho$ and $(b,a) \in \rho$ imply $a = b$.

The *relational product* $\rho \circ \sigma$ of two binary relations ρ and σ on A is given by $(a,b) \in \rho \circ \sigma$ iff for some c in A, one has $(a,c) \in \rho$ and $(c,b) \in \sigma$. For a binary relation ρ on a set A, the relation $\rho \cup (\rho \circ \rho) \cup (\rho \circ \rho \circ \rho) \cup \ldots$ is the smallest transitive relation on A containing ρ. It is called the *transitive closure* of A.

0.3 Functions (mappings, maps). A *function* $f : A \to B$ from a set A to a set B is a relation between A and B such that for each a in A there is exactly one $b \in B$ with $(a,b) \in f$. In this case we write $b = af$ or $b = f(a)$ or $b = a^f$ or $f : a \mapsto b$. We usually specify such a function f as follows:

$$f : A \to B; \ a \mapsto af$$

(or $f : A \to B; \ a \mapsto f(a)$).

The set of all functions from A to B is denoted by B^A or $\underline{\mathrm{Set}}(A,B)$. The *identity function* 1_A on A is defined by $1_A : A \to A; \ a \mapsto a$. For $A \subseteq B$, the *inclusion function* is $\iota : A \to B; \ a \mapsto a$. A function $f : A \to B$ is *one-to-one* or *injective* or an *embedding*, if $af = bf$ implies $a = b$. It is *onto* or *surjective* if for each $b \in B$ there is an $a \in A$ with $af = b$. (Sometimes one uses the notation $f : A \rightarrowtail B$ and $f : A \twoheadrightarrow B$ for injections and surjections respectively.) A function $f : A \to B$ with non-empty domain A is injective iff it has a *right inverse* $f' : B \to A$, i.e. $ff' = 1_A$. A function $f : A \to B$ is surjective iff it has a left inverse $f' : B \to A$, i.e. $f'f = 1_B$. A function $f : A \to B$ is *bijective* or an *isomorphism* iff it is injective and surjective, i.e. if it has a (necessarily unique) *inverse* $f^{-1} : B \to A$ with $ff^{-1} = 1_A$ and $f^{-1}f = 1_B$. The set of all bijections from A to A is denoted by $A!$. If $A' \subseteq A$ and $B' \subseteq B$, then we write

$$A'f = f(A') = \{af \mid a \in A'\} \text{ and}$$
$$B'f^{-1} = f^{-1}(B') = \{a \in A \mid af \in B'\}$$

for the image and inverse image of A' and B', respectively. A *family* of sets A_i indexed by a set I is a function consisting of all pairs (i, A_i) for $i \in I$. The *cartesian product* $\prod_{i \in I} A_i$ of a family A_i, for $i \in I$, is the set

$$\prod_{i \in I} A_i = \{f : I \to \bigcup_{i \in I} A_i \mid if \in A_i\}.$$

For each $i \in I$, the mapping $\pi_i : \prod_{i \in I} A_i \to A_i$; $f \mapsto if$ is called the i-th *projection function*. If for each $i \in I$ one has $A_i = A$, then $\prod_{i \in I} A_i$ is called a *cartesian power* of A and is written in the form A^I (consistent with the earlier notation for sets of functions between given sets). If one of the factors A_i is empty, then the cartesian product $\prod A_i$ is empty. If I is empty, then the cartesian product $\prod A_i$ consists of precisely one element, the empty function. The *kernel* $\ker f$ of a function $f : A \to B$ is the binary relation $\{(a, b) \in A^2 \mid af = bf\}$.

For functions $f : A \to B$ and $g : B \to C$, their relational product fg is called the *composition* of f and g. In particular, one has $fg : A \to B \to C$; $a \mapsto af \mapsto afg$.

A diagram of sets and functions *commutes* or is *commutative* if composites along pairs of paths with common origin and destination agree. For example, commutativity of the diagram below means $f = gh$:

$$\begin{array}{ccc} A & \xrightarrow{f} & B \\ g \downarrow & & \parallel \\ C & \xrightarrow{h} & B. \end{array}$$

0.4 Equivalence relations. A binary relation ρ on a set A is an *equivalence relation* if it is reflexive, symmetric and transitive. For an equivalence relation ρ on a set A, the *equivalence class* of an element a of A under ρ is the subset

$$a^\rho := \{b \in A \mid a\rho b\}$$

of A. (Note that many writers use notations such as a/ρ or $[a]\rho$ in place of a^ρ.) The set of equivalence classes

$$A^\rho := \{a^\rho \mid a \in A\}$$

is called the *quotient* of the set by the equivalence relation ρ (leading some writers to use the notation A/ρ). The set A^ρ can be considered as a *partition* of A, i.e. $A = \bigcup \{a^\rho \mid a \in A\}$ with $a^\rho \cap b^\rho = \emptyset$ for $a^\rho \neq b^\rho$. The map

$$\operatorname{nat}\rho : A \to A^\rho; \ a \mapsto a^\rho$$

is called the *(natural) projection* of A onto A^ρ. The smallest equivalence relation on A is the *diagonal*

$$\widehat{A} = \{(a,a) \mid a \in A\},$$

which represents the equality relation. The largest is the direct square of A, the *universal* relation $A^2 = \{(a,b) \mid a,b \in A\}$.

First Isomorphism Theorem 0.4.1. *Let $f : A \to B$ be a function. Then the mapping $i : A^{\ker f} \to Af;\ a^{\ker f} \mapsto af$ is an isomorphism. Moreover, the function f factorizes as $f = (\operatorname{nat}\ker f)i\iota$, where ι is the inclusion of Af in B, i.e. the following diagram commutes:*

$$\begin{array}{ccc} A & \xrightarrow{f} & B \\ {\scriptstyle \operatorname{nat}\ker f}\downarrow & & \uparrow{\scriptstyle \iota} \\ A^{\ker f} & \xrightarrow[i]{} & Af \end{array} \quad . \quad \square$$

0.5 Order relations. A binary relation $\leq$ on a set A is an *order relation* (or a *partial order*) if it is reflexive, transitive and antisymmetric. The set A together with such an order is often denoted by $(A, \leq)$ and called an *ordered set* or a *partially ordered set* or (by an acronym) a *poset*. An ordering $\leq$ on a set A induces an ordering $\leq \cap\, B^2$ on each subset B of A. An ordering $\leq$ on a set C is said to be *total* or *linear* if for all a, b in C, one has $a \leq b$ or $b \leq a$. A linearly ordered set $(C, \leq)$ is also called a *chain*, or a *flag*. An element u of a poset $(A, \leq)$ is an *upper bound* for a subset S of A if $s \leq u$ for all s in S. If the whole set A has an upper bound, then this upper bound is uniquely defined and is sometimes called the *greatest* element of $(A, \leq)$. An element M of a poset $(A, \leq)$ is said to be *maximal* if $M \leq a \in A$ implies $M = a$. Lower bounds and minimal elements of $(A, \leq)$ are defined dually. In particular, an element ℓ in $(A, \leq)$ is a *lower bound* of S if $\ell \leq s$ for all s in S. If the set A has a lower bound, then it is uniquely defined and is called the *least element* of $(A, \leq)$. An element m of a poset $(A, \leq)$ is *minimal* if $a \leq m$ implies $a = m$. An ordered set $(A, \leq)$ in which each chain has an upper bound is called *inductive*.

Kuratowski-Zorn Lemma 0.5.1. *Each inductive poset $(A, \leq)$ has a maximal element. More precisely, for each a in A there is a maximal element M_a with $a \leq M_a$.* □

A *well-ordered set* is a chain $(A, \leq)$ such that each non-empty subset B of A has a least element. An ordered set $(A, \leq)$ is *directed* if each two-element subset of A has an upper bound. For elements a, b of an ordered set $(A, \leq)$, one says that a is *covered by* b, or that b *covers* a, if $a < b$ (i.e. $a \leq b$ and $a \neq b$) and $a \leq c < b$ implies $a = c$. One may represent a (finite) poset $(A, \leq)$ by its so-called *Hasse diagram*, a directed graph having elements of A as vertices and a (unique) arrow from an element a to an element $b \neq a$ precisely if b covers a. The symbol $a \| b$ means that the elements a and b of a poset $(A, \leq)$ are incomparable. We also use the following symbols for intervals of a poset

$(A, \leq)$:

$$[a, b] := \{x \in A \mid a \leq x \leq b\}, \ (a, b] := \{x \in A \mid a < x \leq b\},$$
$$[a, b) := \{x \in A \mid a \leq x < b\} \text{ and } (a, b) := \{x \in A \mid a < x < b\}.$$

A subset L of an ordered set $(A, \leq)$ is said to be a *lower set* or a *down-set* if $a \leq l \in L \Rightarrow a \in L$. The dual concepts are called *upper sets* or *up-sets*.

For two ordered sets $(A, \leq)$ and $(B, \leq)$, a mapping $h : A \to B$ is *order-preserving* (or *monotone*) if $a \leq a'$ in A implies $ah \leq a'h$ in B. The mapping is *order-reversing* (or *antitone*) if $a \leq a'$ implies $a'h \leq ah$. The mapping is an *isomorphism of ordered sets* if it bijects, and both h and h^{-1} are order-preserving. Finally, it is a *dual isomorphism* if it bijects, and both h and h^{-1} are order-reversing.

0.6 Quasi-order relations. A binary relation $\preceq$ on a set A is a *quasi-order* (or a *pre-order*) relation if it is reflexive and transitive. The set A together with such a quasi-order is denoted $(A, \preceq)$, and is called a *quasi-ordered* (or *pre-ordered*) set. A quasi-order $\preceq$ on a set A is *full* if $a \preceq b$ and $b \preceq a$ for all a and b in A. A quasi-ordered set $(A, \preceq)$ is *directed* if for each pair a and b of elements of A, there is c in A with $a \preceq c$ and $b \preceq c$.

Theorem 0.6.1. *Let $(A, \preceq)$ be a quasi-ordered set. Let α be the binary relation on A defined by:*

$$(a, b) \in \alpha \text{ iff } a \preceq b \text{ and } b \preceq a.$$

Then the relation α is an equivalence relation. The relation $\leq$ on the set A^α of equivalence classes of α defined by

$$a^\alpha \leq b^\alpha \text{ iff } a \preceq b$$

is an order relation. □

In other words, a quasi-ordered set is a disjoint sum of (fully quasi-ordered) equivalence classes of α, and the set of equivalence classes is an ordered set.

0.7 Cardinals. The sets A and B are *equipollent* if there is a bijection from A onto B. The equipollence of sets is reflexive, symmetric and transitive. The equivalence classes (that are proper classes) are called *cardinal numbers* or *cardinals*. For any set A, the *cardinality* or *power* $|A|$ of A is the equivalence class containing A. The equivalence class containing an n-element set is denoted by n, and is usually identified with the set $\{0, 1, \ldots, n-1\}$. The power set $\mathcal{P}(A)$ of a set A has the same cardinality as 2^A. For a pairwise disjoint family of sets A_i, where $i \in I$ and $|A_i| = m_i$, one has $\sum_{i \in I} m_i = |\bigcup_{i \in I} A_i|$ and $\prod_{i \in I} m_i = |\prod_{i \in I} A_i|$. If $|A| = m$ and $|B| = n$, and there is an injection $\varphi : A \to B$, then one writes $m \leq n$. If n is less than the cardinality of the set of natural numbers, then n is a *finite cardinal*. Otherwise, it is *infinite*. The set of all cardinals is well-ordered with respect to the relation $\leq$.

A *multiset* is a set in which each element occurs with a finite multiplicity that may be greater than 1. For example, the multisets $\langle 1, 1, 3, 2, 3, 3 \rangle$ and $\langle 1, 3, 2 \rangle$ are considered different. If X is the set of elements of a multiset $M(X)$ in which each element x occurs with multiplicity m_x, then $M(X)$ can be considered as the disjoint union of simple multisets $m(x) = \langle x, \ldots, x \rangle$, each consisting of all occurrences of x in $M(X)$. The *cardinality* of $M(X)$ is then the sum of the multiplicities m_x of all the elements x in X.

For a given set X, the multisets defined on subsets of X form an ordered set. For $A, B \subseteq X$, one defines $M(A) \subseteq M(B)$ iff $A \subseteq B$ and for each a in A, one has $|m(a)|$ in $M(A)$ not greater than $|m(a)|$ in $M(B)$. By convention, a set may be considered as a multiset in which each element has multiplicity 1.

0.8 Ordinals. Two well-ordered sets $(A, \leq)$ and $(B, \leq)$ have the same *order type* if they are isomorphic. The equivalence classes obtained this way are called *ordinals*. The equivalence class containing the empty set is denoted by 0. The equivalence class containing the chain $0 < 1 < \cdots < n-1$ is denoted by n. The order type of the chain $0 < 1 < \cdots < n < \ldots$ is usually denoted by ω.

If the order type of $(A, \leq)$ is α, the order type of $(B, \leq)$ is β and $\varphi : A \to B$ is a monotone injective mapping, then $\alpha \leq \beta$. For any set A of ordinals, $(A, \leq)$ is a well-ordered set and the type of $(\{\gamma \mid \gamma < \alpha\}, \leq)$ is α. Note that in many axiomatic set theories, an ordinal equals the set of smaller ordinals. In this approach $\alpha \leq \beta$ whenever $\alpha \subseteq \beta$. Any ordinal α has one of three forms: either $\alpha = 0$, or $\alpha = \beta + 1$ for some ordinal β, or α is the least upper bound of all $\beta < \alpha$. In this last case α is called a *limit ordinal*. A limit ordinal α can also be written as $\alpha = \bigcup \{\beta \mid \beta < \alpha\}$. The Principle of Mathematical Induction is then generalized as follows.

Principle of Transfinite Induction 0.8.1. Let the statement $\varphi(\alpha)$ be defined for all ordinals α. Assume that

if $\varphi(\beta)$ holds for all $\beta < \alpha$, then $\varphi(\alpha)$ holds.

Then $\varphi(\alpha)$ holds for all ordinals α. □

The condition of Theorem 0.8.1 is usually replaced by the following three:

(a) $\varphi(0)$ holds;

(b) If $\varphi(\beta)$ holds, then $\varphi(\beta + 1)$ holds;

(c) If $\alpha = \bigcup \{\beta \mid \beta < \alpha\}$ and $\varphi(\beta)$ holds for $\beta < \alpha$, then $\varphi(\alpha)$ holds.

0.9 Logical notation. We use the following abbreviations:

For universal and existential quantifiers:

$\forall x \in X, \varphi(x)$ denotes that every x satisfies the proposition $\varphi(x)$;

$\exists x \in X, \varphi(x)$ denotes that there exists some x which satisfies $\varphi(x)$.

The symbols $\Leftrightarrow$ and $\Rightarrow$ denote an equivalence and an implication. To define new objects we frequently use the symbols $:\Leftrightarrow$ and $:=$. If an expression equals a if α holds and otherwise equals b, then we write

$$\text{if} \quad \alpha \quad \text{then} \quad a \quad \text{else} \quad b.$$

CHAPTER I

ALGEBRAS

This chapter introduces basic algebraic concepts, basic constructions of algebras, and mappings preserving algebraic structures. The central idea is the concept of an algebra. It is accompanied by such notions as subalgebras, direct products, homomorphisms, congruence relations, reducts and subreducts, clones of operations, and identities satisfied in an algebra. These notions lie at the heart of universal algebra, but encompass also most of the well-known algebraic structures, as well as numerous lesser known algebras, in particular the algebras of prime interest for this book. We present here, and in Chapter 3, only that part of universal algebra that suits our specific needs. Some topics are covered more thoroughly in our book [PMA]. The reader who is interested in pursuing "traditional" universal algebra should consult one of the standard books [Mal'cev 1973], [Cohn 1981], [Gratzer 1979], [Burris and Sankappanavar 1981], [McKenzie, McNulty, Taylor 1987], as well as the more recent [Wechler 1992]. However, it should be noted that these books often make assumptions that are unsuitable for current purposes, such as the exclusion of empty algebras or the inclusion of fictitious variables.

1.1. Algebras and subalgebras

Let A be a set, and let n be a natural number. Then an *n-ary operation on the set A* is a function $\omega : A^n \to A$ from the n-th direct power of A to A. The number n is called the *arity* of the operation ω. For $n = 0, 1, 2, 3, 4$ respectively, the term "n-ary" becomes: *nullary, unary, binary, ternary, quaternary*. A nullary operation ω or *constant* on a (necessarily non-empty) set A is a function $\omega : \{\varnothing\} \to A$ from a singleton $\{\varnothing\}$ to A selecting the element $\varnothing\omega$ in A. For example, the identity element e of a group G is selected by the nullary operation $e : G^0 \to G : \varnothing \mapsto e$.

Given an n-ary operation $\omega : A^n \to A$ on a set A, various notations are used for elements of A^n and their images under ω. A typical element $(a_1, \ldots, a_n)$ of A^n is sometimes written as a "word" $a_1 \ldots a_n$, and its image under ω as $a_1 \ldots a_n \omega$. This form of writing is called *postfix notation*, or occasionally "reverse Polish notation," and in many situations is much more convenient than the clumsier form $\omega(a_1, \ldots, a_n)$. Then the n-ary operation ω may be described as

$$\omega : A^n \to A; \ (a_1, \ldots, a_n) \mapsto a_1 \ldots a_n \omega.$$

However, familiar operations of small arity may be written in infix, prefix or nofix notation. For example, addition and negation in an abelian group A are written as

$$+ : A^2 \to A; \ (a, b) \mapsto a + b,$$
$$- : A \to A; \ a \mapsto -a,$$

and multiplication in a semigroup A as $(a, b) \mapsto a \cdot b$ or $(a, b) \mapsto ab$.

Definition 1.1.1. A *(finitary) type* is a function $\tau : \Omega \to \mathbb{N}$. The domain Ω of the type τ is called its *operator domain*, and the elements of Ω are called *operators* (or *operation symbols*). The image multiset $\langle \omega\tau \mid \omega \in \Omega \rangle$ of τ is called the *signature* of τ. For ω in Ω, the natural number $\omega\tau$ is called the *arity* of the operator ω. □

Definition 1.1.2. Given a (finitary) type $\tau : \Omega \to \mathbb{N}$, a *$\tau$-algebra* or an *$\Omega$-algebra* or an *algebra (A, Ω) of type τ* or of *signature $\langle \omega\tau \mid \omega \in \Omega \rangle$* is defined to be a set A equipped with an operation $\omega : A^{\omega\tau} \to A$ corresponding to each operator or element ω of the domain Ω of τ. □

The operations ω determined by operators ω in Ω are referred to as *basic operations*. The type τ is often described by (or identified with) its graph $\{(\omega, \omega\tau) \in \Omega \times \mathbb{N} \mid \omega \in \Omega\}$. For instance, an abelian group $(A, +, -, 0)$ is an algebra of type $\{(+, 2), (-, 1), (0, 0)\}$. An algebra of signature $\langle 2 \rangle$, i.e. with a single binary operation, is called a *groupoid* (or *magma* or *binar*). Note, however, that the term "groupoid" is also used with quite a different meaning in category theory. (It denotes a category in which each morphism is an isomorphism.) Note also that if $\tau^{-1}\{0\}$ is nonempty, then a τ-algebra (A, Ω) cannot be empty. If not indicated otherwise, all "non-classical" algebras considered in this book have a *plural* finitary type, i.e. a type $\tau : \Omega \to \mathbb{N}$ such that $\omega\tau \geq 1$ for each ω in Ω, and for at least one ω, one has $\omega\tau > 1$. An algebra (A, Ω) consisting of precisely one element, say $A = \{a\}$, is called *trivial*. Note that in such an algebra, for each ω in Ω, one has $\omega : A^n \to A$; $(a, \ldots, a) \mapsto a$. To denote Ω-algebras, we will use the symbol (A, Ω) or, if there is no danger of confusion, we will write simply A to denote both the algebra and its set of elements. The class of all τ-algebras is denoted by $\underline{\tau}$. Algebras of the same type in a given class $\underline{K}$ are sometimes called $\underline{K}$-*algebras*.

We recall now the definitions of some "classical" algebras. Note that the different kinds of algebras listed below are distinguished from each other by their basic operations and the fact that they satisfy certain laws.

Examples

1.1.3 (Semigroups and monoids). A *semigroup* is a groupoid $(A, \cdot)$ with an associative multiplication, i.e. an algebra of type $\{(\cdot, 2)\}$ such that

(1.1.1) $$a \cdot (b \cdot c) = (a \cdot b) \cdot c \quad \text{for all } a, b, c \in A.$$

We will sometimes use the following "nofix" notation for an expression as in (1.1.1):

$$a \cdot bc = ab \cdot c.$$

A semigroup is *commutative* if the multiplication is commutative, i.e.

(1.1.2) $$ab = ba \quad \text{for all } a, b \text{ in } A.$$

A *monoid* $(A, \cdot, 1)$ is an algebra of type $\{(\cdot, 2), (1, 0)\}$ such that $(A, \cdot)$ is a semigroup and 1 is an *identity* or *neutral* element for $\cdot$, i.e.

(1.1.3) $$a1 = 1a = a \quad \text{for each } a \text{ in } A.$$

The typical example of a semigroup is given by the semigroup of all functions from a set X into X with the composition of functions as multiplication. If we consider the identity mapping on X as an identity, we get the monoid of all functions from X to X. The "simplest" semigroups are *left zero* or *left trivial* semigroups $(A, \cdot)$ such that

(1.1.4) $$ab = a \quad \text{for all } a, b \text{ in } A,$$

and *right zero* or *right trivial* semigroups $(A, \cdot)$ such that

(1.1.5) $$ab = b \quad \text{for all } a, b \text{ in } A.$$

1.1.4 (Groups). *Groups* can be defined as algebras $(A, \cdot, ^{-1}, 1)$ of type $\{(\cdot, 2), (^{-1}, 1), (1, 0)\}$ such that $(A, \cdot, 1)$ is a monoid and the operation $^{-1}$ satisfies

(1.1.6) $$a \cdot a^{-1} = a^{-1} \cdot a = 1 \quad \text{for each } a \text{ in } A.$$

A group A is *abelian* or *commutative* if the corresponding monoid is commutative. Abelian groups are usually written as $(G, +, -, 0)$, the image of the negation $-$ applied to an element a usually written in prefix notation as $-a$. Then (1.1.6) is written as

$$(1.1.7) \qquad a - a = -a + a = 0.$$

A typical example of a group is given by the group of all bijections or permutations from a set X onto X, or a group of automorphisms of a vector space.

1.1.5 (Rings). *Rings* are algebras $(A, +, -, 0, \cdot)$ such that $(A, +, -, 0)$ is an abelian group, $(A, \cdot)$ is a semigroup, and the *distributive laws*

$$(1.1.8) \qquad a \cdot (b + c) = (a \cdot b) + (a \cdot c) \text{ for all } a, b, c \text{ in } A$$

and

$$(1.1.9) \qquad (b + c) \cdot a = (b \cdot a) + (c \cdot a) \text{ for all } a, b, c \text{ in } A$$

are satisfied. We will usually write (1.1.8) and (1.1.9) in the simpler form

$$a(b + c) = ab + ac,$$
$$(b + c)a = ba + ca.$$

A *unital ring* or *ring with identity* is an algebra $(A, +, -, 0, \cdot, 1)$ such that $(A, +, -, 0, \cdot)$ is a ring and $(A, \cdot, 1)$ is a monoid. Familiar examples of rings are provided by rings of numbers with the usual operations, and by rings of matrices with entries in a given ring. A ring is *commutative* if its multiplication is commutative. Unless stated otherwise explicitly, all the rings considered in this book are commutative rings with identity. A particular rôle will be played by *integral domains* – unital commutative rings without zero divisors, *fields* – commutative rings with identity such that each non-zero element is invertible, and *polynomial rings* with commuting indeterminates.

1.1.6 (Semilattices). A *semilattice* is a commutative semigroup $(A, \cdot)$ satisfying the following *idempotent law*:

$$(1.1.10) \qquad aa = a \text{ for all } a \text{ in } A.$$

Semilattices are closely associated with ordered sets having either greatest lower bounds, or least upper bounds, of all pairs of elements. If $(A, \leq)$ has a greatest lower bound $glb\{a, b\}$ for each subset $\{a, b\}$ of A, then $glb\{a, b\} = a \cdot b$ is a semilattice operation on A, and $(A, \cdot)$ is called a *meet semilattice*. On the other hand, defining

$$(1.1.11) \qquad a \leq_{\cdot} b \Leftrightarrow a \cdot b = a$$

on a semilattice $(A, \cdot)$ yields an ordered set $(A, \leq_{\cdot})$ with greatest lower bounds given by products in the semilattice. Dually, one obtains a *join semilattice* $(A, +)$ with

$$(1.1.12) \qquad a \leq_+ b \Leftrightarrow a + b = b,$$

the "sum" operation + on the semilattice corresponding to the least upper bound in $(A, \leq_+)$. A typical example of a meet semilattice is provided by taking A to be the collection of all subsets of a given set, with the operation of intersection. An example of a join semilattice is given by the set of integers, with the operation of taking the maximum of any two numbers.

1.1.7 (Lattices). *Lattices* are algebras $(A, +, \cdot)$ such that $(A, +)$ is a join semilattice, $(A, \cdot)$ is a meet semilattice, and the following *absorption law* holds:

$$(1.1.13) \qquad a \cdot (a + b) = a + (a \cdot b) = a \text{ for all } a, b \text{ in } A,$$

usually written as

$$a(a + b) = a + ab = a.$$

In a lattice, both semilattice orderings coincide, and one has $\leq := \leq_{\cdot} = \leq_{+}$. In particular, each lattice $(A, +, \cdot)$ is an ordered set $(A, \leq)$ having both greatest lower bounds and least upper bounds of all pairs of elements. On the other hand, each ordered set with such a property is a lattice, where the "sum" or "join" operation corresponds to the least upper bound and the "product" or "meet" operation corresponds to the greatest lower bound in $(A, \leq)$. Typical examples are given by the collection of subsets of a given set with the operations of union and intersection, or by the set of integers with the operations of taking the maximum and minimum of two numbers. Occasionally, the lattice join is also denoted by $\vee$ and the lattice meet by $\wedge$.

A *complete lattice* is an ordered set $(A, \leq)$ having both greatest lower bounds and least upper bounds of all subsets of A. The greatest lower bound, infimum, or meet of a subset B of A is denoted by $\prod B$ or $\inf B$. The least upper bound, supremum, or join of B is denoted by $\sum B$ or $\sup B$. Note that each complete lattice is a lattice, and that each finite lattice is complete.

A *bounded lattice* is an algebra $(A, +, \cdot, 0, 1)$ such that $(A, +, \cdot)$ is a lattice and both $(A, +, 0)$ and $(A, \cdot, 1)$ are monoids. Note that in the ordered set $(A, \leq)$, the element 0 is the *least element* or the *lower bound* of $(A, \leq)$ and the element 1 is the *greatest element* or the *upper bound* of $(A, \leq)$. A lattice is *distributive* if it satisfies the distributive laws (1.1.8) [and hence also (1.1.9)] and

(1.1.14) $\qquad a + bc = (a+b)(a+c)$ for all a, b, c in A.

A bounded lattice is *complemented* if for each a in A, there is an element a' in A such that

$$a + a' = 1 \text{ and } aa' = 0.$$

In a distributive bounded lattice, each element has at most one complement.

1.1.8 (Boolean algebras). A *Boolean algebra* $(A, +, \cdot, ', 0, 1)$ is a bounded complemented distributive lattice. The collection of all subsets of a set X, with the operations of union, intersection, complementation, and two constants, the empty set and the set X, forms a typical example of a Boolean algebra. □

Definition 1.1.9. Let (A, Ω) be a τ-algebra, and let B be a subset of A.

(a) If for each ω in Ω,

$$(\forall 1 \leq i \leq \omega\tau, b_i \in B) \Rightarrow (b_1 \ldots b_{\omega\tau}\omega \in B),$$

then the subset B is said to form a *subalgebra* (B, Ω) of (A, Ω). This is denoted by $(B, \Omega) \leq (A, \Omega)$ or simply $B \leq A$. One also says that (A, Ω) is a *superalgebra* or an *extension* of (B, Ω).

(b) If for each ω in Ω

$$(\forall 1 \leq i \leq \omega\tau, b_i \in B) \Leftrightarrow (b_1 \ldots b_{\omega\tau}\omega \in B),$$

then the subalgebra (B, Ω) of (A, Ω) is said to be a *wall* of (A, Ω).

(c) If for each ω in Ω and all $a_1, \ldots, a_{\omega\tau}$ in A

$$(\exists 1 \leq i \leq \omega\tau \,.\, a_i \in B) \Rightarrow (a_1 \ldots a_{\omega\tau}\omega \in B),$$

then B is said to be a *sink* of (A, Ω). □

Note that a subalgebra (B, Ω) of (A, Ω) is itself a τ-algebra with

$$\omega : B^{\omega\tau} \to B; (b_1, \ldots, b_{\omega\tau}) \mapsto b_1 \ldots b_{\omega\tau}\omega$$

for each ω in Ω. In particular, a subalgebra contains each (element selected by a) nullary operation of Ω. Note also that a sink is a subalgebra. The unique element of a one-element sink of an algebra (A, Ω) is sometimes called a *zero* of the algebra. Note however that the element 0 of a ring or lattice as above is not a zero in this sense.

Example 1.1.10. Consider the closed unit interval $I := [0,1] = \{x \in \mathbb{R} \mid 0 \leq x \leq 1\}$ with the operation $\circ : I \times I \to I; (x,y) \mapsto (x+y)/2$. It is easy to see that the open unit interval $I^\circ =]0,1[= \{x \in \mathbb{R} \mid 0 < x < 1\}$ is a subalgebra of $(I, \circ)$, and is a sink. The singletons $\{0\}$ and $\{1\}$ are also subalgebras, and form walls. □

Definition 1.1.11. Let (A, Ω) be a τ-algebra.

(a) The *subalgebra poset* $Sb(A, \Omega)$ or SbA of (A, Ω) is the subset of the power set $\mathcal{P}(A)$ comprising the subalgebras of (A, Ω), with the order induced from $(\mathcal{P}(A), \subseteq)$.

(b) The *wall poset* $Wl(A, \Omega)$ or WlA of (A, Ω) is the subset of the power set $\mathcal{P}(A)$ comprising the walls of (A, Ω), with the order induced from $(\mathcal{P}(A), \subseteq)$.

(c) The *sink poset* $Sk(A, \Omega)$ or SkA of (A, Ω) is the subset of the power set $\mathcal{P}(A)$ comprising the sinks of (A, Ω), with the order induced from $(\mathcal{P}(A), \subseteq)$. □

Note that if A is a set with an empty set $\varnothing$ of operations, then $Sb(A, \varnothing) = Wl(A, \varnothing) = Sk(A, \varnothing) = \mathcal{P}(A)$.

If a singleton subset $\{e\}$ of an algebra (A, Ω) is a subalgebra $(\{e\}, \Omega)$, then the element e is said to be *idempotent*. For example, the identity of a group is the unique idempotent element in the group. An algebra (A, Ω) is said to be *idempotent* if each element of A is an idempotent of (A, Ω). For example, semilattices and lattices are idempotent algebras. Idempotent semigroups are also called *bands*. Idempotence of an algebra may be expressed by the laws

$$a \ldots a\omega = a \text{ for all } a \text{ in } A,$$

for each ω in Ω. Note that an algebra (A, Ω) with a nullary operation ω is idempotent precisely if $|A| = 1$.

Lemma 1.1.12. *Let (B_i, Ω), for $i \in I$, be a family of subalgebras of (A, Ω). Then the intersection $B = \bigcap_{l \in I} B_i$ also forms a subalgebra of (A, Ω).* □

The proof is left as Exercise 1.1A.

Since intersections of subalgebras are subalgebras, each subset X of A determines a smallest subalgebra $\langle X \rangle$ of (A, Ω) containing X, the intersection of all subalgebras of (A, Ω) containing X. It is known as the *subalgebra generated by X*. The elements of X are called *generators* of the algebra $(\langle X \rangle, \Omega)$. If the set X is finite, then the algebra $(\langle X \rangle, \Omega)$ is called *finitely generated*. If (A, Ω) is idempotent, then $\langle a \rangle = \{a\}$ for each element a of A. Note that the intersection of a family of subalgebras may be empty, as is the case with the subalgebras $(\{0\}, \circ)$ and $(\{1\}, \circ)$ of Example 1.1.10. Unions of subalgebras are not subalgebras in general. Again Example 1.1.10 provides an example. The subalgebra of (A, Ω) generated by the union of a family of subalgebras is called the *join* of these subalgebras. In fact, one can easily show the following.

Proposition 1.1.13. *For a τ-algebra (A, Ω), each of the posets SbA, WlA and SkA is a complete lattice.* □

The proof is left as Exercise 1.1D.

Now let Ψ be a subset of Ω, and let $\sigma : \Psi \to \mathbb{N}$ be the corresponding restriction of τ. Given a τ-algebra (A, Ω), one then obtains an algebra (A, Ψ) of type σ. The algebra (A, Ψ) is said to be a *(basic) reduct* or *impoverishment* of the algebra (A, Ω). One also says that (A, Ω) is an *augment* or *enrichment* of the algebra (A, Ψ). For example, a lattice $(A, +, \cdot)$ is an enrichment of its two semilattice reducts $(A, +)$ and $(A, \cdot)$. Note that reducts of a given algebra may have many more subalgebras than the original algebra. For example, $\mathbb{N}$ is a subalgebra of the semigroup reduct $(\mathbb{Z}, +)$ of the group $(\mathbb{Z}, +, -, 0)$, but not a subgroup.

Example 1.1.14. Any algebra (A, Ω) can be considered as a reduct of the algebra with the set of all possible finitary operations. To be more specific, let $\underline{\operatorname{Set}}(A^n, A)$ denote the set of all n-ary operations on the set A. Then $(A, \underline{\operatorname{Set}}(A^n, A))$ is an algebra of type $\tau : \underline{\operatorname{Set}}(A^n, A) \to \{n\}$, and

$$(A, \bigcup_{n \in \mathbb{N}} \underline{\operatorname{Set}}(A^n, A))$$

is an algebra of type $\tau_A : \bigcup_{n \in \mathbb{N}} \underline{\operatorname{Set}}(A_n, A) \to \mathbb{N}$. □

Example 1.1.15. Let $(A, \cdot, 1)$ be a monoid. For $a_1, \ldots, a_n$ in A define the product $\prod_{i=1}^{n} a_i$ inductively by

$$\prod_{i=1}^{0} a_i := 1 \text{ and } \prod_{i=1}^{r} a_i := \left(\prod_{i=1}^{r-1} a_i\right) a_r$$

for $r \leq n$. Let $w_n : A^n \to A; (a_1, \ldots, a_n) \mapsto \prod_{i=1}^{n} a_i$ for $n \in \mathbb{N}$. Then $(A, \{w_n \mid n \in \mathbb{N}\})$ is an algebra of type $\tau : w_n \mapsto n$, and the monoid $(A, \cdot, 1)$ is a reduct of $(A, \{w_n \mid n \in \mathbb{N}\})$. Given a type $\tau : \Omega \to \mathbb{N}$, one then obtains a τ-algebra (A, Ω) with Ω-operations defined by

$$\omega : A^{\omega\tau} \to A; (a_1, \ldots, a_{\omega\tau}) \mapsto \prod_{i=1}^{\omega\tau} a_i.$$

Obviously, the algebra (A, Ω) is a reduct of the algebra $(A, \{w_n \mid n \in \mathbb{N}\})$. If Ω contains nullary and binary operations, then (A, Ω) is an enrichment of the monoid $(A, \cdot, 1)$. However, as we will see later, the nature of this enrichment differs essentially from that obtained in Example 1.1.14.

In a similar way, one can obtain an Ω-algebra from a semigroup $(A, \cdot)$. Such a semigroup will be called an Ω-*semigroup*. □

Exercises

1.1A. Prove Lemma 1.1.12.

1.1B. Show that each complete lattice $(A, \leq)$ is bounded, in particular $\prod \varnothing = 1$ and $\sum \varnothing = 0$.

1.1C. Show that, for an ordered set $(A, \leq)$, the following conditions are equivalent:

(a) $(A, \leq)$ is a complete lattice;

(b) the least upper bound $\sum B$ exists for every subset B of A;

(c) the greatest lower bound exists for every subset B of A.

1.1D. Prove Proposition 1.1.13.

1.1E. Consider the closed unit interval I of Example 1.1.10. For each r in $I°$, define the binary operation $\underline{r} : I \times I \to I$; $x \mapsto x(1-r) + yr$. Let $\underline{I}° = \{\underline{r} \mid r \in I°\}$.

(a) Describe all subalgebras of the algebra $(I, \underline{I}°)$.

(b) Describe all subalgebras of the algebra $(I, \circ)$ of Example 1.1.10.

1.1F. Consider the set $\mathbb{N}$ of natural numbers.

(a) Show that $\mathbb{N}$ with its usual ordering is a distributive lattice.

(b) Show that $\mathbb{N}$ ordered by the divisibility relation is a distributive lattice. In particular, the meet of two numbers i and j is their greatest common divisor $GCD(i, j)$, while the join of i and j is their least common multiple $LCM(i, j)$.

1.2. Homomorphisms and congruences

Let (A_1, Ω) and (A_2, Ω) be τ-algebras. Then the Cartesian product $A_1 \times A_2$ of the sets A_1 and A_2 becomes an algebra $(A_1 \times A_2, \Omega)$ of the same type τ, the *direct product* of (A_1, Ω) and (A_2, Ω) with operations ω in Ω defined ("componentwise") as follows:

(1.2.1) $\qquad (a_1, b_1) \ldots (a_{\omega\tau}, b_{\omega\tau})\omega := (a_1 \ldots a_{\omega\tau}\omega, b_1 \ldots b_{\omega\tau}\omega).$

Direct products of semigroups, groups and lattices are known examples of such constructions. The mappings

(1.2.2) $\qquad\qquad \pi_i : A_1 \times A_2 \to A_i;\ (a_1, a_2) \mapsto a_i$

for $i = 1, 2$ are known as *projections on the i-th coordinate*. A function $f : A \to B$ is said to be an Ω-*homomorphism* or a τ-*homomorphism*, or just a *homomorphism*, if the set

(1.2.3) $$\{(a, b) \in A \times B \mid af = b\}$$

is a subalgebra of $(A \times B, \Omega)$. The condition (1.2.3) means that for each ω in Ω and $a_1, \ldots, a_{\omega\tau}$ in A and $b_1 = a_1 f, \ldots, b_{\omega\tau} = a_{\omega\tau} f$ in B, one has

$$(a_1, b_1) \ldots (a_{\omega\tau}, b_{\omega\tau})\omega$$
$$= (a_1 \ldots a_{\omega\tau}\omega, b_1 \ldots b_{\omega\tau}\omega)$$
$$= (a_1 \ldots a_{\omega\tau}\omega, a_1 \ldots a_{\omega\tau}\omega f).$$

In other words, (1.2.3) says that

(1.2.4) $$(a_1 \ldots a_{\omega\tau}\omega)f = (a_1 f) \ldots (a_{\omega\tau} f)\omega$$

for each ω in Ω, or that f *preserves* the operation ω. If $A = B$, then f is called an *endomorphism*. If f is one-to-one, then it is called an *embedding*. A one-to-one and onto homomorphism f is called an *isomorphism*, and if in addition $A = B$, then f is called an *automorphism*. An automorphism is *proper*, if it is not the identity map. If f is an isomorphism, then f^{-1} is also an isomorphism [Exercise 1.2A], and the algebras A and B are *isomorphic*. This is usually denoted by $A \cong B$. However, we will frequently identify isomorphic algebras to such an extent that sometimes we will even write $A = B$ for isomorphic algebras, even when this is not quite precise, if the context is clear. The symbols $\Omega(A, B)$ or $\underline{\tau}(A, B)$ denote the set of all homomorphisms from (A, Ω) to (B, Ω). Note that the identity mapping 1_A on A is a homomorphism. Homomorphisms of semigroups, groups, rings and lattices, as well as linear mappings of vector spaces, are all familiar special cases of homomorphisms as defined above. Other easy examples of homomorphisms are given by the projection maps (1.2.2) [Exercise 1.2B]. To differentiate between mappings and homomorphisms, we will sometimes write $f : (A, \Omega) \to (B, \Omega)$ for a homomorphism, leaving $f : A \to B$ for maps. Elementary properties of homomorphisms are collected in Exercise 1.2C. The reader unfamiliar with these concepts is encouraged to prove all of them carefully.

Suppose that $f : (A, \Omega) \to (B, \Omega)$ is a homomorphism between τ-algebras. Consider the equivalence relation

(1.2.5) $$\ker f = \{(a,b) \in A \times A \mid af = bf\},$$

the *kernel* of f. It follows from the defining property of homomorphisms that $\ker f$ is a subalgebra of $(A \times A, \Omega)$ [Exercise 1.2D]. In general, an equivalence relation θ on the set A of elements of an algebra (A, Ω) for which θ is a subalgebra of $(A \times A, \Omega)$ is called a *τ-congruence* or an *Ω-congruence relation*, or briefly a *congruence* on (A, Ω). It is easy to see that θ is a congruence on (A, Ω) if it satisfies the following *compatibility property* for each ω in Ω and a_i, b_i in A:

(1.2.6) $$\left.\begin{array}{r}(a_1, b_1) \in \theta \\ \vdots \\ (a_{\omega\tau}, b_{\omega\tau}) \in \theta\end{array}\right\} \Rightarrow (a_1 \ldots a_{\omega\tau}\omega, b_1 \ldots b_{\omega\tau}\omega) \in \theta.$$

Then obviously, for a τ-homomorphism $f : (A, \Omega) \to (B, \Omega)$, the relation $\ker f$ is a congruence on (A, Ω). For each a in A, define $a^\theta := \{b \in A \mid (a,b) \in \theta\}$. The set a^θ is called the *congruence class of a under θ* or briefly a *θ-class of a*. The set $\{a^\theta \mid a \in A\}$ of all congruence classes of θ is denoted by A^θ. It is again a τ-algebra under Ω-operations defined by

(1.2.7) $$a_1^\theta \ldots a_{\omega\tau}^\theta \omega := (a_1 \ldots a_{\omega\tau}\omega)^\theta$$

for ω in Ω and a_i in A. Note that the condition (1.2.6) gives a guarantee that the definition (1.2.7) is good. The algebra (A^θ, Ω) is called a *quotient* of

(A, Ω). (Note that some authors use the symbols a/θ and A/θ to denote the congruence class of θ and the set of all congruence classes of θ respectively.) A τ-algebra (A, Ω) always has the *improper* (or *universal*) Ω-congruence A^2 and the *trivial* congruence $\widehat{A} := \{(a, a) \mid a \in A\}$ that can be identified with the equality relation on A, and sometimes is also called the *diagonal* of (A, Ω). If a non-empty, non-trivial algebra (A, Ω) has no proper, non-trivial congruences, then it is said to be *simple*. A simple algebra whose only proper subalgebras are empty or singletons is described as *plain* or *strictly simple*.

Since the intersection of any set of congruence relations on (A, Ω) is a congruence relation [Exercise 1.2H], it follows that for a binary relation β on A, i.e. for a subset β of $A \times A$, there is a smallest congruence $\langle\beta\rangle_{cg}$ or $\langle\beta\rangle$ containing β, the intersection of all congruences containing β. For $\varnothing \neq X \subseteq A$, the smallest congruence containing $X \times X$ is called the *congruence generated by X* and is denoted by cgX. In particular, if X is a two-element subset $\{a, b\}$, then one writes $cg(a, b)$. Such a congruence is called *principal*.

Definition 1.2.1. Let (A, Ω) be a τ-algebra. Then the *congruence poset* $Cg(A, \Omega)$ or CgA of (A, Ω) is the subset of the power set $\mathcal{P}(A \times A)$ comprising the congruence relations of (A, Ω), with the order on this subset induced from $(\mathcal{P}(A \times A), \subseteq)$. □

Similarly as in the case of subalgebras of (A, Ω), one proves the following.

Proposition 1.2.2. *For a τ-algebra (A, Ω), the poset CgA is a complete lattice.* □

The proof is relegated to Exercise 1.2G.

Familiar examples of congruences are given by the decomposition of a group as the disjoint union of cosets of a normal subgroup or the decomposition of a ring as the disjoint union of cosets of an ideal [Exercises 1.2M, 1.2N]. Note, however, that for a group or a ring homomorphism $h: A \to B$, the symbol

$\mathrm{Ker}\, h = \{a \in A \mid ah = 0\}$ denotes the inverse image of 0, i.e. the normal subgroup or the ideal, whereas the symbol $\ker h$ denotes the corresponding congruence.

For a congruence θ on (A, Ω), the function

(1.2.8) $$\mathrm{nat}\,\theta : A \to A^\theta;\ a \mapsto a^\theta,$$

mapping an element a of A to its congruence class a^θ, is a homomorphism [Exercise 1.2E], and is called the *natural projection*. Note that $\ker\mathrm{nat}\,\theta = \theta$. On the other hand, one has the following.

First Isomorphism Theorem 1.2.3. *Let $f : (A, \Omega) \to (B, \Omega)$ be a τ-homomorphism. Then there is a unique isomorphism $i : A^{\ker f} \to Af$ such that*

(1.2.9) $$(\mathrm{nat}\,\ker f)i\iota = f,$$

where $\iota : Af \to B$ is the inclusion of the image of f. □

The proof is left as Exercise 1.2O. The condition (1.2.9) is best illustrated by the following diagram

(1.2.10)
$$\begin{array}{ccc} A & \xrightarrow{f} & B \\ \mathrm{nat}\,\ker f \downarrow & & \uparrow \iota \\ A^{\ker f} & \xrightarrow{i} & Af \end{array}$$

There are two more classical isomorphism theorems to accompany the first. Their proofs are also left as Exercises 1.2P and 1.2Q.

Second Isomorphism Theorem 1.2.4. *Let (B, Ω) be a subalgebra of a τ-algebra (A, Ω). Then for $\theta \in CgA$, one has the following.*

(a) *The set $B\theta := \{a \in A \mid \exists b \in B\,.\, a\theta b\}$ is a subalgebra of (A, Ω);*
(b) $((B\theta)^\theta, \Omega) \cong B^{\theta \cap B^2}$. □

Let $f : (A, \Omega) \to (B, \Omega)$ be a τ-homomorphism. For $\beta \in CgB$, consider the composition
$$f\,\mathrm{nat}\,\beta : A \to B \to B^\beta; \quad a \mapsto af \mapsto (af)^\beta.$$

Define
$$f^* : CgB \to CgA; \quad \beta \mapsto \ker(f\,\mathrm{nat}\,\beta).$$

In particular, for a congruence $\theta \in CgA$ and the natural projection $\mathrm{nat}\,\theta : (A, \Omega) \to (A^\theta, \Omega)$ and $\beta \in Cg(A^\theta)$, one has

$$\mathrm{nat}\,\theta\,\,\mathrm{nat}\,\beta : A \to (A^\theta)^\beta; \quad a \mapsto a^\theta \mapsto (a^\theta)^\beta$$

and
$$(\mathrm{nat}\,\theta)^* : Cg(A^\theta) \to CgA; \quad \beta \mapsto \ker(\mathrm{nat}\,\theta\,\,\mathrm{nat}\,\beta).$$

Third Isomorphism Theorem 1.2.5. *Let θ be a congruence on a τ-algebra (A, Ω). Then $(\mathrm{nat}\,\theta)^* : Cg(A^\theta) \to CgA$ induces a (lattice) isomorphism*

$$(\mathrm{nat}\,\theta)^* : Cg(A^\theta) \to [\theta, A^2]; \quad \beta \mapsto (\mathrm{nat}\,\theta)^*(\beta)$$

between $Cg(A^\theta)$ and the interval $[\theta, A^2] = \{\gamma \in CgA \mid \theta \leq \gamma \leq A^2\}$ in the poset $(CgA, \subseteq)$. Moreover, for $\beta \in Cg(A^\theta)$, one has

$$(A^\theta)^\beta \cong A^{(\mathrm{nat}\,\theta)^*(\beta)}. \quad \square$$

For a τ-algebra (A, Ω) consider the set $\mathcal{P}(A \times A)$ of binary relations β on the set A. The binary operation $\circ$ of relational product on the set $\mathcal{P}(A \times A)$ is associative and has an identity element $\widehat{A}$. Hence $(\mathcal{P}(A \times A), \circ, \widehat{A})$ is a monoid [Exercise 1.2S], known as a *(binary) relation monoid*. In general, the relational product is not commutative. A set of binary relations $\{\beta_i \mid i \in I\}$ is said to be *permutable*, or a set of *permutable relations*, if

$$\beta_i \circ \beta_j = \beta_j \circ \beta_i$$

for all i and j in I. In particular, if two equivalence relations β_1 and β_2 are permutable, then their product $\beta_1 \circ \beta_2 = \beta_2 \circ \beta_1$ is again an equivalence relation, the smallest equivalence containing both β_1 and β_2 [Exercise 1.2J]. By Exercise 1.2G, the equivalence relations on the set A form a (complete) lattice $(EqvA, \subseteq)$. In this lattice, the join of any two permutable equivalences is just their product.

Now the congruence lattice $(CgA, +, \cdot)$ of (A, Ω) is a sublattice of the equivalence lattice $(EqvA, +, \cdot)$ of the set A. In fact, one can show even more, that the complete lattice $(CqA, \subseteq)$ is a complete sublattice of $(EqvA, \subseteq)$. (See Exercise 1.2K.) One says that the algebra (A, Ω) is *congruence permutable* or that it is a *Mal'cev algebra* if every pair of congruences on (A, Ω) permutes. In a congruence permutable algebra (A, Ω), the join of any two congruences coincides with their product, and CgA is also a submonoid of the relation monoid $(\mathcal{P}(A \times A), \circ, \widehat{A})$.

Exercises

1.2A. Show that the inverse map f^{-1} of an isomorphism $f : (A, \Omega) \to (B, \Omega)$ is an isomorphism $f^{-1} : (B, \Omega) \to (A, \Omega)$.

1.2B. Show that the projection maps (1.2.2) are homomorphisms.

1.2C. Let $g : (A, \Omega) \to (B, \Omega)$ and $h : (B, \Omega) \to (C, \Omega)$ be homomorphisms of τ-algebras. Show the following:

(a) The composition $gh : (A, \Omega) \to (C, \Omega)$ is a homomorphism;

(b) If D is a subalgebra of (A, Ω), then its image Dg $(= g(D))$ is a subalgebra of (B, Ω), the *homomorphic image* of (D, Ω).

(c) If D is a subalgebra of (B, Ω), then its inverse image Dg^{-1} $(= g^{-1}(D))$ is a subalgebra of (A, Ω).

(d) If $A \subseteq B$, then the identity mapping 1_A is a homomorphism iff A is a subalgebra of (B, Ω).

(e) If X is a subset of A, then $\langle X \rangle g$ $(= g(\langle X \rangle)) = \langle g(X) \rangle) = \langle Xg \rangle$.

(f) If $A = \langle X \rangle$ and $f : (A, \Omega) \to (B, \Omega)$ is another Ω-homomorphism such that $xg = xf$ for all x in X, then $g = f$. In other words, a homomorphism $f : (A, \Omega) \to (B, \Omega)$ is specified uniquely by its restriction $f : X \to B$ to X.

1.2D. Show that for a homomorphism $f : (A, \Omega) \to (B, \Omega)$ of τ-algebras, the set ker f is a subalgebra of $(A \times A, \Omega)$.

1.2E. Show that the mapping (1.2.8) is a homomorphism.

1.2F. Show that, under composition, the endomorphisms of an algebra form a monoid, and the automorphisms form a group.

1.2G. Let (A, Ω) be a τ-algebra. Show that

(a) each intersection $\bigcap_{i \in I} \varepsilon_i$ of equivalence relations ε_i on the set A is an equivalence relation;

(b) each intersection $\bigcap_{i \in I} \theta_i$ of congruence relations θ_i on the algebra (A, Ω) is a congruence relation;

(c) the ordered set $(Eqv A, \subseteq)$ of equivalence relations on A and the ordered set $(Cg A, \subseteq)$ of congruence relations on (A, Ω) are complete lattices.

1.2H. Show that if θ_1 and θ_2 are two equivalence relations on a set A, then

$$\theta_1 + \theta_2 = \theta_1 \cup (\theta_1 \circ \theta_2) \cup (\theta_1 \circ \theta_2 \circ \theta_1) \cup (\theta_1 \circ \theta_2 \circ \theta_1 \circ \theta_2)$$

$$\cup \ldots$$

Equivalently, this means that $(a, b) \in \theta_1 + \theta_2$ iff there is a positive integer n and elements $c_1, \ldots, c_n$ in A such that $(c_i, c_{i+1}) \in \theta_1$ or $(c_i, c_{i+1}) \in \theta_2$ for $i = 1, \ldots, n - 1$, with $a = c_1$, $b = c_n$.

1.2I. Let (A, Ω) be a τ-algebra. Consider the set $Eqv A$, and let ε_i, for $i \in I$, be in $Eqv A$. Show that for a, b in A, for $X \subseteq A^2$, and for $\langle X \rangle \in Cg A$, one has

(a) $\sum_{i\in I}\varepsilon_i = \bigcup\{\varepsilon_{i_1} \circ \cdots \circ \varepsilon_{i_n} \mid i_1,\ldots,i_n \in I,\ n \in \mathbb{Z}^+\}$;

(b) $\langle X \rangle = \sum\{cg(a,b) \mid (a,b) \in X\}$.

1.2J. Show that the following conditions are equivalent for any two equivalence relations θ_1 and θ_2 in $EqvA$:

(a) $\theta_1 \circ \theta_2 \subseteq \theta_2 \circ \theta_1$;

(b) $\theta_1 \circ \theta_2 = \theta_2 \circ \theta_1$;

(c) $\theta_1 + \theta_2 = \theta_1 \circ \theta_2$.

1.2K. A subset K of a complete lattice $(L, \subseteq)$ is called a *complete* sublattice of $(L, \subseteq)$ if for every subset A of K the elements $\sum A$ and $\prod A$ are also in K. Show that for a τ-algebra (A, Ω), the complete lattice $(CgA, \subseteq)$ is a complete sublattice of the complete lattice $(EqvA, \subseteq)$.

1.2L. Show that for a directed set $\{\theta_i \mid i \in I\}$ of congruences of a τ-algebra (A, Ω),
$$\bigcup_{i\in I} \theta_i = \sum_{i\in I} \theta_i.$$

1.2M. Let $(A, \cdot, ^{-1}, 1)$ be a group. Show the following:

(a) If $\theta \in CgA$, then 1^θ is a normal subgroup of A, and for $a, b \in A$, one has $(a,b) \in \theta$ iff $a \cdot b^{-1} \in 1^\theta$;

(b) If N is a normal subgroup of A, then the binary relation defined on A by
$$(a,b) \in \theta \text{ iff } a \cdot b^{-1} \in N$$
is a congruence on the group A with $1^\theta = N$.

(c) The mapping $CgA \to \mathcal{P}(A);\ \theta \mapsto 1^\theta$ is an order-preserving bijection between congruences of A and normal subgroups of A.

1.2N. Let $(A, +, -, 0, \cdot, 1)$ be a ring. Show the following:

(a) If $\theta \in CgA$, then 0^θ is an ideal of A, and for $a, b \in A$, one has $(a,b) \in \theta$ iff $a - b \in 0^\theta$;

1. ALGEBRAS

(b) If I is an ideal of A, then the binary relation θ defined on A by

$$(a,b) \in \theta \text{ iff } a - b \in I$$

is a congruence on the ring A with $0^\theta = I$.

(c) The mapping $\text{Cg}A \to \mathcal{P}(A)$; $\theta \mapsto 0^\theta$ is an order-preserving bijection between congruences of A and ideals of A.

1.2O. Prove the First Isomorphism Theorem.

1.2P. Prove the Second Isomorphism Theorem.

1.2Q. Prove the Third Isomorphism Theorem.

1.2R. Let L_1 and L_2 be two lattices. Show that a mapping $h : L_1 \to L_2$ is an isomorphism of ordered sets if and only if it is a lattice isomorphism.

1.2S. Show that for a set A, the algebra $(\mathcal{P}(A \times A), \circ, \widehat{A})$ is a monoid.

1.2T. Let $\theta \in CgA$ for a τ-algebra (A, Ω). Show that (A^θ, Ω) is simple iff θ is a maximal element in the poset $CgA - \{A \times A\}$.

1.2U. Let X be a set and let $(M, \cdot, 1)$ be a monoid. The set X with a monoid homomorphism

$$R : (M, \cdot, 1) \to (X^X, \cdot, 1); \ m \mapsto (R_m : X \to X; \ x \mapsto xm)$$

is called a *(right) action of the monoid M on the set X*, or a *(right) M-action*, or a *(right) M-set*. Show that $\forall m, n \in M$ and $\forall x \in X$,

$$(xm)n = x(mn) \text{ and } x1 = x.$$

An M-set can be considered as an algebra (X, M) with a unary operation R_m for each m in M. The *orbit xM* of an element x of X is the set $\{xm \mid m \in M\}$. An element x of X is a *fixed point* of (X, M) if $xR_m = x$ for each m in M. Show that a subset S of an M-set (X, M) is a subalgebra if it contains the orbit sM of each of its elements s. Show that the set $\text{Fix}(X, M)$ of all fixed points of (X, M) is a subalgebra of (X, M).

1.3. Direct products

The concept of a direct product of two τ-algebras easily extends to a *direct product* $(A_1, \Omega) \times \cdots \times (A_n, \Omega)$, or simply $A_1 \times \cdots \times A_n$, of finitely many τ-algebras. Its set of elements consists of all n-tuples $(a_1, \ldots, a_n)$ with $a_i \in A_i$, and the operations of Ω are defined componentwise, i.e. for each ω in Ω and a_{ij} in A_i,

$$(1.3.1) \quad (a_{11}, \ldots, a_{n1}) \ldots (a_{1\omega\tau}, \ldots, a_{n\omega\tau})\omega$$
$$:= (a_{11} \ldots a_{1\omega\tau}\omega, \ldots, a_{n1} \ldots a_{n\omega\tau}\omega).$$

The corresponding i-th *projections* are

$$(1.3.2) \quad \pi_i : A_1 \times \cdots \times A_n \to A_i; \; (a_1, \ldots, a_n) \mapsto a_i.$$

These are obviously homomorphisms.

Similarly, given an (infinite) family of τ-algebras (A_i, Ω), for $i \in I$, the Cartesian product $\prod_{i \in I} A_i$ becomes an algebra $\prod_{i \in I}(A_i, \Omega)$ or $\prod(A_i, \Omega)$ or $(\prod A_i, \Omega)$, the *direct product* of (A_i, Ω), under componentwise operations, i.e. for each ω in Ω and $a_1, \ldots, a_{\omega\tau}$ in $(\prod A_i)^{\omega\tau}$

$$(1.3.3) \quad i(a_1 \ldots a_{\omega\tau}\omega) := (ia_1) \ldots (ia_{\omega\tau})\omega$$

for each i in I. The i-th *projections* are defined by

$$(1.3.4) \quad \pi_i : \prod_{i \in I} A_i \to A_i; \; (f : I \to \bigcup_{i \in I} A_i) \mapsto if,$$

and are Ω-homomorphisms. If $A_i = A$ for each $i \in I$, then $\prod A_i$ is also denoted by A^I, and the algebra (A^I, Ω) is referred to as a *direct power* of (A, Ω).

Direct products give a very useful way of constructing new τ-algebras from families of given τ-algebras, and representing some algebras as isomorphic to products of some simpler algebras. For example, a well-known theorem from

1. ALGEBRAS

group theory says that each finite abelian group is a direct product of cyclic groups. (More generally, each abelian group embeds into a direct product of cyclic groups.) However, given a τ-algebra (A, Ω), it is not easy to decide if this algebra is isomorphic to a direct product of some other τ-algebras. The following theorems will give two characterizations of algebras that are isomorphic to, or in other words are *decomposable as*, the direct product of two algebras. An easy generalization for algebras decomposable as the product of finitely many algebras is left as Exercise 1.3E. A more general case will be considered in Chapter 3.

Let (A_1, Ω) and (A_2, Ω) be τ-algebras. Consider their direct product $(A_1 \times A_2, \Omega)$ together with the projections π_1 and π_2. First note that the kernels of π_1 and π_2 satisfy the following:

(1.3.5) $\quad \begin{cases} \ker \pi_1 \cap \ker \pi_2 = \widehat{A_1 \times A_2}, \\ \ker \pi_1 + \ker \pi_2 = (A_1 \times A_2)^2, \\ \ker \pi_1 \circ \ker \pi_2 = \ker \pi_2 \circ \ker \pi_1. \end{cases}$

[Exercise 1.3B]. In general, if two congruences θ_1 and θ_2 on a τ-algebra satisfy the condition (1.3.5), i.e. if

(1.3.6) $\quad \begin{cases} \theta_1 \cap \theta_2 = \widehat{A}, \\ \theta_1 + \theta_2 = A^2, \\ \theta_1 \circ \theta_2 = \theta_2 \circ \theta_1, \end{cases}$

then one says that they form a *pair of factor congruences* on (A, Ω).

Theorem 1.3.1. *Two congruences θ_1 and θ_2 on a τ-algebra (A, Ω) form a pair of factor congruences if and only if the mapping*

(1.3.7) $\quad h : A \to A^{\theta_1} \times A^{\theta_2}; \; a \mapsto (a^{\theta_1}, a^{\theta_2})$

is an isomorphism.

Proof. Note that h injects iff $\theta_1 \cap \theta_2 = \widehat{A}$, and surjects iff $\theta_1 \circ \theta_2 = A \times A = \theta_2 \circ \theta_1$. The details of the proof are left as Exercise 1.3C. □

An algebra that is not isomorphic to a direct product of two non-trivial algebras is called (*directly*) *indecomposable*. For example, any simple algebra is directly indecomposable, and any finite algebra with a prime number of elements must be directly indecomposable.

In Exercise 1.3F, the reader is asked to show that *rectangular* bands, i.e. bands $(A, \cdot)$ satisfying the law

(1.3.8) $\qquad abc = ac$ for all a, b, c in A,

are characterized as being isomorphic to direct products of a left zero and a right zero band. The existence of certain rectangular band operations on a τ-algebra (A, Ω) is closely connected with the decompositions of the τ-algebra (A, Ω) into a direct product of two τ-algebras.

Theorem 1.3.2. *A τ-algebra (A, Ω) is isomorphic to a direct product $(A_1, \Omega) \times (A_2, \Omega)$ of τ-algebras (A_1, Ω) and (A_2, Ω) if and only if there is a rectangular band operation $\cdot$ on (A, Ω) such that for each ω in Ω and $a_1, \ldots, a_{\omega\tau}, b_1, \ldots, b_{\omega\tau}$ in A,*

(1.3.9) $\qquad a_1 \ldots a_{\omega\tau}\omega \cdot b_1 \ldots b_{\omega\tau}\omega = (a_1 \cdot b_1) \ldots (a_{\omega\tau} \cdot b_{\omega\tau})\omega.$

Proof. ($\Rightarrow$) Let $(A, \Omega) \cong (A_1 \times A_2, \Omega)$. By Exercise 1.3F, there is a rectangular band operation $\cdot$ on A such that $(A, \cdot) \cong (A_1, \cdot) \times (A_2, \cdot)$, where $(A_1, \cdot)$ is a left zero and $(A_2, \cdot)$ is a right zero band. It remains to show that condition (1.3.9) is satisfied. So let $a_{11}, \ldots, a_{1\omega\tau}, b_{11}, \ldots, b_{1\omega\tau}$ be in A_1 and $a_{21}, \ldots, a_{2\omega\tau}, b_{21}, \ldots, b_{2\omega\tau}$ be in A_2. Then by the definition of $\cdot$ one has:

$$(a_{11}, a_{21}) \ldots (a_{1\omega\tau}, a_{2\omega\tau})\omega \cdot (b_{11}, b_{21}) \ldots (b_{1\omega\tau}, b_{2\omega\tau})\omega$$
$$= (a_{11} \ldots a_{1\omega\tau}\omega, a_{21} \ldots a_{2\omega\tau}\omega) \cdot (b_{11} \ldots b_{1\omega\tau}\omega, b_{21} \ldots b_{2\omega\tau}\omega)$$
$$= (a_{11} \ldots a_{1\omega\tau}\omega, b_{21} \ldots b_{2\omega\tau}\omega)$$
$$= (a_{11}, b_{21}) \ldots (a_{1\omega\tau}, b_{2\omega\tau})\omega$$
$$= ((a_{11}, a_{21}) \cdot (b_{11}, b_{21})) \ldots ((a_{1\omega\tau}, a_{2\omega\tau}) \cdot (b_{1\omega\tau}, b_{2\omega\tau}))\omega,$$

1. ALGEBRAS

as required.

($\Leftarrow$) Now assume that there is a rectangular band operation $\cdot$ on (A, Ω) satisfying (1.3.9) for each ω in Ω. Consider the relations β and γ defined in Exercise 1.3F. By this exercise, β and γ are congruences of $(A, \cdot)$, the quotient $(A^\beta, \cdot)$ is a left zero band, and $(A^\gamma, \cdot)$ is a right zero band. These relations are also Ω-congruences. Indeed, for $a_1, \ldots, a_{\omega\tau}, a'_1, \ldots, a'_{\omega\tau}$ in A such that $(a_i, a'_i) \in \beta$, one has $a_1 \ldots a_{\omega\tau}\omega \cdot a'_1 \ldots a'_{\omega\tau}\omega = (a_1 \cdot a'_1) \ldots (a_{\omega\tau} \cdot a'_{\omega\tau})\omega = a'_1 \ldots a'_{\omega\tau}\omega$. Similarly, for $(a_i, a'_i) \in \gamma$, one has $a_1 \ldots a_{\omega\tau}\omega \cdot a'_1 \ldots a'_{\omega\tau}\omega = a_1 \ldots a_{\omega\tau}\omega$. Now the band isomorphism i defined in Exercise 1.3F is in fact an Ω-homomorphism. This follows by the calculations below:

$$(a_1 \ldots a_{\omega\tau}\omega)i = (a_1 \ldots a_{\omega\tau}\omega^\beta, a_1 \ldots a_{\omega\tau}\omega^\gamma)$$
$$= (a_1^\beta \ldots a_{\omega\tau}^\beta \omega, a_1^\gamma \ldots a_{\omega\tau}^\gamma \omega) = (a_1^\beta, a_1^\gamma) \ldots (a_{\omega\tau}^\beta, a_{\omega\tau}^\gamma)\omega$$
$$= (a_1 i) \ldots (a_{\omega\tau} i)\omega.$$

Consequently, the algebra (A, Ω) is isomorphic to $(A^\beta, \Omega) \times (A^\gamma, \Omega)$. $\square$

Note that the condition (1.3.9) means that the operation $\cdot = h$ is in fact a homomorphism from the algebra $(A \times A, \Omega)$ into the algebra (A, Ω). Indeed, one can rewrite (1.3.9) as

$$((a_1, b_1) \ldots (a_{\omega\tau}, b_{\omega\tau})\omega)h$$
$$= (a_1 \ldots a_{\omega\tau}\omega, b_1 \ldots b_{\omega\tau}\omega)h$$
$$= (a_1, b_1)h \ldots (a_{\omega\tau}, b_{\omega\tau})h\omega.$$

Two operations ω and $\cdot$ satisfying (1.3.9) are said to *commute*. More generally, two operations ω and φ in Ω commute, if for all $a_{11}, \ldots, a_{1\omega\tau}, \ldots, a_{\varphi\tau 1}, \ldots, a_{\varphi\tau\omega\tau}$, one has

(1.3.10)
$$(a_{11} \ldots a_{1\omega\tau}\omega) \ldots (a_{\varphi\tau 1} \ldots a_{\varphi\tau\omega\tau}\omega)\varphi$$
$$= (a_{11} \ldots a_{\varphi\tau 1}\varphi) \ldots (a_{1\omega\tau} \ldots a_{\varphi\tau\omega\tau}\varphi)\omega.$$

Since the operation · is used to decompose the algebra (A, Ω) into a direct product, it is sometimes called a *decomposition* operation. Later, we will also encounter different "decompositions" of algebras and different "decomposition" operations.

Exercises

1.3A. Let $(A_1, \Omega), (A_2, \Omega)$ and (A_3, Ω) be τ-algebras. Show that
   (a) $(A_1 \times A_2, \Omega) \cong (A_2 \times A_1, \Omega)$;
   (b) $(A_1 \times (A_2 \times A_3), \Omega) \cong ((A_1 \times A_2) \times A_3, \Omega)$.

1.3B. Show that the kernels of the projections π_1 and π_2 on the direct product $(A_1 \times A_2, \Omega)$ satisfy the condition (1.3.5).

1.3C. Give details of the proof of Theorem 1.3.1.

1.3D. Show, by induction on the number of elements, that each finite algebra is a direct product of directly indecomposable algebras.

1.3E. Let $\theta_1, \ldots, \theta_n$ be congruences on a τ-algebra (A, Ω). Show that the mapping

$$h : A \to A^{\theta_1} \times \cdots \times A^{\theta_n}; \ a \mapsto (a^{\theta_1}, \ldots, a^{\theta_n})$$

is an isomorphism iff $\{\theta_1, \ldots, \theta_n\}$ is a permutable set of congruences, and for each $i = 1, \ldots, n$, the congruences θ_i and $\bigcap_{j \neq i} \theta_j$ form a pair of factor congruences.

1.3F. (a) Consider the cartesian product $B \times C$ of non-empty sets B and C, together with the binary operation · defined by

$$\cdot : (B \times C)^2 \to B \times C; \ ((b, c'), (b', c)) \mapsto (b, c).$$

Show that $(B \times C, \cdot)$ is a rectangular band, and that $(B \times C, \cdot)$ is (isomorphic to) the direct product of the left zero band $(B, \cdot)$ and the right zero band $(C, \cdot)$.

(b) Given a rectangular band $(A, \cdot)$, define binary relations $\beta = \{(a, a') \mid$

$aa' = a'\}$ and $\gamma = \{(a, a') \mid aa' = a\}$ on A. Show that β and γ are band congruences, and that $(A^\beta, \cdot)$ and $(A^\gamma, \cdot)$ are left zero and right zero bands, respectively. Show that the mappings

$$i : A \to A^\beta \times A^\gamma;\ a \mapsto (a^\beta, a^\gamma)$$

and

$$j : A^\beta \times A^\gamma \to A;\ (b^\beta, c^\gamma) \mapsto bc$$

are mutually inverse band isomorphisms. Conclude that $(A, \cdot)$ is a direct product of a left zero and a right zero band.

1.3G. Show that each of the following operations on A commutes with itself:

(a) $(a, b) \mapsto ah + bg$, where h and g are commuting endomorphisms of a commutative semigroup $(A, +)$;

(b) [Frink 1955] $(a, b) \mapsto ah + bg$, where h and g are arbitrarily fixed elements of a field A;

(c) [Etherington 1958a] $(a, b) \mapsto a\bar{b}$, where A is the field $\mathbb{C}$ of complex numbers;

(d) [Padmanabhan 1969] $(a, b) \mapsto a - b$, where A is an abelian group.

(e) [Aczél 1948, Etherington 1949] the operations of taking the arithmetic and geometric means on $A = \mathbb{R}$;

(f) [Robinson 1962] the operation $(a, b) \mapsto a - b + a$ on a nilpotent group $(A, +, -, 0)$ of class 2.

Which of these operations are idempotent?

1.4. Word algebras

Let X be a set, referred to as a set of *variables* or an *alphabet*. A (*non-empty*) *word in an alphabet* X is a concatenation $x_1 x_2 \ldots x_n$ of (not necessarily distinct) elements of X. Let X^+ denote the set of all non-empty words in X.

Concatenation gives a semigroup multiplication on X^+, in particular

$$(x_1 \ldots x_m, y_1 \ldots y_n) \mapsto x_1 \ldots x_m \cdot y_1 \ldots y_n := x_1 \ldots x_m y_1 \ldots y_n.$$

Under this multiplication, X^+ becomes the so-called *free* semigroup over X. Adjoining an identity element 1 to X^+, called the *empty word*, one obtains the *free* monoid X^* over X. The *length* of a non-empty word $x_1 \ldots x_n$ in X is the number n of "letters" appearing in it. The length of the empty word is 0. Length gives a monoid homomorphism

$$X^* \to (\mathbb{N}, +, 0); \; x_1 \ldots x_n \mapsto n.$$

Note the following "universality property" of the free monoid X^*:

(1.4.1) $\quad \begin{cases} \text{For each monoid } (A, \cdot, 1) \text{ and a set map } f : X \to A, \\ \text{there is a unique monoid homomorphism} \\ \bar{f} : (X^*, \cdot, 1) \to (A, \cdot, 1) \text{ such that } \bar{f}|_X = f. \end{cases}$

[Exercise 1.4A]. The property (1.4.1) may be expressed diagrammatically as

(1.4.2)
$$\begin{array}{ccc} X & \longrightarrow & X^* \\ f \downarrow & & \downarrow \bar{f} \\ A & = & A \end{array}.$$

Now let $\tau : \Omega \to \mathbb{N}$ be a fixed finitary type and let X be a set of variables. Form the free monoid $(X \cup \Omega)^*$ of words in the alphabet $X \cup \Omega$. The free monoid becomes a τ-algebra $((X \cup \Omega)^*, \Omega)$ on defining

(1.4.3) $\quad \omega : ((X \cup \Omega)^*)^{\omega \tau} \to (X \cup \Omega)^*; \; (a_1, \cdots, a_{\omega \tau}) \mapsto a_1 \ldots a_{\omega \tau} \omega$

for each ω in Ω. The (τ-)*word algebra* or *absolutely free* (τ-)*algebra* $X\Omega$ or $(X\Omega, \Omega)$ over X is then defined to be the subalgebra of $(X \cup \Omega)^*$ generated by

X. The elements of $X\Omega$ are called Ω-*words* or τ-*words in* X, or sometimes just *words*. An Ω-word w containing precisely variables $x_1, \ldots, x_n$ is usually denoted by $x_1 \ldots x_n w$ or $w(x_1, \ldots, x_n)$. The *length* of a (non-empty) word is the number of "letters" from $X \cup \Omega$ appearing in it. The Ω-words of length 1 are just the elements of X and the nullary operators of Ω (if Ω contains such operators). Longer Ω-words have the form

(1.4.4) $$w_1 \ldots w_n \omega,$$

where ω is an n-ary operator, $n > 0$ and $w_1, \ldots, w_n$ are shorter Ω-words in X. Note that two Ω-words are equal if they are of the same length and their corresponding letters are the same.

The word algebra $X\Omega$ has a similar "universality property" as the free monoid X^*. [Cf. (1.4.2).]

Proposition 1.4.1. *For each mapping* $f : X \to A$ *from* X *to the underlying set* A *of a* τ-*algebra* (A, Ω), *there is a unique homomorphism* $\bar{f} : (X\Omega, \Omega) \to (A, \Omega)$ *whose restriction to the set* X *is* f:

$$\begin{array}{ccc} X & \longrightarrow & X\Omega \\ f \downarrow & & \downarrow \bar{f} \\ A & = & A \end{array}$$

Proof. Define $\bar{f}$ on words of length 1 by setting $x\bar{f} := xf$ for x in X and $\omega\bar{f} := \omega$ for nullary operators. Define $\bar{f}$ generally by induction on the length of words: for a word $w = w_1 \ldots w_n \omega$ of length greater than 1, define

$$w\bar{f} := (w_1\bar{f}) \ldots (w_n\bar{f})\omega.$$

The mapping $\bar{f}$ is a homomorphism, since for an n-ary operator ω and elements $w_1, \ldots, w_n$ of $X\Omega$, one has

$$w_1 \ldots w_n \omega \bar{f} = (w_1\bar{f}) \ldots (w_n\bar{f})\omega.$$

Finally, $\bar{f}$ is the unique such homomorphism, by Exercise 1.2C(f). □

Consider the power set monoid $(\mathcal{P}(X), \cup, \emptyset)$. As in Example 1.1.15, it may be realized as a τ-algebra $(\mathcal{P}(X), \Omega)$ with $Y_1 \ldots Y_{\omega\tau}\omega = \bigcup_{i=1}^{\omega\tau} Y_i$ for ω in Ω and $Y_i \subseteq X$. There is a function $f : X \to \mathcal{P}(X); x \mapsto \{x\}$ embedding X in $\mathcal{P}(X)$ as the set of singletons. Now the monoid $(\mathcal{P}(X), \cup, \emptyset)$ has the universality property described in Exercise 1.4B, so that the mapping f can be uniquely extended to the Ω-homomorphism

$$\bar{f} : X\Omega \to \mathcal{P}(X); \quad x_1 \ldots x_n\omega \mapsto \{x_1, \ldots, x_n\}.$$

The homomorphism $\bar{f}$ is called the *argument map*, and is written as

(1.4.5) $\qquad\qquad \text{arg} : X\Omega \to \mathcal{P}(X); \; w \mapsto \text{arg}(w).$

Elements of the set $\arg(w)$ are called *arguments* of the word w. In particular, one has $\arg(x) = \{x\}$ and $\arg(x_1 \ldots x_{\omega\tau}\omega) = \{x_1, \ldots, x_{\omega\tau}\}$. For a nullary operator ω, one has $\arg(\omega) = \emptyset$. Note that the image of the argument map consists of finite subsets of the set X.

Define

(1.4.6) $\qquad\qquad \mathfrak{P} = \{x_1 < x_2 < \ldots\}$

as a linearly ordered, countably infinite set of variables. (It will sometimes be identified with the ordered set of positive integers.) Consider the function

(1.4.7) $\qquad\qquad \max \circ \arg : \mathfrak{P}\Omega \to \mathbb{N}; \; x_{i_1} \ldots x_{i_k}w \mapsto i_k,$

where $i_1 \leq \cdots \leq i_k$, selecting the maximum of the (finite) set $\{i_1, \ldots, i_k\}$. If ω is a nullary operator, then $\max \circ \arg(\omega) = 0$. For a function $f : A \to B$ with an ordered codomain $(B, \leq)$, the *epigraph* of f is the set

(1.4.8) $\qquad\qquad \text{epi} f := \{(a, b) \in A \times B \mid af \leq b\}.$

1. ALGEBRAS

Define

(1.4.9) $$\bar{\Omega} := \text{epi max} \circ \text{arg}.$$

Then for $w = x_{i_1} \ldots x_{i_k} w$ in $\mathfrak{P}\Omega$, one has epi max $\circ$ arg $(x_{i_1} \ldots x_{i_k} w) = \{(w, n) \mid i_k \leq n\}$. For example, for the groupoid type $\{(\cdot, 2)\}$ and the groupoid word $x_2 x_4 x_5 w = x_2 \cdot (x_5 \cdot x_4)$, one has max $\circ$ arg $(w) = 5$ and epi max $\circ$ arg $(w) = \{(x_2 \cdot (x_5 \cdot x_4), n) \mid n \geq 5\}$.

Define

(1.4.10) $$\bar{\tau} : \bar{\Omega} \to \mathbb{N}; \ (w, n) \mapsto n.$$

Note that there is an embedding

$$j : \Omega \to \bar{\Omega} : \omega \mapsto (\omega, \omega\tau),$$

so that one can identify Ω with its image Ωj in $\bar{\Omega}$. There is also an embedding

$$\iota : \mathfrak{P}\Omega \to \text{max} \circ \text{arg} \ : x_{i_1} \ldots x_{i_k} w \mapsto (w, i_k)$$

from $\mathfrak{P}\Omega$ to the graph of max $\circ$ arg, a subset of $\bar{\Omega}$, so that one can identify $\mathfrak{P}\Omega$ with its image $\mathfrak{P}\Omega\iota$ in max $\circ$ arg contained in $\bar{\Omega}$. Denote the restriction max $\circ$ arg of $\bar{\tau}$ to $\mathfrak{P}\Omega\iota$ by τ'. The original type $\tau : \Omega \to \mathbb{N}$ thus yields three types:

(1.4.11) $$\begin{cases} (a) & \tau : \Omega \to \mathbb{N}; \ w \mapsto \omega\tau; \\ (b) & \tau' : \mathfrak{P}\Omega \to \mathbb{N}; \ w \mapsto \text{max} \circ \text{arg} \ (w); \\ (c) & \bar{\tau} : \bar{\Omega} \to \mathbb{N}; \ (w, n) \mapsto n. \end{cases}$$

The type $\tau' : \mathfrak{P}\Omega \to \mathbb{N}$ is called the *type derived* from $\tau : \Omega \to \mathbb{N}$. The type $\bar{\tau} : \bar{\Omega} \to \mathbb{N}$ is called the *closure* of the type τ. The words in $\mathfrak{P}\Omega$ are called *derived operators* or *terms*, while the elements of Ω are sometimes described as *basic operators* in this context. A derived operator is *linear* if it does not include any repeated variables. The set $\bar{\Omega}$ will be described as a *closed operator domain*, and specifically as the *closure* of the operator domain Ω. The word algebra $(\mathfrak{P}\Omega, \Omega)$ is also called a "term algebra".

Example 1.4.2. For each operator domain Ω, the closure $\overline{\Omega}$ contains the elements (x_i, n) for $1 \leq i \leq n$. Such elements are called the *n-ary projection operators* π_i^n *onto the i-th factor*. □

Now consider a τ-algebra (A, Ω), and an element (w, n) of $\overline{\Omega}$ with $w = x_{i_1} \ldots x_{i_k} w$. Set $X = \{x_1, \ldots, x_n\}$. Note that $\arg(w) \subseteq X$ and $w \in X\Omega$. Define the n-ary operation $(w, n)_A$ or (w, n) on A by

$$(w, n) : A^n \to A;\ (a_1, \ldots, a_n) \mapsto a_{i_1} \ldots a_{i_n} w.$$

For example, for the groupoid word $x_2 x_4 x_5 w = x_2 \cdot (x_5 \cdot x_4)$, and $n = 6$, one has

$$(w, 6) : A^6 \to A, (a_1, \ldots, a_6) \mapsto a_2 \cdot (a_5 \cdot a_4).$$

This makes the τ-algebra (A, Ω) into an algebra $(A, \overline{\Omega})$ of type $\overline{\tau}$, or an algebra $(A, \mathfrak{P}\Omega)$ of type τ', and determines the following mappings:

$$R_\tau^{\tau'} : (A, \mathfrak{P}\Omega) \mapsto (A, \Omega);$$
$$R_\tau^{\overline{\tau}} : (A, \overline{\Omega}) \mapsto (A, \Omega);$$
$$R_{\tau'}^{\overline{\tau}} : (A, \overline{\Omega}) \mapsto (A, \mathfrak{P}\Omega).$$

The operations of $\mathfrak{P}\Omega$ on A are called the *derived operations* or *term operations* of the τ-algebra (A, Ω). A derived operation is *linear* if the corresponding operator is linear. A basic reduct of $(A, \overline{\Omega})$ is called a *reduct* of (A, Ω). A subalgebra of a reduct of (A, Ω) is called a *subreduct*. The set $\overline{\Omega}$ of operations on A is called the *closed set of operations* or the *clone* of the τ-algebra (A, Ω), and the *closure* of the set Ω of operations on A. The operations from $\overline{\Omega}$ are called the *clone operations* of the τ-algebra (A, Ω).

Example 1.4.3. A real vector space V may be considered as an algebra $(V, +, \mathbb{R})$ of type $(\tau : \Omega \to \mathbb{N}) = \{(+, 2)\} \cup (\mathbb{R} \times \{1\})$. Use infix notation for addition and postfix notation for scalar multiplication, and consider linear combinations $x_1(1-r) + x_2 r$ for $r \in I^\circ :=]0,1[$. For each r in I° define the binary operations $\underline{r}$ on V as follows:

(1.4.12) $\qquad \underline{r} : V^2 \to V;\ (x,y) \mapsto xy\underline{r} := x(1-r) + yr.$

Now the set $\mathfrak{P}\Omega$ of derived operations on V consists of all linear combinations. Denote by $\underline{I}^\circ$ the subset (1.4.12) of operations on V. Consider the reduct $(V, \underline{I}^\circ)$. Then the convex subsets of the vector space V are precisely the subalgebras of the reduct $(V, \underline{I}^\circ)$ of $(V, +, \mathbb{R})$ [Exercise 1.4C and Section 1.6.2]. □

Proposition 1.4.4. *Let (A, Ω) be a τ-algebra. Then the following hold:*

(a) *A subset $B \subseteq A$ is a subalgebra of (A, Ω) if and only if it is a subalgebra of $(A, \bar{\Omega})$;*

(b) *A mapping $h : A \to B$ is an Ω-homomorphism $(A, \Omega) \to (B, \Omega)$ if and only if it is an $\bar{\Omega}$-homomorphism;*

(c) *The algebra (A, Ω) is (isomorphic to) a direct product $\left(\prod_{i \in I} A_i, \Omega \right)$ if and only if the algebra $(A, \bar{\Omega})$ is (isomorphic to) the direct product $\left(\prod_{i \in I} A_i, \bar{\Omega} \right)$.*

(d) *A relation ρ is a congruence on (A, Ω) if and only if it is a congruence on $(A, \bar{\Omega})$.*

Proof. The proofs of (a), (b) and (c) are relegated to Exercise 1.4D. We will prove (d) by induction on the length of the clone operations. Let θ be in Cg A, and let $a_1, \ldots, a_n, b_1, \ldots, b_n$ be in A with $(a_i, b_i) \in \theta$ for $i = 1, \ldots, n$. If $w = x_i$ is a variable, then $a_1 \ldots a_n(w, n) = a_1 \ldots a_n \pi_i = a_i\ \theta\ b_i = b_1 \ldots b_n \pi_i = b_1 \ldots b_n(w, n)$. Suppose that the statement has been proved for operations $(w_1, n), \ldots, (w_{w\tau}, n)$, and consider $w = w_1 \ldots w_{w\tau} \omega$ for ω in Ω. Then

$$a_1 \ldots a_n(w, n) = (a_1 \ldots a_n(w_1, n)) \ldots (a_1 \ldots a_n(w_{w\tau}, n))\omega$$
$$\theta\ (b_1 \ldots b_n(w_1, n)) \ldots (b_1 \ldots b_n(w_{w\tau}, n))\omega = b_1 \ldots b_n(w, n),$$

as required. □

Proposition 1.4.5. *The subalgebra $\langle X \rangle$ of a τ-algebra (A, Ω) generated by a subset X of A is the set of images $x_1 \ldots x_n w$ of elements of X under derived operations.*

Proof. Let $f : X \to A$ be the inclusion mapping of X in A. Then by the definition of derived operations, the image $X\Omega \bar{f}$ of $X\Omega$ under the corresponding homomorphism $\bar{f}$ given by Proposition 1.4.1 is the set of images of elements of X under derived operations. By Exercise 1.2C, it follows that $X\Omega \bar{f}$ is a subalgebra of (A, Ω), so $\langle X \rangle \subseteq X\Omega \bar{f}$. Conversely, the subalgebra $\langle X \rangle$ is closed under the derived operations, so that $X\Omega \bar{f} \subseteq \langle X \rangle$. □

Given a τ-algebra (A, Ω), consider the corresponding algebra $(A, A \cup \Omega)$ of type

$$(\tau_A : A \cup \Omega \to \mathbb{N}) = (A \to \{0\}) \cup (\tau : \Omega \to \mathbb{N}).$$

This is the algebra (A, Ω) with adjoined basic nullary operations $a : A^0 \to A$; $\varnothing \mapsto a$ corresponding to each element of the underlying set of the algebra. Then the operations of the clone $\overline{A \cup \Omega}$ are called *polynomial operations* of the algebra (A, Ω). For example, polynomials of a commutative, unital ring R are polynomial operations of the ring R.

Definition 1.4.6. Let (A, Ω) and (A, Ψ) be two algebras with the same set of elements, but possibly of different type.

(a) The algebras (A, Ω) and (A, Ψ) are *(clonally) equivalent* if $(A, \bar{\Omega}) = (A, \bar{\Psi})$.

(b) The algebras (A, Ω) and (A, Ψ) are *polynomially equivalent* if $(A, \overline{A \cup \Omega}) = (A, \overline{A \cup \Psi})$.

[Note that some authors use the term "term equivalent" for (a).] □

Example 1.4.7. A typical example of equivalent algebras is given by a Boolean algebra and a Boolean ring defined on the same set A. A commutative, unital ring A is Boolean if for each a in A, one has $a \cdot a = a$. If $(A, +, -, 0, \cdot, 1)$ is a Boolean ring, then by defining

$$a \vee b := a + b + a \cdot b,$$
$$a \wedge b := a \cdot b,$$
$$a' := 1 + a,$$

one obtains a Boolean algebra $(A, \vee, \wedge, ', 0, 1)$. If $(A, \vee, \wedge, ', 0, 1)$ is a Boolean algebra, then by defining

$$a + b := (a \wedge b') \vee (a' \wedge b),$$
$$a \cdot b := a \wedge b,$$
$$-a := a,$$

one obtains a Boolean ring $(A, +, -, 0, \cdot, 1)$ [Exercise 1.4E]. It follows that for the operator domains $\Omega = \{\vee, \wedge, ', 0, 1\}$ and $\Psi = \{+, -, 0, \cdot, 1\}$, one has $(A, \overline{\Omega}) = (A, \overline{\Psi})$, and hence the Boolean algebra A and the Boolean ring A are (clonally) equivalent. □

In general, a set Ω of operations on a set A is said to be *closed* or a *clone* on A if $\Omega = \overline{\Omega}$. Examples are provided by the clone $\overline{\Omega}$ of the clone operations and the clone $\overline{A \cup \Omega}$ of the polynomial operations of a τ-algebra (A, Ω). For n-ary operations $\omega_1, \ldots, \omega_k$ and a k-ary operation φ on a set A, their *composition* is the n-ary operation

$$\omega_1 \ldots \omega_k \varphi : A^n \to A; \quad (a_1, \ldots, a_n) \mapsto (a_1 \ldots a_n \omega_1) \ldots (a_1 \ldots a_n \omega_k) \varphi.$$

A clone on A can also be described as a set of operations on A containing the projection operations and closed under all compositions. The clone of all operations on A is sometimes denoted by Clo$A$. The clone of a τ-algebra (A,Ω) is obviously contained in Clo$A$, and is the smallest clone on A containing the basic operations of (A,Ω) or *generated* by the operations of Ω. For more information see [Szendrei 1986] and [McKenzie, McNulty and Taylor 1987]. In particular, for minimal clones of commuting operations, see [Lévai and Palfy 1996], [Kearnes and Szendrei 1998].

Exercises

1.4A. Prove the universal property (1.4.1).

[Hint: Define $\bar{f} : X^* \to A$; $x_1 \ldots x_n \mapsto x_1 f \ldots x_n f$.]

1.4B. Consider the monoid $(\mathcal{P}_{<\infty}(X), \cup, \varnothing)$ of finite subsets of a set X. Show that it has the following universality property: For each (join) semilattice $(A, +, 0)$ with 0 and a set map $f : X \to A$, there is a unique monoid homomorphism $\bar{f} : (\mathcal{P}_{<\infty}(X), \cup, \varnothing) \to (A, +, 0)$ such that $\bar{f}|_X = f$.

1.4C. Show that the convex subsets of a real vector space V are precisely the subalgebras of the reduct $(V, \underline{I}^\circ)$ of $(V, +, \mathbb{R})$.

1.4D. Prove Proposition 1.4.4(a), (b) and (c).

1.4E. Show that a Boolean algebra and its corresponding Boolean ring are equivalent.

1.5. Identities and equationally definable classes

Recall that most of the examples of algebras provided in Section 1.1 were defined as "satisfying" certain "laws" for all their elements. Such laws were written in the form $w = t$, where w and t were words of the appropriate type. And a law $w = t$ was "satisfied" in an algebra if it became an identity after substituting any choice of elements of the algebra for the variables of w and t. The concept of a law satisfied in an algebra can then be formalized as follows. Fix a finitary type $\tau : \Omega \to \mathbb{N}$. Let $\mathfrak{P}$ be defined as in (1.4.6).

Definition 1.5.1. An *identity* (or an *equation* or a *law*) $u = v$ of type τ is a pair (u, v) of derived operators, i.e. an element of the direct square $\mathfrak{P}\Omega \times \mathfrak{P}\Omega$ of the word algebra $\mathfrak{P}\Omega$ on the set $\mathfrak{P}$. A τ-algebra (A, Ω) is said to *satisfy the identity* $u = v$ if the n-ary clone operations $(u, n)_A$ and $(v, n)_A$ on A coincide, where $n = \max(\arg(u) \cup \arg(v))$. A class $\underline{K}$ of τ-algebras *satisfies* an identity $u = v$ if each algebra in $\underline{K}$ satisfies it. □

One also says that the identity $u = v$ *holds* in (A, Ω) or is *true* in (A, Ω), and that the words u and v are *synonymous* in A. For example, the groupoid words $u = (x_1 \cdot x_2) \cdot x_3$ and $v = x_1 \cdot (x_2 \cdot x_3)$ are synonymous in each semigroup $(A, \cdot)$. Indeed, the operations $(u, 3) : A^3 \to A; (a_1, a_2, a_3) \mapsto (a_1 \cdot a_2) \cdot a_3$ and $(v, 3) : A^3 \to A; (a_1, a_2, a_3) \mapsto a_1 \cdot (a_2 \cdot a_3)$ coincide. If there is no danger of confusion, the consecutive variables in an Ω-word are also sometimes denoted by x, y, z, etc. instead of $x_1, x_2, \ldots$.

Definition 1.5.2. If Σ is a set of identities of type $\tau : \Omega \to \mathbb{N}$, then the class $V(\Sigma)$ *defined* or *axiomatized* by $\sum$ is the class of all τ-algebras which satisfy all the identities in Σ. One says that $V(\Sigma)$ is an *equationally definable class* (or briefly *equational class*) or a *primitive class*. □

One writes
$$(A, \Omega) \models u = v \text{ or } A \models u = v$$
if the algebra (A, Ω) satisfies $u = v$. Similarly, if $\underline{K}$ is a class of τ-algebras, then the symbol
$$\underline{K} \models u = v$$
means that the class $\underline{K}$ satisfies $u = v$. Note that an equational class $\underline{K} = V(\Sigma)$ may also satisfy identities not contained in Σ. For instance, the class of semilattices is defined by idempotence, associativity and commutativity, but satisfies also e.g. the identity $(x \cdot yz)x = xy \cdot z$. The set Σ is called a *set of axioms for* $\underline{K}$ or a *basis for the identities true in* $\underline{K}$ or an *equational basis for*

$\underline{K}$. If $\underline{K} = V(\Sigma)$ for some finite set Σ of identities, then $\underline{K}$ is said to be *finitely based*, and to have Σ as a *finite equational basis*. Note that the laws defining semigroups, monoids, groups, rings, semilattices, lattices, and Boolean algebras given in Section 1.1 provide (finite) equational bases for the (equational) classes of semigroups, monoids, groups, rings, semilattices, lattices, Boolean algebras, respectively. Given a type $\tau : \Omega \to \mathbb{N}$, the symbol $\mathrm{Id}(\tau)$ or $\mathrm{Id}(\Omega)$ denotes the set of all identities of type τ. Given a class $\underline{K}$ of τ-algebras, the symbol $\mathrm{Id}(\underline{K})$ denotes the set of identities true in $\underline{K}$, i.e.

$$\mathrm{Id}(\underline{K}) := \{u = v \in \mathrm{Id}(\tau) \mid \underline{K} \models u = v\}.$$

One also writes $\mathrm{Id}(A_1, \ldots, A_n)$ if the class $\underline{K}$ consists of the algebras $A_1, \ldots, A_n$.

Proposition 1.5.3. *Let (A, Ω) be a τ-algebra and let $u = v$ be an identity of type τ. Then (A, Ω) satisfies $u = v$ if and only if for each homomorphism $h : (\mathfrak{P}\Omega, \Omega) \to (A, \Omega)$, one has $uh = vh$, i.e.*

$$A \models u = v \Leftrightarrow \forall h \in \underline{\tau}(\mathfrak{P}\Omega, A), uh = vh.$$

Proof. ($\Rightarrow$) Let n be the maximal among the indices in $\mathrm{arg}(u) \cup \mathrm{arg}(v)$. Set $X = \{x_1, \ldots, x_n\}$. If A satisfies $u = v$ and $h \in \underline{\tau}(\mathfrak{P}\Omega, A)$, then

$$uh = (x_1 \ldots x_n(u, n)_A)h = (x_1 h) \ldots (x_n h)(u, n)_A$$
$$= (x_1 h) \ldots (x_n h)(v, n)_A = (x_1 \ldots x_n(v, n)_A)h = vh.$$

($\Leftarrow$) Suppose that for each $h \in \underline{\tau}(\mathfrak{P}\Omega, A)$, one has $uh = vh$. If A is empty, then there are no nullary derived operations, and for $n > 0$, one has $(u, n)_A = (v, n)_A$ vacuously. Otherwise, let a_0 be an element of A. Given $f_X \in \underline{\mathrm{Set}}(X, A)$, define $f := (f_X : X \to A) \cup ((\mathfrak{P} - X) \to \{a_0\})$. Extend f to the homomorphism $\bar{f} : (\mathfrak{P}\Omega, \Omega) \to (A, \Omega)$ defined in Proposition 1.4.1. One then has

$$(x_1 f) \ldots (x_n f)(u, n)_A = u\bar{f} = v\bar{f} = (x_1 f) \ldots (x_n f)(v, n)_A. \quad \square$$

An identity $u = v$ of type τ is said to be *regular* if $\arg(u) = \arg(v)$, and *irregular* otherwise. For example, the associativity and commutativity of a binary operation are regular identities, whereas the absorption laws of lattices are not.

Example 1.5.4. Let $(A, \cdot)$ be a semilattice and $x_1 \ldots x_n w$ a groupoid word. By applying the axioms of semilattices, one easily sees that the identity

$$(1.5.1) \qquad x_1 \ldots x_n w = x_1 \cdot \ldots \cdot x_n$$

holds in $(A, \cdot)$. E.g. $(x_1 \cdot x_2) \cdot (x_2 x_1 \cdot x_3) = x_1 x_2 x_2 x_1 x_3 = x_1 x_1 x_2 x_2 x_3 = x_1 x_2 x_3$. One sometimes says that the right hand side of (1.5.1) is a *standard* or *normal* form of the word $x_1 \ldots x_n w$. It follows that any identity

$$x_1 \ldots x_m u = x_1 \ldots x_n v,$$

where $m \leq n$, may be written in the form

$$(1.5.2) \qquad x_1 \cdot \ldots \cdot x_m = x_1 \cdot \ldots \cdot x_m \cdot \ldots \cdot x_n.$$

Now if (1.5.2) holds in $(A, \cdot)$ and $m \neq n$, then also

$$(1.5.3) \qquad x = x \cdot \ldots \cdot x = x \cdot \ldots \cdot x \cdot y \cdot \ldots \cdot y = x \cdot y$$

holds in $(A, \cdot)$. Thus (1.5.3) and commutativity imply that

$$(1.5.4) \qquad x = x \cdot y = y \cdot x = y$$

also holds in $(A, \cdot)$. But (1.5.4) means that $(A, \cdot)$ is trivial. It follows that a non-trivial semilattice $(A, \cdot)$ satisfies precisely the regular identities of groupoid type, and any derived operation of $(A, \cdot)$ has the form

$$A^n \to A; \ (a_1, \ldots, a_n) \mapsto a_1 \cdot \ldots \cdot a_n.$$

Now let $\tau : \Omega \to \mathbb{N}$ be any (plural) type of algebras. As in Example 1.1.15, one can define an Ω-semilattice (A, Ω) by setting

$$\omega : A^{\omega\tau} \to A; \quad (a_1, \ldots, a_{\omega\tau}) \mapsto a_1 \cdot \ldots \cdot a_{\omega\tau}$$

for each ω in Ω. Conversely, given an Ω-semilattice (A, Ω), one may define a binary operation $\cdot$ on A by

$$x \cdot y := xy \ldots y\omega$$

for each n-ary ω in Ω with $n \geq 2$. Note that an Ω-semilattice (A, Ω) satisfies precisely the regular identities of type τ. $\square$

Given an equational class $\underline{K}$ of τ-algebras, the class $\underline{\widetilde{K}}$ of τ-algebras defined by the regular identities satisfied in $\underline{K}$ is called the *regularization* of $\underline{K}$. For example, the regularization of the class of semilattices is the same class. On the other hand, if $\underline{K}$ satisfies an irregular identity, $\underline{K}$ cannot coincide with its regularization. A class $\underline{K}$ of τ-algebras is *regular* if the only identities satisfied by all algebras in $\underline{K}$ are regular. Otherwise, $\underline{K}$ is *irregular*. Note that different authors have used different names for the identities we call "regular" and "irregular". Semigroup theorists call regular identities "homotypical" and irregular ones "heterotypical". Regular identities have also been called "variable-uniform" and "normal".

An identity $u = v$ is said to be *linear* if the multiplicities of each argument in u and v are at most 1. An identity is said to be *balanced* if it is both regular and linear. Thus commutativity and associativity are balanced, whereas idempotence is balanced only if the operation ω occurring there is unary.

Example 1.5.5. Let $(A, \cdot)$ be a semilattice. Define the *complex product* of subsemilattices A_1 and A_2 of $(A, \cdot)$ to be the set

(1.5.5) $$A_1 \cdot A_2 := \{a_1 a_2 \mid a_1 \in A_1, a_2 \in A_2\}.$$

Since for a_1a_2 and b_1b_2 in $A_1 \cdot A_2$, one has $a_1a_2 \cdot b_1b_2 = a_1b_1 \cdot a_2b_2 \in A_1 \cdot A_2$, it follows that $A_1 \cdot A_2$ is a subsemilattice of $(A, \cdot)$ as well. Let AS be the set of non-empty subsemilattices of $(A, \cdot)$. Thus (1.5.5) makes AS into an algebra $(AS, \cdot)$. Now let $x_1 \ldots x_n w = x_1 \ldots x_n v$ be a balanced identity of groupoid type. Then for non-empty subsemilattices $A_1, \ldots, A_n$ one has $A_1 \ldots A_n w = \{a_1 \ldots a_n w \mid a_i \in A_i\} = \{a_1 \ldots a_n v \mid a_i \in A_i\} = A_1 \ldots A_n v$, so that each balanced identity is satisfied in $(AS, \cdot)$. One can show even more. In fact, $A_1 A_1 = \{a_1 a_2 \mid a_1, a_2 \in A_1\} \subseteq A_1$, and since $(A, \cdot)$ is idempotent, $A_1 = \{a \mid a \in A_1\} = \{aa \mid a \in A_1\} \subseteq A_1 A_1$. Hence $A_1 A_1 = A_1$, and consequently $(AS, \cdot)$ is again a semilattice. □

Satisfaction of identities is preserved by the basic algebraic constructions of subalgebras, homomorphic images and direct products.

Proposition 1.5.6. *Let $\underline{K}$ be an equationally definable class of τ-algebras. Then the following hold:*

(a) *If (A, Ω) is in the class $\underline{K}$ and (B, Ω) is a subalgebra of (A, Ω), then (B, Ω) also lies in $\underline{K}$;*

(b) *If (A, Ω) is in $\underline{K}$ and (B, Ω) is a homomorphic image of (A, Ω), then (B, Ω) also lies in $\underline{K}$;*

(c) *If (A_i, Ω), for $i \in I$, are in $\underline{K}$, then their product $\left(\prod_{i \in I} A_i, \Omega \right)$ also lies in $\underline{K}$.* □

The proof is relegated to Exercise 1.5A. Note that a subalgebra or a homomorphic image of a given algebra (A, Ω) may satisfy more identities than the algebra (A, Ω).

Definition 1.5.7. *Let $\underline{K}$ be a class of τ-algebras.*

(h) *A τ-algebra is a member of the class $H\underline{K}$ iff it is the image of a τ-homomorphism whose domain is a member of $\underline{K}$.*

(i) A τ-algebra is a member of the class $I\underline{K}$ iff it is isomorphic to a member of $\underline{K}$. If $\underline{K} = I\underline{K}$, one also says that $\underline{K}$ is *abstract*.

(p) A τ-algebra is a member of the class $P\underline{K}$ iff it is isomorphic to the direct product of a multiset of members of $\underline{K}$.

(s) A τ-algebra is a member of the class $S\underline{K}$ iff it is isomorphic to a subalgebra of a member of $\underline{K}$. □

Note that H, I, P and S can all be considered as operators mapping classes of τ-algebras into classes of τ-algebras. Some further operators on classes of algebras will also be introduced later. The image of such an operator O acting on a class $\underline{K}$ will usually be denoted by $O(\underline{K})$ or simply $O\underline{K}$, or, if $\underline{K}$ consists of a single algebra A, just by OA. A class $\underline{K}$ is said to be *closed* under O if $O\underline{K} \subseteq \underline{K}$.

Corollary 1.5.8. *For any class $\underline{K}$ of τ-algebras, the classes $\underline{K}, I\underline{K}, H\underline{K}, P\underline{K}$ and $S\underline{K}$ satisfy the same identities.*

Corollary 1.5.9. *If $\underline{K}$ is an equational class of τ-algebras, then:*

$(h) \quad H\underline{K} = \underline{K};$

$(p) \quad P\underline{K} = \underline{K};$

$(s) \quad S\underline{K} = \underline{K}.$ □

Lemma 1.5.10. *Let $\underline{K}$ be a class of τ-algebras. Then:*

$(a) \quad PS\underline{K} \subseteq SP\underline{K};$

$(b) \quad PH\underline{K} \subseteq HP\underline{K};$

$(c) \quad SH\underline{K} \subseteq HS\underline{K}.$

Moreover, each of the operators H, S and P is idempotent.

Proof. (a) Suppose that for each i in I, one has $A_i \leq B_i \in \underline{K}$. Then $\prod_{i \in I} A_i \leq \prod_{i \in I} B_i \in P\underline{K}$.

(b) Suppose that for each i in I, the algebra B_i is the image of a surjective homomorphism $h_i : A_i \to B_i$ whose domain A_i lies in $\underline{K}$. Then the algebra $B := \prod_{i \in I} B_i = \prod_{i \in I} h_i(A_i)$ is in $P_H\underline{K}$. Now it is easy to check that the mapping

$$h : \prod_{i \in I} A_i \to \prod_{i \in I} B_i; (f : i \mapsto a_i) \mapsto (h(f) : i \mapsto a_i h_i)$$

is an epimorphism, i.e. $h\left(\prod_{i \in I} A_i\right) = \prod_{i \in I} B_i$. Hence $B \in HP\underline{K}$.

(c) Suppose that $A \in SH\underline{K}$. Then for some $B \in \underline{K}$ and a surjective homomorphism $h : B \to C$, one has $h^{-1}(A) \leq B$. Since $h(h^{-1}(A)) = A$, it follows that $A \in HS\underline{K}$.

Finally, it is very easy to check that $HH\underline{K} = H\underline{K}, SS\underline{K} = S\underline{K}$ and $PP\underline{K} = P\underline{K}$. □

Definition 1.5.11. Let $\underline{K}$ be a (non-empty, abstract) class of τ-algebras. Then $\underline{K}$ is said to be a *prevariety* if

$$P\underline{K} \subseteq \underline{K} \text{ and } S\underline{K} \subseteq \underline{K}.$$

A prevariety $\underline{K}$ is called a *variety* if additionally

$$H\underline{K} \subseteq \underline{K}. \quad □$$

Evidently, each variety is a prevariety. The class $\underline{\tau}$ of all τ-algebras forms a variety. If a type τ contains nullary operators, then the class of trivial τ-algebras forms a *trivial* variety. For a plural type τ, the class of trivial and empty τ-algebras forms a *trivial* variety. It is usually denoted by $\underline{\text{Tr}}$. As the intersection of a class of (pre)varieties of type τ is again a (pre)variety, one easily sees that for each class $\underline{K}$ of τ-algebras there is a smallest prevariety $R\underline{K}$ containing $\underline{K}$ or *generated by* $\underline{K}$, and a smallest variety $V\underline{K}$ containing $\underline{K}$ or *generated by* $\underline{K}$. One simply writes $V(A_1, \ldots, A_n)$ if $\underline{K}$ consists of algebras $A_1, \ldots, A_n$. One says that a prevariety is *finitely generated* if it is of the form $R\underline{K}$ for some finite set $\underline{K}$ of finite algebras. Similarly, one says that a variety is *finitely generated* if it is of the form $V\underline{K}$ for some finite set $\underline{K}$ of finite algebras.

Proposition 1.5.12. *Let $\underline{K}$ be a (non-empty, abstract) class of τ-algebras. Then:*

(a) $R\underline{K} = SP\underline{K}$;

(b) $V\underline{K} = HSP\underline{K}$.

Proof. (a) By Lemma 1.5.10, one has that $SSP\underline{K} = SP\underline{K}$ and $PSP\underline{K} \subseteq SPP\underline{K} = SP\underline{K}$, whence $SP\underline{K}$ is a prevariety. Obviously $SP\underline{K} \subseteq R\underline{K}$, and since $R\underline{K}$ is the smallest prevariety containing $\underline{K}$, one also has $SP\underline{K} = R\underline{K}$.

(b) As in (a), Lemma 1.5.11 implies that $HHSP\underline{K} = HSP\underline{K}$, $SHSP\underline{K} \subseteq HSSP\underline{K} = HSP\underline{K}$ and $PHSP\underline{K} \subseteq HPSP\underline{K} \subseteq HSPP\underline{K} = HSP\underline{K}$, whence $HSP\underline{K}$ is a variety. It is obvious that $HSP\underline{K} \subseteq V\underline{K}$, and since $V\underline{K}$ is the smallest variety containing $\underline{K}$, one has $V\underline{K} = HSP\underline{K}$. □

Note an important consequence of Proposition 1.5.6.

Proposition 1.5.13. *Each equational class is a variety.* □

In Chapter 3 we will prove a theorem, attributed to G. Birkhoff, showing that the converse is also true.

Exercises

1.5A. Prove Proposition 1.5.6.

1.5B. An identity $u = v$ is called *left regular* if the leftmost variables of u and v are the same. It is called *right regular* if the rightmost variables of u and v are the same.

    (a) Show that a non-trivial left zero band $(A, \cdot)$ satisfies precisely the left regular identities of groupoid type, and that any derived operation of $(A, \cdot)$ has the form

$$A^n \to A; \ (a_1, \ldots, a_n) \mapsto a_1.$$

    (b) Show that a non-trivial right zero band $(A, \cdot)$ satisfies precisely the right regular identities of groupoid type, and that any derived operation of $(A, \cdot)$ has the form

$$A^n \to A; \quad (a_1, \ldots, a_n) \mapsto a_n.$$

1.5C. Let $\tau : \Omega \to \mathbb{N}$ be any (plural) type. Let $(A, \cdot)$ be a left zero band. Define an Ω-left zero band (A, Ω) by setting

$$\omega : A^{\omega\tau} \to A; \quad (a_1, \ldots, a_{\omega\tau}) \mapsto a_1 \cdot \ldots \cdot a_{\omega\tau} = a_1$$

for each ω in Ω. Show that (A, Ω) satisfies precisely the left regular identities of type τ. Formulate and prove a corresponding statement for right zero bands.

1.5D. Show that an identity $u = v$ is satisfied in a rectangular band iff u and v have the same leftmost and rightmost variables.

1.5E. A band is called *normal* if it satisfies the *entropic* identity

$$xy \cdot zt = xz \cdot yt.$$

Show that an identity $u = v$ holds in a normal band iff it is regular, and u and v have the same leftmost and rightmost variables.

1.5F. A band is called *left normal* if it satisfies the identity

$$x \cdot yz = x \cdot zy,$$

and is called *right normal* if it satisfies the identity

$$xy \cdot z = yx \cdot z.$$

Show that an identity $u = v$ holds in a left (right) normal band iff it is regular, and u and v have the same left(right)most variables.

1.5G. Show that each rectangular band satisfies the identities:

(a) $xyx = x$;

(b) $xyzt = xt = xzyt$.

Show that each band satisfying (a) or (b) is rectangular, and that each idempotent groupoid satisfying (b) is a rectangular band.

1.5H. Generalize rectangular bands $(A, \cdot)$ to idempotent algebras (A, d) with one n-ary *diagonal* operation d satisfying the diagonal identity

$$(x_{11} \ldots x_{1n}d) \ldots (x_{n1} \ldots x_{nn}d)d = x_{11}x_{12} \ldots x_{nn}d.$$

Prove the following generalization of Theorem 1.3.2: A τ-algebra (A, Ω) is isomorphic to a direct product $(A_1, \Omega) \times \cdots \times (A_n, \Omega)$ of τ-algebras (A_i, Ω) iff there is an idempotent diagonal n-ary operation on (A, Ω) that commutes with each ω in Ω, i.e. (A, Ω) satisfies the identities

$$(a_{11} \ldots, a_{1\omega\tau}\omega) \ldots (a_{n1} \ldots a_{n\omega\tau}\omega)d$$
$$=(a_{11} \ldots a_{n1}d) \ldots (a_{1\omega\tau} \ldots a_{n\omega\tau}d)\omega.$$

Moreover, for each $i = 1, \ldots, n$, the algebra (A_i, Ω) satisfies the identity

$$x_1 \ldots x_i \ldots x_n d = x_i.$$

1.6. Further examples of algebras

In this section, the list of algebras from Section 1.1 is extended by the provision of several further kinds of algebras that will play an essential rôle in this book. We will first recall the definitions of quasigroups, modules, semirings and semi-modules. Then we will describe certain algebraizations of affine spaces and convex sets.

1.6.1. Quasigroups

A *quasigroup* $(Q, \cdot)$ is a groupoid with a binary operation called *multiplication* such that, in the equation

$$x \cdot y = z,$$

knowledge of any two of x, y, z specifies the third uniquely. Let us write the unique solution x of $x \cdot y = z$ for given y and z as z/y. Then $/$ may be regarded as a new binary operation on Q called *right division*. Similarly, the unique solution y of $x \cdot y = z$ for given x and z may be written as $x \backslash z$. The binary operation $\backslash$ thus defined on Q is called *left division*. The algebra $(Q, \cdot, /, \backslash)$ satisfies the following identities for all x, y in Q:

(1.6.1.1) $$x(x \backslash y) = y = (y/x) \cdot x,$$

(1.6.1.2) $$x \backslash (x \cdot y) = y = (y \cdot x)/x.$$

Proposition 1.6.1.1. *If $(Q, \cdot)$ is a quasigroup, then $(Q, \cdot, /, \backslash)$ satisfies the identities (1.6.1.1) and (1.6.1.2) for all x, y in Q. Conversely, if $(Q, \cdot, /, \backslash)$ is an algebra with three binary operations satisfying these identities, then $(Q, \cdot)$ is a quasigroup.*

Proof. ($\Rightarrow$) If $(Q, \cdot)$ is a quasigroup, then y/x is a solution of $(y/x)x = x$. Two solutions z of $z \cdot x = y \cdot x$ for given x and y are $(yx)/x$ and y. By the uniqueness of z, one has $(yx)/x = y$. The proof that the remaining identities hold is similar.

($\Leftarrow$) If the algebra $(Q \cdot, /, \backslash)$ satisfies (1.6.1.1) and (1.6.1.2), then (1.6.1.1) gives the existence of solutions z to the equations $xz = y$ and $zx = y$, while (1.6.1.2) gives the uniqueness of such solutions. $\square$

Proposition 1.6.1.1 shows that one may consider a quasigroup either as a groupoid $(Q, \cdot)$ satisfying the solution property, or equivalently as an algebra $(Q, \cdot, /, \backslash)$ satisfying the identities (1.6.1.1) and (1.6.1.2) for all x, y in Q. Note, however, that the algebraic properties of $(Q, \cdot)$ and $(Q, \cdot, /, \backslash)$ are not necessarily the same. Exercise 1.6A shows that a non-empty quasigroup with associative multiplication is a group. However, for a group $(G, \cdot, ^{-1}, 1)$, a subgroupoid $(H, \cdot)$ of the groupoid $(G, \cdot)$ is not necessarily a subgroup of

$(G, \cdot, ^{-1}, 1)$. Note that an empty quasigroup is not considered as a group, even though it is associative, since it does not have an identity element.

If we are given the multiplication table of a finite groupoid $(Q, \cdot)$, then it is easy to check whether or not $(Q, \cdot)$ is a quasigroup.

Proposition 1.6.1.2. *A finite groupoid $(Q, \cdot)$ is a quasigroup if and only if each element of Q appears exactly once in each row and in each column of the multiplication table of $(Q, \cdot)$.* □

The proof is left as Exercise 1.6E. Now an $n \times n$ matrix (q_{ij}) whose entries are elements of an n-element set Q, such that each member of Q appears exactly once in each row and each column, is called a *Latin square* of order n. Proposition 1.6.1.2 shows that the body of the multiplication table of a quasigroup is a Latin square. On the other hand, each Latin square of order n may be the body of the multiplication table of (at most $n! \times n!$) different quasigroups defined on the same set.

A fairly encyclopedic survey of quasigroup theory is available in [Chein, Pflugfelder and Smith 1990]. For earlier stages, see [Belousov 1967] and [Bruck 1971].

1.6.2. Modules over a (fixed, unital) ring

Let S be a given (unital) ring. A *(right) module over the ring S*, or briefly an *S-module* $(M, +, S)$, is defined as an abelian group $(M, +, -, 0)$ together with a ring homomorphism

(1.6.2.1) $\qquad R : S \to \text{End } M; \; s \mapsto (R_s : M \to M; \; m \mapsto ms)$

from the ring S to the ring of all endomorphisms of the group $(M, +, -, 0)$. The mapping R assigns to each element s of S the right "multiplication" of elements of M by the element s, i.e. $mR_s =: ms$, the image of m under R_s.

The fact that each R_s is a group homomorphism means that the identity

(M1) $$(m_1 + m_2)s = m_1s + m_2s$$

holds for all m_1, m_2 in M and s in S. The fact that R is a ring homomorphism means that the identities

(M2) $$m(s_1 + s_2) = ms_1 + ms_2,$$
(M3) $$(ms_1)s_2 = m(s_1s_2),$$
(M4) $$m1 = m$$

are satisfied for all m in M and s_1, s_2 in S. It follows that the class of all S-modules $(M, +, S)$, considered as algebras with three basic group operations and the unary operations, one for each element of the ring, can also be described by satisfaction of certain identities. If the mapping R injects, the S-module $(M, +, S)$ is called *faithful*. In this case, any two elements s and t of the ring S determine two different operations R_s and R_t. On the other hand, if $(M, +, S)$ is not faithful, some of the operations R_s may coincide. However, one can always turn the module $(M, +, S)$ into a faithful one. Define the *annihilator* $AnnM$ of $(M, +, S)$ by

$$AnnM := \operatorname{Ker} R$$
$$= \{s \in S \mid Ms = \{ms \mid m \in M\} = \{0\}\}.$$

Obviously $AnnM$ is an ideal of the ring S, and $(M, +, S)$ is faithful if and only if $AnnM = \{0\}$. Then the following holds.

Proposition 1.6.2.1. *Given any S-module $(M, +, S)$, let $S' := S/AnnM$. Then the S'-module $(M, +, S')$ with unary basic operations $R_s := R_{s+AnnM}$ defined by*

$$m \mapsto m(s + AnnM) := ms$$

is faithful. □

The proof is left as Exercise 1.6G. On the other hand, given an S-module $(M, +, S)$ and a ring homomorphism $h : T \to S$, one can always make the module $(M, +, S)$ into a T-module by defining the operations R_t for t in T as follows:

(1.6.2.2) $$mt := m(th)$$

[Exercise 1.6H]. Note that any two of the three modules $(M, +, S)$, $(M, +, S/AnnM)$ and $(M, +, T)$ are equivalent.

Let us recall some elementary examples of modules. If the ring S is a field, then the S-modules are just the vector spaces over this field. If the ring S is the ring $\mathbb{Z}$ of integers, then S-modules are just abelian groups. The operations R_s are defined by $m \mapsto ms$, where ms is calculated in the group. Any ring T with a subring $S \leq T$ is an S-module. The operations R_s are the restrictions of the multiplication in T. In particular, any ring S is an S-module. For more information about modules, see [Anderson and Fuller 1992] and [Golan and Head 1991].

1.6.3. Semirings and semimodules

A *semiring* is defined here to be an algebra $(S, +, \cdot)$ such that both $(S, +)$ and $(S, \cdot)$ are semigroups, the semigroup $(S, +)$ is commutative, and the following distributive laws hold:

$$x \cdot (y + z) = x \cdot y + x \cdot z;$$
$$(x + y) \cdot z = x \cdot z + y \cdot z.$$

One or both of the semigroups above may have an identity element. If $1 \in S$ is an identity of $(S, \cdot)$, then $(S, \cdot, 1)$ is a monoid, and the semiring S is said to have an *identity*. If $0 \in S$ is a unit for $(S, +)$, then it is called a *zero*.

In this case $(S, +, 0)$ is a monoid, and the semiring S is said to have a *zero*. Note however that the zero of a semiring is not a zero in the sense defined in Section 1.1. However, it may happen that the zero of a semiring is a zero for the semigroup $(S, \cdot)$, i.e. the identity

$$0 \cdot x = x \cdot 0 = 0$$

is satisfied. In this case, one also says that 0 is an *absorbing zero*. A semiring is *commutative* if its multiplicative reduct is commutative. Standard examples of semirings are rings, the sets of positive integers, rationals, or reals under addition and multiplication, and distributive lattices.

Example 1.6.3.1. Examples of semirings are given by endomorphisms of a commutative semigroup. Let $(T, +)$ be such a semigroup. Consider the set $\text{End}(T, +)$ of the endomorphisms of $(T, +)$, together with the following two operations of addition and multiplication:

$$t(\varphi + \psi) := t\varphi + t\psi,$$
$$t(\varphi\psi) := (t\varphi)\psi$$

for all $\varphi, \psi \in \text{End}(T, +)$. It is easy to see that $(\text{End}(T, +), +, \cdot)$ is a semiring. The identity map on T is the identity of $\text{End}(T, +)$. If one considers a monoid $(T, +, 0)$ instead of a semigroup, then the mapping $0 : T \to T;\ t \mapsto 0$ is a zero of the semiring $\text{End}(T, +, 0)$. □

A *semimodule* over a semiring S is a commutative semigroup $(T, +)$ together with a semiring homomorphism

$$R : (S, +, \cdot) \to (\text{End}(T, +), +, \cdot);\ s \mapsto (R_s : T \to T;\ t \mapsto tR_s =: ts).$$

The semimodule T may equivalently be defined as an algebra $(T, +, S)$ such that $(T, +)$ is a commutative semigroup, and each $s \in S$ defines a unary operation R_s satisfying the following laws:

$$(t_1 + t_2)s = t_1 s + t_2 s \quad (R_s \text{ is a semigroup homomorphism}),$$
$$t(s_1 + s_2) = ts_1 + ts_2,$$
$$t(s_1 s_2) = (ts_1)s_2 \quad (R \text{ is a semiring homomorphism}).$$

If the semiring S has an identity 1, then additionally one has

$$t1 = t.$$

Sometimes, one defines a semimodule as a monoid $(T, +, 0)$ rather than a semigroup $(T, +)$. A typical example of a semimodule is given by a monoid $(T, +, 0)$ with a semiring of endomorphisms of $(T, +, 0)$ as the semiring S. For more information, see [Golan 1999].

1.6.4. Affine spaces

Traditionally, an *affine space over a field F* is defined, more or less explicitly, as a "space of points" A associated with a vector space V of "free vectors" over F such that the additive group of the vector space V *acts strongly transitively on the set A*, i.e. there is a group homomorphism

(1.6.4.1) $$V \to A!; \; v \mapsto (t_v : A \to A; \; p \mapsto p + v)$$

onto the group of bijections t_v, usually called (*right*) *translations* of A, and for any two $p, q \in A$ there is exactly one $v \in V$ with $p + v = q$. In particular, (1.6.4.1) means that for all vectors $v, w \in V$ and all points $p \in A$,

$$p + 0 = p, \; (p + v) + w = p + (v + w).$$

The unique vector v with $p + v = q$ is denoted by $q - p$. One proves that

(1.6.4.2) $$p + (q - p) = q, \; (p - q) + (q - r) = p - r,$$
$$p - q = 0 \Leftrightarrow p = q,$$

and

(1.6.4.3) $\quad (p-q)+v = (p+v)-q, \ (p+v)-(q+w) = (p-q)+(v-w)$

[Cf. Exercise 1.6K].

Let $0 \in A$ be any "basis" point of A. Then the mapping

(1.6.4.4) $\quad\quad\quad\quad \alpha : A \to V; \ p \mapsto p - 0$

is a bijection between the set of points and the set of vectors. By defining

(1.6.4.5) $\quad\quad\quad p_1 f_1 + p_2 f_2 := 0 + (p_1 - 0)f_1 + (p_2 - 0)f_2$

on the set A, for all $f_1, f_2 \in F$, one introduces a vector space structure on A, and one obtains a vector space A^* isomorphic to V [Exercise 1.6L] so that in what follows we can identify the space V with A^*. This conveniently allows us to consider elements of A either as points or as vectors. One proves that the definition of $p_1 f_1 + p_2 f_2$ does not depend on the choice of 0. For every $n \in \mathbb{Z}^+$, and $f_1, \ldots, f_n \in F$ with $f_1 + \cdots + f_n = 1$, one defines an n-ary operation of *weighted mean* on A:

$$A^n \to A; \ (p_1, \ldots, p_n) \mapsto p_1 f_1 + \cdots + p_n f_n.$$

Especially useful weighted means are the binary ones

$$\underline{f} : A^2 \to A; (p_1, p_2) \mapsto p_1(1-f) + p_2 f =: p_1 p_2 \underline{f}$$

defined for each $f \in F$, and the ternary *Mal'cev parallelogram operation*

(1.6.4.6) $\quad P : A^3 \to A; \ (p_1, p_2, p_3) \mapsto p_1 1 + p_2(-1) + p_3 1 = p_1 - p_2 + p_3$

giving the fourth vertex of the parallelogram with vertices p_1, p_2 and p_3.

Lemma 1.6.4.1. For $n > 2$, each n-ary weighted mean operation is a composition of binary weighted means and the parallelogram operation P.

Proof. First note that for $p_1, \ldots, p_n \in A$ and $f_1, \ldots, f_n$ in F with $f_1 + \cdots + f_n = 1$, one has

$$
\begin{aligned}
p_1 f_1 + \cdots + p_n f_n &= p_1 f_1 - p_2 f_1 + p_2 f_1 + (p_2 f_2 + \cdots + p_n f_n) \\
&= p_1 - (p_1(1 - f_1) + p_2 f_1) + (p_2(f_1 + f_2) + p_3 f_3 + \cdots + p_n f_n) \\
&= p_1 - p_1 p_2 \underline{f}_1 + (p_2(f_1 + f_2) + p_3 f_3 + \cdots + p_n f_n).
\end{aligned}
\tag{1.6.4.7}
$$

The proof of the lemma thus proceeds by induction on n. For $n = 3$, one has

$$
\begin{aligned}
p_1 f_1 + p_2 f_2 + p_3 f_3 &= p_1 - p_1 p_2 \underline{f}_1 + (p_2(f_1 + f_2) + p_3 f_3) \\
&= p_1 - p_1 p_2 \underline{f}_1 + (p_2(1 - f_3) + p_3 f_3) \\
&= p_1 - p_1 p_2 \underline{f}_1 + p_2 p_3 \underline{f}_3.
\end{aligned}
$$

Now if the lemma holds for $n - 1$, then the third summand of (1.6.4.7) is an $n - 1$-ary weighted mean, and hence is a composition of binary weighted means and the operation P. Consequently $p_1 f_1 + \cdots + p_n f_n$ has the same property. □

Note that the operation P is needed in Lemma 1.6.4.1 only in the case where the characteristic of the field F is 2, in particular for $F = GF(2^n)$. In other fields, $2 = 1 + 1$ is invertible, and the operation P can be built up from binary weighted means:

$$p_1 - p_2 + p_3 = p_2\ p_3 p_1 \underline{2}^{-1}\ \underline{2}.$$

The coefficients $f_1, \ldots, f_n$ in $p_1 f_1 + \cdots + p_n f_n$ are sometimes called *weights*, and the point $p_1 f_1 + \cdots + p_n f_n$ is called a *center of gravity* of the points $p_1, \ldots, p_n$. The name is motivated by the physical example of a system of weights $f_1, \ldots, f_n$ located at respective points $p_1, \ldots, p_n$ in Euclidean space E^k (for $k = 1, 2, 3$). The center of gravity of the system is then located at the point $p_1\frac{f_1}{f} + \cdots + p_n\frac{f_n}{f}$, where $f = f_1 + \cdots + f_n$. By Lemma 1.6.4.1, it is clear that the set A together with the weighted mean operations can be defined as an algebra $(A, P, \underline{F})$ with the operation P and the set $\underline{F}$ of binary operations $\underline{f}$ for each f in F. The next proposition recalls the well-known fact that subalgebras of $(A, P, \underline{F})$ are precisely the affine subspaces of the affine space A. We will see later that it is fully justified to identify the algebras $(A, P, \underline{F})$ with the affine spaces A. Now recall that a subset B of an affine space A is called an *affine subspace* of A if there is a point $p \in A$ and a vector subspace W of the vector space V such that $B = p + W := \{p + w \mid w \in W\}$.

Proposition 1.6.4.2. *Let B be a non-empty subset of an affine space A, and let $p \in A$. The following conditions are equivalent.*

(a) *B is a non-empty affine subspace of A, i.e. $B = p + W$ for some vector subspace W of V.*

(b) *B is a subalgebra of the algebra $(A, P, \underline{F})$.*

Proof. (a) $\Rightarrow$ (b) Let $B = p + W$, whence $W = B - p = \{b - p \mid b \in B\}$ is a subspace of the vector space V. It suffices to consider the ternary operations $p_1 f_1 + p_2 f_2 + p_3 f_3$ for $p_i \in A$ and $f_i \in F$. Using (1.6.4.5), one shows that $p_1 f_1 + p_2 f_2 + p_3 f_3 \in B$ for $p_1, p_2, p_3 \in B$ [Exercise 1.6M].

(b) $\Rightarrow$ (a) Let B be a nonempty subalgebra of $(A, P, \underline{F})$. One proves that for each $p \in A$, the set $W := B - p$ forms a subspace of the vector space V [Exercise 1.6M]. Hence $b = p + W$. □

Each affine subspace B of A is closed under taking all the centers of gravity of points in B. Identifying points of A with vectors of A^*, one can say that non-empty affine subspaces of the space A are precisely the cosets of vector subspaces of A^*. Moreover, we can consider the non-empty algebras $(A, P, \underline{F})$ as reducts of vector spaces $(A, +, \underline{F})$. The operations P and $\underline{f}$ for $f \in F$, and more generally all weighted means, are obviously all the idempotent linear combinations of the vector space A^*.

Traditionally, an *affine map* $a : A \to A'$ from an affine space A associated with a vector space V over F to an affine space A' associated with a vector space V' over F is defined as a set mapping $a : A \to A'$ such that there is a linear map $a^* : V \to V'$ and a point $0 \in A$ with $(0+v)a = 0a + va^*$. The next, very well-known, proposition shows that such maps are actually homomorphisms from $(A, P, \underline{F})$ to $(A', P, \underline{F})$.

Proposition 1.6.4.3. *Let A and A' be non-empty affine spaces as above. Let $a : A \to A'$ be a set mapping. Then the following conditions are equivalent:*

(a) *The mapping $a : A \to A'$ is a homomorphism from the algebra $(A, P, \underline{F})$ to the algebra $(A', P, \underline{F})$;*

(b) *The mapping $a : A \to A'$ is an affine map from the affine space A to the affine space A'.*

Proof. (a) $\Rightarrow$ (b) First note that homomorphisms of $(A, P, \underline{F})$ preserve all weighted means. Choose a point $0 \in A$, and define

$$a^* : V \to V'; \quad v \mapsto (0+v)a - 0a.$$

Then the mapping a^* is a linear map. Indeed, for f and g in F, one has

$$(vf + wg)a^* = (0 + (vf + wg))a - 0a$$
$$= (0(1 - f - g) + (0 + v)f + (0 + w)g)a - 0a$$
$$= (0a - 0a) + 0a(-f - g) + (0 + v)af + (0 + w)ag$$
$$= ((0 + v)a - 0a)f + ((0 + w)a - 0a)g = va^*f + wa^*g.$$

(b) $\Rightarrow$ (a) For any point $p \in A$ and any vector $v \in V$, one has $(p + v)a = (0 + (p - 0) + v)a = 0a + ((p - 0) + v)a^* = 0a + (p - 0)a^* + va^* = (0 + (p - 0))a + va^* = pa + va^*$. Use (1.6.4.3) and (1.6.4.4) to show that the mapping a preserves all the operations p and $\underline{f}$ for $f \in F$ [Exercise 1.6R]. □

A related approach to affine geometry may be found in [Mac Lane and Birkhoff 1967], [Białynicki-Birula 1987], and [Mal'cev 1975].

The definition of $(A, P, \underline{F})$ and Propositions 1.6.4.2 and 1.6.4.3 show that an affine space A over a field F can be considered as an algebra $(A, P, \underline{F})$. So in what follows we will call the algebra $(A, P, \underline{F})$ an *affine space over F* or *an affine F-space*. Let $\underline{F}$ be the class of all affine F-spaces, including the empty set. A surprising fact is that the class $\underline{F}$ is a variety.

Proposition 1.6.4.4. *For any field F, the class $\underline{F}$ of affine F-spaces is a variety.*

Proof. By Propositions 1.6.4.2 and 1.6.4.3, the class $\underline{F}$ is closed under the operators S and H. Now if A_i, for $i \in I$, are affine F-spaces associated with vector spaces V_i, then $\prod_{i \in I} A_i$ is an affine F-space associated with the vector space $\prod_{i \in I} V_i$. Hence the class $\underline{F}$ is also closed under the operator P. □

The considerations of this example may easily be extended to the case where, instead of a field F, one takes a commutative ring R with unity. One obtains algebras $(A, P, \underline{R})$ that we will also call *affine R-spaces*. Some authors call them "affine modules."

1.6.5. Convex sets

Let E be any real vector space. For each real number p in the open unit interval $I^\circ = \,]0,1[$ in $\mathbb{R}$, consider the binary weighted mean operations $\underline{p}$ given by

$$\underline{p} : E \times E \to E; \; (x,y) \mapsto xy\underline{p}$$

and the algebra $(E, \underline{I^\circ}) = (E, \{\underline{p} \mid p \in I^\circ\})$. Note that the set $\{xy\underline{p} \mid p \in \mathbb{R}\}$ is the set of points of a line through the points x and y, while the set $\{xy\underline{p} \mid p \in I^\circ\}$ is just the line segment joining the points x and y.

Proposition 1.6.5.1. *The subalgebras of the algebra $(E, \underline{I}^\circ)$ are the convex subsets of the real vector space E.*

Proof. A subset B of the set E is a subalgebra of $(E, \underline{I}^\circ)$ iff for all x, y in B and p in I°, the element $xy\underline{p} = x(1-p) + yp$ lies in B. This happens precisely when the line segment joining the points x and y is contained in B. And this is precisely the criterion that B be a convex subset of E. □

Convex sets as subalgebras of $(E, \underline{I}^\circ)$ provide a useful description of the proportions of component parts making up a mixture, such as points of different primary colours being mixed to produce a certain shade, or chemical elements being mixed to produce an alloy. Other instances of such circumstances include mixture of prospects or commodities in decision theory and mathematical economics, and the mixture of pure states in quantum mechanics. (See [Gudder 1973, 1977, 1978] and [Morgenstern and Neumann 1953].) The component parts of a mixture may be represented by the set X of elements $e_0 = (1, 0, \ldots, 0), e_1 = (0, 1, \ldots, 0), \ldots, e_n = (0, \ldots, 0, 1)$ of the $n+1$-dimensional real vector space $E = \mathbb{R}^{n+1}$. The subalgebra $(\Delta_n, \underline{I}^\circ)$ of $(E, \underline{I}^\circ)$ generated by X is called the *n-dimensional simplex*. For example, if e_0 represents red paint, e_1 represents yellow paint and e_2 represents blue paint, then Δ_2 represents the colours of paints obtained by mixing these. The point $e_0 e_1 \underline{\tfrac{1}{2}}$ would represent a 2:1 mixture of red and yellow, say orange. The point $e_0 e_2 \underline{\tfrac{1}{2}}$ would represent a 2:1 mixture of red and blue, say violet. The point $(e_0 e_1 \underline{\tfrac{1}{2}})(e_0 e_2 \underline{\tfrac{1}{2}})\underline{\tfrac{1}{2}}$ would represent a shade between orange and violet.

Example 1.6.5.2. The closed unit interval $I = [0, 1]$ of $\mathbb{R}$ is a convex subset of the vector space $\mathbb{R}$, and hence can be considered as an algebra $(I, \underline{I}^\circ)$. Let $A = \{a, b, c\}$ and let $(A, \cdot)$ be a semilattice with $a < b, a < c$ and $b \| c$. For each $p \in I^\circ$ and x, y in A, define $xy\underline{p} := x \cdot y$, and consider the algebra $(A, \underline{I}^\circ)$, the

$\underline{I}^\circ$-semilattice. Now put

$$h: I \to A; \quad r \mapsto \text{if } r = 0 \text{ then } b \text{ else if } r = 1 \text{ then } a \text{ else } c.$$

The reader is asked to check that h is an $\underline{I}^\circ$-homomorphism. Note that the homomorphic image $h(I) = Ih$ of I is not a convex set. □

Consider the class $\underline{C}$ of convex subsets $(A, \underline{I}^\circ)$ of real vector spaces. Exercise 1.6W shows that $\underline{C}$ is closed under the taking of subalgebras and direct products. However, Example 1.6.5.2 shows that it is not closed under the taking of homomorphic images. Hence, it is not a variety. The smallest variety containing $\underline{C}$ consists of the homomorphic images of convex sets. Algebras in this variety are called *barycentric algebras*. Note that the variety $\underline{B}$ of barycentric algebras contains the class $\underline{C}$ of convex sets and the class of $\underline{I}^\circ$-semilattices. The significance of barycentric algebras is that they provide a general algebraic framework for the study of convexity. Moreover, convexity is closely connected with a (semilattice) order, and barycentric algebras give a uniform treatment of convex sets and semilattices. □

Exercises

1.6A. Show that a non-empty quasigroup $(Q, \cdot)$ with associative multiplication is a group.

[Hint: Obtain the identity element e by solving $xe = x$ for a particular x in Q, then show that it satisfies $ye = y = ey$ for any y in Q.]

1.6B. Show by examples that a subgroupoid $(H, \cdot)$ of a quasigroup $(Q, \cdot, /, \backslash)$ is not necessarily closed under the divisions, i.e. is not necessarily a subquasigroup.

1.6C. Define a *right quasigroup* to be a groupoid $(Q, \cdot)$ such that the equation $x \cdot a = b$ has a unique solution x in Q. Define the right division $/$ as in the case of quasigroups. Show the following: If $(Q, \cdot)$ is a right quasigroup, then $(Q, \cdot, /)$ satisfies the identities $y = (y/x)x$ and

$y = (yx)/x$ for all x, y in Q. Conversely, if an algebra $(Q, \cdot, /)$ with two binary operations satisfies these identities, then $(Q, \cdot)$ is a right quasigroup.

1.6D. Define left quasigroups, and formulate a counterpart of 1.6C for left quasigroups.

1.6E. Prove Proposition 1.6.1.2.

1.6F. Classify all quasigroups of order 3 up to isomorphism.

1.6G. Prove Proposition 1.6.2.1.

1.6H. Prove that for an S-module $(M, +, S)$ and a ring homomorphism $h : T \to S$, the condition (1.6.2.2) defines a T-module on the set M.

1.6I. Show that an S-module $(M, +, S)$ is faithful iff for each $0 \neq s \in S$ there is $m \in M$ such that $ms \neq 0$.

1.6J. Show that if the ring S is commutative, then each mapping R_s, for s in S, is in fact a homomorphism of the module $(M, +, S)$, i.e. for each t in $S, (mt)R_s = (mR_s)t$. Additionally, if s is invertible, show that R_s is bijective and a module isomorphism.

1.6K. Prove that each quasigroup satisfies the identities (1.6.1.2) and (1.6.1.3).

1.6L. Prove that A^* is a vector space isomorphic to V. Extend the definition (1.6.4.5) to the linear combination $p_1 f_1 + \cdots + p_n f_n$ for $n \geq 2$.

1.6M. Prove all claims of Proposition 1.6.4.2. Prove that for any $p, q \in B$, one has $B - p = B - q$.

1.6N. Let A be a vector space over a finite field $GF(2^n)$. Consider the algebra $(A, \underline{GF(2^n)})$ with the binary weighted means. Show that non-empty subalgebras of $(A, \underline{GF(2^n)})$ are not necessarily affine subspaces of the affine space A.

1.6O. Let A be an affine space over a field F of characteristic not equal to 2. Describe all 2-generated subalgebras of (A, F), and show that for distinct generators, they are precisely the lines through the generators.

Show that a non-empty subset B of A is an affine subspace of A precisely when it contains, with any two distinct points, the line through these points.

1.6P. Show that the mapping $\alpha : A \to V$ defined by (1.6.4.4) is a linear map from the vector space A^* to the vector space V.

1.6Q. Prove the implication $(b) \Rightarrow (a)$ of Proposition 1.6.4.3.

1.6R. Let A and A' be affine spaces with associated vector spaces V and V'. Let $p \in A$ and $p' \in A'$, and let $h : V \to V'$ be a linear map. Show that there is a unique affine map $a : A \to A'$ with $pa = p'$ and $a^* = h$. Show that for any $q \in A$, one has $qa = p' + (q - p)a^*$.

1.6S. Let A and A' be affine spaces, and let $a : A \to A'$ be an affine map. Show that for all $p, q \in A$, one has $(p - q)a^* = pa - qa$.

1.6T. (a) Find all homomorphisms from $(I, \underline{I}^\circ)$ to $\underline{I}^\circ$-semilattices.

(b) Find all congruence relations corresponding to the above homomorphisms.

(c) Show that there are no other proper non-trivial congruences of $(I, \underline{I}^\circ)$.

1.6U. Show that the class $\underline{C}$ of real convex sets is closed under the operators S and P.

1.6V. Find examples of barycentric algebras that are neither convex sets nor semilattices.

1.7. Quasi-identities and equational implications

Let $\tau : \Omega \to \mathbb{N}$ be a fixed (finitary) type of algebras. A formula of the form

$$(q) \qquad (w_1 = w_1' \,\&\, \ldots \,\&\, w_n = w_n') \to (w = w'),$$

where $w_1, \ldots, w_n, w_1', \ldots, w_n', w, w'$ are derived operators of type τ (cf. Section 1.4), is called a *quasi-identity*. Let $\arg(q)$ denote the (finite) set $\arg(w_1) \cup$

$\cdots \cup \arg(w_n) \cup \arg(w'_1) \cup \cdots \cup \arg(w'_n) \cup \arg(w) \cup \arg(w')$ of variables appearing in q. A τ-algebra (A, Ω) is said to *satisfy the quasi-identity* q if for each mapping $f \colon \arg(q) \to A$ and the (uniquely determined) homomorphism $\bar{f} \colon (\arg(q)\Omega, \Omega) \to (A, \Omega)$ extending f, whenever $w_1\bar{f} = w'_1\bar{f}, \ldots, w_n\bar{f} = w'_n\bar{f}$, then also $w\bar{f} = w'\bar{f}$. A class of τ-algebras *satisfies the quasi-identity* q, if each algebra in this class does.

Definition 1.7.1. If Σ is a set of quasi-identities of type τ, then the class $Q(\Sigma)$ *defined or axiomatized by* Σ is the class of all τ-algebras which satisfy all the quasi-identities in Σ. One says that $Q(\Sigma)$ is *defined by the quasi-identities* Σ or that it is a *quasi-equationally definable class* (or briefly a *quasi-equational class*). □

To denote that an algebra (A, Ω) or a class $\underline{K}$ of τ-algebras satisfy a quasi-identity q or a set Σ of quasi-identities, one respectively writes

$$A \models q, \ \underline{K} \models q, \ A \models \Sigma, \ \underline{K} \models \Sigma.$$

The equations $w_1 = w'_1, \ldots, w_n = w'_n$ in q are usually called the *premises*, and the equation $w = w'$ is called the *conclusion*. If q is satisfied in (A, Ω) or in $\underline{K}$, then one also says that q is *true* or *holds* in (A, Ω) or $\underline{K}$, respectively.

Remark 1.7.2. The set $\{w_1 = w'_1, \ldots, w_n = w'_n\}$ of premises of a quasi-identity q may be empty. In this case the quasi-identity q is considered to be the identity $w = w'$. By Proposition 1.5.3, it is clear that such a quasi-identity is satisfied in an algebra precisely if it is satisfied as an identity. It follows that identities form a special case of quasi-identities. Similarly, equational classes form a special case of quasi-equational classes. □

Example 1.7.3. The definition of a quasigroup $(Q, \cdot)$ implies that the *right multiplication* $R(q)$ by q in Q,

$$R(q) \colon Q \to Q; \ x \mapsto xq,$$

and the *left multiplication* $L(q)$ by q,

$$L(q) : Q \to Q; \ x \mapsto qx ,$$

are bijections of the underlying set Q. Their injectivity may be expressed by *right and left cancellation* quasi-identities:

(1.7.1) $\qquad\qquad\qquad (xz = yz) \to (x = y);$

(1.7.2) $\qquad\qquad\qquad (xy = xz) \to (y = z).$

They are satisfied in all quasigroups, and in particular, in all groups. Note however that the class of quasigroups, as well as the class of groups, can be defined purely by identities. □

Example 1.7.4. The *modular law* for lattices is the quasi-identity

(1.7.3) $\qquad\qquad\qquad (x + z = x) \to x(y + z) = xy + z.$

Exercise 1.7A shows that this quasi-identity is in fact equivalent to an identity, i.e. if either one holds in a lattice, then the other holds, too. Examples of modular lattices are provided by distributive lattices, lattices of normal subgroups of groups, and lattices of subspaces of vector spaces [Exercises 1.7C and 1.7D]. Exercise 1.7E shows that each Mal'cev algebra has a modular congruence lattice, i.e. it is *congruence modular*. □

Example 1.7.5. Example 1.7.3 may be generalized further. The cancellation quasi-identities (1.7.1) and (1.7.2) may also be considered in groupoids, semigroups and monoids. In this way, one obtains quasi-equational classes of (right, left, and two-sided) cancellative groupoids, semigroups and monoids. The name "cancellative" will be reserved for the case when both right and left cancellation quasi-identities hold. Note that the class of cancellative monoids is not equational. Consider the multiplicative monoid $(\mathbb{Z}, \cdot, 1)$ of integers. It is cancellative. However, for each n that is not prime, its homomorphic image $\mathbb{Z}_n = \mathbb{Z}/n\mathbb{Z}$ is not cancellative. By Proposition 1.5.6, the class of cancellative monoids cannot be equational. □

Proposition 1.7.6. *Each quasi-equational class of τ-algebras is a prevariety.* □

The proof is relegated to Exercises 1.7G and 1.7H.

Similarly, as identities form a special case of quasi-identities, so quasi-identities form a special case of more general formulas admitting an infinite number of premises and an arbitrary set of variables. Such an "infinitely long implication"

$$(1.7.4) \qquad (w_1 = w_1' \ \& \ w_2 = w_2' \ \& \ldots) \to (w = w'),$$

where $w_1, \ldots, w_n, w_1', \ldots, w_n', w, w'$ are Ω-words over a fixed set X of arbitrary cardinality, will be called an *equational implication*. More general equational implications with an arbitrary set M of premises have the form

$$(i) \qquad (\&_{m \in M} \ w_m = w_m') \to (w = w').$$

Satisfaction of an implication (i) in an algebra or a class of τ-algebras is defined as for quasi-identities. Other concepts concerning quasi-identities also have their implicational counterparts. In particular, one can speak about classes of τ-algebras satisfying given implications, or *implicational classes*. Exercises 1.7G and 1.7H can also be generalized.

The next lemma gives a simple condition for a τ-algebra A to satisfy an implication i. Let P denote the set of premises of i, and let C be the set consisting of the conclusion of i. One can think of an implication i as a pair (P, C) of subsets of $X\Omega \times X\Omega$, where $X = \arg(i)$. Then A satisfies i if for each mapping $f : X \to A$ with corresponding Ω-homomorphism $\bar{f} : X\Omega \to A$

1. ALGEBRAS

extending f, one has

$$(\forall m \in M, (w_m, w'_m) \in \ker \bar{f}) \Rightarrow ((w, w') \in \ker \bar{f}).$$

This condition can be rewritten as follows:

$$(P \subseteq \ker \bar{f}) \Rightarrow (C \subseteq \ker \bar{f}).$$

Now for the congruences $\varphi = \langle P \rangle_{cg}$ and $\psi = \langle P \cup C \rangle_{cg}$ on $X\Omega$ there is an Ω-homomorphism

$$h_i : X\Omega^\varphi \to X\Omega^\psi; \; t^\varphi \mapsto t^\psi.$$

Evidently, $X\Omega^\psi \models i$.

Lemma 1.7.7. *An implication i is satisfied in an Ω-algebra A if and only if for every Ω-homomorphism $h : X\Omega^\varphi \to A$, there is an Ω-homomorphism $g : X\Omega^\psi \to A$ with $h = h_i g$:*

$$\begin{array}{ccc} X\Omega^\varphi & \xrightarrow{h_i} & X\Omega^\psi \\ h \downarrow & & \downarrow g \\ A & = & A \end{array}$$

Proof. ($\Rightarrow$) Assume that $A \models i$. Then for the Ω-homomorphism $\bar{f} : X\Omega \to A$ extending the mapping

$$f : X \to X\Omega \to X\Omega^\varphi \to A; \; x \mapsto x^\varphi h,$$

one has

$$(\&_{m \in M} \; w_m \bar{f} = w'_m \bar{f}) \Rightarrow (w\bar{f} = w'\bar{f}).$$

Note that $\bar{f} = \mathrm{nat}\varphi \; h$, and consider the following diagram:

$$\begin{array}{ccccccc} A & = & A & = & A & = & A \\ f \uparrow & & \bar{f} \uparrow & & h \uparrow & & g \uparrow \\ X & \to & X\Omega & \xrightarrow{\mathrm{nat}\varphi} & X\Omega^\varphi & \xrightarrow{h_i} & X\Omega^\psi \end{array}$$

It is clear that

$$(\&_{m \in M} \ w_m^\varphi h = w'^\varphi_m h) \Rightarrow (w^\varphi h = w'^\varphi h).$$

Define $g : X\Omega^\psi \to A;\ t^\psi \mapsto t^\varphi h$:

$$\begin{array}{ccccc}
t\bar{f} & = & t^\varphi h & = & t^\psi g \\
\bar{f} \uparrow & & \uparrow h & & \uparrow g \\
t & \xrightarrow{\text{nat}\varphi} & t^\varphi & \xrightarrow{h_i} & t^\psi
\end{array}$$

Clearly, g is a well defined Ω-homomorphism, and $h = h_i g$.

($\Leftarrow$) Let $f : X \to A$ be any mapping such that the homomorphism $\bar{f} : X\Omega \to A$ satisfies $w_m \bar{f} = w'_m \bar{f}$ for each $m \in M$. Then obviously $\bar{f} = \text{nat}\varphi\ h$, where $h : X\Omega^\varphi \to A;\ t^\varphi \mapsto t\bar{f}$. By the hypothesis, there is a homomorphism $g : X\Omega^\psi \to A$ with $h = h_i g$. Thus $\bar{f} = \text{nat}\varphi\ h_i\ g = \text{nat}\psi\ g$. This implies $w\bar{f} = w^\psi g = w'^\psi g = w'\bar{f}$, and hence $A \models i$. □

If the condition of Lemma 1.7.7 is satisfied, one says that the algebra A is *injective with respect to* h_i, or is *i-injective*.

Exercises

1.7A. Let L be a lattice. Show that the following quasi-identities are pairwise equivalent in L:
   (a) $(x + z = x) \to (x(y + z) = xy + z)$;
   (b) $(x + z = x) \to (x(y + z) = xy + xz)$;
   (c) $x(y + xz) = xy + xz$.

1.7B. Show that the monoid $(\mathbb{Z}_n, \cdot, 1)$, where n is not prime, does not satisfy the cancellation quasi-identities.

1.7C. Show that the lattice of normal subgroups of a group is modular.

1.7D. Show that the lattice of subspaces of a vector space is modular.

1.7E. [Birkhoff 1967] Show that each Mal'cev algebra is congruence modular.

1.7F. Show that each convex set $(A, \underline{I}^\circ)$ satisfies the cancellation quasi-identity for each $\underline{p} \in \underline{I}^\circ$.

1.7G. Show that if a quasi-identity is satisfied in a τ-algebra (A, Ω), then it is satisfied in each subalgebra of (A, Ω).

1.7H. Show that if a quasi-identity is satisfied in the τ-algebras (A_i, Ω), for i in I, then it is satisfied in the direct product $\left(\prod_{i \in I} A_i, \Omega \right)$.

CHAPTER II

CATEGORIES OF ALGEBRAS

In many areas of mathematics, we often have to consider sets with a certain structure, together with mappings between the sets which preserve the structure. Examples are given by algebras of a given type and homomorphisms between them, topological spaces and continuous maps, or relational structures and structure-preserving mappings. A general setting that encompasses all these situations, and many others, is given by the concept of a category discussed in Section 2.1. This concept allows one to comprehend an entire class of algebras, or other sets with structure, together with appropriate mappings. It also provides a common language allowing one to find analogies between different classes of objects with their corresponding mappings. Different categories are compared using special types of mappings called functors, discussed in Section 2.2. Finally, Section 2.3 provides the basic categorical constructions used in this book: products, coproducts, and certain types of limits.

The current chapter presents the basic notions needed later in this book. Occasionally, in a few places, we refer the reader to more advanced notions and results of category theory from other sources, usually our monograph [PMA] or [Mac Lane 1971, 1998]. See also [Herrlich and Strecker 1973] for a good basic introduction to category theory, and [Bergman 1998] for a discussion of universal constructions in algebra.

2.1. Categories and subcategories

Definition 2.1.1. A *category* $\mathcal{C}$ consists of two classes, a class Ob$\mathcal{C}$ whose members are called *objects* or *points* of $\mathcal{C}$, and a class Mor$\mathcal{C}$ whose members are called *morphisms* or *arrows* of $\mathcal{C}$.

For each arrow $f \in$ Mor$\mathcal{C}$ there is a unique pair (a,b) of objects of $\mathcal{C}$, where a is called the *domain* or *tail* of f and b is called the *codomain* or *head* of f. For each pair $(a,b) \in$ Ob$\mathcal{C} \times$ Ob$\mathcal{C}$, the class $\mathcal{C}(a,b)$ of arrows with tail a and head b is a set.

For each $a \in$ Ob$\mathcal{C}$, there is an *identity* morphism $1_a \in \mathcal{C}(a,a)$. For each triple a,b,c of objects of $\mathcal{C}$, there is a *composition*

$$\mathcal{C}(a,b) \times \mathcal{C}(b,c) \to \mathcal{C}(a,c); (f,g) \mapsto fg$$

such that, for all $a,b \in$ Ob$\mathcal{C}$ and all $f \in \mathcal{C}(a,b)$,

$$1_a f = f = f 1_b ,$$

and for all $a,b,c,d \in$ Ob$\mathcal{C}$, all $f \in \mathcal{C}(a,b)$, all $g \in \mathcal{C}(b,c)$ and all $h \in \mathcal{C}(c,d)$,

$$(fg)h = f(gh). \quad \square$$

An element f of $\mathcal{C}(a,b)$ may be represented pictorially as $a \xrightarrow{f} b$ or $f : a \to b$. In this convention, the composition fg appears as

$$\begin{array}{ccc} a & = & a \\ {\scriptstyle f} \downarrow & & \downarrow {\scriptstyle fg} \\ b & \xrightarrow{g} & c \end{array}$$

Example 2.1.2 (Sets). The category $\underline{\mathrm{Set}}$ of sets has the class of all sets as the class of objects. For two sets A and B, the set $\underline{\mathrm{Set}}(A,B)$ is the set B^A of functions from A to B, and the class Mor$\underline{\mathrm{Set}}$ of arrows of $\underline{\mathrm{Set}}$ is the (disjoint) union of the sets $\underline{\mathrm{Set}}(A,B)$ for all sets A and B. $\quad \square$

2. CATEGORIES OF ALGEBRAS

Example 2.1.3 (τ-algebras). The category $\underline{\tau}$ or (Ω) of algebras of a given type $\tau : \Omega \to \mathbb{N}$ has the class of all τ-algebras as the class of objects. For two τ-algebras A and B, the set $\underline{\tau}(A, B)$ is the set of all τ-homomorphisms from A to B. The class $\text{Mor}\underline{\tau}$ is the (disjoint) union of the sets $\underline{\tau}(A, B)$ for all τ-algebras A and B. □

Definition 2.1.4. A category $\mathcal{C}_1$ is a *subcategory* of a category $\mathcal{C}$ if $\text{Ob}\mathcal{C}_1 \subseteq \text{Ob}\mathcal{C}$, $\text{Mor}\mathcal{C}_1 \subseteq \text{Mor}\mathcal{C}$, and moreover $\mathcal{C}_1(a, b) \subseteq \mathcal{C}(a, b)$ for all $a, b \in \text{Ob}\mathcal{C}_1$. If for $f, g \in \text{Mor}\mathcal{C}_1$, the composition fg is defined, then it is the same in both categories. Moreover, for each $a \in \text{Ob}\mathcal{C}_1$, the identity morphism 1_a is the same in both categories. If for all $a, b \in \text{Ob}\mathcal{C}_1$ one has $\mathcal{C}_1(a, b) = \mathcal{C}(a, b)$, then the subcategory $\mathcal{C}_1$ is said to be a *full* subcategory of $\mathcal{C}$. □

Example 2.1.5. The categories of Example 2.1.3 have many full subcategories with τ-homomorphisms as morphisms. Here is a list of some of them:

$\underline{Q}$ - quasigroups and quasigroup homomorphisms;

$\underline{\text{Mon}}$ - monoids and monoid homomorphisms;

$\underline{\text{CMon}}$ - commutative monoids, a full subcategory of $\underline{\text{Mon}}$;

$\underline{\text{Gp}}$ - groups, a full subcategory of $\underline{\text{Mon}}$;

$\underline{\text{AGp}}$ - abelian groups, a full subcategory of $\underline{\text{Gp}}$;

$\underline{\text{Sgp}}$ - semigroups and semigroup homomorphisms;

$\underline{\text{CSgp}}$ - commutative semigroups, a full subcategory of $\underline{\text{Sgp}}$;

$\underline{\text{Ring}}$ - unital rings and ring homomorphisms;

$\underline{\text{CRing}}$ - commutative rings, a full subcategory of $\underline{\text{Ring}}$;

$\underline{\text{Mod}}_S$ (for a fixed unital ring S) - unital right S-modules and S-module homomorphisms. □

If there is no danger of confusion, we usually denote a class $\underline{K}$ of algebras of a given type and the corresponding category of $\underline{K}$-algebras and $\underline{K}$-homomorphisms by the same symbol.

Definition 2.1.6. Let $\mathcal{C}$ be a category. A category $\mathcal{C}^{op}$ is called the *opposite* or *dual* of the category $\mathcal{C}$ if $\text{Ob}\mathcal{C}^{op} = \text{Ob}\mathcal{C}$, $\text{Mor}\mathcal{C}^{op} = \text{Mor}\mathcal{C}$, but the arrows of $\mathcal{C}$ are reversed (turned head to tail) to yield the arrows of $\mathcal{C}^{op}$. Moreover, for $a, b, c \in \text{Ob}\mathcal{C}$

$$\mathcal{C}^{op}(a,b) = \mathcal{C}(b,a),$$

and for $f \in \mathcal{C}^{op}(a,b)$ and $g \in \mathcal{C}^{op}(b,c)$, the composition $f \circ g$ or gf of f and g in $\mathcal{C}^{op}$ is defined by

$$\mathcal{C}^{op}(a,b) \times \mathcal{C}^{op}(b,c) \to \mathcal{C}^{op}(a,c); \ (f,g) \mapsto f \circ g := gf. \quad \square$$

Example 2.1.7. Let $(A, \leq)$ be a poset. Define a category (A) or $\leq_A$ as follows. Take $\text{Ob}\leq_A := A$, and for all $a, b \in A$, define $\leq_A(a,b) =$ **if** $a \leq b$ **then** $\{(a,b)\}$ **else** $\varnothing$. Define the identity morphism 1_a to be (a,a). The composition of morphisms is then defined uniquely by the transitivity of $\leq$. The dual of the category $\leq_A$ is the ordered set $(A, \geq)$. $\quad \square$

Note that in the category $\leq_A$ both the class of objects and the class of morphisms are actually sets. Such categories are called *small*. Note that the category $\leq_A$ is different from the category of the next example.

Example 2.1.8 (Ordered sets). The category <u>Ord</u> of ordered sets has the class of ordered sets as its class of objects and the class of order preserving mappings as its class of morphisms. $\quad \square$

Let $e : a \to b$ be a morphism in a category $\mathcal{C}$. Let $c \in \text{Ob}\mathcal{C}$. If

$$\forall f : b \to c, \ \forall g : b \to c, \ ef = eg \Rightarrow f = g,$$

then e is said to be an *epimorphism* in $\mathcal{C}$. Note that surjective homomorphisms in the category <u>τ</u> are all epimorphisms.

To obtain the dual of a concept in a category $\mathcal{C}$, one may interpret the concept in the dual category $\mathcal{C}^{op}$. The dual of the concept "epimorphism" is the concept "monomorphism". Thus a *monomorphism* $m : b \to a$ in $\mathcal{C}$ is a morphism such that

$$\forall f : c \to b, \ \forall g : c \to b, \ fm = gm \Rightarrow f = g.$$

In particular, injective homomorphisms in the category $\underline{\tau}$ are all monomorphisms.

Suppose that for two morphisms $r : a \to b$ and $s : b \to a$ of a category $\mathcal{C}$ one has

(2.1.1) $$sr = 1_b.$$

Then s is a *left inverse* of r and r is a *right inverse* of s. If (2.1.1) holds, then r is an epimorphism and s is a monomorphism. (Note however that epimorphisms do not need to be right inverses.) A morphism in a category is *invertible* or an *isomorphism* if it is both a left inverse and a right inverse. The coincident right and left inverses of an isomorphism f are denoted by f^{-1}, the *inverse* of the isomorphism f. One writes $a \cong b$, if there is an isomorphism $f : a \to b$.

Example 2.1.9. Let $(M, \cdot, 1)$ be a monoid. Define $\mathrm{Ob}\mathcal{M} := \{1\}$ and $\mathrm{Mor}\mathcal{M} := M$. Let $\mathcal{M}(1,1) := M$. Define the composition

$$\mathcal{M}(1,1) \times \mathcal{M}(1,1) \to \mathcal{M}(1,1); \ (m,n) \mapsto mn$$

by monoid multiplication. Thus $\mathcal{M}$ forms a small category, with a single object. If the monoid M is a group, then each morphism of $\mathcal{M}$ is an isomorphism. □

Definition 2.1.10. (a) An object t of a category $\mathcal{C}$ is *terminal* if $\mathcal{C}(a,t)$ is a singleton for each object a of $\mathcal{C}$.
(b) An object $\bot$ of a category $\mathcal{C}$ is an *initial object* of $\mathcal{C}$ if $\mathcal{C}(\bot,a)$ is a singleton for each object a of $\mathcal{C}$.
(c) An object 0 of a category $\mathcal{C}$ is a *zero object* if it is both initial and terminal. □

Initial objects in a category $\mathcal{C}$ are terminal in the dual category $\mathcal{C}^{op}$. Similarly, terminal objects in $\mathcal{C}$ are initial in $\mathcal{C}^{op}$. So the concepts of initial and terminal objects are dual to each other. In the category <u>Set</u>, the empty set is initial and each one-element set is terminal. In a poset category $\leq_A$, the least element is initial and the greatest element is terminal. The zero group $\{0\}$ is a zero object in the category <u>AGp</u> of abelian groups. In the category <u>Ring</u>, the ring $\mathbb{Z}$ is initial and the zero ring is terminal. A category may have no initial or terminal objects.

Proposition 2.1.11.

(a) Any two terminal objects of a category $\mathcal{C}$ are isomorphic.

(b) Any two initial objects of a category $\mathcal{C}$ are isomorphic.

(c) Any two zero objects of a category $\mathcal{C}$ are isomorphic.

Proof. Suppose that $\mathcal{C}$ has two terminal objects t_1 and t_2. Then there is a unique morphism $h : t_1 \to t_2$ and a unique morphism $h' : t_2 \to t_1$. Since $hh' : t_1 \to t_1$ and $h'h : t_2 \to t_2$ are uniquely defined, and $1_{t_1} : t_1 \to t_1$ and $1_{t_2} : t_2 \to t_2$, it follows that $hh' = 1_{t_1}$ and $h'h = 1_{t_2}$, so that t_1 and t_2 are isomorphic.

(b) is the dual of (a), and (c) is a special case of (a). □

Terminal and initial objects play an important rôle in the specification of categorical constructions. (See Section III.2 of [PMA].)

Exercises

2.1A. (a) Generalize Example 2.1.7 to show that each quasi-order yields a small category.

(b) Suppose that for each pair (a, b) of objects of a small category $\mathcal{C}$, the set $\mathcal{C}(a, b)$ has at most one element. Show that Mor$\mathcal{C}$ is a quasi-order on Ob$\mathcal{C}$.

2.1B. Let f be a morphism in a category $\mathcal{C}$. Show that f is an isomorphism iff it has a unique left inverse and a unique right inverse that coincide.

2.1C. (a) Show that composition in a small category with a single object forms a monoid.

(b) Show that if $\mathcal{C}$ is a category with a single object, and such that each morphism is an isomorphism, then $\mathcal{C}$ forms a group.

2.2. Functors

There is a special type of function between two categories that allows one to compare the classes of objects and the classes of morphisms.

Definition 2.2.1. Let $\mathcal{C}$ and $\mathcal{D}$ be categories. Then a *(covariant) functor* $F : \mathcal{C} \to \mathcal{D}$ is a pair of mappings (F_o, F_m), an *object part* $F_o : \mathrm{Ob}\mathcal{C} \to \mathrm{Ob}\mathcal{D}$ and a *morphism part* $F_m : \mathrm{Mor}\mathcal{C} \to \mathrm{Mor}\mathcal{D}$, with the following properties:

(a) $\forall a, b, \in \mathrm{Ob}\mathcal{C},\ \mathcal{C}(a,b)F_m \subseteq \mathcal{D}(aF_o, bF_o)$;

(b) $\forall a \in \mathrm{Ob}\mathcal{C},\ 1_a F_m = 1_{aF_o}$;

(c) $\forall a, b, c \in \mathrm{Ob}\mathcal{C},\ \forall f \in \mathrm{Mor}(a,b),\ \forall g \in \mathrm{Mor}(b,c),\ (fg)F_m = (fF_m)(gF_m)$.

One usually suppresses the indices o and m of Definition 2.2.1, and writes simply $F : \mathrm{Ob}\mathcal{C} \to \mathrm{Ob}\mathcal{C}$ and $F : \mathrm{Mor}\mathcal{C} \to \mathrm{Mor}\mathcal{C}$. Note the analogy relating homomorphisms between algebras and functors between categories. In particular, for each category $\mathcal{C}$, the pair $1_\mathcal{C} := (1_{\mathrm{Ob}\mathcal{C}}, 1_{\mathrm{Mor}\mathcal{C}})$ of identities on $\mathrm{Ob}\mathcal{C}$ and $\mathrm{Mor}\mathcal{C}$ is a functor. Moreover, for two functors $F : \mathcal{C} \to \mathcal{D}$ and $G : \mathcal{D} \to \mathcal{E}$, their *composition* $FG = (F_o G_o, F_m G_m)$ is a functor, too. One may thus form a *category* <u>Cat</u> of *small categories*, with Ob<u>Cat</u> as the proper class of small categories. Given small categories c and d, the set <u>Cat</u>(c,d) consists of functors from c to d. Composition in <u>Cat</u> is defined by composition of functors.

Example 2.2.2. Let $\mathcal{C}_1$ be a subcategory of a category $\mathcal{C}$. Then there are two inclusions $\mathrm{Ob}\mathcal{C}_1 \rightarrowtail \mathrm{Ob}\mathcal{C}$ and $\mathrm{Mor}\mathcal{C}_1 \rightarrowtail \mathrm{Mor}\mathcal{C}$ yielding the *inclusion functor*

$J : \mathcal{C}_1 \to \mathcal{C}$. □

Example 2.2.3. Let $\mathcal{C}$ be a subcategory of the category $\underline{\tau}$ of algebras of a given type $\tau : \Omega \to \mathbb{N}$. Let $\Omega' \subseteq \Omega$. For each (A, Ω) in $\mathcal{C}$, let $(A, \Omega)F_o$ be the reduct (A, Ω') of (A, Ω). And for each τ-homomorphism $f : (A, \Omega) \to (B, \Omega)$, let fF_m be the same mapping, but considered as an Ω'-homomorphism $(A, \Omega') \to (B, \Omega')$. In this way one obtains the *restriction functor* $F = (F_o, F_m)$. This functor "forgets" a part of the structure of an object, "remembering" only the Ω'-structure. Functors of this type are called *forgetful functors*. In particular, it may happen that $\Omega' = \emptyset$. In this case, one obtains the *underlying set* functor $\mathcal{C} \to \underline{\text{Set}}$ that forgets the structure of each algebra, remembering only the underlying set. If $\mathcal{C} = \underline{\tau}$, then the underlying set functor functor is written as $U : \underline{\tau} \to \underline{\text{Set}}$. □

Example 2.2.4. There is also an important functor $\Omega : \underline{\text{Set}} \to \underline{\tau}$, the *word algebra functor*, going in the opposite direction. Its object part is defined by $X \mapsto X\Omega$. To define its morphism part note that, by Proposition 1.4.1, for each mapping $f : X \to Y$, the composition $f' : X \to Y \rightarrowtail Y\Omega$ has a unique extension $\overline{f} : (X\Omega, \Omega) \to (Y\Omega, \Omega)$ such that $\overline{f}|_X = f$, i.e. the following diagram commutes:

$$\begin{array}{ccc} X & \longrightarrow & X\Omega \\ f\downarrow & & \downarrow \overline{f} \\ Y & \longrightarrow & Y\Omega \end{array}.$$

In particular, for $w = w_1 \ldots w_n \omega$ in $X\Omega$, one has $w\overline{f} = (w_1\overline{f})\ldots(w_n\overline{f})\omega$. The morphism part of the functor Ω is defined by $(f : X \to Y) \mapsto (\overline{f} : X\Omega \to Y\Omega)$. □

Example 2.2.5. Let $(A, \leq)$ and $(B, \leq)$ be poset categories (A) and (B), as in Example 2.1.7. Then the functors from (A) to (B) are just the monotonic or order-preserving functions $f : (A, \leq) \to (B, \leq)$. □

A *contravariant* functor $F : \mathcal{C} \to \mathcal{D}$ from a category $\mathcal{C}$ to a category $\mathcal{D}$ is a functor $F : \mathcal{C} \to \mathcal{D}^{op}$, or equivalently $F : \mathcal{C}^{op} \to \mathcal{D}$.

Example 2.2.6. For poset categories as in Example 2.2.5, the contravariant functors are just the order-reversing functions from $(A, \leq)$ to $(B, \leq)$. □

Example 2.2.7. Let A be a given set. We will define two functors $\underline{\text{Set}}(A, \_) : \underline{\text{Set}} \to \underline{\text{Set}}$ and $\underline{\text{Set}}(\_, A) : \underline{\text{Set}} \to \underline{\text{Set}}$ as follows:

$$\begin{array}{ccccc} X & & \underline{\text{Set}}(A, X) & \ni & \alpha \\ f \downarrow & \xrightarrow{\underline{\text{Set}}(A,\_)} & \downarrow & & \downarrow \\ Y & & \underline{\text{Set}}(A, Y) & \ni & \alpha f \end{array}$$

where $\alpha(f\underline{\text{Set}}(A, \_)) = \alpha f$. Similarly

$$\begin{array}{ccccc} X & & \underline{\text{Set}}(X, A) & \ni & f\alpha \\ f \downarrow & \xrightarrow{\underline{\text{Set}}(\_,A)} & \uparrow & & \uparrow \\ Y & & \underline{\text{Set}}(Y, A) & \ni & \alpha \end{array}$$

where $\alpha(f\underline{\text{Set}}(\_, A)) = f\alpha$. The first functor is covariant, the second is contravariant. □

For arbitrary categories $\mathcal{C}$ and $\mathcal{D}$, a functor $F : \mathcal{C} \to \mathcal{D}$ is an *isomorphism* if there is a functor $G : \mathcal{D} \to \mathcal{C}$ with $FG = 1_\mathcal{C}$ and $GF = 1_\mathcal{D}$. In this context, the categories $\mathcal{C}$ and $\mathcal{D}$ are *isomorphic*, a relation denoted by $\mathcal{C} \cong \mathcal{D}$. The functor G is the *(two-sided) inverse* F^{-1} of F.

Example 2.2.8. Let $\underline{\text{BAlg}}$ be the category of Boolean algebras and homomorphisms. Let $\underline{\text{BRing}}$ be the category of Boolean rings and homomorphisms. Define two functors $F : \underline{\text{BAlg}} \to \underline{\text{BRing}}$ and $G : \underline{\text{BRing}} \to \underline{\text{BAlg}}$ as follows. For each Boolean algebra $(B, \vee, \wedge, ', 0, 1)$, define BF to be the Boolean ring $(B, +, -, 0, \cdot, 1)$ as in Example 1.4.7. Given a Boolean algebra homomorphism $h : A \to B$, define hF to be the same mapping, and observe that hF is a

Boolean ring homomorphism. For each Boolean ring $(B, +, -, 0, \cdot, 1)$, define the Boolean algebra $(B, \vee, \wedge, ', 0, 1)$ as in Example 1.4.7. Given a Boolean ring homomorphism $f : A \to B$, define fG to be the same mapping, and observe that fG is a Boolean algebra homomorphism. Then it is easy to see that the functors F and G are mutually inverse isomorphisms between the category of Boolean algebras and the category of Boolean rings. □

Example 2.2.8 can be generalized as follows. Let $\underline{V}$ be an equational class of algebras of type $\tau : \Omega \to \mathbb{N}$. Let $\underline{W}$ be an equational class of algebras of type $\sigma : \Psi \to \mathbb{N}$. Assume additionally that both types τ and σ are plural (cf. Section 1.1). (Such a restriction is not absolutely necessary, but does make things simpler. This book is mostly concerned with algebras of plural type.) Assume that there are Ω-words w_ψ, where $\psi \in \Psi$, such that by defining $\psi := w_\psi$ in any $\underline{V}$-algebra (A, Ω), one obtains a reduct (A, Ψ) in the variety $\underline{W}$. Similarly, assume that there are Ψ-words t_ω, where $\omega \in \Omega$, such that by defining $\omega := t_\omega$ in any $\underline{W}$-algebra (B, Ψ), one obtains a reduct (B, Ω) in the variety $\underline{V}$. Define the functor $F : \underline{V} \to \underline{W}$ by

$$\begin{array}{ccc} (A, \Omega) & & (A, \Psi) \\ h \downarrow & \xrightarrow{F} & \downarrow hF \\ (B, \Omega) & & (B, \Psi) \end{array},$$

where $hF = h$ considered as a Ψ-homomorphism. Define the functor $G : \underline{W} \to \underline{V}$ by

$$\begin{array}{ccc} (A, \Psi) & & (A, \Omega) \\ f \downarrow & \xrightarrow{G} & \downarrow fG \\ (B, \Psi) & & (B, \Omega) \end{array},$$

where $fG = f$ considered as an Ω-homomorphism. Then the equational classes $\underline{V}$ and $\underline{W}$ are (*clonally*) *equivalent* if F and G are mutually inverse isomorphisms between the category of $\underline{V}$-algebras and $\underline{V}$-homomorphisms and the category of $\underline{W}$-algebras and $\underline{W}$-homomorphisms. This relation is denoted by $\underline{V} \simeq \underline{W}$. The functors F and G are sometimes called *interpretations* of $\underline{W}$ in $\underline{V}$ and of $\underline{V}$ in $\underline{W}$, respectively. Exercise 2.2B shows that the class $\underline{V}$ is equivalent to the class $\underline{V}_{\mathfrak{P}\Omega}$ of $\underline{V}$-algebras considered as algebras of the derived type $\tau' : \mathfrak{P}\Omega \to \mathbb{N}$. This allows one to consider any equational class $\underline{V}$ of τ-algebras as the category $\underline{V}_{\mathfrak{P}\Omega}$ (in fact, also an equational class) of algebras of the derived type $\tau' : \mathfrak{P}\Omega \to \mathbb{N}$, where one does not choose any particular set of basic operators. We will sometimes identify these two categories, and suppress the index $\mathfrak{P}\Omega$ in $\underline{V}_{\mathfrak{P}\Omega}$.

The relation of isomorphism between categories is often too restrictive, and so is replaced by a broader relation of "equivalence" which indicates that the categories are essentially the same. Some new concepts are needed to discuss this more general notion of equivalence.

The first of these concepts, that of a natural transformation, enables different functors between the same pair of categories to be compared.

Definition 2.2.9. Let $S : \mathcal{C} \to \mathcal{D}$ and $T : \mathcal{C} \to \mathcal{D}$ be functors with common domain and codomain. Then a *natural transformation* $\tau : S \longrightarrow T$ between S and T is a function

$$\tau : \mathrm{Ob}\mathcal{C} \to \mathrm{Mor}\mathcal{D}; \quad x \mapsto (\tau_x : xS \to xT)$$

such that, for each morphism $f : x \to y$ of $\mathcal{C}$, the diagram

$$\begin{array}{ccc} xS & \xrightarrow{\tau_x} & xT \\ {\scriptstyle fS}\downarrow & & \downarrow{\scriptstyle fT} \\ yS & \xrightarrow[\tau_y]{} & yT \end{array}$$

commutes. The $\mathcal{D}$-morphism $\tau_x : xS \to xT$ is called the *component* of τ at the object x of $\mathcal{C}$. A natural transformation is called a *natural isomorphism* if each of its components is an isomorphism of $\mathcal{D}$. This is denoted by $\tau : S \cong T$ or $S \cong T$. □

Given two natural transformations $\sigma : S \longrightarrow T$ and $\tau : T \longrightarrow U$ between functors $S : \mathcal{C} \to \mathcal{D}$, $T : \mathcal{C} \to \mathcal{D}$, and $U : \mathcal{C} \to \mathcal{D}$, the *composite* $\sigma\tau : S \longrightarrow U$ is the natural transformation defined by its components $(\sigma\tau)_x := \sigma_x \tau_x$ at objects x of S. This composition is associative. Given a functor $S : \mathcal{C} \to \mathcal{D}$, there is a natural isomorphism $1_S : S \longrightarrow S$, called the *identity transformation*, with components $1_{xS} : xS \to xS$.

Definition 2.2.10. For arbitrary categories $\mathcal{C}$ and $\mathcal{D}$, a functor $F : \mathcal{C} \to \mathcal{D}$ is an *equivalence* (and the categories $\mathcal{C}$ and $\mathcal{D}$ are *equivalent*), if there is a functor $G : \mathcal{D} \to \mathcal{C}$ such that $FG \cong 1_\mathcal{C}$ and $GF \cong 1_\mathcal{D}$. In this case $G : \mathcal{D} \to \mathcal{C}$ is also an equivalence of categories. □

Example 2.2.11. Let $\mathcal{C}$ be a category. A full subcategory $\mathcal{D}$ of $\mathcal{C}$, such that each object of $\mathcal{C}$ is isomorphic to exactly one object of $\mathcal{D}$, is called a *skeleton* of $\mathcal{C}$. Then $\mathcal{D}$ is equivalent to $\mathcal{C}$, and the inclusion $\mathcal{T} : \mathcal{D} \to \mathcal{C}$ is an equivalence of categories. Indeed, for each object c of $\mathcal{C}$, one can select an isomorphism $i_c : c \to cT$ with $cT \in \text{Ob}\mathcal{D}$. Then define a functor $T : \mathcal{C} \to \mathcal{D}$ in the unique possible way so that i will become a natural isomorphism $TJ \cong 1$. Moreover, $JT \cong 1$ too, so J is indeed an equivalence. In fact, a category is equivalent to any one of its skeletons. □

A functor $F : \mathcal{C} \to \mathcal{D}$ is *full*, if for each pair x and y of objects of $\mathcal{C}$, the function $\mathcal{C}(x,y) \to \mathcal{D}(xF, yF)$; $f \mapsto fF$ surjects. Note that a subcategory $\mathcal{C}$ of a category $\mathcal{D}$ is full if the inclusion functor $J : \mathcal{C} \to \mathcal{D}$ is full.

A functor $F : \mathcal{C} \to \mathcal{D}$ is *faithful* if, for each pair x and y of objects of $\mathcal{C}$, the function $\mathcal{C}(x,y) \to \mathcal{D}(xF, yF)$; $f \mapsto fF$ injects. For example, the inclusion functor $J : \mathcal{C} \to \mathcal{D}$ from any subcategory $\mathcal{C}$ of $\mathcal{D}$ is faithful.

Finally, a functor $F : \mathcal{C} \to \mathcal{D}$ is *dense*, if each object of $\mathcal{D}$ is isomorphic to

2. CATEGORIES OF ALGEBRAS

the image of an object of $\mathcal{C}$ under F.

Theorem 2.2.12. *For arbitrary categories $\mathcal{C}$ and $\mathcal{D}$, a functor $F : \mathcal{C} \to \mathcal{D}$ is an equivalence if and only if it is full, faithful and dense.*

For a proof, we refer the reader to [Maclane 1971, IV.4, Theorem 1] or [PMA, Theorem 3.2.3].

Example 2.2.13. Consider again the functors $U : \underline{\tau} \to \underline{\text{Set}}$ and $\Omega : \underline{\text{Set}} \to \underline{\tau}$ of Examples 2.2.3 and 2.2.4. Proposition 1.4.1 shows that there is a bijection

(2.2.1) $$\underline{\text{Set}}(X, AU) \cong \underline{\tau}(X\Omega, A).$$

More explicitly, the bijection

(2.2.2) $$\varphi_A^X : \underline{\tau}(X\Omega, A) \to \underline{\text{Set}}(X, AU)$$

is the component at A of a natural isomorphism

(2.2.3) $$\varphi^X : \underline{\tau}(X\Omega, \_) \cong \underline{\text{Set}}(X, \_U),$$

and the component at X of a natural isomorphism

(2.2.4) $$\varphi_A : \underline{\tau}(\_\Omega, A) \cong \underline{\text{Set}}(\_, AU),$$

where the four functors appearing in (2.2.3) and (2.2.4) are defined as follows. For $k : A \to A'$ in $\text{Mor}\underline{\tau}$, one has

$$\begin{array}{ccccc} A & & \underline{\tau}(X\Omega, A) & \ni & g \\ k \downarrow & \xrightarrow{\underline{\tau}(X\Omega,\_)} & \downarrow & & \downarrow \\ A' & & \underline{\tau}(X\Omega, A') & \ni & gk \end{array},$$

and

$$\begin{array}{ccccc} A & & \underline{\text{Set}}(X, AU) & \ni & h \\ k \downarrow & \xrightarrow{\underline{\text{Set}}(X,\_\Omega)} & \downarrow & & \downarrow \\ A' & & \underline{\text{Set}}(X, A'U) & \ni & h(kU) \end{array}.$$

For $f : X' \to X$ in Mor$\underline{\text{Set}}$, one has

$$\begin{array}{ccccc}
X' & \underline{T}(X'\Omega, A) & \ni & (f\Omega)g \\
f\downarrow & \xrightarrow{\underline{T}(\_\Omega, A)} & \uparrow & & \uparrow \\
X & \underline{T}(X\Omega, A) & \ni & g
\end{array},$$

and

$$\begin{array}{ccccc}
X' & \underline{\text{Set}}(X', AU) & \ni & fh \\
f\downarrow & \xrightarrow{\underline{\text{Set}}(\_, A\Omega)} & \uparrow & & \uparrow \\
X & \underline{\text{Set}}(X, AU) & \ni & h
\end{array}.$$

The corresponding natural isomorphisms are described by the diagrams below.

$$\begin{array}{ccc}
\underline{T}(X\Omega, A) & \xrightarrow{\varphi_A^X} & \underline{\text{Set}}(X, AU) \\
\downarrow & & \downarrow \\
\underline{T}(X\Omega, A') & \xrightarrow{\varphi_{A'}^X} & \underline{\text{Set}}(X, A'U)
\end{array} \qquad \begin{array}{ccc}
\underline{T}(X'\Omega, A) & \xrightarrow{\varphi_A^{X'}} & \underline{\text{Set}}(X', AU) \\
\uparrow & & \uparrow \\
\underline{T}(X\Omega, A) & \xrightarrow{\varphi_A^X} & \underline{\text{Set}}(X, AU)
\end{array} \quad \square$$

The relationship (2.2.1) between the underlying set functor U and the word algebra functor Ω is an example of a phenomenon that is very common in mathematics, and particularly in algebra.

Definition 2.2.14. Let $\mathcal{C}$ and $\mathcal{D}$ be categories. Let $F : \mathcal{D} \to \mathcal{C}$ and $G : \mathcal{C} \to \mathcal{D}$ be functors. Then F is a *left adjoint* of G, and G is a *right adjoint* of F, if there is a natural isomorphism

(2.2.5) $$\mathcal{C}(XF, Y) \cong \mathcal{D}(X, YG)$$

for each object X of $\mathcal{D}$ and each object Y of $\mathcal{C}$. In other words, the isomorphism (2.2.5) or

(2.2.6) $$\varphi_Y^X : \mathcal{C}(XF, Y) \to \mathcal{D}(X, YG)$$

is the component at Y of a natural isomorphism

(2.2.7) $$\varphi^X : \mathcal{C}(XF, \_) \cong \mathcal{D}(X, \_G),$$

and the component at X of a natural isomorphism

(2.2.8) $$\varphi_Y : \mathcal{C}(\_F, Y) \cong \mathcal{D}(\_, YG).$$

The four functors of (2.2.7) and (2.2.8) are defined similarly as in Example 2.2.13. The relationship between F and G is described as *adjunction*. $\square$

Proposition 2.2.15. *Consider the adjunction (2.2.6).*

(a) *There is a natural transformation* $\eta : 1_\mathcal{D} \longrightarrow FG$ *whose component at a $\mathcal{D}$-object X is $1_{XF}\varphi^X_{XF}$.*

(b) *There is a natural transformation* $\varepsilon : GF \longrightarrow 1_\mathcal{C}$ *whose component at an $\mathcal{C}$-object Y is $1_{YG}(\varphi^{YG}_Y)^{-1}$.* $\square$

See Section 3.1 of [PMA] for the proof. Note that $1_{XF}\varphi^X_{XF}$ is a $\mathcal{D}$-morphism $\eta_X : X \to XFG$. Similarly $1_{YG}(\varphi^{YG}_Y)^{-1}$ is a $\mathcal{C}$-morphism $\varepsilon_Y : YGF \to Y$. The natural transformation η is called the *unit* of the adjunction (2.2.6). The natural transformation ε is called its *counit*. As shown e.g. in [PMA, Section 3.1], the existence of a left adjoint of $G : \mathcal{C} \to \mathcal{D}$ is equivalent to the fact that for each object X of $\mathcal{D}$, the morphism η_X is an initial object of a certain category called the *comma category* (X, G). A dual result holds for the existence of a right adjoint of G. In particular the unit η and the counit ε are uniquely determined by the functors F and G. In the same section of [PMA], it is shown that F is a left adjoint of G if and only if there are natural transformations $\eta : 1_\mathcal{D} \longrightarrow FG$ and $\varepsilon : GF \longrightarrow 1_\mathcal{C}$ satisfying

$$(\eta F)(F\varepsilon) = 1 \quad \text{and} \quad (G\eta)(\varepsilon G) = 1.$$

Here ηF is a natural transformation $1_\mathcal{D} F \longrightarrow FGF$ with components $(\eta F)_X : XF \to XFGF$, and $F\varepsilon$ is a natural transformation $FGF \longrightarrow F1_\mathcal{C}$ with components $(F\varepsilon)_X : XFGF \to XF$. In view of these results, one may specify the adjunction (2.2.6) by the quadruple $(F, G, \eta, \varepsilon)$ consisting of the left adjoint, right adjoint, unit, and counit.

Note that an isomorphism $F : \mathcal{D} \to \mathcal{C}$ between categories $\mathcal{D}$ and $\mathcal{C}$ can be described as an adjunction $(F, G, \eta, \varepsilon)$, where G is the two-sided inverse of F, the unit η is the identity transformation $1 : 1_\mathcal{D} \xrightarrow{\sim} FG$, and the counit ε is the identity transformation $1 : GF \xrightarrow{\sim} 1_\mathcal{C}$. In this adjunction, F is a left adjoint of G. But there is also an adjunction between G and F whose unit is the identity transformation $1 : 1_\mathcal{C} \xrightarrow{\sim} GF$ and whose counit is the identity transformation $1 : FG \xrightarrow{\sim} 1_\mathcal{D}$. In this adjunction, F is a right adjoint of G. More generally, one has the following.

Theorem 2.2.16. *Let $F : \mathcal{D} \to \mathcal{C}$ be a functor. Then F is an equivalence if and only if it is a part of an adjunction $(F, G, \eta, \varepsilon)$ such that the unit and counit are natural isomorphisms.* □

For a proof see [MacLane 1971, IV.4, Theorem 1] or [PMA, Theorem 3.2.3].

Exercises

2.2A. Let $\underline{Sl}$ be the category of semilattices and semilattice homomorphisms. Let $\underline{Sl}_\Omega$ be the category of Ω-semilattices and Ω-homomorphisms. (Cf. Examples 1.1.15 and 1.5.4.) Show that the categories $\underline{Sl}$ and $\underline{Sl}_\Omega$ are isomorphic. (We usually suppress the index Ω and write $\underline{Sl}$ instead of $\underline{Sl}_\Omega$.)

2.2B. Let $\underline{V}$ be an equational class of Ω-algebras of a given (plural) type $\tau : \Omega \to \mathbb{N}$. Let $\underline{V}_{\mathfrak{P}\Omega}$ be the category of $\underline{V}$-algebras considered as algebras of the derived type $\tau' : \mathfrak{P}\Omega \to \mathbb{N}$ with $\mathfrak{P}\Omega$-homomorphisms. (Cf. Section 1.4.) Show that the categories $\underline{V}$ and $\underline{V}_{\mathfrak{P}\Omega}$ are isomorphic, and that the varieties $\underline{V}$ and $\underline{V}_{\mathfrak{P}\Omega}$ are equivalent. (As in Exercise 2.2A, we usually denote both categories by $\underline{V}$.)

2.2C. Let $\underline{V}$ be an equational class of algebras of a (plural) type $\tau : \Omega \to \mathbb{N}$,

2. CATEGORIES OF ALGEBRAS

and let $\underline{W}$ be an equational class of algebras of a (plural) type $\sigma : \Psi \to \mathbb{N}$. Show that the classes $\underline{V}$ and $\underline{W}$ are equivalent if and only if the corresponding classes $\underline{V}_{\mathfrak{P}\Omega}$ and $\underline{W}_{\mathfrak{P}\Psi}$ are equivalent.

2.2D. Let $\underline{\text{Aff}}$ be the category comprising the empty set together with all affine spaces over a field F, the morphisms being the insertions of the empty set, and the affine maps defined in the traditional geometrical fashion. (Cf. Section 1.6.4.) Show that the category $\underline{\text{Aff}}$ is isomorphic to the category $\underline{F}$ of algebras $(A, P, \underline{F})$ with corresponding homomorphisms.

2.2E. Let S be a ring, and let T be the ring of 2×2-matrices over S. Show that the equational classes $\underline{\text{Mod}}_S$ of S-modules and $\underline{\text{Mod}}_T$ of T-modules are equivalent.

2.3. Some categorical constructions

Lemma 2.3.1. *Let $\tau : \Omega \to \mathbb{N}$ be a given type, and let (A_i, Ω), for $i \in I$, be τ-algebras. Then the direct product $(A, \Omega) = (\prod A_i, \Omega)$ together with the projections $\pi_i : \prod A_i, \to A_i$, satisfies the following universal property:*

(2.3.1)
$$\begin{cases} \text{For any } \tau\text{-algebra } (B, \Omega) \text{ and } \Omega\text{-homomorphisms } h_i : B \to A_i, \\ \text{for } i \in I, \text{ there is a unique } \Omega\text{-homomorphism } h : B \to A \text{ such that} \\ h\pi_i = h_i \text{ for each } i \text{ in } I. \end{cases}$$

(2.3.2)
$$\begin{array}{ccc} \prod_{i \in I} A_i & \xrightarrow{\pi_i} & A_i \\ h \uparrow & & \uparrow h_i \\ B & = & B \end{array}$$

Proof. The map
$$h : B \to \prod A_i; \; b \mapsto bh,$$

where $i(bh) := bh_i$, is the unique Ω-homomorphism $B \to \prod A_i$ such that $h\pi_i = h_i$. □

The homomorphism h of Lemma 2.3.1 is usually denoted by $\prod_{i \in I} h_i$.

The universality property (2.3.1) characterizes direct products up to isomorphism.

Proposition 2.3.2. *A τ-algebra (A', Ω), together with Ω-homomorphisms $\pi'_i : A' \to A_i$, satisfies the universal property (2.3.1) if and only if (A', Ω) is isomorphic to the direct product $(\prod A_i, \Omega)$.*

Proof. It is enough to show that given a τ-algebra A' and homomorphisms π'_i, the algebras (A', Ω) and $(\prod A_i, \Omega)$ are isomorphic. First note that the universal property for A' and π'_i implies that there is a unique homomorphism $j' : \prod_{i \in I} A_i \to A'$ with $j'\pi'_i = \pi_i$. On the other hand, the universal property for A and π_i implies that there is a unique homomorphism $j : A' \to A$ with $j\pi_i = \pi'_i$. Hence $j'j\pi_i = j'\pi'_i = \pi_i$. Since also $1_A\pi_i = \pi_i$, it follows that $j'j = 1_A$. Similarly, one shows that $jj' = 1_{A'}$, which implies that j and j' are mutually inverse isomorphisms. □

The definition of a product in any category is modelled on the characterization of (direct) products of algebras given in Proposition 2.3.2.

Definition 2.3.3. Let I be a set and let $\mathcal{C}$ be a category. Let $A : I \to \mathrm{Ob}\mathcal{C}$; $i \mapsto A_i$ be a function. Then the *product* π or π^A of A is an object $\prod_{i \in I} A_i$ (or $\prod A_i$) of $\mathcal{C}$ together with morphisms $\pi_i : \prod A_i \to A_i$, known as *projections*, for each i in I, having the following *universal property*:

(2.3.3) $\begin{cases} \text{For each object } B \text{ of } \mathcal{C} \text{ and morphisms } h_i : B \to A_i, \text{ for } i \in I, \\ \text{there is a unique morphism } \prod_{i \in I} h_i : B \to \prod A_i \\ \text{(or } \prod h_i\text{) such that } (\prod h_i)\pi_i = h_i \text{ for each } i \text{ in } I. \end{cases}$

2. CATEGORIES OF ALGEBRAS

$$\begin{array}{ccc} \prod A_i & \xrightarrow{\pi_i} & A_i \\ \prod h_i \uparrow & & \uparrow h_i \\ B & = & B \end{array} \quad \square$$

Usually, the object $\prod_{i \in I} A_i$ itself is called the *product*, leaving tacit the rôle of the projections. Obviously direct products of τ-algebras are products in the category $\underline{\tau}$, and cartesian products of sets are products in the category $\underline{\text{Set}}$.

Example 2.3.4. Let $(X, \leq)$ be an ordered set, construed as a category. Let $A : I \to X$; $i \mapsto x_i$ be a function. Then the product $\prod^A$ is the greatest lower bound $\prod_{i \in I} x_i$ of the set $\{x_i \mid i \in I\}$. Such a greatest lower bound may or may not exist. $\square$

Products in the dual $\mathcal{C}^{op}$ of a category $\mathcal{C}$ are called "coproducts". Formally, one has the following definition.

Definition 2.3.5. Let I be a set and let $\mathcal{C}$ be a category. Let $A : I \to \text{Ob}\mathcal{C}$; $i \mapsto A_i$ be a function. Then the *coproduct* ι or ι^A of A is an object $\sum_{i \in I} A_i$ or $\coprod_{i \in I} A_i$ (written also as $\sum A_i$ or $\coprod A_i$) together with morphisms $\iota_i : A_i \to \sum A_i$, known as *insertions*, for each i in I, having the following *universal property*:

(2.3.4) $\begin{cases} \text{For each object } B \text{ of } \mathcal{C} \text{ and morphisms } h_i : A_i \to B, \text{ for } i \in I, \\ \text{there is a unique morphism } \sum h_i : \sum_{i \in I} A_i \to B \text{ (or } \sum h_i) \\ \text{such that } \iota_i(\sum h_i) = h_i \text{ for each } i \text{ in } I. \end{cases}$

$$\begin{array}{ccc} \sum A_i & \xleftarrow{\iota_i} & A_i \\ \sum h_i \downarrow & & \downarrow h_i \\ B & = & B \end{array} \quad \square$$

Again, one sometimes refers to $\sum A_i$ alone as the *coproduct*, not explicitly mentioning the insertions.

Example 2.3.6. In the category Set, the coproduct $\sum A_i$ of an arbitrary function $A : I \to \mathrm{ObSet}$ from a set I is given by the disjoint union of the sets A_i. The insertions ι_i are embeddings of A_i into the disjoint sum. □

Example 2.3.7. In the category AGp of abelian groups, the coproduct $\sum A_i$ is the direct sum of the groups A_i. The insertions ι_i are embeddings of the groups A_i into the group $\sum A_i$. □

Example 2.3.8. Let $(X, \leq)$ be an ordered set, construed as a category. Let $A : I \to X$; $i \mapsto x_i$ be a function. Then the coproduct ι^A is the least upper bound $\sum_{i \in I} x_i$ of the set $\{x_i \mid i \in I\}$. Such a least upper bound may or may not exist. From this example and Example 2.3.4, it is clear that an ordered set $(X, \leq)$ has all products and coproducts (or is a category *closed under products and coproducts*) iff it is a complete lattice. □

A third categorical construction we want to discuss is that of a directed colimit. We define it at first for the category $\underline{\tau}$ of τ-algebras.

Let $(I, \leq)$ be a directed ordered set considered as a category (I). Let $F : (I) \to (\Omega)$ be a functor to the category of τ-algebras. Then for $\varphi_{i,j} := (i \to j)F : (A_i, \Omega) \to (A_j, \Omega)$, where $i \leq j$, one has

(2.3.5) $$\begin{cases} \varphi_{i,i} = 1_{A_i}, \\ \varphi_{i,j}\varphi_{j,k} = \varphi_{i,k} \text{ for } i \leq j \leq k. \end{cases}$$

The algebras A_i, for $i \in I$, together with the homomorphisms $\varphi_{i,j}$ are said to form a *directed system* $((A_i)_{i \in I}, (\varphi_{i,j})_{i \leq j})$ of τ-algebras.

Consider the disjoint sum $\bigcup_{i \in I} A_i$ of the sets A_i. Define the following binary relation δ on the set $\bigcup A_i$. For a_i in A_i and b_j in A_j

(2.3.6) $\qquad (a_i, b_j) \in \delta :\Leftrightarrow \exists k \geq i, j \text{ with } a_i \varphi_{i,k} = b_j \varphi_{j,k}$.

It is easy to see that δ is an equivalence relation. Denote by $\varinjlim_{i \in I} A_i$ the set $(\bigcup A_i)^\delta$. And for each $i \in I$ define the mapping

(2.3.7) $\qquad\qquad \varphi_i : A_i \to \varinjlim_{i \in I} A_i;\ a_i \mapsto a_i^\delta.$

2. CATEGORIES OF ALGEBRAS

For each (n-ary) ω in Ω, define the operation ω on the set $\varinjlim A_i$ by

(2.3.8) $\qquad a_{i_1}^\delta \ldots a_{i_n}^\delta \omega := (a_{i_1}\varphi_{i_1,j} \ldots a_{i_n}\varphi_{i_n,j}\omega)^\delta,$

where j is any upper bound of $i_1, \ldots, i_n$ and $a_{i_k} \in A_{i_k}$. Exercise 2.3B shows that ω is well-defined.

Definition 2.3.9. The algebra $(\varinjlim_{i \in I} A_i, \Omega)$ with the Ω-operations defined by (2.3.8), together with the mappings φ_i defined by (2.3.7), is called a *directed colimit of the directed system* $((A_i)_{i \in I}, (\varphi_{i,j})_{i \leq j})$ or *of the functor* $F : (I) \to (\Omega)$, and is denoted $\varinjlim F$. □

Directed colimits are also called "direct limits" or "inductive limits". By abuse of notation, the algebra $\varinjlim A_i$ itself is also called a directed colimit.

Example 2.3.10. Let I be a (join) semilattice. Evidently I is a directed ordered set. Any two elements i and j of I have their join $i+j$ as an upper bound. Consider a functor $F : (I) \to (\Omega)$, and the corresponding directed system $((A_i)_{i \in I}, (\varphi_{i,j})_{i \leq j})$. On the disjoint sum $A := \bigcup_{i \in I} A_i$ of the sets A_i, define the operation ω for each (n-ary) ω in Ω and a_{i_r} in A_{i_r} as follows:

$$a_{i_1} \ldots a_{i_n} \omega := a_{i_1}\varphi_{i_1,i} \ldots a_{i_n}\varphi_{i_n,i}\,\omega,$$

where $i = i_1 + \cdots + i_n$. Then the equivalence relation δ defined on $\bigcup A_i$ by (2.3.6) is a congruence on the algebra (A, Ω). To show this, let $(a_{i_r}, b_{j_r}) \in \delta$ for $r = 1, \ldots, n$. This means that there are indices $k_r \geq i_r$ and $k_r \geq j_r$ such that $a_{i_r}\varphi_{i_r,k_r} = b_{j_r}\varphi_{j_r,k_r}$. Let $i = i_1 + \cdots + i_n$ and $j = j_1 + \cdots + j_r$. Let k be any upper bound of i and j. Then for each $r = 1, \ldots, n$, one has $i_r, j_r \leq i_r + j_r \leq i + j \leq k_1 \cdots + k_n =: k$, and by (2.3.5)

$$a_{i_r}\varphi_{i_r,k} = a_{i_r}\varphi_{i_r,k_r}\varphi_{k_r,k} = b_{j_r}\varphi_{j_r,k_r}\varphi_{k_r,k} = b_{j_r}\varphi_{j_r,k}.$$

It follows that for each (n-ary) ω in Ω one has:

$$a_{i_1}\ldots a_{i_n}\omega\ \varphi_{i,k} = a_{i_1}\varphi_{i_1,i}\ldots a_{i_n}\varphi_{i_n,i}\omega\ \varphi_{i,k}$$
$$= a_{i_1}\varphi_{i_1,i}\varphi_{i,k}\ldots a_{i_n}\varphi_{i_n,i}\varphi_{i,k}\ \omega$$
$$= a_{i_1}\varphi_{i_1,k}\ldots a_{i_n}\varphi_{i_n,k}\ \omega$$
$$= b_{j_1}\varphi_{j_1,k}\ldots b_{j_n}\varphi_{j_n,k}\ \omega$$
$$= b_{j_1}\varphi_{j_1,j}\varphi_{j,k}\ldots b_{j_n}\varphi_{j_n,j}\varphi_{j,k}\ \omega$$
$$= b_{j_1}\varphi_{j_1,j}\ldots b_{j_n}\varphi_{j_n,j}\omega\ \varphi_{j,k}$$
$$= b_{j_1}\ldots b_{j_n}\omega\ \varphi_{j,k}.$$

This shows that $(a_{i_1}\ldots a_{i_n}\omega, b_{j_1}\ldots b_{j_n}\omega) \in \delta$, whence δ is a congruence relation on (A,Ω). It follows that, if I is a semilattice, the directed colimit of the functor F is just the quotient (A^δ,Ω) of the algebra (A,Ω) by the congruence δ. □

Similarly as products, directed colimits may also be defined in categorical terms. To do this, one needs the following lemma.

Lemma 2.3.11. *The following hold for a directed system* $((A_i)_{i\in I}, (\varphi_{i,j})_{i\leq j})$:

(a) *For* $i \leq j$, *one has* $\varphi_i = \varphi_{i,j}\varphi_j$;

(b) *For each* $i \in I$, *the mapping* φ_i *is a* τ-*homomorphism*;

(c) *The directed colimit* $\varinjlim A_i$, *together with the mappings* $\varphi_i : A_i \to \varinjlim A_i$, *satisfies the following universal property:*

(2.3.9)
$$\begin{cases} \text{If for each } i \in I, \text{ a } \tau\text{-homomorphism } \psi_i : A_i \to B \text{ is given such that} \\ \psi_i = \varphi_{i,j}\psi_j, \text{ where } i \leq j, \text{ then there is precisely one } \Omega\text{-homomorphism} \\ \varphi : \varinjlim A_i \to B \text{ with } \varphi_i\varphi = \psi_i \text{ for all } i \text{ in } I. \end{cases}$$

2. CATEGORIES OF ALGEBRAS

$$\begin{array}{ccc}
\varinjlim A_i & \xrightarrow{\varphi} B & = B \\
\| & & \| \\
\varinjlim A_i & \xleftarrow{\varphi_i} A_i \xrightarrow{\psi_i} B \\
\| & \downarrow{\varphi_{i,j}} & \| \\
\varinjlim A_i & \xleftarrow{\varphi_j} A_j \xrightarrow{\psi_j} B
\end{array}$$

Proof. (a) Since $a_i \varphi_{i,j} = a_i \varphi_{i,j} \varphi_{j,j}$, it is clear that $(a_i, a_i \varphi_{i,j}) \in \delta$.

(b) Let $a_{i_1} \ldots, a_{i_n}$ be in A_i, let ω be an n-ary operator in Ω, and let $i \leq j$. Then the following hold directly from the corresponding definitions and (a):

$$a_{i_1}\varphi_i \ldots a_{i_n}\varphi_i \, \omega = a_{i_1}^\delta \ldots a_{i_n}^\delta \omega$$
$$= (a_{i_1}\varphi_{i,j} \ldots a_{i_n}\varphi_{i,j} \, \omega)^\delta$$
$$= (a_{i_1} \ldots a_{i_n} \omega \, \varphi_{i,j})^\delta$$
$$= a_{i_1} \ldots a_{i_n} \omega \, \varphi_{i,j}\varphi_j = a_{i_1} \ldots a_{i_n} \omega \, \varphi_i.$$

(c) Define the mapping φ as follows:

$$\varphi : \varinjlim A_i \to B; \ a_i^\delta \mapsto a_i \psi_i.$$

The mapping φ is well defined, since $a_i^\delta = b_j^\delta$ implies that there is $k \in I$ with $a_i\varphi_{i,k} = b_j\varphi_{j,k}$, whence $a_i\psi_i = a_i\varphi_{i,k}\psi_k = b_j\varphi_{j,k}\psi_k = b_j\psi_j$. It is an Ω-homomorphism, since for a_{i_k} in A_{i_k} one has

$$a_{i_1}^\delta \varphi \ldots a_{i_n}^\delta \varphi \, \omega = a_{i_1}\psi_{i_1} \ldots a_{i_n}\psi_{i_n} \omega$$
$$= (a_{i_1}\varphi_{i_1,j}\psi_j) \ldots (a_{i_n}\varphi_{i_n,j}\psi_j)\omega$$
$$= a_{i_1}\varphi_{i_1,j} \ldots a_{i_n}\varphi_{i_n,j}\omega \, \psi_j$$
$$= (a_{i_1}\varphi_{i_1,j} \ldots a_{i_n}\varphi_{i_n,j}\omega)^\delta \varphi$$
$$= (a_{i_1}^\delta \ldots a_{i_n}^\delta \omega)\varphi.$$

Finally, if $\alpha : \varinjlim A_i \to B$ is another Ω-homomorphism with $\psi_i = \varphi_i \alpha$ for each $i \in I$, then

$$a_i^\delta \alpha = a_i \varphi_i \alpha = a_i \psi_i = a_i^\delta \varphi,$$

whence $\alpha = \varphi$. $\square$

Proposition 2.3.12. *A τ-algebra (A, Ω), together with the Ω-homomorphisms $\sigma_i : A_i \to A$, satisfies the universal property (2.3.9) if and only if (A, Ω) is isomorphic to $\varinjlim A_i$.*

Proof. It suffices to show that the implication ($\Rightarrow$) holds. The universal property for A and σ_i implies that there is a unique homomorphism $\sigma : A \to \varinjlim A_i$ with $\varphi_i = \sigma_i \sigma$. Then the universal property for $\varinjlim A_i$ and φ_i implies that there is a unique homomorphism $\varphi : \varinjlim A_i \to A$ with $\sigma_i = \varphi_i \varphi$ for each i in I. Hence $\varphi_i = \varphi_i \varphi \sigma$ and $\sigma_i = \sigma_i \sigma \varphi$. Since also $\varphi_i = \varphi_i 1_{\varinjlim A_i}$ and $\sigma_i = \sigma_i 1_A$, it follows that $\varphi \sigma = 1_{\varinjlim A_i}$ and $\sigma \varphi = 1_A$. This implies that φ and σ are mutually inverse isomorphisms. $\square$

In particular, Proposition 2.3.12 shows that the universality property (2.3.9) characterizes $\varinjlim A_i$ up to isomorphism, and becomes the basis for the following generalization.

Definition 2.3.13. Let I be a directed ordered set and let $\mathcal{C}$ be a category. Then the *directed colimit* $\varinjlim F$ of a functor $F : (I) \to \mathcal{C}$ is an object $\varinjlim_{i \in I} A_i$ (or $\varinjlim A_i$) of $\mathcal{C}$ together with morphisms $\varphi_i : A_i \to \varinjlim A_i$, for each i in I, having the universal property (2.3.9), where all A_i and B are objects of $\mathcal{C}$ and the word "homomorphism" is replaced by the word "morphism". $\square$

By omitting the word "directed" in Definition 2.3.13, one obtains the definition of the colimit $\varinjlim F$ of the functor $F : (I) \to \mathcal{C}$ from the ordered set I to the category $\mathcal{C}$.

The notion of a (directed) limit is obtained dually.

2. CATEGORIES OF ALGEBRAS

Definition 2.3.14. Let I be a (directed) ordered set and let C be a category. Then the (*directed*) *limit* $\varprojlim F$ of a functor $F : (I) \to C$ is an object $\varprojlim_{i \in I} A_i$ (or $\varprojlim A_i$) of C, together with morphisms $\varphi_i : \varprojlim A_i \to A_i$ for each i in I, satisfying the *universal property* (2.3.9^{op}) dual to (2.3.9), where all A_i and B are objects of C.

(2.3.9^{op})
$$\begin{cases} \text{If for each } i \in I, \text{ a morphism } \psi_i : B \to A_i \text{ of } C \text{ is given such that} \\ \psi_i \varphi_{i,j} = \psi_j, \text{ where } i \leq j, \text{ then there is precisely one morphism} \\ \varphi : B \to \varprojlim A_i \text{ of } C \text{ with } \varphi \varphi_i = \psi_i \text{ for all } i \text{ in } I. \end{cases}$$

$$\begin{array}{ccccc}
\varprojlim A_i & \xleftarrow{\varphi} & B & = & B \\
\| & & & & \| \\
\varprojlim A_i & \xrightarrow{\varphi_i} & A_i & \xleftarrow{\psi_i} & B \qquad \square \\
\| & & \downarrow{\varphi_{i,j}} & & \| \\
\varprojlim A_i & \xrightarrow{\varphi_j} & A_j & \xleftarrow{\psi_j} & B
\end{array}$$

Note that if I is a totally unordered poset, then the colimit $\varinjlim F$ reduces to the coproduct, and the limit $\varprojlim F$ reduces to the product. The concepts of colimit and limit may be generalized further to the case where I is an arbitrary small category. In this case also, the objects $\varinjlim F$ and $\varprojlim F$ are defined uniquely up to isomorphism. (See [Mac Lane 1971] and [PMA].)

One more special case of the general concept of a colimit will be needed later.

Definition 2.3.15. Let C be a category.
(a) A *parallel pair* $A_1 \underset{f_2}{\overset{f_1}{\rightrightarrows}} A_2$ of arrows in C is a pair (f_1, f_2) of arrows of C with common tail A_1 and common head A_2.
(b) Let $A_1 \underset{f_2}{\overset{f_1}{\rightrightarrows}} A_2$ be a parallel pair of arrows of C. Then the *coequalizer* of the pair (f_1, f_2) is an object Q of C together with a morphism $q : A_2 \to Q$ of C

satisfying $f_1 q = f_2 q$ and the following *universal property*:

(2.3.10) $\left\{ \begin{array}{l} \text{For each } \mathcal{C}\text{-morphism } f : A_2 \to B \text{ with } f_1 f = f_2 f, \\ \text{there is a unique morphism } \widetilde{f} : Q \to B \text{ such that } q\widetilde{f} = f : \end{array} \right.$

$$\begin{array}{ccc} A_1 \underset{f_2}{\overset{f_1}{\rightrightarrows}} A_2 & \xrightarrow{q} & Q \\ & {\scriptstyle f}\downarrow & \downarrow {\scriptstyle \widetilde{f}} \\ & B = \!\!= B & \end{array} \quad . \quad \Box$$

As in the case of coproducts, one often refers to the object Q alone as the coequalizer. There is a corresponding dual concept of "equalizer". (See e.g. [PMA].)

Example 2.3.16. Let (A, Ω) be a τ-algebra of a given type τ. Let θ be a congruence on (A, Ω). Consider θ as a subalgebra of (A^2, Ω). The direct product $A \times A$ comes equipped with the two projections $\pi_i : A^2 \to A;\ (a_1, a_2) \mapsto a_i$ for $i = 1, 2$, each of which restricts to the subset θ of A^2 yielding the following commuting diagram:

$$\begin{array}{ccc} \theta & \xrightarrow{\pi_2} & A \\ {\scriptstyle \pi_1}\downarrow & & \downarrow {\scriptstyle \mathrm{nat}\theta} \\ A & \xrightarrow[\mathrm{nat}\theta]{} & A^\theta \end{array} \quad .$$

Moreover, given an Ω-homomorphism $h : A \to B$ with $\pi_1 h = \pi_2 h$, there is a unique, well-defined Ω-homomorphism $\widetilde{h} : A^\theta \to B;\ a^\theta \mapsto ah$ such that $(\mathrm{nat}\theta)\widetilde{h} = h$. This shows that the quotient algebra A^θ together with the homomorphism $\mathrm{nat}\theta$ is the coequalizer of the pair (π_1, π_2). $\Box$

As proved in [PMA, Theorem 2.2.3], any prevariety $\underline{\underline{K}}$ of τ-algebras is closed under all limits $\varprojlim F$ and colimits $\varinjlim F$ of functors $F : (I) \to \mathcal{C}$ from a small category (I).

Exercises

2.3A. Show that the relation δ defined by (2.3.6) is an equivalence relation.

2.3B. Show that the definition (2.3.8) does not depend on the particular choice of an upper bound j.

2.3C. Note that the finitely generated subalgebras of an algebra (A, Ω) form a (join) semilattice. Such subalgebras together with the inclusion of subalgebras form a directed system. Use these observations to show that each algebra is a directed colimit of its finitely generated subalgebras.

2.3D. Consider a directed colimit $\varinjlim A_i$ of algebras A_i.

    (a) Show that all the $\varphi_{i,j}$ are injective if and only if all the φ_i are injective;

    (b) Show that if all the $\varphi_{i,j}$ are surjective, then all the φ_i are surjective. Show that the converse is false.

2.3E. Show that a subalgebra of a directed colimit $\varinjlim A_i$ of algebras A_i is a directed colimit of subalgebras of A_i.

2.3F. Let I be a directed ordered set. Let $I_0 := I \cup \{0\}$, where $0 \notin I$. Extend the order of I to I_0 by setting $0 < i$ for each $i \in I$. Extend the functor $F : (I) \to (\Omega)$ to $F_0 : (I_0) \to (\Omega)$ by setting $0F$ to be an initial object in (Ω) (if such exists). Show that

$$\varinjlim F \cong \varinjlim F_0.$$

2.3G. Let $\underline{K}$ be a category of τ-algebras and τ-homomorphisms. For $i \in I$, let $\iota_i : A_i \to A$ be morphisms in $\underline{K}$ such that if $h_i : A_i \to B$, for $i \in I$, are also morphisms in $\underline{K}$, then there is a morphism $h : A \to B$ in $\underline{K}$ such that $\iota_i h = h_i$ for all i in I. Show that the algebra A together with the morphisms ι_i is a coproduct $\coprod A_i$ if and only if $\bigcup \iota_i(A_i)$ generates A.

2.3H. Let $\mathcal{C}$ be a category of τ-algebras. Extend the definition of the directed colimit $\varinjlim F$ of a functor $F : (I) \to \mathcal{C}$ whose domain is a directed ordered set $(I, \leq)$ to the case where the domain $(I, \leq)$ of the functor is a directed quasi-ordered set. Show that $\varinjlim F$ can be constructed in the same way as in the original case.

2.3I. Consider the rational unit interval $I_{\mathbb{Q}} := \{x \in \mathbb{Q} \mid 0 \leq x \leq 1\}$. Define operations $\oplus$ and $\ominus$ on $I_{\mathbb{Q}}$ as follows:

$$x \oplus y = \text{if } x+y < 1 \text{ then } x+y \text{ else } x+y-1,$$

$$\ominus x = \text{if } x \neq 0 \text{ then } 1-x \text{ else } 0.$$

Show that:
- (a) $(I_{\mathbb{Q}}, \oplus, \ominus, 0)$ is an abelian group.
- (b) Each finitely generated subgroup of the group $I_{\mathbb{Q}}$ is finite.
- (c) Each finite n-element subgroup C_n of the group $I_{\mathbb{Q}}$ is cyclic (and is generated by $\frac{1}{n}$).
- (d) Show that the group $I_{\mathbb{Q}}$ is the directed colimit $\varinjlim_{n \in \mathbb{Z}^+} C_n$, where $\mathbb{Z}^+$ is ordered by

$$m \leq n \text{ iff } m \mid n,$$

and all the colimit morphisms $\varphi_{m,n} : C_m \to C_n$ are embeddings.

CHAPTER III

VARIETIES, PREVARIETIES and QUASIVARIETIES

Chapter 2 considered classes of algebras as categories. This chapter considers classes of algebras of a given type defined by certain formulae, namely equational, quasi-equational and implicational classes. It also considers classes of algebras defined by closure under products, homomorphic images, and the like. The three main results of the chapter show that these two types of definitions are interconnected even more closely than was described in Chapter 1. Prevarieties are characterized as implicational classes, and varieties as equational classes. Also, quasi-equational classes are characterized by closure under certain operators. Moreover, some other properties of (pre-, quasi-) varieties and the relationship between them will be described. All this demands a deeper knowledge of congruences of algebras and certain special algebras in such classes. The first two sections are devoted to a further study of lattices of congruences and closely related closure operators, a new construction of subdirect products of algebras, and the representation of algebras in prevarieties by means of such constructions. The third section considers replicas and corresponding constructions of free algebras. The three main results are presented in the subsequent sections.

3.1. Subdirect product representation

Definition 3.1.1. A subalgebra A of a direct product $\prod_{i \in I} A_i$ of Ω-algebras A_i is called a *subdirect product of A_i*, written as $A \leq_s \prod_{i \in I} A_i$, if for each projection π_i, with $i \in I$, one has $A\pi_i = A_i$. □

The corresponding embedding $A \to \prod_{i \in I} A_i$ is called *subdirect*. If $I = \varnothing$, then A is trivial. Definition 3.1.1 implies that

$$(3.1.1) \qquad A \leq_s \prod_{i \in I} A_i \Rightarrow \bigcap_{i \in I} \ker \pi_i = \widehat{A}.$$

Lemma 3.1.2. *If φ and φ_i, for $i \in I$, are in $Cg\,A$ and $\varphi = \bigcap_{i \in I} \varphi_i$, then*

$$A^\varphi \leq_s \prod_{i \in I} A^{\varphi_i}.$$

Proof. The mapping

$$A^\varphi \to \prod_{i \in I} A^{\varphi_i}; \ a^\varphi \mapsto (I \to \bigcup_{i \in I} A^{\varphi_i}; \ i \mapsto a^{\varphi_i})$$

is a well defined subdirect embedding. □

Corollary 3.1.3. *Let A and A_i, for $i \in I$, be Ω-algebras. Let $A \leq \prod_{i \in I} A_i$. Then*

$$A \leq_s \prod_{i \in I} A_i \Leftrightarrow \bigcap_{i \in I} \ker \pi_i = \widehat{A}. \quad \square$$

The equivalence of Corollary 3.1.3 may also be formulated as follows:

$$A \leq_s \prod_{i \in I} A_i \text{ iff there is a set of congruences}$$
$$\theta_i, \text{ with } i \in I, \text{ of } A \text{ such that } \bigcap_{i \in I} \theta_i = \widehat{A}$$

[Exercise 3.1B]. In particular, every subset of $Cg\,A$ that intersects to $\widehat{A}$ determines A as a subdirect product. Thus A may usually be represented as a subdirect product in many different ways.

Example 3.1.4. Consider the unit closed disc $C = \{(x_1, x_2) \mid x_1^2 + x_2^2 \leq 1\}$ in the Euclidean space $\mathbb{R}^2$ as a barycentric algebra $(C, \underline{I}^\circ)$ (cf. Section 1.6.5). Clearly $(C, \underline{I}^\circ) \leq ([-1,1]^2, \underline{I}^\circ) \leq (\mathbb{R}^2, \underline{I}^\circ)$, and the images $C\pi_1$ and $C\pi_2$ of C under the projections π_1 to the x_1-axis and π_2 to the x_2-axis are both isomorphic to $([-1,1], \underline{I}^\circ)$. Hence $(C, \underline{I}^\circ)$ is a subdirect product of two copies of $([-1,1], \underline{I}^\circ)$. To illustrate Corollary 3.1.3, note that the $\ker \pi_1$-classes of C are segments of C parallel to the x_2-axis. Similarly, the $\ker \pi_2$-classes of C are segments of C parallel to the x_1-axis. Clearly $\ker \pi_1 \cap \ker \pi_2 = \widehat{C}$. Now taking projections to rotated axes will give a different subdirect embedding of C into $([-1,1]^2, \underline{I}^\circ)$. □

Example 3.1.5. Obviously each algebra A isomorphic to a direct product $\prod_{i \in I} A_i$ is at the same time a subdirect product. Note that by Corollary 3.1.3 and Exercise 3.1C, the mapping

$$e : A \to \prod_{i \in I} A^{\varphi_i}; \ a \mapsto \left(I \to \bigcup_{i \in I} A^{\varphi_i}; \ i \mapsto a^{\varphi_i} \right)$$

is a subdirect embedding iff $\bigcap_{i \in I} \varphi_i = \widehat{A}$, i.e. iff e is injective. Then e is also surjective if additionally for every $f : I \to A; \ i \mapsto a_i$, there is an element $a \in A$ such that $(a, a_i) \in \varphi_i$ for all $i \in I$ [Exercise 3.1D]. □

Let $\underline{K}$ be a prevariety of τ-algebras. For a τ-algebra (A, Ω) or A, the congruences α with A^α in the prevariety $\underline{K}$ are called $\underline{K}$-*congruences* on A. The subposet of $Cg(A, \Omega)$ induced on the set of $\underline{K}$-congruences is denoted by $Cg_K(A, \Omega)$ or $Cg_K A$. Note that the singleton τ-algebra, as the product of an empty set of $\underline{K}$-algebras, lies in $\underline{K}$. Thus $A^2 \in Cg_K A$ for each prevariety $\underline{K}$. The trivial congruence $\widehat{A}$ belongs to $Cg_K A$ only if A is in $\underline{K}$.

Proposition 3.1.6. *Let $\underline{K}$ be a prevariety of τ-algebras. Then for each τ-algebra A, the poset $Cg_K A$ is a complete lattice, and a meet subsemilattice of CgA and $\mathcal{P}(A^2)$.*

Proof. Let $\{\alpha_i \mid i \in I\}$ be a set of $\underline{K}$-congruences on A. Then the kernel of $\prod_{i \in I} \text{nat } \alpha_i$ is $\bigcap_{i \in I} \alpha_i$. By the First Isomorphism Theorem 1.2.3, one has

$$A^{\bigcap \alpha_i} = A\left(\text{nat} \bigcap_{i \in I} \alpha_i\right) \lesssim \prod_{i \in I}(A \text{ nat } \alpha_i) = \prod A^{\alpha_i} \in P\underline{K} = \underline{K}.$$

Thus $\bigcap_{i \in I} \alpha_i \in Cg_K A$. □

Definition 3.1.7. Let $\underline{K}$ be a prevariety. A (non-empty) non-trivial $\underline{K}$-algebra A is said to be *subdirectly irreducible relative to $\underline{K}$*, or just *relatively subdirectly irreducible*, if for every subdirect embedding $e : A \to \prod_{i \in I} A_i$ into the direct product of $\underline{K}$-algebras A_i, there is an $i \in I$ such that $e\pi_i : A \xrightarrow{e} \prod_{i \in I} A_i \xrightarrow{\pi_i} A_i$ is an isomorphism. If $\underline{K} = \underline{\tau}$, the class of all algebras of a type τ, the algebra A is called *subdirectly irreducible*. □

In what follows the symbol $cg_K(a, b)$, for a, b in A, denotes the $\underline{K}$-congruence generated by $\{a, b\}$. It is called a *principal $\underline{K}$-congruence*. If $\underline{K} = \underline{\tau}$, then one simply writes $cg(a, b)$, as before.

Proposition 3.1.8. *Let A be an algebra in a prevariety $\underline{K}$. Then the following are equivalent:*

(a) *The algebra A is subdirectly irreducible relative to $\underline{K}$;*

(b) *The congruence lattice $Cg_K A$ has a smallest non-trivial element;*

(c) *There exist elements $a, b \in A$ such that for every pair $(c, d) \in A \times A$, one has $(a, b) \in cg_K(c, d)$ iff $c \neq d$.*

Proof. (a) $\Rightarrow$ (b) Let Θ be the set of all non-trivial $\underline{K}$-congruences of A, and let $\beta := \bigcap_{\theta \in \Theta} \theta$. If $\beta \neq \widehat{A}$, then obviously β is the smallest non-trivial $\underline{K}$-congruence of A, and (b) holds. If $\beta = \bigcap_{\theta \in \Theta} \theta = \widehat{A}$, then the map $A \to \prod_{\theta \in \Theta} A^\theta; a \mapsto \left(\Theta \to \bigcup_{\theta \in \Theta} A^\theta; \theta \mapsto a^\theta\right)$ is a subdirect embedding. The projections $\pi_\theta : A \to A^\theta$ cannot be one-to-one. This contradicts the subdirect irreducibility of A. Thus (b) holds.

(b) $\Rightarrow$ (c). Let β be the smallest non-trivial $\underline{K}$-congruence of A, and let $(a,b) \in \beta - \widehat{A}$. If c and d are distinct elements of A, then $cg_K(c,d) \neq \widehat{A}$, whence $(a,b) \in \beta \subseteq cg_K(c,d)$. If $c = d$, then $cg_K(c,d) = \widehat{A}$ and $(a,b) \notin cg_K(c,d)$. Thus (c) holds.

(c) $\Rightarrow$ (a) Let $A \to \prod_{i \in I} A_i$; $a \mapsto \left(I \to \bigcup_{i \in I} A_i;\ i \mapsto a\pi_i \right)$ be a subdirect embedding into a product of $\underline{K}$-algebras A_i. By Exercise 3.1C, the mappings π_i separate points, i.e. there is $i \in I$ such that $a\pi_i \neq b\pi_i$. Thus $(a,b) \notin \ker \pi_i$, and by (c), $\ker \pi_i$ must be $\widehat{A}$. Hence π_i is injective, and since it is certainly surjective, it is an isomorphism. $\square$

In particular, if $\underline{K} = \underline{\tau}$ for any type τ, then Proposition 3.1.8 gives a characterization of subdirectly irreducible algebras. The least non-trivial congruence of a (relatively) subdirectly irreducible algebra A is called its (*relative*) *monolith*. Note that every subdirectly irreducible algebra A that lies in a prevariety $\underline{K}$ is subdirectly irreducible relative to $\underline{K}$. More generally, if $\underline{K}$ is a subclass of a prevariety $\underline{K}'$ that is itself a prevariety, i.e. $\underline{K}$ is a subprevariety of $\underline{K}'$, then each relatively subdirectly irreducible $\underline{K}'$-algebra that is a member of $\underline{K}$ is subdirectly irreducible relative to $\underline{K}$. The converse does not hold in general. (See e.g. [Pigozzi 1988].)

Example 3.1.9. Each non-empty simple Ω-algebra, i.e. algebra with just two congruences, is obviously subdirectly irreducible. The monolith is the universal congruence. $\square$

Example 3.1.10. Each two-element algebra is simple, and hence subdirectly irreducible. $\square$

Example 3.1.11. The two-element semilattice is the only subdirectly irreducible semilattice. To show this, let $(S, \cdot)$ be a subdirectly irreducible semilattice, with monolith μ. Let $(a,b) \in \mu$. Without loss of generality assume that $b \neq ab$. Let $f : (S, \cdot) \to \underset{\sim}{2}$ be the homomorphism onto the two-element meet semilattice $(\{0,1\}, \cdot)$, with $0 < 1$, defined by $xf =$ **if** $x \geq b$ **then** 1 **else** 0. Obviously $(a, b) \notin \ker f$, whence $\ker f = \widehat{S}$. Since $\ker f$ has precisely two congruence classes, it follows that $(S, \cdot)$ has two elements a and b with $a < b$. □

One can show that the two-element lattice $\underset{\sim}{2}$ is the unique subdirectly irreducible distributive lattice, and the two-element Boolean algebra is the unique subdirectly irreducible Boolean algebra. (See e.g. [McKenzie, McNulty, Taylor 1987].) Further examples are provided in the exercises.

Now it will be shown that any τ-algebra may be built up from subdirectly irreducible τ-algebras.

Birkhoff's Subdirect Representation Theorem 3.1.12. *Every non-trivial algebra is a subdirect product of subdirectly irreducible homomorphic images.*

Proof. Let A be a non-trivial algebra. For $a, b \in A$, let $\theta_{a,b}$ be a maximal congruence on A such that $(a, b) \notin \theta_{a,b}$. Such a congruence exists by the Kuratowski-Zorn Lemma 0.5.1. Indeed, for any chain C of congruences $\varphi \in Cg\, A$ such that $(a, b) \notin \varphi$, Exercise 1.2L implies that $\sum_{\varphi \in C} \varphi = \bigcup_{\varphi \in C} \varphi$, whence $(a, b) \notin \sum_{\varphi \in C} \varphi$. By the Third Isomorphism Theorem 1.2.5, the congruence lattice $Cg\, A^{\theta_{a,b}}$ is isomorphic to the lattice $[\theta_{a,b}, A^2]$. Since $\theta_{a,b}$ is maximal subject to $(a, b) \notin \theta_{a,b}$, it follows that for any congruence θ of A with $\theta > \theta_{a,b}$, the pair (a, b) is in θ, whence $\theta \geq cg(a, b)$. Consequently $(a, b) \in \theta_{a,b} + cg(a, b)$, and by Proposition 3.1.8, the algebra $A^{\theta_{a,b}}$ is subdirectly irreducible. As clearly $\bigcap_{(a,b) \in A^2} \theta_{a,b} = \widehat{A}$, Exercise 3.1B shows that A is subdirectly embeddable in the product $\prod_{(a,b) \in A^2} A^{\theta_{a,b}}$ for all pairs (a, b) with $a \neq b$. □

Corollary 3.1.13. *Every finite algebra is a subdirect product of a finite number of subdirectly irreducible finite algebras.* □

Corollary 3.1.14. *Let $\underline{V}$ be a variety. Then each $\underline{V}$-algebra A is a subdirect product of subdirectly irreducible $\underline{V}$-algebras.* □

It follows that a variety is determined by its subdirectly irreducible members.

Exercises

3.1A. Prove Lemma 3.1.2.

3.1B. Prove that $A \leq_s \prod_{i \in I} A_i$ iff there is a set $\{\theta_i \mid i \in I\}$ of congruences of A such that $\bigcap_{i \in I} \theta_i = \widehat{A}$.

3.1C. Let $\alpha_i : A \to A_i$, for $i \in I$, be homomorphisms of algebras. Then α_i *separates points* if for all $a_1, a_2 \in A$ with $a_1 \neq a_2$, there is an α_i such that $a_1 \alpha_i \neq a_2 \alpha_i$. Show that the following are equivalent:

(a) The maps α_i separate points;

(b) The homomorphism $e : A \to \prod_{i \in I} A_i;\ a \mapsto \left(I \to \bigcup_{i \in I} A_i;\ i \mapsto a\alpha_i \right)$ is injective;

(c) $\bigcap_{i \in I} \ker \alpha_i = \widehat{A}$.

3.1D. Prove the claims of Example 3.1.5.

3.1E. Show that a vector space is subdirectly irreducible iff its dimension is equal to 1.

3.1F. Show that each of the following lattices is subdirectly irreducible:

(N_5)

(M_3)

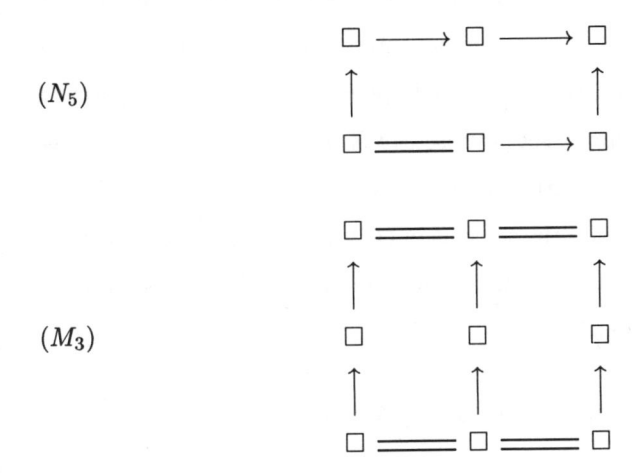

3.1G. Show that a finite abelian group G is subdirectly irreducible iff it is cyclic and has order p^n for some prime p.

[Hint: Each finite abelian group is a direct product of cyclic groups.]

3.1H. The *quasi-cyclic* group $\mathbb{Z}_{p^\infty}$, for a prime p, is the multiplicative group of complex numbers α satisfying $\alpha^{p^k} = 1$ for some $k \geq 0$. Verify that $\mathbb{Z}_{p^\infty}$ is subdirectly irreducible.

3.1I. Prove that every subdirectly irreducible abelian group is cyclic or quasi-cyclic.

3.1J. A congruence θ on an algebra (A, Ω) is said to be *completely meet irreducible* if for $\theta_i \in Cg\,A$ with $\theta = \bigcap_{i \in I} \theta_i$, one has $\theta = \theta_i$ for some $i \in I$. Show that A^θ is subdirectly irreducible iff θ is completely meet irreducible.

3.1K. Show that for a field F, the affine space $(F, P, \underline{F})$ is simple. (Cf. Section 1.6.4.)

3.1L. Show that the barycentric algebras $(I^\circ, \underline{I}^\circ)$ and $(I, \underline{I}^\circ)$ are subdirectly irreducible. (Cf. Section 1.6.5 and Exercise 1.1E.)

3.2. Relative subdirect product representation

Let X be a set, and let $\mathcal{L}$ be a subset of the power set $\mathcal{P}(X)$. The set $\mathcal{L}$ is called a *closure system* if $\mathcal{L}$ is closed under intersections, i.e. for $\mathcal{K} \subseteq \mathcal{L}$, one has $\bigcap \mathcal{K} \in \mathcal{L}$. In particular, the set X itself appears in each closure system $\mathcal{L}$ as the intersection of the empty set of subsets of X [Exercise 1.1B]. By Exercise 1.1C, the set $\mathcal{L}$ is a complete lattice with respect to the ordering by inclusion. The greatest lower bounds are the same as in $(\mathcal{L}, \subseteq)$. However, the least upper bounds are different. The join $\sum_{i \in I} A_i$ of a subset $\{A_i \mid i \in I\}$ of $\mathcal{L}$ is not given by set union in general, but by

$$\sum_{i \in I} A_i = \bigcap \{B \in \mathcal{L} \mid \bigcup_{i \in I} A_i \subseteq B\}.$$

Examples of closure systems are provided by the system of all subalgebras of an algebra, the system of all congruences of an algebra, and the system of all closed sets of a topological space.

A mapping $C : \mathcal{P}(X) \to \mathcal{P}(X)$ is called a *closure operator* on X if for $A, B \subseteq X$, it satisfies the following:

(C1) $A \subseteq A^C$;

(C2) $A^{CC} = A^C$;

(C3) $A \subseteq B$ implies $A^C \subseteq B^C$.

A subset A of X is *closed* if $A^C = A$. Under set inclusion, the closed subsets of A form an ordered set. The mapping

(3.2.1) $\qquad \langle \ \rangle : \mathcal{P}(X) \to \mathcal{P}(X); \ A \mapsto \langle A \rangle,$

where (X, Ω) is a τ-algebra and $\langle A \rangle$ denotes the subalgebra generated by A, is an example of a closure operator. The closed sets are the subalgebras of (X, Ω).

The notions of a closure system and a closure operator turn out to be equivalent.

Theorem 3.2.1. *Every closure system $\mathcal{L}$ on a set X defines a closure operator $C = C_\mathcal{L}$ on X by*

$$A^C = \bigcap \{B \in \mathcal{L} \mid B \supseteq A\}.$$

Conversely, every closure operator C on A defines a closure system $\mathcal{L} = \mathcal{L}_C$ on X by

$$\mathcal{L} = \{A \subseteq X \mid A^C = A\}.$$

The mappings $\mathcal{L} \mapsto C_\mathcal{L}$ and $C \mapsto \mathcal{L}_C$ are mutually inverse bijections. □

The proof is left as Exercise 3.2A.

If $\mathcal{L}$ is a closure system, then $(\mathcal{L}, \subseteq)$ is a complete lattice. On the other hand, one has the following.

Proposition 3.2.2. *Every complete lattice L is isomorphic to the lattice of closed subsets of L with respect to the closure operator C on the set L defined by*
$$A^C = \{a \in L \mid a \leq \sum A\}. \quad \square$$

The proof is left as Exercise 3.2B.

To investigate congruences of τ-algebras, further properties of closure operators and complete lattices are needed. A closure operator C on a set X is said to be *algebraic* if for any $A \subseteq X$ and $a \in X$, each containment $a \in A^C$ implies $a \in B^C$ for some finite subset B of A.

Proposition 3.2.3. *Let C be a closure operator on a set X, and let $\mathcal{L}_C$ be the associated closure system. Then the following are equivalent:*

(a) *C is an algebraic closure operator;*

(b) *for all $A \subseteq X$,*
$$A^C = \bigcup \{B^C \mid B \subseteq A \text{ and } B \text{ is finite}\};$$

(c) *for every directed subset $\{A_i \mid i \in I\}$ of $\mathcal{P}(X)$,*

(3.2.2)
$$\left(\bigcup_{i \in I} A_i\right)^C = \bigcup_{i \in I} A_i^C;$$

(d) *for every directed subset $\mathcal{K}$ of $\mathcal{L}_C$, one has $\bigcup \mathcal{K} \in \mathcal{L}_C$, and hence $\sum \mathcal{K} = \bigcup \mathcal{K}$.*

Proof. (a) $\Leftrightarrow$ (b) is Exercise 3.2C.

(b) $\Rightarrow$ (c) Let $\{A_i \mid i \in I\}$ be a directed subset of $\mathcal{P}(X)$. Hence $\left(\bigcup_{i \in I} A_i\right)^C =$ $\bigcup \{B^C \mid B \subseteq \bigcup_{i \in I} A_i \text{ and } B \text{ is finite}\} = \bigcup \{B^C \mid B \subseteq A_k \text{ for some } k \in I \text{ and finite } B\} \subseteq \bigcup_{i \in I} A_i^C$. By Exercise 3.2D, the condition (3.2.2) holds.

(c) $\Rightarrow$ (d) is obvious, since $\mathcal{L}_C$ consists of closed sets.

(d) $\Rightarrow$ (a) Let $A \subseteq X$. Then the set $\mathcal{D} = \{B^C \mid B \subseteq A \text{ and } B \text{ is finite}\}$ is directed. Hence $\bigcup \mathcal{D} \in \mathcal{L}_C$. Since for $x \in A$, one has $x \in \{x\} \subseteq \{x\}^C \subseteq \bigcup \mathcal{D}$, it follows that $A \subseteq \bigcup \mathcal{D}$. Now if $a \in A^C \subseteq (\bigcup \mathcal{D})^C = \bigcup \mathcal{D} = \bigcup \{B^C \mid B \subseteq A$ and B is finite$\}$, then there is a finite subset $B \subseteq A$ with $a \in B^C$. □

Condition (d) implies that for every chain $\mathcal{D}$ in $\mathcal{L}_C$, one has $\bigcup \mathcal{D} \in \mathcal{L}_C$, and thus $\sum \mathcal{D} = \bigcup \mathcal{D}$. In fact, the converse is true as well. See e.g. Chapter I of [Cohn 1981]. A closure system satisfying any of the equivalent conditions of Proposition 3.2.3 is called *algebraic*.

Let L be a complete lattice. An element $a \in L$ is called *compact* if and only if, whenever $a \leq \sum A$ for $A \subseteq L$, then $a \leq \sum B$ for some finite $B \subseteq A$. The set of compact elements of L is denoted by $K(L)$. A complete lattice L is said to be *algebraic* if for each $a \in L$,

$$a = \sum \{k \in K(L) \mid k \leq a\}.$$

Theorem 3.2.4. *(a) Let C be an algebraic closure operator on a set X, and let $\mathcal{L}_C$ be the corresponding algebraic closure system. Then $\mathcal{L}_C$ is an algebraic lattice, and the compact elements of this lattice are precisely of the form Y^C for finite sets $Y \subseteq X$.*
(b) Every algebraic lattice is isomorphic to some algebraic closure system.

Proof. (a) By Proposition 3.2.3, it suffices to show that the compact elements are the closures of the finite sets. Then Proposition 3.2.3 will imply that $\mathcal{L}_C$ is an algebraic lattice. We will use the equivalent definition of compact elements given in Exericse 3.2E(b).

Let Y be a finite subset of X, and let $A = Y^C$. We want to show that if $A \subseteq \sum \mathcal{D} = \bigcup \mathcal{D}$ for a directed subset $\mathcal{D}$ of $\mathcal{L}_C$, then there is B in $\mathcal{D}$ with $A \subseteq B$. Indeed, since $Y \subseteq Y^C = A \subseteq \bigcup \mathcal{D}$, and since Y is finite and $\mathcal{D}$ directed, there is $B \in \mathcal{D}$ such that $Y \subseteq B$, and hence $A = Y^C \subseteq B^C = B$. Hence A is compact.

Now assume that $A \in \mathcal{L}_C$ is compact. Since $A = \sum \{Y^C \mid Y \subseteq A$ and Y is finite$\}$ and A is compact, it follows that there is a finite set $Y \subseteq A$ such that $A \subseteq Y^C$. The reverse inclusion holds since $Y \subseteq A$ implies $Y^C \subseteq A^C = A$.

(b) Let L be an algebraic lattice. For each $a \in L$, define $D_a = \{k \in K(L) \mid k \leq a\}$. The proof that $\mathcal{L} = \{D_a \mid a \in L\}$ is a closure system is left as Exercise 3.2F(a). The map $\varphi : L \to \mathcal{L}$; $a \mapsto D_a$ is an isomorphism [Exercise 3.2F(b)]. Now let $\mathcal{D} = \{D_b \mid b \in B\}$ be a directed subset of $\mathcal{L}$. Since $D_b \subseteq D_c$ iff $b \leq c$, the indexing set B is a directed subset of L. Set $a = \sum B$. Then

$$k \in D_a \Leftrightarrow k \in K(L) \text{ and } k \leq a = \Sigma B$$
$$\Leftrightarrow k \in K(L) \text{ and } k \leq b \text{ for some } b \in B$$
$$\Leftrightarrow k \in D_b \text{ for some } b \in B$$
$$\Leftrightarrow k \in \bigcup \mathcal{D}.$$

Hence $\mathcal{L}$ is closed under directed unions, and so is algebraic. □

Example 3.2.5.

(a) Each finite lattice is algebraic.

(b) The lattice of subsets of a set is algebraic. The compact elements are precisely the finite sets.

(c) The closure operator (3.2.1) is algebraic. The compact elements are the finitely generated subalgebras [Exercise 3.2.6]. Hence the lattice of subalgebras of any algebra is algebraic. □

For more information about algebraic lattices and closure operators, see [Davey and Priestley 1990].

The algebraic lattices of most interest here arise as lattices of congruences of algebras. For each τ-algebra (A, Ω), consider the map

(3.2.3) $\qquad \langle\ \rangle_{cg} : \mathcal{P}(A^2) \to \mathcal{P}(A^2) : \beta \mapsto \langle \beta \rangle_{cg}$

assigning to each $\beta \subseteq A^2$ the smallest congruence $\langle \beta \rangle_{cg}$ containing it.

3. VARIETIES

Proposition 3.2.6. *For each τ-algebra (A, Ω), the map (3.2.3) is an algebraic closure operator. The closed sets are the congruences of (A, Ω). The poset $Cg\, A$ is an algebraic lattice.*

Proof. Consider the direct product $(A \times A, \Omega)$. On $A \times A$, define the following further operations: nullary operations (a, a) for each $a \in A$, a unary operation s defined by $(a, b)s = (b, a)$, and a binary operation t defined by $(a, b)(c, d)t =$ if $b = c$ **then** (a, d) **else** (a, b).

The reader may verify that B is a subalgebra of the algebra

$$(A \times A, \Omega, s, t, \{(a, a) | a \in A\})$$

iff B is a congruence of (A, Ω). Hence the map $\langle\ \rangle_{cg}$ coincides with the algebraic closure operator $\langle\ \rangle$ of Example 3.2.5(c). In particular, $Cg\, A$ is an algebraic lattice. □

Note that the compact elements of $Cg\, A$ are the congruences generated by finite sets.

As was shown in Proposition 3.1.6, if (A, Ω) is a τ-algebra, and $\underline{\underline{K}}$ is a prevariety of τ-algebras, then the lattice $Cg_K A$ is a complete lattice, and a meet subsemilattice of the lattice $Cg\, A$. If $\underline{\underline{K}}$ is a quasi-equational class, then $Cg_K A$ turns out to be algebraic. This was already proved for the case $\underline{\underline{K}} = \underline{\underline{\tau}}$ in Proposition 3.2.6.

Now let $\underline{\underline{K}}$ be a quasi-equational class $\underline{\underline{Q}}$ different from $\underline{\underline{\tau}}$. Recall that $\underline{\underline{Q}}$ is a prevariety. For each τ-algebra (A, Ω), $Cg_Q A$ is a (complete) meet subsemilattice of $(Cg\, A, \bigcap)$, i.e. meets in $Cg_Q A$ and $Cg\, A$ are the same. The join in $Cg_Q A$ of any set of $\underline{\underline{Q}}$-congruences is the least $\underline{\underline{Q}}$-congruence containing all summands, and is usually different from its join in $Cg\, A$. To prove that the lattice $Cg_Q A$ is algebraic, we will show that there is an appropriate algebraic closure operator.

Lemma 3.2.7. *Let $\underline{\underline{Q}}$ be a quasi-equational class of τ-algebras, and let A be a τ-algebra. Then the mapping*

(3.2.4) $$\langle\ \rangle_{cg\underline{\underline{Q}}} : \mathcal{P}(A^2) \to \mathcal{P}(A^2);\ X \mapsto \langle X\rangle_{cg\underline{\underline{Q}}}$$

assigning to each subset X of A^2 the smallest $\underline{\underline{Q}}$-congruence containing X, is an algebraic closure operator on $A \times A$.

Proof. The easy proof that $\langle\ \rangle_{cg\underline{\underline{Q}}}$ is a closure operator is left as Exercise 3.2I. The closed subsets of A^2 are precisely the $\underline{\underline{Q}}$-congruences of A. To prove that $\langle\ \rangle_{cg\underline{\underline{Q}}}$ is algebraic, it is enough to show that the condition (d) of Proposition 3.2.3 is satisfied, i.e. for any directed set Φ of $\underline{\underline{Q}}$-congruences on A, the union $\alpha = \bigcup_{\varphi \in \Phi} \varphi$ is again a $\underline{\underline{Q}}$-congruence. First note that by Exercise 1.2L, α is a congruence on A, and that $\sum_{\varphi \in \Phi} \varphi = \alpha = \bigcup_{\varphi \in \Phi} \varphi$. We have to show that if

(q) $$(\&_{i=1,\ldots,n}\ w_i = w'_i) \to (w = w')$$

is any quasi-identity true in $\underline{\underline{Q}}$, then for the homomorphism $\bar{f} : \arg(q)\Omega \to A$ extending any map $f : \arg(q) \to A$, if $\{(w_1\bar{f}, w'_1\bar{f}), \ldots, (w_n\bar{f}, w'_n\bar{f})\} \subseteq \alpha$, then $(w\bar{f}, w'\bar{f}) \in \alpha$. So suppose that for each $i = 1, \ldots, n$, one has $(w_i\bar{f}, w'_i\bar{f}) \in \alpha = \bigcup_{\varphi \in \Phi} \varphi$. This means that there are $\varphi_1, \ldots, \varphi_n \in \Phi$ with $(w_i\bar{f}, w'_i\bar{f}) \in \varphi_i$. Since Φ is directed, the congruences $\varphi_1, \ldots, \varphi_n$ have a common bound, say φ, in Φ. Obviously all the $(w_i\bar{f}, w'_i\bar{f})$ are in φ. Since φ is also a $\underline{\underline{Q}}$-congruence, one has $(w\bar{f}, w'\bar{f}) \in \varphi$, and hence $(w\bar{f}, w'\bar{f}) \in \alpha$. This means that $A^\alpha \models q$ for any quasi-identity q true in $\underline{\underline{Q}}$, and hence $\alpha = \bigcup_{\varphi \in \Phi} \varphi$ is a $\underline{\underline{Q}}$-congruence. Consequently, $\langle\ \rangle_{cg\underline{\underline{Q}}}$ is an algebraic closure operator. $\square$

Corollary 3.2.8. *If A is a τ-algebra, and $\underline{\underline{Q}}$ is a quasi-equational class of τ-algebras, then the lattice $Cg_{\underline{\underline{Q}}}A$ of $\underline{\underline{Q}}$-congruences of A is algebraic.*

Proof. The corollary follows immediately by Theorem 3.2.4. $\square$

3. VARIETIES

Note that the compact elements of Cg_QA are precisely the finitely generated $\underline{Q}$-congruences of A. Moreover, by Proposition 3.2.3 (b), for $X \subseteq A^2$, the congruence $\langle X \rangle_{cg_Q}$ is the union of finitely generated $\underline{Q}$-congruences $\langle Y \rangle_{cg_Q}$ for $Y \subseteq X$, i.e.

$$\langle X \rangle_{cg_Q} = \bigcup \{\langle Y \rangle_{cg_Q} \mid Y \subseteq X \text{ and } Y \text{ is finite}\}.$$

An element a of a complete lattice L is called *completely meet-irreducible* (cf. Exercise 3.1J) if, for every subset $A \subseteq L$, the equation $a = \prod A$ implies $a \in A$.

Lemma 3.2.9. *Each element of an algebraic lattice L is a meet of completely meet-irreducible elements.*

Proof. Let $a \in L$ and let $L_a = \{b \in L \mid a \leq b \text{ and } b \text{ is completely meet irreducible}\}$. It will be shown that $a = \prod L_a$. It is clear that $a \leq \prod L_a$. To prove that $a = \prod L_a$, it is enough to show that for each compact element c of L, if $c \leq \prod L_a$, then $c \leq a$. Suppose on the contrary that there is a compact element c such that $c \leq \prod L_a$ but $c \not\leq a$. Let $X = \{x \in L \mid a \leq x \text{ and } c \not\leq x\}$. Obviously $a \in X$, and since c is compact, the join of any chain in X is again in X. By the Kuratowski-Zorn Lemma 0.5.1, the set X has a maximal element, say m. Since by the maximality of m the inequality $m < d$ implies $m + c \leq d$, it follows that $m \in L_a$, and obviously $\prod L_a \leq m$. This gives a contraction, since $c \leq \prod L_a$ but $c \not\leq m$. □

Mal'cev's Relative Subdirect Representation Theorem 3.2.10. *Let $\underline{Q}$ be a quasi-equational class of τ-algebras. Then every algebra in $\underline{Q}$ is a subdirect product of relatively subdirectly irreducible algebras in $\underline{Q}$.*

Proof. Let A be a non-trivial $\underline{Q}$-algebra. By Corollary 3.2.8, Cg_QA is an algebraic lattice. Hence by Lemma 3.2.9, $\widehat{A} = \bigcap_{i \in I} \theta_i$ for some completely meet irreducible $\underline{Q}$-congruences θ_i of A. Now by Corollary 3.1.3, one has $A \leq_s$

$\prod_{i \in I} A^{\theta_i}$. To show that the algebras A^{θ_i} are relatively subdirectly irreducible, first note that by the Third Isomorphism Theorem 1.2.3, one has $C_{g_Q} A^{\theta_i} \cong [\theta_i, A^2]$. Since each congruence θ_i is completely meet irreducible in $C_{g_Q} A$, it follows that $\widehat{A^{\theta_i}}$ is completely meet irreducible in $C_{g_Q} A^{\theta_i}$. By Exercise 3.2M, the algebras A^{θ_i}, for i in I, are relatively subdirectly irreducible. □

Exercises

3.2A. Prove Theorem 3.2.1.

3.2B. Prove Proposition 3.2.2.

3.2C. Prove the equivalence $(a) \Leftrightarrow (b)$ in Proposition 3.2.3.

3.2D. Prove that for any closure operator C on a set X, and $A \subseteq X$,

$$A^C \supseteq \bigcup \{B^C \mid B \subseteq A \text{ and } B \text{ is finite}\}.$$

3.2E. [Davey, Priestley 1990] Let L be a complete lattice, and let S be a subset of L.

(a) Show that $D = \{\sum T \mid T \subseteq S \text{ and } T \text{ is finite}\}$ is a directed subset of L with $\sum D = \sum S$.

(b) Show that an element k of L is compact if and only if, for every directed set D in L,

$$(k \leq \sum D) \Rightarrow (k \leq d \text{ for some } d \in D).$$

3.2F. (a) Show that the poset $\mathcal{L}$ in the proof of Theorem 3.2.4(b) is a closure system.

(b) Show that the mapping φ in the proof of Theorem 3.2.4(b) is an isomorphism.

[Hint: The injectivity of φ follows from the fact that L is algebraic.]

3.2G. Show that for a τ-algebra (A, Ω), the map $\langle \ \rangle : \mathcal{P}(A) \to \mathcal{P}(A); B \mapsto \langle B \rangle$ is an algebraic closure operator.

[Hint: Use Proposition 1.4.5.]

3. VARIETIES

3.2H. Let A be a τ-algebra, and let $\underline{\underline{Q}}$ be a quasi-equational class of τ-algebras. Let $a_1, \ldots, a_n, b_1, \ldots, b_n$ be in A and $\varphi \in Cg_Q A$. Show the following:

(a) $\langle\{(a_1, b_1), \ldots, (a_n, b_n)\}\rangle_{cg_Q} = \sum_{i=1}^{n} \langle(a_i, b_i)\rangle_{cg_Q} = \sum_{i=1}^{n} cg_Q(a_i, b_i)$
    (cf. Exercise 1.2I);

(b) $\varphi = \bigcup \{cg_Q(a,b) \mid (a,b) \in \varphi\} = \sum\{cg_Q(a,b) \mid (a,b) \in \varphi\}$.

(c) $\varphi = \bigcup \{\langle\{(a_1, b_1), \ldots, (a_n, b_n)\}\rangle_{cg_Q} \mid (a_i, b_i) \in \varphi, n \geq 1\}$.

3.2I. Show that the mapping $\langle\ \rangle_{cg_Q}$ defined by (3.2.1) is a closure operator.

3.2J. Show that the normal subgroups of a group form an algebraic lattice.

3.2K. Show that the ideals of a ring form an algebraic lattice.

3.2L. Show that for any quasi-equational class $\underline{\underline{Q}}$ and any $\underline{\underline{Q}}$-algebra A, the restriction $\langle\ \rangle_{cg_Q} : Cg\,A \to Cg_Q A$ is a complete join homomorphism, i.e. for congruences φ_i of A, with $i \in I$, one has

$$\langle \sum_{i \in I} \varphi_i \rangle_{cg_Q} = \sum_{i \in I} \langle\varphi_i\rangle_{cg_Q}.$$

3.2M. Let $\underline{\underline{Q}}$ be a quasi-equational class of τ-algebras, and let A be a $\underline{\underline{Q}}$-algebra. Show that A is subdirectly irreducible relative to $\underline{\underline{Q}}$ iff $\widehat{A}$ is completely meet irreducible in $Cg_Q A$.

3.3. Replicas and free algebras

Let $\underline{K}$ be a prevariety of τ-algebras for a given type $\tau : \Omega \to \mathbb{N}$, and let (A, Ω) be a τ-algebra. Define Θ to be the set $\{\theta \in Cg\,A \mid A^\theta \in \underline{K}\}$ of $\underline{K}$-congruences of the algebra (A, Ω). Then the universality property (2.3.3) for products implies that for each τ-algebra (B, Ω) and set of τ-homomorphisms

$f_\theta : B \to A^\theta$, there is a unique τ-homomorphism $r := \prod_{\theta \in \Theta} (f_\theta : B \to \prod_{\theta \in \Theta} A^\theta)$ such that the following diagram commutes:

(3.3.1)
$$\begin{array}{ccc} \prod_{\theta \in \Theta} A^\theta & \xrightarrow{\pi_\theta} & A^\theta \\ {\scriptstyle r}\uparrow & & \uparrow{\scriptstyle f_\theta} \\ B & = & B \end{array}$$

i.e. $r\pi_\theta = f_\theta$ for each congruence θ in the set Θ. In particular, there is a unique homomorphism $r := \prod_{\theta \in \Theta} (\mathrm{nat}\,\theta : A \to \prod_{\theta \in \Theta} A^\theta)$ such that the diagram below commutes:

(3.3.2)
$$\begin{array}{ccc} \prod_{\theta \in \Theta} A^\theta & \xrightarrow{\pi_\theta} & A^\theta \\ {\scriptstyle r}\uparrow & & \uparrow{\scriptstyle \mathrm{nat}\,\theta} \\ A & = & A \end{array}$$

Note that $ar = br$ iff $a^\theta = b^\theta$ for each $\theta \in \Theta$ iff $(a,b) \in \bigcap_{\theta \in \Theta} \theta$. Hence $\rho := \ker r = \bigcap_{\theta \in \Theta} \theta$ is the least $\underline{K}$-congruence of the algebra (A, Ω). Since all the A^θ are in $\underline{K}$, and $\underline{K}$ is a prevariety, it follows that $\prod_{\theta \in \Theta} A^\theta$ is in $\underline{K}$ and $r(A) \leq \prod_{\theta \in \Theta} A^\theta$ is in $\underline{K}$, too. Now First Isomorphism Theorem 1.2.3 implies that (A^ρ, Ω) and $(r(A), \Omega)$ are isomorphic. The $\underline{K}$-algebra (A^ρ, Ω), or briefly A^ρ, is said to be the *replica of* (A, Ω) *in the prevariety* $\underline{K}$, or briefly the $\underline{K}$-*replica of* (A, Ω). Note that

$$\pi_\theta(r(A)) = \{a^\theta \mid a \in A\} = A^\theta,$$

so that each projection π_θ is surjective. Hence the $\underline{K}$-replica A^ρ is a subdirect product of the algebras A^θ. (Cf. Section 3.1.) The congruence ρ is called the $\underline{K}$-*replica congruence of* (A, Ω), and the natural homomorphism $\mathrm{nat}\,\rho$ is called the $\underline{K}$-*replica morphism*.

In Section 2.2, we showed that the functors $\Omega : \underline{\mathrm{Set}} \to \underline{\tau}$ and $U : \underline{\tau} \to \underline{\mathrm{Set}}$ are adjoint to each other. Now it will be shown that the assignment of the

3. VARIETIES

$\underline{K}$-replica to each τ-algebra is the object part of a functor called a "replica functor", and that this functor is left adjoint to a certain forgetful functor. The following lemma is needed to define the morphism part.

Lemma 3.3.1 (**Universality property of replication**). *Let $\underline{K}$ be a prevariety of τ-algebras. For each homomorphism $h : (A, \Omega) \to (B, \Omega)$ into a $\underline{K}$-algebra (B, Ω), there is a unique τ-homomorphism $\bar{h} : (A^\rho, \Omega) \to (B, \Omega)$ such that*

(3.3.3) $$(\operatorname{nat} \rho) \bar{h} = h,$$

i.e. the following diagram commutes:

$$\begin{array}{ccc} A^\rho & \xrightarrow{\bar{h}} & B \\ {\scriptstyle \operatorname{nat} \rho} \uparrow & & \uparrow {\scriptstyle h} \\ A & = & A \end{array}$$

Proof. This follows by (3.3.2) and the First Isomorphism Theorem 1.2.3, and can be summarized in the following commuting diagram:

$$\begin{array}{ccccccc} & & & \bar{h} & & & \\ & & & & & & \downarrow \\ \prod_{\theta \in \Theta} A^\theta \geq A^\rho & \xrightarrow{\pi_\theta} & A^\theta & \longrightarrow & h(A) & \rightarrowtail & B \\ {\scriptstyle \operatorname{nat} \rho} \uparrow & & {\scriptstyle \operatorname{nat} \theta} \uparrow & & \uparrow {\scriptstyle h} & & \\ A & = & A & = & A & & \end{array}$$

where $\theta := \ker h$, and π_θ is the restriction to A^ρ of the projection onto A^θ. □

Using Lemma 3.3.1, one can define the *replica* functor $R : \underline{\tau} \to \underline{K}$ (or R_K) as follows. To each object (A, Ω) of $\underline{\tau}$, one assigns its $\underline{K}$-replica $AR := (A^\rho, \Omega)$. To each τ-homomorphism $h : (A, \Omega) \to (B, \Omega)$, one assigns the $\underline{K}$-morphism

$hR : AR \to BR$, i.e.

$$\begin{array}{ccc} (A, \Omega) & & (AR, \Omega) \\ h \downarrow & \stackrel{R}{\longmapsto} & \downarrow hR \\ (B, \Omega) & & (BR, \Omega), \end{array}$$

where hR is determined by the following diagram:

$$hR = \overline{h \, \text{nat} \rho'}$$

$$\begin{array}{ccccc} AR = A^\rho & \stackrel{\pi_\theta}{\longrightarrow} & A^\theta & \longrightarrow & B^{\rho'} = BR \\ \text{nat}\rho \uparrow & & \uparrow \text{nat}\theta & & \uparrow \text{nat}\rho' \\ A & = & A & \stackrel{}{\underset{h}{\longrightarrow}} & B, \end{array}$$

in which $\theta = \ker(h \, \text{nat}\rho')$. Note that for $a \in A$, one has $a^\rho hR = (ah)^{\rho'}$.

Now let $J : \underline{K} \to \underline{\tau}$ (or J_K) be the inclusion functor, "forgetting" that $\underline{K}$-algebras are members of the prevariety $\underline{K}$. The universality property of Lemma 3.3.1 shows that for each τ-homomorphism $h : (A, \Omega) \to (B, \Omega)$ into a $\underline{K}$-algebra (B, Ω), there is precisely one $\underline{K}$-morphism $\bar{h} : (AR, \Omega) \to (B, \Omega)$. Hence one can readily see that there is a bijection

$$(3.3.4) \qquad \underline{\tau}(A, BJ) \cong \underline{K}(AR, B).$$

This yields the following.

Proposition 3.3.2. *The replica functor* $R : \underline{\tau} \to \underline{K}$ *is left adjoint to the inclusion functor* $J : \underline{K} \to \underline{\tau}$. □

Proposition 3.3.2 provides an adjunction $(R, J, \eta, \varepsilon)$. The component η_A of the unit η of this adjunction at a τ-algebra (A, Ω) is the natural projection of (A, Ω) into its $\underline{K}$-replica $AR = (A^\rho, \Omega)$.

Using Lemma 3.3.1, one can easily extend the concepts of $\underline{K}$-replica and $\underline{K}$-replica morphism to any class $\underline{K}$ of τ-algebras of type $\tau : \Omega \to \mathbb{N}$.

Definition 3.3.3. An Ω-homomorphism $r : A \to Ar$ onto a $\underline{K}$-algebra Ar is called a $\underline{K}$-*replica morphism*, and the algebra Ar is called a *replica* of A in $\underline{K}$ or a $\underline{K}$-*replica* of A, if they satisfy the universality property of Lemma 3.3.1, i.e. if for each Ω-homomorphism $h : A \to B$ into a $\underline{K}$-algebra B, there is a unique homomorphism $\bar{h} : Ar \to B$ with $r\bar{h} = h$:

$$\begin{array}{ccc} Ar & \xrightarrow{\bar{h}} & B \\ r\uparrow & & \uparrow h \quad \square \\ A & =\!=\!= & A. \end{array}$$

A τ-algebra A may not have homomorphic images in the class $\underline{K}$, and hence may not have a $\underline{K}$-replica. If A has a replica in $\underline{K}$, it is uniquely defined up to isomorphism [Exercise 3.3C], and can be realized as the quotient A^ρ by the smallest $\underline{K}$-congruence $\rho = \ker r$ of A.

Now let X be a set, and let $\underline{K}$ be a prevariety of τ-algebras as before. The absolutely free τ-algebra $X\Omega$ over X is an object of the category $\underline{\tau}$, and its $\underline{K}$-replica $X\Omega R$ is an object of the category $\underline{K}$. Let $f : X \to B$ be a mapping of X into the underlying set of a $\underline{K}$-algebra (B,Ω), and let $\iota : X \to X\Omega$; $x \mapsto x$ be the inclusion of X into $X\Omega$. Then the universality property for $X\Omega$ yields the unique τ-homomorphism $\bar{f} : (X\Omega, \Omega) \to (B, \Omega)$ with $f = \iota\bar{f}$, while the universality property for $X\Omega R$ yields the unique $\underline{K}$-homomorphism $\bar{\bar{f}} : (X\Omega R, \Omega) \to (B, \Omega)$ with $f = (\iota \operatorname{nat}\rho)\bar{\bar{f}}$. (Here ρ is the $\underline{K}$-replica congruence of $X\Omega$.) Note that one also has $\bar{f} = (\operatorname{nat}\rho)\bar{\bar{f}}$. Indeed, for an Ω-word $x_1 \ldots x_n w$, one has

$$\begin{aligned} x_1 \ldots x_n w \bar{f} &= x_1 f \ldots x_n f w \\ &= (x_1^\rho \bar{\bar{f}}) \ldots (x_n^\rho \bar{\bar{f}}) w \\ &= (x_1^\rho \ldots x_n^\rho w) \bar{\bar{f}} \\ &= (x_1 \ldots x_n w)^\rho \bar{\bar{f}}. \end{aligned}$$

This shows that the following diagram commutes:

(3.3.5)
$$\begin{array}{ccc} X & \xrightarrow{\iota} X\Omega & \xrightarrow{\mathrm{nat}\rho} X\Omega R \\ {\scriptstyle f}\downarrow & {\scriptstyle \bar{f}}\downarrow & {\scriptstyle \bar{\bar{f}}}\downarrow \\ B & = B & = B. \end{array}$$

Definition 3.3.4. A $\underline{K}$-algebra XK is called (*relatively*) *free* in the prevariety $\underline{K}$, or the *free $\underline{K}$-algebra over X*, if it satisfies the following *universality property* for $\underline{K}$-algebras XK and mappings $\eta_X : X \to XK$:

(3.3.6) $\quad\begin{cases} \text{For each mapping } f : X \to B \text{ into the underlying set of a} \\ \underline{K}\text{-algebra } (B, \Omega), \text{ there is a unique homomorphism} \\ \bar{\bar{f}} : (XK, \Omega) \to (B, \Omega) \text{ such that } \eta_X \bar{\bar{f}} = f \end{cases}$

$$\begin{array}{ccc} X & \xrightarrow{\eta_X} & XK \\ {\scriptstyle f}\downarrow & & {\scriptstyle \bar{\bar{f}}}\downarrow \quad \square \\ B & = & B. \end{array}$$

It is clear that the algebra $X\Omega R$, together with the mapping $\iota\,\mathrm{nat}\,\rho$, satisfies the universality property (3.3.6), so that it is a free K-algebra.

Proposition 3.3.5. *The free algebra XK is isomorphic to the $\underline{K}$-replica $X\Omega R$ of the Ω-word algebra $X\Omega$.*

Proof. The universality properties for $X\Omega R$ and XK yield the following commuting diagrams:

(3.3.7)
$$\begin{array}{ccccccccc} X & \xrightarrow{\iota} & X\Omega & \xrightarrow{\mathrm{nat}\rho} & X\Omega R & & X & \xrightarrow{\eta_X} & XK \\ {\scriptstyle \eta_X}\downarrow & & {\scriptstyle \bar{\eta}_X}\downarrow & & {\scriptstyle \bar{\bar{\eta}}_X}\downarrow & & {\scriptstyle \iota}\downarrow & & {\scriptstyle \overline{\iota\,\mathrm{nat}\rho}}\downarrow \\ XK & = & XK & = & XK & & X\Omega & \xrightarrow{\mathrm{nat}\rho} & X\Omega R \end{array}$$

3. VARIETIES

Note that

$$\iota \, \mathrm{nat}\rho = \eta_X \overline{\overline{\iota \mathrm{nat}\rho}} = \iota \, \bar{\bar{\eta}}_X \overline{\overline{\iota \mathrm{nat}\rho}} = \iota \, \mathrm{nat}\rho \, \bar{\bar{\eta}}_X \overline{\overline{\iota \mathrm{nat}\rho}}.$$

Similarly

$$\eta_X = \iota \bar{\bar{\eta}}_X = \iota \, \mathrm{nat}\rho \, \bar{\bar{\eta}}_X = \eta_X \overline{\overline{\iota \mathrm{nat}\rho}} \, \bar{\bar{\eta}}_X.$$

Since evidently $\iota \, \mathrm{nat}\rho = \iota \, \mathrm{nat}\rho \, 1_{X\Omega R}$ and $\eta_X = \eta_X 1_{XK}$, it follows, by the requirements of uniqueness, that

$$\bar{\bar{\eta}}_X \overline{\overline{\iota \mathrm{nat}\rho}} = 1_{X\Omega R} \quad \text{and} \quad \overline{\overline{\iota \mathrm{nat}\rho}} \, \bar{\bar{\eta}}_X = 1_{XK},$$

whence $\bar{\bar{\eta}}_X$ and $\overline{\overline{\iota \mathrm{nat}\rho}}$ are mutually inverse isomorphisms. $\square$

In particular, Proposition 3.3.5 shows that any free algebra XK in the trivial variety of a plural type is a one-element algebra. Moreover, for any type, each $\underline{K}$-algebra is a homomorphic image of a suitable free $\underline{K}$-algebra. Indeed, for a $\underline{K}$-algebra (B, Ω), the free $\underline{K}$-algebra $BK \cong B\Omega R$ over the underlying set B of (B, Ω) will do the job:

$$\begin{array}{ccc} B & \xrightarrow{\eta_B} & BK \\ {\scriptstyle 1_B}\downarrow & & \downarrow{\scriptstyle \bar{\bar{1}}_B} \\ B & = & B. \end{array}$$

One obtains thus a functor K from the category $\underline{\mathrm{Set}}$ to the category $\underline{K}$ as the composite of the functor $\Omega : \underline{\mathrm{Set}} \to \underline{\tau}$ defined in Example 2.2.4 and the replica functor $R : \underline{\tau} \to \underline{K}$. This functor is called the *free $\underline{K}$-algebra* functor. It is left adjoint to the underlying set functor

$$U : \underline{K} \to \underline{\mathrm{Set}}; \ (f : (A, \Omega) \to (B, \Omega)) \mapsto (f : A \to B).$$

The component $\eta_X : X \to XKU$ of the unit η of the adjunction $(K, U, \eta, \varepsilon)$ is the composite of the unit $\eta_X^\tau : X \to X\Omega U$ of the adjunction $(\Omega : \underline{\mathrm{Set}} \to$

$\underline{\tau}$, $U : \underline{\tau} \to \underline{\mathrm{Set}}$, $\eta^\tau, \varepsilon^\tau$) with the unit $\eta^K_{X\Omega} : X\Omega \to X\Omega RJ$ of the adjunction $(R : \underline{\tau} \to \underline{K}, J : \underline{K} \to \underline{\tau}, \eta^K, \varepsilon^K)$.

Under rather mild restrictions, one can define free $\underline{K}$-algebras in more "algebraic" terms.

Proposition 3.3.6. *Let $\underline{K}$ be a non-trivial prevariety, and let X be a non-empty set. Let $e : X \to A$ be an embedding into the underlying set of a $\underline{K}$-algebra (A, Ω). Then the algebra (A, Ω) is (isomorphic to) a free $\underline{K}$-algebra XK if and only if it satisfies the following properties:*

(a) *(A, Ω) is generated by a subset (isomorphic to) X;*

(b) *for each mapping $h : X \to B$ into the underlying set B of a $\underline{K}$-algebra (B, Ω), there is a (unique) homomorphism $\overline{\overline{h}} : (A, \Omega) \to (B, \Omega)$ extending h.*

Proof. ($\Leftarrow$) By the defining property of free $\underline{K}$-algebras it is clear that the algebra A together with the embedding $\eta_X = e : X \to A$; $x \mapsto xe$ is a free $\underline{K}$-algebra over X with $\overline{\overline{h}} : A \to B$ defined by $a_1 \ldots a_n w \overline{\overline{h}} = a_1 h \ldots a_n h w$, for $a_1, \ldots, a_n$ in X.

($\Rightarrow$) Now assume that A is the free $\underline{K}$-algebra XK over X with $\eta_X = e$. Consider X as a subset of A. It will be shown that the set X generates A. Let B be the subalgebra of A generated by X, and let $j : B \to A$; $a \mapsto a$ be the corresponding inclusion. We want to show that j is also surjective. Consider the inclusion map $\iota : X \to B$, $x \mapsto x$. Then the universality property for XK yields a unique homomorphism $\overline{\overline{\iota}} : A \to B$ with $\iota = e\overline{\overline{\iota}}$. Hence $\iota j = e\overline{\overline{\iota}} j$. The same universality property gives the following commuting diagram

$$\begin{array}{ccc} X & \xrightarrow{e} & A \\ \iota \downarrow & & \downarrow \overline{\overline{\iota}} j \\ B & \xrightarrow{j} & A \end{array}$$

with $\iota j = e\iota j$. Since $\overline{\overline{\iota j}}$ is uniquely defined, one has $\overline{\overline{\iota j}} = \overline{\iota} j$. On the other hand, $\overline{\overline{\iota j}} = 1_{XK}$, and hence $\overline{\iota} j = 1_{XK}$. It follows that j is surjective. □

In the case where X generates XK, we also say that the set X is a set of *free generators* for XK, and that XK is *freely generated* by X. Such a freely generated algebra XK is determined uniquely up to isomorphism by the cardinality of the generating set X.

It is easy to see that the free semigroups X^+ and monoids X^* over X defined in Section 1.4 are indeed free in the variety of semigroups and in the variety of monoids, respectively, with the sets X as sets of free generators.

The following propositions summarize key properties of free $\underline{K}$-algebras.

Proposition 3.3.7. *A prevariety $\underline{K}$ of τ-algebras satisfies an identity $u = v$ with $\arg(u) \cup \arg(v) = X$ if and only if the free $\underline{K}$-algebra XK satisfies it.*

Proof. It is clear that if a prevariety $\underline{K}$ satisfies an identity $u = v$, then the free $\underline{K}$-algebra XK also satisfies it. To prove the converse, suppose that XK satisfies $u = v$. We can assume that $X \subseteq \mathfrak{P}$, and that $u = x_{i_1} \ldots x_{i_m} u$ and $v = x_{j_1} \ldots x_{j_n} v$. Moreover, we can identify XK with $X\Omega R = X\Omega^\rho$. Let $f : \mathfrak{P}\Omega \to A$ be an Ω-homomorphism into a $\underline{K}$-algebra A. One has to show that $uf = vf$. By the universality property for free $\underline{K}$-algebras, there is a homomorphism $g : XK \to A$ such that $x_i f = x_i^\rho g$ for all $x_i \in X$. Let h be any homomorphism $\mathfrak{P}\Omega \to XK$ such that $x_i h = x_i^\rho$ for $x_i \in X$. Note that $f = hg$, since these maps agree on $X\Omega$.

$$\begin{array}{ccc} X\Omega & \xrightarrow{h} & X\Omega^\rho = XK \\ {\scriptstyle f}\downarrow & & \downarrow{\scriptstyle g} \\ A & = & A \end{array}$$

Note also that, by assumption,

$$x_{i_1}^\rho \ldots x_{i_m}^\rho u = x_{j_1}^\rho \ldots x_{j_n}^\rho v$$

in XK. Hence it follows that

$$uf = (x_{i_1} \ldots x_{i_m} u)hg = ((x_{i_1}h) \ldots (x_{i_m}h))u)g$$
$$= (x_{i_1}^\rho \ldots x_{i_m}^\rho u)g = (x_{j_1}^\rho \ldots x_{j_n}^\rho v)g$$
$$= ((x_{j_1}h) \ldots (x_{j_n}h)v)g = (x_{j_1} \ldots x_{j_n}v)hg = vf.$$

Consequently, each $\underline{K}$-algebra satisfies the identity $u = v$. □

Proposition 3.3.8. *Let $\underline{K}$ be a prevariety of Ω-algebras. Consider the variety $v\underline{K}$ generated by $\underline{K}$. If the free $\underline{K}$-algebra XK is freely generated by X, then*

$$XK \cong X(vK).$$

Proof. It is enough to prove that XK satisfies the condition (b) of Proposition 3.3.6 for any $v\underline{K}$-algebra B. Let A be a $\underline{K}$-algebra, and let $h : A \to B$ be a homomorphism onto a $v\underline{K}$-algebra B. Let $f : X \to B$ be any map. By the Axiom of Choice, there is a mapping $g : X \to A$ such that $xg \in h^{-1}(xf)$ for each $x \in X$. By the universality property for XK, there is a homomorphism $\bar{g} : XK \to A$ extending g:

$$\begin{array}{ccc} X & \longrightarrow & XK \\ g \downarrow & & \downarrow \bar{g} \\ A & = & A. \end{array}$$

Define $\tilde{g} := \bar{g}h$,

$$\tilde{g} : XK \xrightarrow{\bar{g}} A \xrightarrow{h} B.$$

Then for each $x \in X$ one has

$$x\tilde{g} = x\bar{g}h = xgh = xf,$$

so that $\tilde{g}$ is the (unique) homomorphism extending f. [Cf. Exercise 1.2C(f).] □

3. VARIETIES

We conclude this section with an explicit description of Ω-semilattice replicas of τ-algebras. The Ω-semilattice replica AR of an algebra (A, Ω) of type $\tau : \Omega \to \mathbb{Z}^+$ is the quotient (A^ρ, Ω) of (A, Ω) by the semilattice replica congruence ρ, so that any Ω-homomorphism $h : (A, \Omega) \to (S, \Omega)$ from (A, Ω) to an Ω-semilattice (S, Ω) factors as $h = (\mathrm{nat}\,\rho)\bar{h}$ through a unique homomorphism $\bar{h} : (A^\rho, \Omega) \to (S, \Omega)$. Thus the congruence ρ identifies precisely those elements of A which are identified in all Ω-homomorphisms from (A, Ω) to Ω-semilattices.

The semilattice replica of an Ω-algebra may be given an explicit description in terms of walls of the algebra. (Cf. Definition 1.1.9.) For an algebra (A, Ω) and an element a of A, let $[a]$ be the *principal wall* generated by a, i.e. the smallest wall containing a.

Lemma 3.3.9. *For any $a_1, \ldots, a_n$ in A and n-ary ω in Ω, one has*

$$[a_1] + \cdots + [a_n] = [a_1 \ldots a_n \omega].$$

Proof. On the one hand $a_1 \ldots a_n \omega \in [a_1 \ldots a_n \omega]$ implies $a_1, \ldots, a_n \in [a_1 \ldots a_n \omega]$ by the definition of a wall, so $[a_1], \ldots, [a_n] \subseteq [a_1 \ldots a_n \omega]$ and $[a_1] + \cdots + [a_n] \leq [a_1 \ldots a_n \omega]$. On the other hand, since $a_1, \ldots, a_n$ are contained in the subalgebra $[a_1] + \cdots + [a_n]$, one has $a_1 \ldots a_n \omega \in [a_1] + \cdots + [a_n]$, whence $[a_1 \ldots a_n \omega] \leq [a_1] + \cdots + [a_n]$. $\square$

In particular, it follows that for a and b in A and n-ary ω in Ω, one has $[a] + [b] = [a] + [b] + \cdots + [b] = [ab \ldots b\omega]$. This shows that the principal walls of the algebra (A, Ω) form a subsemilattice $(\mathcal{W}, +)$ of the join semilattice of all walls. Moreover, it shows that the mapping

(3.3.8) $$h : A \to \mathcal{W};\ a \mapsto [a]$$

is an Ω-homomorphism from the algebra (A, Ω) onto the Ω-semilattice $(\mathcal{W}, \Omega)$ of principal walls. Set $\sigma := \ker h$. Thus $(a, b) \in \sigma$ iff $[a] = [b]$.

An Ω-algebra is said to be *algebraically open* if it has no proper non-empty walls. The name is motivated by the fact (proven in Chapter 7) that a convex subset of a real space is open in its affine hull, i.e. in the smallest affine space containing it, iff it is algebraically open.

Proposition 3.3.10. *An algebra (A, Ω) is algebraically open if and only if there is no epimorphism $(A, \Omega) \to \underset{\sim}{2} = (\{0, 1\}, \Omega)$ onto the two-element (join) semilattice with $0 < 1$.*

Proof. The equivalence is clear for trivial algebras, so assume that A has more than 1 element. If (A, Ω) has no proper non-empty walls, then there is no Ω-epimorphism of (A, Ω) onto the two-element semilattice. Indeed, a proper non-empty wall W furnishes an Ω-epimorphism $(W \to \{0\}) \cup ((A - W) \to \{1\})$. Conversely, if there is no such epimorphism, then (A, Ω) has no proper non-empty walls. Indeed, the preimage of 0 under an epimorphism $(A, \Omega) \to (\{0, 1\}, \Omega)$ is a proper non-empty wall of (A, Ω). $\square$

Decomposition Theorem 3.3.11. *The semilattice replica of an algebra (A, Ω) is its semilattice of principal walls. The corresponding ρ-classes are algebraically open subalgebras of (A, Ω).*

Proof. First note that the kernel σ of the homomorphism h defined by (3.3.8) contains the semilattice replica congruence ρ. The ρ-classes are subalgebras of (A, Ω), and have no non-trivial semilattice quotients. Thus they are algebraically open by Proposition 3.3.10. It remains to show that σ actually coincides with ρ.

Now for b in A, one has

(3.3.9) $$\forall a \in b^\rho, \; [a] = [b^\rho].$$

Certainly $a \in b^\rho$ implies $[a] \leq [b^\rho]$. Conversely, note that $[a] \cap b^\rho$ is a non-empty wall of b^ρ. Indeed, $a_1 \ldots a_n \omega \in [a] \cap b^\rho$ with $a_1, \ldots, a_n$ in b^ρ implies

$a_1, \ldots, a_n \in [a]$, so $a_1, \ldots, a_n \in [a] \cap b^\rho$. But b^ρ, being algebraically open, has no proper non-empty walls. Thus $[a] \cap b^\rho = b^\rho$, whence $b^\rho \subseteq [a]$ and $[b^\rho] \leq [a]$, completing the verification of (3.3.9).

Also, considering the semilattice replica (A^ρ, Ω) as a join semilattice $(A^\rho, \leq)$, one has

$$(3.3.10) \qquad [b^\rho] = \bigcup \{c^\rho \mid c^\rho \leq b^\rho\}$$

for each b in A. Certainly $c^\rho \leq b^\rho$ implies $cb \ldots bw \in c^\rho + b^\rho = b^\rho \subseteq [b^\rho]$, whence $c \in [b^\rho]$ and $[b^\rho] \supseteq \bigcup \{c^\rho \mid c^\rho \leq b^\rho\}$. On the other hand, the right hand side of (3.3.10) is the preimage in (A, Ω) under $\mathrm{nat}\rho$ of the principal wall $\{c^\rho \mid c^\rho \leq b^\rho\}$ of the semilattice (A^ρ, Ω) generated by b^ρ. Now preimages of walls under epimorphisms are walls, so that $[b^\rho]$ is contained in the right hand side of (3.3.10).

To complete the proof, assume that $(a, b) \in \sigma$. Then by (3.3.9) one has $[a^\rho] = [a] = [b] = [b^\rho]$. The expression (3.3.10) then shows that $a^\rho \leq b^\rho$ and $b^\rho \leq a^\rho$, whence $a^\rho = b^\rho$ or $(a, b) \in \rho$. Thus σ is also contained in ρ, so that ρ and σ do indeed coincide. □

We will see later (cf. Section 7.5) that in the case of idempotent algebras with commuting basic operations, it is possible to replace the concept of "wall" by the concept of "sink" in the Decomposition Theorem.

Exercises

3.3A. Find an example of a class $\underline{K}$ of τ-algebras and a τ-algebra (A, Ω) such that (A, Ω) has no replica in $\underline{K}$.

3.3B. Let $\underline{K}'$ be a subclass of a prevariety $\underline{K}$ that also is a prevariety. Let $R_K : \underline{\tau} \to \underline{K}$ and $R_{K'} : \underline{\tau} \to \underline{K}'$ be replica functors. Show that for each τ-algebra A one has $A R_K R'_K = A R_{K'}$.

3.3C. Let $\underline{K}$ be a class of τ-algebras. Show that $\underline{K}$-replicas of isomorphic algebras are isomorphic.

3.3D. Show that if B is non-trivial, the mapping η_X in Definition 3.3.4 is injective.

3.3E. Prove that if a $\underline{K}$-algebra (A, Ω) satisfies conditions (a) and (b) of Proposition 3.3.6, then the homomorphism $\overline{\overline{h}}$ is uniquely defined.

3.3F. Show that a prevariety $\underline{K}$ of Ω-algebras satisfies an identity $u = v$ iff $(u, v) \in \rho$, where ρ is the $\underline{K}$-replica congruence of $\mathfrak{P}\Omega$.

3.3G. Show that Proposition 3.3.7 holds also for any free $\underline{K}$-algebra YK with $|Y| \geq |X|$.

3.3H. Let $\underline{K}$ be a prevariety of τ-algebra. Let X be a non-empty set, and let $Y \supseteq X$ be an infinite set. Show that $\{u = v \mid (u, v) \in X\Omega^2, \underline{K} \models u = v\} = \{u = v \mid (u, v) \in X\Omega^2,\ YK \models u = v\}$. In particular, for $(u, v) \in X\Omega^2$, one has $\underline{K} \models u = v$ iff $PK \models u = v$.

3.3I. Show that the semilattice $(\mathcal{P}_{0<\infty}(X), \cup)$ of finite non-empty subsets of a set X is a free semilattice over X. (Cf. Exercise 1.4B.)

3.3J. Find all walls and sinks of the (join) semilattice given by the diagram

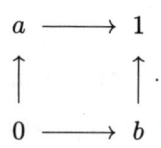

$$\begin{array}{ccc} a & \longrightarrow & 1 \\ \uparrow & & \uparrow \\ 0 & \longrightarrow & b \end{array}.$$

3.3K. Find all walls and sinks of the unit interval $(I, \underline{I}^\circ)$ of Exercise 1.1E. Find the $\underline{I}^\circ$-semilattice replica of $(I, \underline{I}^\circ)$.

3.3L. Let $\underline{K}$ be a non-trivial prevariety of idempotent τ-algebras of a given type

$\tau : \Omega \to \mathbb{Z}^+$. The *free product* in $\underline{K}$ of non-empty $\underline{K}$-algebras A_i, for $i \in I$, is the coproduct $\coprod_{i \in I} A_i$ in $\underline{K}$ of the algebras A_i with injective insertions ι_i. Show the following.

(a) There is a $\underline{K}$-algebra B with injective homomorphisms $\varphi_i : A_i \to B$ for all i in I.

(b) If for $i \neq j$ one has $A_i\varphi_i \cap A_j\varphi_j = \varnothing$, then also $A_i\iota_i \cap A_j\iota_j = \varnothing$, and $\coprod_{i\in I} A_i$ is generated by the disjoint union $\bigcup_{i\in I} A_i\iota_i$ of isomorphic copies of A_i.

3.3M. Let $\underline{K}$ be a class of Ω-algebras of a given type $\tau : \Omega \to \mathbb{N}$. Show that a $v\underline{K}$-algebra F is free over a subset X iff each mapping $f : X \to A$ of X into the underlying set of a $\underline{K}$-algebra A may be extended to a homomorphism $\bar{f} : (F, \Omega) \to (A, \Omega)$.

3.4. Prevarieties

This section provides two characterizations of prevarieties. In the previous section it was shown that each τ-algebra has a replica in any prevariety $\underline{K}$ of τ-algebras. Here, it will be shown that the converse also holds. If a class $\underline{K}$ of τ-algebras contains a replica of each τ-algebra, then it is a prevariety. The second characterization goes back to classes of τ-algebras defined by implications. Section 1.7 showed that such classes are prevarieties. Here, it will be shown that each prevariety can be defined by implications. The final theorem of the section provides a simple condition for a prevariety to be a variety.

A class $\underline{K}$ of τ-algebras is said to be *closed under replicas* if any τ-algebra A has a replica in $\underline{K}$ (as specified in Definition 3.3.3).

Theorem 3.4.1. *An abstract class $\underline{K}$ of τ-algebras is closed under replicas if and only if it is a prevariety.*

Proof. ($\Leftarrow$) The proof was given in Section 3.3.

($\Rightarrow$) Now suppose that each τ-algebra A has a replica Ar in the class $\underline{K}$. Let B be a $\underline{K}$-algebra, and let C be a subalgebra with corresponding inclusion $\iota : C \to B$ and corresponding $\underline{K}$-replica morphism $r : C \to Cr$. It will be shown that C is also a member of $\underline{K}$. First note that there is an Ω-homomorphism $\bar{\iota} : Cr \to B$ with $r\bar{\iota} = \iota$:

$$\begin{CD} Cr @>\iota>> B \\ @AArA @AA\iota A \\ C @= C \end{CD}$$

Since ι injects, r does, too. And since r is also surjective, it follows that r is an isomorphism. Hence $C \cong Cr$ is in $\underline{K}$.

It remains to show that $\underline{K}$ is closed under products. For $i \in I$, let A_i be $\underline{K}$-algebras, and let $A = \prod_{i \in I} A_i$ be the direct product of the A_i. If I is empty, then A is the trivial one-element algebra. Its $\underline{K}$-replica is trivial, too. Hence A is in $\underline{K}$. Now assume that I is non-empty. Consider the $\underline{K}$-replica morphism

$$r : A \to Ar,$$

and for all i in I, the projections

$$\pi_i : A \to A_i.$$

By the universality property for replicas, there are homomorphisms $\bar{\pi}_i : Ar \to A_i$, for each i in I, such that $r\bar{\pi}_i = \pi_i$:

(3.4.1)
$$\begin{CD} Ar @>\bar{\pi}_i>> A_i \\ @AArA @AA\pi_i A \\ A @= A \end{CD}$$

By the universality property for direct products, there is a homomorphism $h := \prod \bar{\pi}_i : Ar \to A$ such that $h\pi_i = \bar{\pi}_i$:

(3.4.2)
$$\begin{CD} A @>\pi_i>> A_i \\ @AAhA @AA\bar{\pi}_i A \\ Ar @= Ar \end{CD}$$

By (3.4.1) and (3.4.2), it follows that $\pi_i = r\bar{\pi}_i = rh\pi_i$. This implies that $rh = 1_A$, and consequently that r is injective, whence it is an isomorphism. It follows that $A \cong Ar$ is in $\underline{K}$. □

Banaschewski-Herrlich Theorem 3.4.2. *An abstract class $\underline{K}$ of τ-algebras is an implicational class if and only if it is a prevariety.*

Proof. ($\Rightarrow$) The proof goes like the proof of Exercises 1.7G and 1.7H.
($\Leftarrow$) Let $\tau : \Omega \to \mathbb{N}$ be a given type. By Theorem 3.4.1, each Ω-algebra A has a $\underline{K}$-replica $AR = A^\rho$ with corresponding $\underline{K}$-replica morphism $R : A \to AR$. Let $A\Omega$ be an absolutely free Ω-algebra over the set A, and let $\bar{1} : A\Omega \to A$ be the unique Ω-homomorphism extending the identity map $1 : A \to A$. Define $\{(w_m, w'_m) \mid m \in M\}$ to be $\ker \bar{1}$. Let $I(A)$ consist of all implications

$$(3.4.3) \qquad (\&_{m \in M} \ w_m = w'_m) \to (w = w')$$

for Ω-words $w, w' \in A\Omega$ such that $w\bar{1}R = w'\bar{1}R$. Finally, $I(K)$ is defined to be the union of all the $I(A)$, where A ranges over all τ-algebras. It will be shown that $\underline{K}$ is the class of Ω-algebras satisfying exactly the implications of $I(K)$.

First assume that B is any $\underline{K}$-algebra. We will prove that B satisfies all the implications (3.4.3) for any Ω-algebra A. Let $\alpha : A \to B$ be any mapping, and let $\bar{\alpha} : A\Omega \to B$ be its unique Ω-homomorphic extension. We want to show that if $(w_m, w'_m) \in \ker \bar{\alpha}$ for all $m \in M$, then also $(w, w') \in \ker \bar{\alpha}$. So assume that $(w_m, w'_m) \in \ker \bar{\alpha}$ for all $m \in M$. Since $\ker \bar{1} = \{(w_m, w'_m) \mid m \in M\}$, it follows that $\ker \bar{1} \subseteq \ker \bar{\alpha}$. Then $\bar{\alpha} : A\Omega \twoheadrightarrow A\Omega\bar{\alpha} \rightarrowtail B$ can be written as $\bar{\alpha} = fg$ with surjective $f : A\Omega \to A\Omega\bar{\alpha}$ and injective $g : A\Omega\bar{\alpha} \to B$. It follows that $\ker \bar{1} \subseteq \ker \bar{\alpha} \subseteq \ker f$. Hence there is a unique homomorphism $h : A \to A\Omega\bar{\alpha}$ with $f = \bar{1}h$:

$$(3.4.4) \qquad \begin{array}{ccccccc} A & \longrightarrow & A\Omega & \xrightarrow{f} & A\Omega\bar{\alpha} & \xrightarrow{g} & B \\ {\scriptstyle 1}\downarrow & & {\scriptstyle \bar{1}}\downarrow & & \uparrow{\scriptstyle h} & & \uparrow{\scriptstyle \bar{h}} \\ A & = & A & = & A & \xrightarrow{R} & AR \end{array}$$

Since B is a $\underline{K}$-algebra and $\underline{K}$ is a prevariety, the universality property for $\underline{K}$-replicas yields a unique homomorphism $\bar{h} : AR \to B$ with $hg = R\bar{h}$. Thus

$$\bar{\alpha} = fg = \bar{1}hg = \bar{1}R\bar{h}.$$

Hence $\ker(\bar{1}R) \subseteq \ker \bar{\alpha}$. By the definition of $I(A)$, one has $(w, w') \in \ker(\bar{1}R)$. This implies that also $(w, w') \in \ker \bar{\alpha}$, and consequently the algebra B satisfies each implication (3.4.3) of $I(A)$ for any τ-algebra A.

Now assume that an Ω-algebra A satisfies all the implications of $I(K)$. To prove that A is a member of $\underline{K}$, it suffices to show that the replica morphism $R : A \to AR$ onto AR is also injective, and hence that R is an isomorphism. Note in particular that A satisfies all the implications (3.4.3) of $I(A)$, i.e. if $(w_m, w'_m) \in \ker \bar{1}$ for all $m \in M$, then $(w, w') \in \ker \bar{1}$ as well. By the definition of $I(A)$, it is clear that $\ker(\bar{1}R) \subseteq \ker \bar{1}$, and hence $\ker(\bar{1}R) = \ker \bar{1}$. It follows that R is injective and A is isomorphic to AR. □

It should be noted that the implications defining the class $\underline{K}$ in the proof of Theorem 3.4.2 form a proper class. The question as to whether this class may be replaced by a set is a very subtle one. In fact, the answer depends on the underlying axiomatization of set theory. This was shown by E.R. Fisher [1977] and J. Adámek [1990].

Example 3.4.3. The class $\underline{TF}$ of torsion free abelian groups can be defined by the infinite sequence of quasi-identities

$$(3.4.5) \qquad (nx = 0) \to (x = 0),$$

where $n = 2, 3, \ldots$. It is not a variety. The group $(\mathbb{Z}, +, -, 0)$ of integers provides a counterexample. Obviously, this group itself is torsion free. However, for any natural number $n \geq 2$, the homomorphic image $\mathbb{Z}_n = \mathbb{Z}/n\mathbb{Z}$ of $\mathbb{Z}$ contains an element that does not satisfy some of the quasi-identities (3.4.5). For example, $x = 1$ does not satisfy the quasi-identity (3.4.5) in $\mathbb{Z}_n$. The class $\underline{TF}$ has a subclass $\underline{T}$ defined by the implication

$$(3.4.6) \qquad (\&_{n \geq 2} \, x = nx_n) \to (x = 0).$$

Note that both the additive groups $\mathbb{Z}$ and $\mathbb{Q}$ are torsion free. However, $\mathbb{Z}$ satisfies (3.4.6), while $\mathbb{Q}$ does not. Hence the class $\underline{T}$ is a proper subclass of $\underline{TF}$. □

The next lemma provides a condition, formulated in the language of $\underline{K}$-congruences, for a prevariety $\underline{K}$ to satisfy a given implication. Let i be the implication of (3.4.3), and let X be a set of variables containing $\arg(i)$. Denote elements of the free $\underline{K}$-algebra XK by their representatives in $X\Omega$. Define P, C and φ as in Section 1.7. Let $\bar{\varphi} = \langle P \rangle_{cg_K}^{X\Omega} = \langle \varphi \rangle_{cg_K}^{X\Omega}$ be the least congruence of $X\Omega$ containing P, and let $\tilde{\varphi} = \langle P \rangle_{cg_K}^{XK}$ be the least congruence of XK containing P.

Lemma 3.4.4 [Quackenbush, Raftery]. *The following conditions are equivalent for a prevariety $\underline{K}$:*

(a) $\underline{K} \models i$;
(b) $(w, w') \in \bar{\varphi}$
(c) $(w, w') \in \tilde{\varphi}$.

Proof. (a) $\Rightarrow$ (b) Note that all (w_m, w'_m) are in $\bar{\varphi}$ and $X\Omega^{\bar{\varphi}}$ belongs to $\underline{K}$. It follows that (w, w') is an element of $\bar{\varphi}$.

(b) $\Rightarrow$ (a) If i fails in $\underline{K}$, then it fails in some member A of $\underline{K}$ generated by $|X|$ elements. This means that for generators of A the values of w_m and w'_m are equal, whereas the values of w and w' are not. Hence A is a quotient of $X\Omega^{\varphi}$, and since A is a quotient of $X\Omega^{\bar{\varphi}}$ as well, it follows that $(w, w') \notin \bar{\varphi}$.

(b) $\Leftrightarrow$ (c) This part of the proof is relegated to Exercise 3.4A. □

Theorem 3.4.5 [Raftery]. *Let $\underline{K}$ be a (non-trivial) prevariety of τ-algebras. Then $\underline{K}$ is a variety if and only if $Cg_K A$ is a complete sublattice of the lattice CgA for all $\underline{K}$-algebras A.*

Proof. If $\underline{K}$ is a variety then clearly $Cg_K A = Cg\, A$ for each $\underline{K}$-algebra A. Now assume that $\underline{K}$ is a prevariety such that for each $\underline{K}$-algebra A, the lattice

$Cg_K A$ is a complete sublattice of the lattice $Cg\, A$. Let I be a set of implications defining $\underline{K}$ and let the implication i be in I. We have to show that the variety $\underline{V} := v(\underline{K})$ generated by $\underline{K}$ satisfies the implication i.

Let $Y = \{y_m \mid m \in M\}$ be a set of variables disjoint from the set X, and let $Z := X \cup Y$. Consider the free $\underline{K}$-algebra ZK. By Lemma 3.4.4, it is clear that $(w, w') \in \widetilde{\varphi}$, where the congruence $\widetilde{\varphi}$ is defined on the algebra ZK similarly as for XK.

For each m in M, define the map $h_m : Z \to ZK$ by

$$zh_m := \text{if } z = z_m \text{ then } w_m \text{ else } z.$$

Extend h_m to the (uniquely defined) endomorphism $\bar{h}_m : ZK \to ZK$. Then clearly the image $ZK\bar{h}_m$ is in $\underline{K}$, and is isomorphic to the quotient algebra $ZK^{\ker \bar{h}_m}$. Recall that by Proposition 3.3.8, the algebra ZK is also free in the variety $\underline{V}$, so that $ZK = ZV$. Define α_m to be the congruence of ZV generated by (y_m, w_m). Note that $\alpha_m \subseteq \ker \bar{h}_m$. The converse inclusion holds as well. Indeed, for any τ-words u and v in $Z\Omega$, if $(u, v) \in \ker \bar{h}_m$ then the variety $\underline{V}$ satisfies the implication

$$(y_m = w_m) \to (u = v).$$

By Lemma 3.4.4 (applied to the variety $\underline{V}$), one obtains that $(u, v) \in \alpha_m$. Thus α_m is a $\underline{K}$-congruence of ZV. Similarly, one proves that the congruence β_m of ZV generated by (y_m, w'_m), is also a $\underline{K}$-congruence. Therefore, by assumption, the congruence

$$\mu := \langle \bigcup_{m \in M} (\alpha_m \cup \beta_m) \rangle_{cg}$$

of ZV generated by the union of all the α_m and β_m is also a $\underline{K}$-congruence. Note that for each $m \in M$, one has $(w_m, w'_m) \in \alpha_m \circ \beta_m \subseteq \mu$. Hence $\widetilde{\varphi} \subseteq \mu$, and therefore also $(w, w') \in \mu$. Applying Lemma 3.4.4 again to the variety $\underline{V}$, one obtains that

$$\underline{V} \models (\&_{m \in M}\ y_m = w_m\ \&\ y_m = w'_m) \to (w = w'),$$

or equivalently, that $\underline{V}$ satisfies the implication i. □

Exercises

3.4A Complete the proof of Lemma 3.4.4.

3.4B [Raftery 2000] Let $\underline{K}$ be a quasi-equational class. Show that $\underline{K}$ is a variety if and only if $Cg_K A$ is a sublattice of the lattice $Cg\,A$ for all $\underline{K}$-algebras A.

3.5. Varieties

This section provides basic information about varieties of algebras. Since varieties are prevarieties, each variety $\underline{V}$ contains free $\underline{V}$-algebras XV for sets X of any cardinality. This fact influences all the results presented in this section. Some direct consequences of the existence of free $\underline{V}$-algebras are presented in the first part. This is followed by the central result, G. Birkhoff's characterization of varieties as equationally defined classes. The remaining part is devoted to a study of the poset of *subvarieties* of a variety $\underline{V}$, i.e. the poset of subclasses of $\underline{V}$ that are themselves varieties.

Free algebras in familiar varieties are usually well-known. (See e.g. [McKenzie, McNulty, Taylor 1987].) Below are some examples that will play a rôle in this book.

Example 3.5.1.

(a) The free semilattice $X\text{Sl}$ over X is described (up to isomorphism) as the (join) semilattice of finite non-empty subsets of the set X [Exercise 3.3I]. Note that $X\text{Sl}$ can be obtained from the free semigroup X^+ over X as the quotient $(X^+)^\rho$ by the semilattice replica congruence ρ. Since semilattices satisfy the idempotent and commutative laws, the order of letters in each semigroup word can be rearranged so that no repetition of letters occurs. It follows that each element of $(X^+)^\rho$ depends only on its set of elements.

(b) The free commutative monoid $X^{*\kappa}$ over X can be described as the quotient $(X^*)^\rho$ of the free monoid X^* by the commutative monoid replica ρ. The commutative law allows one to rearrange letters in each monoid word, so that each element of $X^{*\kappa}$ depends on its set of elements and the multiplicity of each letter in the word. This leads to a description of $X^{*\kappa}$ (up to isomorphism) as the set of finite multisubsets of the set X. The identity element of $X^{*\kappa}$ is represented as the empty multiset. The multiplication in $X^{*\kappa}$ corresponds to a disjoint union of multisets, e.g. the element $(x_1 x_2 x_1 x_3)^\rho$ of $X^{*\kappa}$ can be represented as $x_1 x_1 x_2 x_3$ and as the multiset $\langle 1, 1, 2, 3 \rangle$, then the element $(x_1 x_2 x_1 x_3)^\rho \cdot (x_3 x_2 x_1)^\rho$ can be represented as $x_1 x_1 x_2 x_3 \cdot x_1 x_2 x_3 = x_1 x_1 x_1 x_2 x_2 x_3 x_3$ and as the multiset $\langle 1, 1, 1, 2, 3 \rangle \mathbin{\biguplus} \langle 1, 2, 3 \rangle = \langle 1, 1, 1, 1, 2, 2, 3, 3 \rangle$.

(c) The free commutative ring with unit over X is described as the ring $\mathbb{Z}[X]$ of all polynomials with integer coefficients constructed from commuting indeterminates in X.

(d) The free commutative semiring over X is described as the semiring $\mathbb{N}[X]$ of all polynomials with natural coefficients constructed from commuting indeterminates in X. One constructs the free (additive) commutative semigroup $\mathbb{N}X^{*\kappa}$ over the set $X^{*\kappa}$ as in Example (b). The multiplication in the monoid $X^{*\kappa}$ extends by distributivity to $\mathbb{N}X^{*\kappa}$, making it the free commutative semitring (with a multiplicative identity) over X [Exercise 3.5G]. The elements of $\mathbb{N}X^{*\kappa}$ may be considered as polynomials in commuting indeterminates in X with natural number coefficients. $\square$

3. VARIETIES

One of the consequences of the existence of free algebras in varieties is that varieties always contain simple algebras.

Magari's Theorem 3.5.2. *Each nontrivial variety $\underline{V}$ of τ-algebras contains a nontrivial simple algebra.*

Proof. Let $X = \{x, y\}$. Consider the free $\underline{V}$-algebras XV and $xV := \{x\}V$. Let θ be a congruence on XV belonging to the interval $[cg(xV), XV \times XV]$ of $Cg(XV)$. First note that

$$\theta = XV \times XV \text{ iff } (x, y) \in \theta.$$

To see this, it is enough to show that if $(x, y) \in \theta$, then $(t, x) \in \theta$ for each $t \in X\Omega$. And indeed, if $t = t(x)$, then obviously $(t, x) \in cg(xV) \subseteq \theta$, while if $t = t(x, y)$, then

$$xyt \ \theta \ xxt \ cg(xV) \ x,$$

whence $(t, x) \in \theta$.

Now suppose that $cg(xV) \neq XV \times XV$. By the Kuratowski-Zorn Lemma 0.5.1, the (ordered) set $[cg(xV), XV \times XV)$ has a maximal element θ_m. By Exercise 1.2T, the algebra XV^{θ_m} is a simple $\underline{V}$-algebra.

Next, assume that $cg(xV) = XV \times XV$. By Proposition 3.2.6, cg is an algebraic closure operator, so $(x, y) \in cg(F)$ for some finite subset F of xV. As $\underline{V}$ is non-trivial, one has $x \neq y$ in XV. Since $(x, y) \in cg(xV)$, it follows that xV is non-trivial. We will show $cg(F) = xV \times xV$ in xV. Let $\bar{h} : XV \to xV$ be the (unique) homomorphism extending the mapping $h : x \mapsto x, y \mapsto t = t(x)$ for some arbitrarily fixed element t in xV. Note that $F\bar{h} = F$. Since $(x, y) \in cg(F)$ in XV, Exercise 3.5A implies that $(x\bar{h}, y\bar{h}) = (x, t) \in cg(F)$ in xV. Hence $cg(F) = xV \times xV$. Since the congruence $xV \times xV$ is finitely generated, it follows by the Kuratowski-Zorn Lemma that the poset $Cg(xV) - \{xV \times xV\}$ has a maximal element φ. Using Exercise 1.2T again, one concludes that $(xV)^\varphi$ is a simple algebra. □

Another application of free algebras is given by the following. A τ-algebra A is *locally finite* if every finitely generated subalgebra of A is finite. A class $\underline{K}$ of τ-algebras is *locally finite* if every member of $\underline{K}$ is locally finite.

Proposition 3.5.3. *A variety $\underline{V}$ of τ-algebras is locally finite if and only if the free XV-algebra is finite for each finite set X.*

Proof. Only the implication ($\Leftarrow$) requires a proof. So let A be a $\underline{V}$-algebra generated by a finite set Y. Let $f : X \to Y$ be a bijection, and $\bar{f} : XV \to A$ the homomorphic extension of f. Since $XV\bar{f}$ is a subalgebra of A containing Y, it must coincide with A. As XV is finite, so is A. □

Note also that each non-trivial, locally finite variety contains a plain algebra [Section 1.2], namely a non-trivial algebra of minimal cardinality. A variety of $\underline{V}$ is called *minimal* if it is non-trivial, but every variety properly contained in $\underline{V}$ is trivial.

Corollary 3.5.4. *Every minimal, locally finite variety is generated by a plain algebra.* □

By Proposition 3.3.7 and Exercise 3.3H, it follows that a prevariety $\underline{K}$ of τ-algebras satisfies an identity $u = v$ if and only if the free $\underline{K}$-algebra $\mathfrak{P}K$ satisfies it. By Proposition 1.5.3, this holds precisely if for each homomorphism $h : \mathfrak{P}\Omega \to \mathfrak{P}K$, one has $uh = vh$. One can reformulate this in the context of the universality property by saying that for each homomorphism $\bar{\eta}_\mathfrak{P} := h$ extending $\eta_\mathfrak{P} := h|_\mathfrak{P}$ in the commuting diagram

(3.5.1)
$$\begin{array}{ccc} \mathfrak{P} & \longrightarrow & \mathfrak{P}\Omega \\ \eta_\mathfrak{P} \downarrow & & \downarrow \bar{\eta}_\mathfrak{P} \\ \mathfrak{P}K & =\!=\!= & \mathfrak{P}K \cong \mathfrak{P}\Omega R, \end{array}$$

one has $u\bar{\eta}_\mathfrak{P} = v\bar{\eta}_\mathfrak{P}$, i.e. $(u,v) \in \ker \bar{\eta}_\mathfrak{P}$. This proves the following:

Lemma 3.5.5. *A prevariety $\underline{K}$ satisfies an identity $u = v$ iff $(u,v) \in \ker \bar{\eta}_\mathfrak{P}$.* □

Note however that by Proposition 3.3.8, if the prevariety $\underline{K}$ is not a variety, then there may be τ-algebras not in $\underline{K}$ that also satisfy identities true in $\underline{K}$.

Proposition 1.5.13 shows that each equational class is a variety. The following theorem presents a classical converse to this proposition, and shows that given a variety $\underline{K}$, the set

$$(3.5.2) \qquad I := \{u = v \mid (u,v) \in \ker \bar{\eta}_\mathfrak{P}\}$$

is precisely the set of identities true in $\underline{K}$, i.e. $I = \mathrm{Id}(\underline{K})$.

Birkhoff's Variety Theorem 3.5.6. *Let $\tau : \Omega \to \mathbb{N}$ be a (finitary) type. Let $\underline{K}$ be a class of τ-algebras. Then $\underline{K}$ is a variety if and only if it is an equational class.*

Proof. First note that trivial $\underline{K}$-algebras satisfy all identities of type τ. (Empty algebras satisfy identities of type τ vacuously.) In what follows, assume that $\underline{K}$ is non-trivial.

($\Leftarrow$) This follows by Proposition 1.5.13.

($\Rightarrow$) Assume that $\underline{K}$ is a variety. Then by Lemma 3.5.5 the variety $\underline{K}$ satisfies the set I of identities. Let $\underline{V}$ be the equational class defined by the set I. Then obviously $\underline{K} \subseteq \underline{V}$.

Now we will show that the converse inclusion also holds. Let A be a $\underline{V}$-algebra. Consider the following diagram:

$$\begin{array}{ccccc} A & \longrightarrow & A\Omega & \xrightarrow{\bar{\eta}_A} & AK \\ {\scriptstyle 1_A}\downarrow & & {\scriptstyle \bar{1}_A}\downarrow & & \\ A & = & A & & \end{array},$$

where $\bar{1}_A$ is the (unique) homomorphism extending the identity map 1_A. It will be shown that $\ker \bar{\eta}_A =: \rho \leq \ker \bar{1}_A$, so that a well-defined homomorphism

$$\bar{\bar{1}}_A : AK \cong A\Omega^\rho \to A;\ w^\rho \mapsto w\bar{1}_A$$

exists, making the diagram commute and displaying A as a homomorphic image of the $\underline{K}$-algebra AK. Let $(w_1, w_2) \in \ker \bar{\eta}_A$, and let $\arg(w_1) \cup \arg(w_2)$ be the set $B = \{a_1, \ldots, a_n\}$. Define $j : B \to P$; $a_i \mapsto x_i$. The mapping j extends to injections $j\Omega : B\Omega \to \mathfrak{P}\Omega$ and $jK : BK \to \mathfrak{P}K$ such that the following diagram commutes:

$$\begin{array}{ccc} B\Omega & \xrightarrow{\bar{\eta}_B} & BK \\ {\scriptstyle j\Omega}\downarrow & & \downarrow{\scriptstyle jK} \\ \mathfrak{P}\Omega & \xrightarrow[\bar{\eta}_{\mathfrak{P}}]{} & \mathfrak{P}K \;. \end{array}$$

Since $(w_1, w_2) \in \bar{\eta}_B$, one has $(u, v) := (w_1 j\Omega, w_2 j\Omega) \in \ker \bar{\eta}_{\mathfrak{P}}$. Define

$$a : \{x_1, \ldots, x_n\} \to A; \; x_i \mapsto a_i.$$

Since A satisfies the identity $u = v$, one obtains

$$w_1 \bar{1}_A = (x_1 a) \ldots (x_n a)(u, n)_A = (x_1 a) \ldots (x_n a)(v, n)_A = w_2 \bar{1}_A.$$

Thus $(w_1, w_2) \in \ker \bar{1}_A$. $\square$

The proof of Birkhoff's Variety Theorem 3.5.6 shows that each variety $\underline{K}$ of τ-algebras, with replication $(R, J, \eta, \varepsilon)$, is specified uniquely as the class of τ-algebras satisfying the set I of identities of (3.5.2), i.e. $I = \mathrm{Id}(\underline{K})$. One usually identifies the set $\mathrm{Id}(\underline{K})$ with the relation

$$Eq(\underline{K}) := \ker(\bar{\eta}_{\mathfrak{P}}^K : \mathfrak{P}\Omega \to \mathfrak{P}K \cong \mathfrak{P}\Omega R_K)$$

of (3.5.1), where $R_K = R$ and $\bar{\eta}_{\mathfrak{P}}^K = \bar{\eta}_{\mathfrak{P}}$. This relation $Eq(\underline{K})$ is called the *equational theory* of $\underline{K}$. Now let $\underline{L}$ be a subvariety of the variety $\underline{K}$, with corresponding replica functor R_L. As in (3.3.5), one obtains the following commuting diagram:

(3.5.3)
$$\begin{array}{ccccc}
\mathfrak{P} & \xrightarrow{\eta_{\mathfrak{P}}^K} & \mathfrak{P}K & = & \mathfrak{P}K \\
\| & & & & \| \\
\mathfrak{P} & \xrightarrow{\iota} & \mathfrak{P}\Omega & \xrightarrow{k:=\bar{\eta}_{\mathfrak{P}}^K} & \mathfrak{P}K \cong \mathfrak{P}\Omega R_K \\
\eta_{\mathfrak{P}}^L \downarrow & & l:=\bar{\eta}_{\mathfrak{P}}^L \downarrow & & \downarrow r:=\bar{\bar{\eta}}_{\mathfrak{P}}^L \\
\mathfrak{P}L & = & \mathfrak{P}L & = & \mathfrak{P}L \cong \mathfrak{P}\Omega R_L.
\end{array}$$

Define $Eq(\underline{L};\underline{K}) := \ker \bar{\bar{\eta}}_{\mathfrak{P}}^L$, and note that $Eq(\underline{L};\underline{\tau}) = Eq(\underline{L})$. A natural question arises: Which congruences on $\mathfrak{P}K$ are of the form $Eq(\underline{L};\underline{K})$ for a subvariety $\underline{L}$ of a variety $\underline{K}$? To answer this question, consider a τ-algebra (A, Ω) and the monoid $\underline{\tau}(A, A)$ of endomorphisms of (A, Ω). Consider these endomorphisms as unary operations of (A, Ω). Then the congruences of the algebra $(A, \Omega \cup \underline{\tau}(A, A))$ are called *fully invariant congruences* of the algebra (A, Ω).

Theorem 3.5.7. *Let $\tau : \Omega \to \mathbb{N}$ be a (finitary) type. Let $\underline{K}$ be a variety of τ-algebras with replica functor R_K. Then a congruence on the free $\underline{K}$-algebra $\mathfrak{P}K \cong \mathfrak{P}\Omega R_K$ is fully invariant if and only if it is of the form $Eq(\underline{L};\underline{K})$ for a subvariety $\underline{L}$ of $\underline{K}$.*

Proof. ($\Leftarrow$) Consider an identity $u = v$ with (uk, vk) in $Eq(\underline{L};\underline{K})$, using the notation of (3.5.3), and an endomorphism t of $\mathfrak{P}K$. By the universality property for replicas (Lemma 3.3.1), there is an endomorphism $t' := \overline{ktr}$ of $\mathfrak{P}L \cong \mathfrak{P}\Omega R_L$:

$$\begin{array}{ccc}
\mathfrak{P}\Omega R_L & \xrightarrow{t':=\overline{ktr}} & \mathfrak{P}\Omega R_L \\
l=kr \uparrow & & \uparrow ktr \\
\mathfrak{P}\Omega & = & \mathfrak{P}\Omega
\end{array}$$

with $ktr = lt'$. Since $(ul, vl) \in Eq(\underline{L})$, one has $uktr = ult' = vlt' = vktr$. Thus $(ukt, vkt) \in Eq(\underline{L};\underline{K})$, so that $Eq(\underline{L};\underline{K})$ is fully invariant.

($\Rightarrow$) To prove the converse, let θ be a fully invariant congruence of $\mathfrak{P}K$. Let $\underline{L}$ be the variety of all τ-algebras satisfying the set of identities

$$\mathrm{Id}(L) = \{u = v \mid (u,v) \in \mathfrak{P}\Omega^2 \text{ with } (uk, vk) \in \theta\}.$$

It will be shown that $\theta = Eq(\underline{L}; \underline{K})$. Note that by Proposition 3.3.7 and Lemma 3.3.5, the algebra $\mathfrak{P}L$ satisfies precisely the identities $\mathrm{Id}(L)$ and

$$Eq(\underline{L}; \underline{K}) = \ker r = \{(uk, vk) \mid ul = vl\} = \{(uk, vk) \mid uk \operatorname{nat}\theta = vk \operatorname{nat}\theta\}.$$

Hence certainly $\theta \leq Eq(\underline{L}; \underline{K})$.

To show the converse, consider the algebra $\mathfrak{P}K^\theta$. Let $(u = v) \in \mathrm{Id}(L)$, i.e. $(uk, vk) \in \theta$. Extend a mapping $f' : \arg(u) \cup \arg(v) \to \mathfrak{P}K^\theta$ to a mapping $f : \mathfrak{P} \to \mathfrak{P}K^\theta$. The universality property for free algebras yields the following commuting diagram:

$$\begin{array}{ccccc} \mathfrak{P} & \xrightarrow{\iota} & \mathfrak{P}\Omega & \xrightarrow{k} & \mathfrak{P}K \cong \mathfrak{P}\Omega R_K \\ {\scriptstyle f}\downarrow & & {\scriptstyle \bar{f}}\downarrow & & {\scriptstyle \bar{\bar{f}}}\downarrow \\ \mathfrak{P}K^\theta & = & \mathfrak{P}K^\theta & = & \mathfrak{P}K^\theta. \end{array}$$

Similarly, there is an endomorphism $h : \mathfrak{P}K \to \mathfrak{P}K$ extending a mapping $h' : \mathfrak{P} \to \mathfrak{P}K$ defined (using the axiom of choice) by $x_n h \in x_n \bar{\bar{f}}$ for $x_n \in \mathfrak{P}$. Note that $h \operatorname{nat}\theta = \bar{\bar{f}} : \mathfrak{P}K \to \mathfrak{P}K^\theta$. Since θ is fully invariant, one has that $(ukh, vkh) \in \theta$. Then $u\bar{f} = uk\bar{\bar{f}} = (ukh)(\operatorname{nat}\theta) = (vkh)(\operatorname{nat}\theta) = vk\bar{\bar{f}} = v\bar{f}$, so that $\mathfrak{P}K^\theta$ satisfies the identity $u = v$. Thus $\mathfrak{P}K^\theta$ is in the variety $\underline{L}$. If $(uk, vk) \in (\mathfrak{P}K \times \mathfrak{P}K) - \theta$, then $\mathfrak{P}K^\theta$ does not satisfy $u = v$. Indeed, for the mapping $f = \iota k \operatorname{nat}\theta : \mathfrak{P} \to \mathfrak{P}K^\theta$ and the corresponding homomorphism $\bar{f} : \mathfrak{P}\Omega \to \mathfrak{P}K^\theta$, one has $u\bar{f} = uk \operatorname{nat}\theta \neq vk \operatorname{nat}\theta = v\bar{f}$. It follows that $(uk, vk) \in (\mathfrak{P}K \times \mathfrak{P}K) - Eq(\underline{L}; \underline{K})$. Thus $Eq(\underline{L}; \underline{K}) \leq \theta$, and the required equality $Eq(\underline{L}; \underline{K}) = \theta$ is obtained. $\square$

Theorem 3.5.7 shows that fully invariant congruences of $\mathfrak{P}K$ classify subvarieties of the variety $\underline{K}$. The following corollary shows which binary relations on $\mathfrak{P}K$ provide equational theories of subvarieties of $\underline{K}$.

Corollary 3.5.8 (**Completeness Theorem of Equational Logic**). *A relation Σ on $\mathfrak{P}\Omega$ is an equational theory if and only if it is closed under the following deduction rules:*

(a) $\forall w \in \mathfrak{P}\Omega$, $(w, w) \in \Sigma$;

(b) $((u, v) \in \Sigma) \Rightarrow ((v, u) \in \Sigma)$;

(c) $((u, v) \in \Sigma$ and $(v, u) \in \Sigma) \Rightarrow ((u, w) \in \Sigma)$;

(d) $\forall \omega \in \Omega$,

$$(\forall\, 1 \le i \le \omega\tau,\ (u_i, v_i) \in \Sigma) \Rightarrow ((u_1 \ldots u_{\omega\tau}\omega, v_1 \ldots v_{\omega\tau}\omega) \in \Sigma);$$

(e) $\forall (u, v) \in \Sigma$, with $n = \max(\arg(u) \cup \arg(v))$, *one has*

$$(w_1 \ldots w_n(u, n), w_1 \ldots w_n(v, n)) \in \Sigma.$$

Proof. The relation Σ is closed under the deduction rules iff it is a fully invariant congruence of $\mathfrak{P}\Omega$. Closure under (c) corresponds to transitivity of Σ, and the conditions (a)-(d) show that Σ is a congruence. Closure under (e) corresponds to closure under an endomorphism f of $\mathfrak{P}\Omega$ with $x_i f = w_i$ for $1 \le i \le n$. □

Exercises 3.5H shows that the fully invariant congruences of $\mathfrak{P}K$ form a complete sublattice $Cg_{fi}\mathfrak{P}K$ of the lattice $Cg\mathfrak{P}K$. In fact, this sublattice is also algebraic. (Cf. [Burris, Sankappanavar 1981].) On the other hand, the subvarieties of the variety $\underline{K}$ are ordered under inclusion. By Theorem 3.5.7, the assignment of an appropriate subvariety of $\underline{K}$ to a fully invariant congruence on $\mathfrak{P}K$ is a bijection and a dual isomorphism between the poset $Cg_{fi}\mathfrak{P}K$ and the poset $\mathcal{L}(\underline{K})$ of subvarieties of $\underline{K}$. This implies that $\mathcal{L}(\underline{K})$ is also a complete lattice. The meet of any set of subvarieties is their intersection. In particular, the meet $\underline{V} \cap \underline{W}$ of two subvarieties $\underline{V}$ and $\underline{W}$ corresponds to the least fully invariant congruence $Eq(\underline{V}; \underline{K}) + Eq(\underline{W}; \underline{K})$ containing $Eq(\underline{V}; \underline{K})$

and $Eq(\underline{W};\underline{K})$. The join of two subvarieties $\underline{V}$ and $\underline{W}$ is the smallest subvariety $\underline{V}+\underline{W}$ of $\underline{K}$ containing both $\underline{V}$ and $\underline{W}$, and corresponds to the intersection $Eq(\underline{V};\underline{K}) \cap Eq(\underline{W};\underline{K})$ of the corresponding fully invariant congruences.

The following corollary shows that each subvariety $\underline{V}$ of $\underline{K}$ contains a minimal subvariety, so that minimal subvarieties of $\underline{K}$ are *atoms* of the lattice $\mathcal{L}(\underline{K})$, i.e. they cover the trivial subvariety of $\underline{K}$. If $\underline{W}$ is a minimal variety contained in a variety $\underline{V}$, then obviously $Eq(\underline{V}) \subseteq Eq(\underline{W})$. This is why minimal varieties are also called *equationally complete*.

Corollary 3.5.9. *Let $\underline{V}$ be a non-trivial variety of type $\tau : \Omega \to \mathbb{N}$. Then $\underline{V}$ contains a minimal subvariety.*

Proof. By Theorem 3.5.7, the variety $\underline{V}$ satisfies precisely the set $\mathrm{Id}(\underline{V}) = \{u = v \mid (u,v) \in Eq(\underline{V};\underline{\tau}) = Eq(\underline{V})\}$ of identities, where $Eq(\underline{V})$ is a fully invariant congruence on $\mathfrak{P}\Omega$. The relation $Eq(\underline{V})$ is certainly not the universal relation. By the Kuratowski-Zorn Lemma 0.5.1, each chain of fully invariant proper congruences on $\mathfrak{P}\Omega$ containing $Eq(\underline{V})$ has a maximal fully invariant proper congruence as its supremum, say θ. Then θ corresponds to a minimal subvariety of $\underline{V}$. □

We conclude this section with some remarks concerning joins of subvarieties of a variety. In general, there is no obvious method for describing algebras in such joins, even if the structure of algebras in the summand varieties is quite clear. However, there is a special case where there is a transparent description of algebras in the join of any finite number of varieties of the same type.

Varieties $\underline{V}_1, \ldots, \underline{V}_n$ of a given type $\tau : \Omega \to \mathbb{N}$ are *independent* if there is an Ω-word $x_1 \ldots x_n d$, called a *decomposition word*, such that the identity $x_1 \ldots x_n d = x_i$ holds in $\underline{V}_i$ for $i = 1, 2, \ldots, n$. Then the class $\underline{V}_1 \times \cdots \times \underline{V}_n$ is defined to be the class of all Ω-algebras which are isomorphic to an algebra $A_1 \times \cdots \times A_n$ with A_i in $\underline{V}_i$. Naturally, $\underline{V}_1 \times \cdots \times \underline{V}_n \subseteq \underline{V}_1 + \cdots + \underline{V}_n$.

3. VARIETIES

Lemma 3.5.10. *If the varieties $\underline{V}_1, \ldots, \underline{V}_n$ of τ-algebras are independent, then*
$$\underline{V}_1 + \cdots + \underline{V}_n = \underline{V}_1 \times \cdots \times \underline{V}_n.$$

Proof. Let $x_1 \ldots x_n d$ be a decomposition word for varieties $\underline{V}_1, \ldots, \underline{V}_n$. Then for each $i = 1, \ldots, n$, the variety $\underline{V}_i$ satisfies the identity $x_1 \ldots x_n d = x_i$. This implies that $\underline{V}_i$ also satisfies the identities

(3.5.4) $$x \ldots x d = x,$$

(3.5.5)
$$(x_{11} \ldots x_{1\omega\tau}\omega) \ldots (x_{n1} \ldots x_{n\omega\tau}\omega)d = (x_{11} \ldots x_{n1}d) \ldots (x_{1\omega\tau} \ldots x_{n\omega\tau}d)\omega$$

for each ω in Ω and

(3.5.6) $$(x_{11} \ldots x_{1n}d) \ldots (x_{n1} \ldots x_{nn}d)\, d = x_{11} \ldots x_{nn}d.$$

It follows that the variety $\underline{V}_1 + \cdots + \underline{V}_n$ satisfies these identities, too. By Exercise 1.5H, this means that any algebra A in $\underline{V}_1 + \cdots + \underline{V}_n$ is of the form $A_1 \times \cdots \times A_n$ with A_i in $\underline{V}_i$, and hence $\underline{V}_1 + \cdots + \underline{V}_n \subseteq \underline{V}_1 \times \cdots \times \underline{V}_n$. Since the converse inclusion holds as well, one obtains that $\underline{V}_1 + \cdots + \underline{V}_n = \underline{V}_1 \times \cdots \times \underline{V}_n$. □

If the varieties $\underline{V}_1, \ldots, \underline{V}_n$ are independent, their join $\underline{V}_1 + \cdots + \underline{V}_n$ is called an *independent join*. Note that some authors call such joins "products" or "direct sums" of varieties.

Theorem 3.5.11. *Let $\underline{V}$ be a variety of τ-algebras. Then $\underline{V}$ is the independent join of subvarieties $\underline{V}_1, \ldots, \underline{V}_n$, each $\underline{V}_i$ satisfying the identity $x_1 \ldots x_n d = x_i$, if and only if $\underline{V}$ satisfies the identities (3.5.4), (3.5.5) and (3.5.6).*

Proof. This follows by Exercise 1.5H and Lemma 3.5.10. □

The simplest example of independent varieties is provided by the varieties $\underline{Lz}$ of left zero bands and $\underline{Rz}$ of right zero bands, with decomposition word $x_1 x_2 d = x_1 \cdot x_2$. The independent join $\underline{Lz} + \underline{Rz}$ is the variety $\underline{Re}$ of rectangular bands.

Exercises

3.5A. Let $h : A \to B$ be a τ-homomorphism, and let $X \subseteq A$. Show that if $(a, b) \in cg(X)$, then $(ah, bh) \in cg(Xh)$.

3.5B. Let $A_1, \ldots, A_k$ be finite Ω-algebras.
   (a) Show that $v(A_1, \ldots, A_k) = v(A_1 \times \cdots \times A_k)$.
   (b) Set $\underline{V} := v(A_1, \ldots, A_k)$. Define A to be the product algebra $A_1 \times \cdots \times A_k$. Show that the free $\underline{V}$-algebra $\{x_1, \ldots, x_k\}V$ embeds into $(A^{|A|^k}, \Omega)$.
   (c) Use Proposition 1.5.12(b) to show that $\underline{V}$ is a locally finite variety.

3.5C. Let $\underline{V}$ be a variety of τ-algebras. Show that if $|X| \leq |Y|$, then XV embeds into YV in a natural way.

3.5D. Show that if $\underline{V}_1 \subseteq \underline{V}_2$ are varieties of τ-algebras, then XV_1 is a homomorphic image of XV_2.

3.5E. Let $\underline{V}$ be a variety of τ-algebras. Let X be an infinite set. Show that
$$\underline{V} = HSP(XV).$$

3.5F. Show that for any variety $\underline{V}$ of τ-algebras,
$$\underline{V} = v(\{X_n V \mid n = 1, 2, \ldots, |X_n| = n\})$$

3.5G. Show that the semiring of Example 3.5.1(d) is a free commutative semiring with identity over X.

3.5H. Show that the fully invariant congruences on an algebra A form a complete sublattice of the lattice $Cg(A)$.

3.5I. Let $\underline{V}_1, \ldots, \underline{V}_n$ be independent varieties of τ-algebras, with each $\underline{V}_i$ satisfying the identity $x_1 \ldots x_n d = x_i$. Let each $\underline{V}_i$ be defined by identities $t_j^i = w_j^i$ for $j = 1, \ldots, k_i$. Show that the independent join $V_1 + \cdots + V_n$ is the variety of τ-algebras defined by the identities (3.5.4)-(3.5.6) and the identities

$$x_1 \ldots t_j^i \ldots x_n d = x_1 \ldots w_j^i \ldots x_n d$$

for each $i = 1, \ldots, n$ and $j = 1, \ldots, k_i$.

3.5J. [Kagalovskiĭ 1965] Let $\underline{K}$ be a class of τ-algebras. Let $P_S\underline{K}$ be the class of algebras (isomorphic to) subdirect products of $\underline{K}$-algebras. Show that

$$HP_S\underline{K} = HSP\underline{K}.$$

3.5K. Let $\underline{V}$ and $\underline{W}$ be varieties of algebras (not necessarily of the same type). Show that the following statements are equivalent:

(i) The varieties $\underline{V}$ and $\underline{W}$ are equivalent;

(ii) There exist a $\underline{V}$-algebra A and a $\underline{W}$-algebra B such that A and B are (clonally) equivalent, with $\underline{V} = v(A)$ and $\underline{W} = v(B)$;

(iii) The free algebra $\mathfrak{P}W$ is isomorphic to a $\underline{W}$-algebra C equivalent to A. (In such a case, we say that A and B are *weakly isomorphic*.)

3.6. Quasivarieties and directed colimits

In previous sections, we have seen that prevarieties can be characterized as classes of algebras defined by implications, while varieties can be characterized as classes of algebras defined by identities. In this section we will show that a prevariety defined only by quasi-identities is also closed under all directed colimits, and that each prevariety closed under directed colimits is necessarily a quasi-equational class.

We start with some preliminaries.

Remark 3.6.1 (**Presentation of Algebras**). Let (A, Ω) be a τ-algebra in a prevariety $\underline{K}$. Assume that the algebra (A, Ω) is generated by a set X, i.e. $A = \langle X \rangle$, and consider the free $\underline{K}$-algebra XK. Then there is a congruence relation θ on XK such that XK^θ is isomorphic to (A, Ω). (Cf. Sections 3.3 and 3.4.) One obtains the following coequalizer diagram:

$$\theta \underset{\pi_2|_\theta}{\overset{\pi_1|_\theta}{\rightrightarrows}} XK \xrightarrow[\text{nat}\,\theta]{} XK^\theta \cong A.$$

(See Section 2.3.) Now the congruence θ can be generated by a subset σ of the set $XK \times XK$. The pair (X, σ) is called a *presentation* of (A, Ω) *relative to* $\underline{K}$. If $\underline{K} = \underline{\tau}$, then one speaks simply of a *presentation* of (A, Ω). On the other hand, any pair (X, σ) consisting of a set X and a subset $\sigma \subseteq XK \times XK$ *presents* a $\underline{K}$-algebra, namely the quotient $\langle X|\sigma\rangle$ of XK by the congruence $\langle \sigma \rangle_{cg}$ or simply $\langle \sigma \rangle$ generated by σ. In the case $\underline{K} = \underline{\tau}$, this provides the following coequalizer diagram:

$$\langle \sigma \rangle \underset{\pi_2}{\overset{\pi_1}{\rightrightarrows}} X\Omega \xrightarrow[\text{nat}\langle\sigma\rangle]{} X\Omega^{\langle\sigma\rangle} = \langle X|\sigma\rangle.$$

If (A, Ω) is presented by (X, σ), then we write $A = \langle X|\sigma\rangle$. A presentation (X, σ) is *finite* if both X and σ are finite. An algebra (A, Ω) is *finitely presented* if it is isomorphic to $\langle X|\sigma\rangle$ for some finite presentation (X, σ). □

Let B be an Ω-algebra presented by (X, σ). Then the algebra B can be identified with the quotient $\langle X|\sigma\rangle = X\Omega^{\langle\sigma\rangle}$ of the free $\underline{\tau}$-algebra $X\Omega$ by the congruence generated by σ. Let I be the set of all pairs $i := (X_i, \sigma_i)$ in which X_i is a finite subset of X and σ_i is a finite subset of $X_i\Omega \times X_i\Omega$ restricted to $\langle \sigma \rangle$. The set I is ordered by the following ordering relation:

$$(X_i, \sigma_i) \leq (X_j, \sigma_j) :\Leftrightarrow X_i \subseteq X_j \text{ and } \sigma_i \subseteq \sigma_j.$$

Since the pair $(X_i \cup X_j,\ \sigma_i \cup \sigma_j)$ is also in the set I, it is easy to see that this element is the least upper bound $(X_i, \sigma_i) + (X_j, \sigma_j)$ of (X_i, σ_i) and (X_j, σ_j). Hence $(I, \leq)$ is a (join) semilattice. Note that $(I, \leq)$ has a least element $(\varnothing, \varnothing)$.

Now for each i in I, let A_i be an Ω-algebra presented by (X_i, σ_i). As before, we identify A_i with $\langle X_i | \sigma_i \rangle = X_i \Omega^{\langle \sigma_i \rangle}$. Note that $\langle \sigma_i \rangle \leq \langle \sigma \rangle$. For $i \leq j$ in I, define

(3.6.1) $$\varphi_{i,j} : X_i \Omega^{\langle \sigma_i \rangle} \to X_j \Omega^{\langle \sigma_j \rangle}; \; w^{\langle \sigma_i \rangle} \mapsto w^{\langle \sigma_j \rangle}.$$

It is easy to check that the mappings $\varphi_{i,j}$ are well-defined Ω-homomorphisms satisfying the conditions (2.3.5) [Exercise 3.6A]. One obtains the functor $F : (I) \to (\Omega)$ defined by

$$\begin{array}{ccc} j & & A_j \\ \uparrow & \longmapsto & \uparrow \varphi_{i,j} \\ i & & A_i \end{array}$$

and the corresponding semilattice ordered system $((A_i)_{i \in I}, (\varphi_{i,j})_{i \leq j})$ of finitely presented algebras. The directed colimit $\varinjlim F$ (or $\varinjlim A_i$) of the functor F can be realized as a quotient A^δ of the Ω-algebra $A = \bigcup_{i \in I} A_i$ as described in Example 2.3.10.

Lemma 3.6.2. Let $B = \langle X | \sigma \rangle$ and $F : (I) \to (\Omega)$ be the algebra and functor described above. Then

$$B \cong \varinjlim F.$$

Proof. It will be shown that the algebra B is isomorphic to the algebra A^δ isomorphic to $\varinjlim F$. Define

(3.6.2) $$f : A \to B; \; w^{\langle \sigma_i \rangle} \mapsto w^{\langle \sigma \rangle}.$$

Then certainly f is well-defined. Exercise 3.6B shows that f is an Ω-homomorphism. To see the surjectivity of f note that, for each $w^{\langle \sigma \rangle}$, one can take the (finite) set of variables of w as X_i and any finite subset of $(X_i \Omega \times X_i \Omega) \cap \langle \sigma \rangle$ as σ_i to get $(w^{\langle \sigma_i \rangle}) f = w^{\langle \sigma \rangle}$.

To prove the injectivity of f, we will show that $\ker f = \delta$. Let $w^{\langle\sigma_i\rangle} \in A_i$ and $u^{\langle\sigma_j\rangle} \in A_j$. Note that

$$(w^{\langle\sigma_i\rangle}, u^{\langle\sigma_j\rangle}) \in \ker f \text{ iff } (w,u) \in \langle\sigma\rangle.$$

On the other hand,

$$(w^{\langle\sigma_i\rangle}, u^{\langle\sigma_j\rangle}) \in \delta \text{ iff } (w,u) \in \langle\sigma_k\rangle$$

for some $k \in I$ with $i \leq k$ and $j \leq k$. If $(w,u) \in \langle\sigma_k\rangle$, then obviously $(w,u) \in \langle\sigma\rangle$. If $(w,u) \in \langle\sigma\rangle$ then, since $w \in X_i\Omega \subseteq (X_i \cup X_j)\Omega$ and $u \in X_j\Omega \subseteq (X_i \cup X_j)\Omega$, it follows that $(w,u) \in \langle\sigma_k\rangle$, where $\sigma_k = \sigma_i \cup \sigma_j$ [and $k = (X_i \cup X_j, \sigma_i \cup \sigma_j)$]. This proves that $\ker f = \delta$. Now the First Isomorphism Theorem 1.2.3 shows that the algebras A^δ and B are isomorphic. $\square$

The following is an immediate consequence of Lemma 3.6.2.

Proposition 3.6.3. *Each algebra is a directed colimit of finitely presented algebras.* $\square$

Lemma 3.6.4. *Let $A = \varinjlim A_i$ be a directed colimit of a directed system $((A_i)_{i \in I}, (\varphi_{i,j})_{i \leq j})$ of Ω-algebras. Let $B = \langle Y | \sigma \rangle$ be a finitely presented Ω-algebra, and let $f : B \to A$ be an Ω-homomorphism. Then there is an Ω-homomorphism $g : B \to A_l$ such that*

$$f = g\varphi_l$$

for some $l \in I$, where the mapping $\varphi_l : A_l \to A$; $a_l \mapsto a_l^\delta$ is defined as in (2.3.7).

(3.6.3)
$$\begin{array}{ccc} B & \xrightarrow{g} & A_l \\ {\scriptstyle f}\downarrow & & \downarrow{\scriptstyle \varphi_l} \\ A & = & A = \varinjlim A_i \end{array}$$

3. VARIETIES

Proof. Let $Y = \{y_1, \ldots, y_n\}$, and let $\sigma = \{(w_1, w_1'), \ldots, (w_m, w_m')\}$. The algebra B will be identified with the quotient algebra $Y\Omega^{\langle\sigma\rangle}$. Consider the composite

$$Y\Omega \xrightarrow[\text{nat}\langle\sigma\rangle]{} Y\Omega^{\langle\sigma\rangle} \xrightarrow{f} A = \varinjlim A_i \cong \left(\bigcup A_i\right)^\delta.$$

For each y_i with $i = 1, \ldots, n$, there is a_{k_i} in A_{k_i} for $k_i \in I$ and $k \geq k_1, \ldots, k_n$ such that

(3.6.4) $\qquad y_i(\text{nat}\langle\sigma\rangle f) = a_{k_i}^\delta = a_{k_i}\varphi_{k_i} = a_{k_i}\varphi_{k_i,k}\varphi_k =: b_i\varphi_k.$

Hence $Y(\text{nat}\langle\sigma\rangle f) \subseteq A_k\varphi_k$, and more generally $Y\Omega(\text{nat}\langle\sigma\rangle f) \subseteq A_k\varphi_k$.

Now let

(3.6.5) $\qquad\qquad h : Y \to A_k;\ y_i \mapsto b_i = a_{k_i}\varphi_{k_i,i}.$

The mapping h extends (uniquely) to the Ω-homomorphism

$$\bar{h} : Y\Omega \to A_k;\ y_{i_1} \ldots y_{i_j} w \mapsto b_{i_1} \ldots b_{i_j} w.$$

By (3.6.4) and (3.6.5), for each $i = 1, \ldots, n$,

$$y_i(\text{nat}\langle\sigma\rangle f) = y_i \bar{h} \varphi_k,$$

whence

(3.6.6) $\qquad\qquad \text{nat}\langle\sigma\rangle f = \bar{h}\varphi_k.$

Then (3.6.6) implies that for each pair $(w_i, w_i') \in \sigma$ with $i = 1, \ldots, m$,

$$w_i \bar{h} \varphi_k = w_i^{\langle\sigma\rangle} f = w_i'^{\langle\sigma\rangle} f = w_i' \bar{h} \varphi_k$$

with $w_i \bar{h} \in A_k$ and $w_i' \bar{h} \in A_k$. This means that $(w_i\bar{h}, w_i'\bar{h}) \in \delta$. Hence there are indices l_i in I such that

$$w_i \bar{h} \varphi_{k,l_i} = w_i' \bar{h} \varphi_{k,l_i}.$$

Consequently, there is an index $l \in I$ such that $l_1, \ldots, l_m \leq l$ and for each $i = 1, \ldots, m$,

(3.6.7) $$w_i \bar{h} \varphi_{k,l} = w'_i \bar{h} \varphi_{k,l}.$$

Now equation (3.6.7) yields the containment $\langle \sigma \rangle \leq \ker(\bar{h}\varphi_{k,l})$. Hence there is an Ω-homomorphism $g : Y\Omega^{\langle \sigma \rangle} \to A_l$ such that

(3.6.8) $$\mathrm{nat}\langle \sigma \rangle g = \bar{h}\varphi_{k,l},$$

as illustrated in the following diagram:

$$\begin{array}{ccccc}
Y\Omega & \xrightarrow{\mathrm{nat}\langle\sigma\rangle} & B & = & B \\
{\scriptstyle \bar{h}}\downarrow & & \downarrow {\scriptstyle f} & & \downarrow {\scriptstyle g} \\
A_k & \xrightarrow{\varphi_k} & A & & A_l \\
{\scriptstyle \varphi_{k,l}}\downarrow & & \downarrow {\scriptstyle \varphi_l} & & \| \\
A_l & = & A_l & = & A_l.
\end{array}$$

Since $\varphi_k = \varphi_{k,l}\varphi_l$, equations (3.6.6) and (3.6.8) yield

$$\mathrm{nat}\langle\sigma\rangle f = \bar{h}\varphi_k = \bar{h}\varphi_{k,l}\varphi_l = \mathrm{nat}\langle\sigma\rangle g \varphi_l.$$

Since $\mathrm{nat}\langle\sigma\rangle$ is surjective, it follows that $f = g\varphi_l$. $\square$

For a class $\underline{K}$ of τ-algebras, let $D\underline{K}$ denote the class of τ-algebras isomorphic to directed colimits of $\underline{K}$-algebras. If $D\underline{K} \subseteq \underline{K}$, then we will say that $\underline{K}$ is *closed under directed colimits*. If $\underline{K}$ is closed under subalgebras, direct products and directed colimits, i.e.

$$s\underline{K} \subseteq \underline{K}, \quad P\underline{K} \subseteq \underline{K} \quad \text{and} \quad D\underline{K} \subseteq \underline{K},$$

then $\underline{K}$ is called a *quasivariety*. Hence a quasivariety is a prevariety closed under directed colimits.

3. VARIETIES

Theorem 3.6.5. *Each quasi-equational class is a quasivariety.*

Proof. Let $\underline{\underline{Q}}$ be a quasi-equational class of algebras of a given type $\tau : \Omega \to \mathbb{N}$. Let I be a directed ordered set, and let A_i, for each $i \in I$, be a $\underline{\underline{Q}}$-algebra. Consider a directed system $((A_i)_{i \in I}, (\varphi_{i,j})_{i \leq j})$. It is sufficient to show that if each A_i satisfies a quasi-identity

$$(q) \qquad (w_1 = w_1' \ \& \ldots \& \ w_n = w_n') \to (w = w'),$$

then the directed colimit $A := \varinjlim A_i$ also satisfies it.

By Lemma 1.7.7, the satisfaction of q in A_i means that A_i is q-injective. We will show that the colimit $\varinjlim A_i$ is also q-injective. We keep the notation of Lemma 1.7.7. Let $h : X\Omega^\varphi \to A$, where $\varphi = \langle P \rangle$, be an arbitrary Ω-homomorphism. Since $X\Omega^\varphi = \langle X \mid \{(w_1, w_1'), \ldots, (w_n, w_n')\}\rangle$ is finitely presented, Lemma 3.6.4 implies that for some $k \in I$, there is a homomorphism $h_k : X\Omega^\varphi \to A$ with $h = h_k \varphi_k$. Since A_k is q-injective, Lemma 1.7.7 implies that there is a homomorphism $g_k : X\Omega^\psi \to A_k$, where $\psi = \langle P \cup C \rangle$, with $h_k = h_q g_k$. Hence $h = h_q(g_k \varphi_k)$, demonstrating the q-injectivity of A. It follows that A satisfies the quasi-identity q. $\square$

Example 3.6.6. Recall the implication

$$(3.6.9) \qquad (x = 2x_2 \ \& \ x = 3x_3 \ \& \ \ldots) \to (x = 0)$$

of Example 3.4.3. We know that the group $\mathbb{Z}$ of integers satisfies this implication. However, there are abelian groups that are not torsion free, but also satisfy (3.6.9), e.g. the finite cyclic groups and more generally all finite abelian groups. The group $I_\mathbb{Q}$ of Exercise 2.3I is the directed colimit of the finite cyclic groups. However, it does not satisfy the implication (3.6.9). It follows that the implicational class of abelian groups defined by (3.6.9) is not a quasivariety. $\square$

In the remaining part of this section we will prove that each prevariety closed under directed colimits is a quasi-equational class. Recall (Proposition 3.2.3) that the union of a directed set of congruences of an algebra is again a congruence of the algebra.

Lemma 3.6.7. *Let A be an Ω-algebra, and let $\underline{\underline{K}}$ be a prevariety of Ω-algebras. If the θ_i, for $i \in I$, form a directed set of $\underline{\underline{K}}$-congruences on A, then*

$$A \operatorname{nat} \bigcup_{l \in I} \theta_i \cong \varinjlim_{i \in I} A^{\theta_i}.$$

Proof. The directed colimit $\varinjlim A^{\theta_i}$ may be described as follows. A homomorphism $\varphi_{i,j} : A^{\theta_i} \to A^{\theta_j}$ is defined for $\theta_i \leq \theta_j$ by

$$\varphi_{i,j} : A^{\theta_i} \to A^{\theta_j}; \ a^{\theta_i} \mapsto a^{\theta_j}.$$

The equivalence relation δ on the disjoint union $\biguplus_{i \in I} A^{\theta_i}$ is defined by

$$(a^{\theta_i}, b^{\theta_j}) \in \delta \text{ iff } \exists k \geq i,j. \ \exists c^{\theta_k}. \ a^{\theta_i}\varphi_{i,k} = c^{\theta_k} = b^{\theta_j}\varphi_{j,k},$$

i.e. $(a^{\theta_i}, b^{\theta_j}) \in \delta$ iff there is $k \geq i, j$ with $(a, b) \in \theta_k$. As shown in Section 2.3,

$$(\biguplus_{i \in I} A^{\theta_i})^\delta = \varinjlim A^{\theta_i}.$$

Let $\theta := \bigcup_{i \in I} \theta_i$. We will prove that $A^\theta \cong (\biguplus_{i \in I} A^{\theta_i})^\delta$. Define the mapping

$$f : (\biguplus_{i \in I} A^{\theta_i})^\delta \to A^\theta; \ (a^{\theta_i})^\delta \mapsto a^\theta.$$

The mapping f is well defined. Indeed, if $(a^{\theta_i}, b^{\theta_j}) \in \delta$, then there is $k \geq i,j$ with $(a,b) \in \theta_k \subseteq \bigcup_{l \in I} \theta_i = \theta$. Hence $(a,b) \in \theta$. The mapping f injects, since if $(a,b) \in \theta = \bigcup_{l \in I} \theta_i$, then there is $j \in I$ such that $(a,b) \in \theta_j$, and hence $(a^{\theta_j})^\delta = (b^{\theta_j})^\delta$. Obviously f surjects. It is a standard exercise to prove that

f is a homomorphism. To do it, let $\omega \in \Omega$ be n-ary, and let $i, i_1, \ldots, i_n \in I$ with θ_i a common upper bound of all the θ_{i_k}, where $k = 1, \ldots, n$. Then for $a_1, \ldots, a_n$ in A, one has $(a_1^{\theta_{i_1}})^\delta \ldots (a_n^{\theta_{i_n}})^\delta \omega =$

$$(a_1^{\theta_{i_1}} \varphi_{i_1,i} \ldots a_n^{\theta_{i_n}} \varphi_{i_n,i} \omega)^\delta = (a_1^{\theta_i} \ldots a_n^{\theta_i} \omega)^\delta = ((a_1 \ldots a_n \omega)^{\theta_i})^\delta.$$

Hence $(a_1^{\theta_{i_1}})^\delta \ldots (a_n^{\theta_{i_n}})^\delta \omega f =$

$$((a_1 \ldots a_n \omega)^{\theta_i})^\delta f = (a_1 \ldots a_n \omega)^\theta = a_1^\theta \ldots a_n^\theta \omega = (a_1^{\theta_{i_1}})^\delta f \ldots (a_n^{\theta_{i_n}})^\delta f \omega.$$

This proves that f is a homomorphism, and hence an isomorphism. □

Lemma 3.6.8. *Let $\underline{K}$ be a prevariety of Ω-algebras. Then the following two conditions are equivalent:*

(a) *$\underline{K}$ is closed under directed colimits;*

(b) *For any Ω-algebra A and a directed set θ_i, for $i \in I$, of $\underline{K}$-congruences on A, the union $\theta = \bigcup_{i \in I} \theta_i$ is also a $\underline{K}$-congruence on A.*

Proof. (a)⇒(b) Let $\underline{K}$ be closed under D. Let θ_i be a directed set of $\underline{K}$-congruences on A. Note that the union $\theta = \bigcup_{i \in I} \theta_i$ is always a congruence on A, though not necessarily a $\underline{K}$-congruence. By Lemma 3.6.7, $A^\theta \cong \varinjlim A^{\theta_i}$, and since the A^{θ_i} are $\underline{K}$-algebras and $\underline{K}$ is closed under D, it follows that A^θ is in $\underline{K}$, too. Hence θ is a $\underline{K}$-congruence.

(b)⇒(a) If the A_i, for i in I, form a directed set of $\underline{K}$-algebras, then there is a set X with $A_i \cong X\Omega^{\theta_i}$ for some directed set θ_i, where $i \in I$, of $\underline{K}$-congruences on $X\Omega$. By (b), the union $\theta = \bigcup_{i \in I} \theta_i$ is again a $\underline{K}$-congruence. By Lemma 3.6.7, $\varinjlim A_i = \varinjlim X\Omega^{\theta_i} \cong X\Omega^{\bigcup \theta_i}$. Hence $\varinjlim A_i$ is also in $\underline{K}$. □

Lemma 3.6.9. *Let $\underline{Q}$ be a quasi-equational class of Ω-algebras, and let A be a finitely generated $\underline{Q}$-algebra. Then A is a directed colimit of finitely presented $\underline{Q}$-algebras.*

Proof. Let A be a finitely generated $\underline{Q}$-algebra. Then A is isomorphic to $X\Omega^\theta$ for some finite set X and a $\underline{Q}$-congruence θ. By Corollary 3.2.8, the lattice $Cg_{\underline{Q}}X\Omega$ is algebraic. Hence $\theta = \sum_{i \in I} \theta_i$, the join of finitely generated $\underline{Q}$-congruences θ_i of $X\Omega$ contained in θ. (Cf. Proposition 3.2.3 and Exercise 3.2H.) Now the set of such congruences is directed. Hence $\theta = \sum_{i \in I} \theta_i = \bigcup_{i \in I} \theta_i$. Since $\theta = \bigcup_{i \in I} \theta_i$ is a $\underline{Q}$-congruence, Lemma 3.6.7 implies that

$$A \cong X\Omega^\theta = X\Omega^{\bigcup \theta_i} \cong \varinjlim X\Omega^{\theta_i},$$

where each $X\Omega^{\theta_i}$ is a finitely presented $\underline{Q}$-algebra. □

Theorem 3.6.10 [Banaschewski-Herrlich, Shafaat]. *A class $\underline{K}$ of Ω-algebras is quasi-equational if and only if $\underline{K}$ is closed under directed colimits, direct products and subalgebras, i.e. it is a quasivariety.*

Proof. ($\Rightarrow$) This follows by Theorem 3.6.5, but here is another, shorter proof: If $\underline{K}$ is quasi-equational, then by Corollary 3.2.8, for each $\underline{K}$-algebra A, the lattice $Cg_{\underline{K}}A$ is algebraic. Hence the union of any directed set of $\underline{K}$-congruences is again a $\underline{K}$-congruence. By Lemma 3.6.8, $\underline{K}$ is closed under D.

($\Leftarrow$) Assume that $\underline{K}$ is closed under D,P and S. First note that since each Ω-algebra A is a directed colimit of its finitely generated subalgebras (cf. Exercise 2.3D), it follows that A is in $\underline{K}$ if and only if every finitely generated subalgebra of A is in $\underline{K}$. So it is sufficient to look for the quasi-identites satisfied in all finitely generated $\underline{K}$-algebras.

Let $X = \{x_1, x_1, \ldots\}$ be a countable infinite set. Let $X_n = \{x_1, \ldots, x_n\} \subseteq X$, and let $Cg_f X_n \Omega$ be the set of all finitely generated congruences on $X_n\Omega$. For each ρ in $Cg_f X_n\Omega$, say $\rho = \langle P_\rho \rangle_{Cg X_n\Omega}$ with P_ρ a finite subset of $X_n\Omega \times X_n\Omega$, there is a smallest $\underline{K}$-congruence $\bar{\rho}$ on $X_n\Omega$ containing ρ. Consider the set of quasi-identities $\Sigma_\rho =$

$$\{(\&_{i=1,\ldots,k}\, w_i = w'_i) \to (w = w') \mid \{(w_i, w'_i) \mid i = 1, \ldots, k\} = P_\rho \text{ and } (w, w') \in \bar{\rho}\}.$$

Now let
$$\Sigma := \bigcup_{n \in \mathbb{Z}^+} \bigcup_{\rho \in Cg_f X_n \Omega} \Sigma_\rho.$$

Denote by $\underline{K}(\Sigma)$ the quasi-equational class of all Ω-algebras satisfying the quasi-identities Σ. It will be shown that $\underline{K} = \underline{K}(\Sigma)$. It is sufficient to prove that a finitely generated Ω-algebra A is in $\underline{K}$ iff A is in $\underline{K}(\Sigma)$.

Let A be a finitely generated $\underline{K}$-algebra. One may assume that $A \cong X_n\Omega^\rho$ for some $\underline{K}$-congruence ρ on $X_n\Omega$. Let q be a quasi-identity

$$(\&_{i=1,\ldots,k}\ w_i = w'_i) \to (w = w')$$

in Σ, and let $P = \{(w_i, w'_i) \mid w_i = w'_i \text{ for } i = 1, \ldots, k\}$. Suppose that A does not satisfy q. This means that for every $i = 1, \ldots, k$, one has $(w_i, w'_i) \in \rho$ but $(w, w') \notin \rho$. (Cf. Section 1.7.) Since $(w_i, w'_i) \in \rho$, it follows that $\alpha = \langle P \rangle_{Cg X_n \Omega} \leq \rho$. In particular, there is a homomorphism $h : X_n\Omega^\alpha \to X_n\Omega^\rho \cong A$. The homomorphism h factorizes through the $\underline{K}$-replica of $X_n\Omega^\alpha$:

$$\begin{array}{ccc} X_n\Omega^\alpha & \longrightarrow & X_n\Omega^{\bar\alpha} = X_n\Omega^\alpha R \\ h \downarrow & & \downarrow f \\ A & = & A. \end{array}$$

Since q is in Σ, one has $(w, w') \in \bar\alpha$ and hence $(w, w') \in \bar\alpha \leq \bar\rho = \rho$, a contradiction. It follows that A satisfies q and $\underline{K} \subseteq \underline{K}(\Sigma)$.

Now suppose that A is a finitely presented $\underline{K}(\Sigma)$-algebra, say $A = \langle Y | P \rangle$. Let ρ be the $\underline{K}(\Sigma)$-congruence of $Y\Omega$ generated by P. Suppose that A is not in $\underline{K}$. Then $\rho \neq \bar\rho$, and there is $(w, w') \in \bar\rho$ such that $(w, w') \notin \rho$. However, this contradicts the assumption that A satisfies Σ, for

$$\&_{(w_i, w'_i) \in P}\ (w_i = w'_i) \to (w = w')$$

is a quasi-identity in Σ that does not hold in A. Hence A is a $\underline{K}$-algebra. This together with the previous part of the proof shows that a finitely presented

Ω-algebra is a member of $\underline{\underline{K}}$ if and only if it is a member of $\underline{\underline{K}}(\Sigma)$. On the other hand, by Lemma 3.6.9 each finitely generated $\underline{\underline{K}}(\Sigma)$-algebra is a directed colimit of finitely presented $\underline{\underline{K}}(\Sigma)$-algebras. It follows that each finitely generated $\underline{\underline{K}}(\Sigma)$-algebra is a member of $\underline{\underline{K}}$. Consequently, $\underline{\underline{K}}$ contains all finitely generated $\underline{\underline{K}}(\Sigma)$-algebras, and hence $\underline{\underline{K}}(\Sigma) \subseteq \underline{\underline{K}}$. □

Exercises

3.6A. Show that the mappings $\varphi_{i,j}$ of (3.6.1) are well-defined, are Ω-homomorphisms, and satisfy the conditions (2.3.5).

3.6B. Show that the mapping f of (3.6.2) is well-defined, and is an Ω-homomorphism.

3.6C. Find an example of a directed colimit $A = \varinjlim A_i$ of finitely presented Ω-algebras A_i such that the algebras A_i are not subalgebras of the algebra A. (Compare this with Exercise 2.3C.)

3.7. Quasivarieties and reduced products

This section provides another characterization of quasivarieties, namely as classes of Ω-algebras closed under subalgebras and reduced products, certain special homomorphic images of direct products. It is similar in spirit to Theorem 3.6.10 describing quasivarieties as classes of Ω-algebras closed under D,S and P, and also to the theorems characterizing varieties and prevarieties by means of corresponding operators. Historically, the "reduced product" theorem of Mal'cev was the first known characterization of quasivarieties. It was in fact obtained as a corollary of more general results characterizing classes of Ω-algebras, and more generally classes of relational structures satisfying certain formulae expressed in so-called "first order language". Roughly speaking, in the case of algebras such formulae contain a finite number of identities, some propositional connectives, and the quantifiers $\forall$ and $\exists$. Note that infinite implications are not of this type, though identities and quasi-identities are. In such classes, reduced products, and in particular ultraproducts, play an extremely

3. VARIETIES

important rôle. The proof of Mal'cev's theorem presented here is direct, and does not require any knowledge of first order logic. As a corollary, one obtains the description of the smallest quasivariety containing a given class $\underline{K}$ of Ω-algebras, which generalizes the "HSP-theorem" for varieties.

Consider a Boolean algebra $(B, +, \cdot, ', 0, 1)$. A subset F of B is called a *filter* of the Boolean algebra B if:

(a) $1 \in F$;
(b) $(a, b \in F) \Rightarrow (ab \in F)$;
(c) $(a \in F$ and $b \geq a) \Rightarrow (b \in F)$.

For example, for each $a \in B$, the upper set $a\uparrow := \{b \in B \mid a \leq b\}$ of b is a filter, called a *principal filter*. An example of a non-principal filter is given by the set of all cofinite subsets of a set X in the Boolean algebra $P^{\leq \infty}_{\leq \infty}(X)$ consisting of all finite and all cofinite subsets of X. (See Exercise 3.7D below.) For a subset $X \subseteq B$, let $F(X)$ denote the least filter of B containing X. It is called the *filter generated by X*.

Lemma 3.7.1. *For a Boolean algebra B and a subset X of the set B,*

$$F(X) = \{b \in B \mid b \geq b_1 \ldots b_n \text{ for some } b_1, \ldots, b_n \in X\}. \quad \Box$$

The proof is left as Exercise 3.7A.

A filter F of a Boolean algebra B is *proper* if it does not coincide with B, i.e. $0 \notin F$. A filter F of B is called an *ultrafilter* if it is a maximal, proper filter. Let $(Fi(B), \subseteq)$ denote the poset of all filters of B.

Theorem 3.7.2. *Let F be a proper filter in a Boolean algebra B. Then the following are equivalent:*

(a) *F is an ultrafilter;*
(b) *For $a, b \in B$, if $a + b \in F$, then $a \in F$ or $b \in F$;*
(c) *For all $a \in B$, exactly one of a and a' belongs to F.*

Proof. (a)⇒(b) Let F be an ultrafilter in B, and let $a, b \in B$. Assume $a + b \in F$ and $a \notin F$. Define $F_a := \{ac \mid c \in F\}\uparrow := \{d \in B \mid \exists c \in F.\ d \geq ac\}$. Then F_a is the smallest filter containing F and a [Exercise 3.7E]. Since F is an ultrafilter, it follows that $F_a = B$. In particular $0 \in F_a$, and $0 = ad$ for some $d \in F$. Then $bd = bd + ad = (b + a)d \in F$. Since $bd \leq b$, one has $b \in F$.

(b)⇒(c) First note that, for any $a \in B$, one has $a + a' = 1 \in F$. Hence by (b), $a \in F$ or $a' \in F$. If both a and a' belong to F, then $0 = aa' \in F$. This gives a contradiction.

(c)⇒(a) Let G be a filter properly containing F. Let $a \in G - F$. Then $a' \in F \subseteq G$. Hence $0 = aa' \in G$. Therefore $G = B$. It follows that F is maximal. □

Theorem 3.7.3. *Let B be a Boolean algebra. Then:*

(a) *each proper filter F of B is contained in an ultrafilter of B;*

(b) *for each $1 \neq a \in B$, there exists an ultrafilter U of B with $a \notin U$;*

(c) *for each proper filter F of B and $a \in B - F$, there exists an ultrafilter U with $F \subseteq U$ and $a \notin U$.*

Proof. (a) Let F be a proper filter of B. Let $\mathcal{A} := \{X \in Fi(B) \mid F \subseteq X \neq B\}$, and consider the ordered set $(\mathcal{A}, \subseteq)$. Let $C = \{Y_i \mid i \in I\}$ be a chain in $\mathcal{A}$. Let $Y = \bigcup_{i \in I} Y_i$. Certainly $Y \neq B$ and $Y \supseteq F$. If $x, y \in Y$, then there are $i, j \in I$ with $x \in Y_i$ and $y \in Y_j$. One may assume without loss of generality that $Y_i \subseteq Y_j$. Then $x, y \in Y_j$. Hence $xy \in Y_j \subseteq Y$. And if $x \in Y_i \subseteq Y$ and $x \leq y$, then obviously $y \in Y_i \subseteq Y$. It follows that $Y \in \mathcal{A}$. By the Kuratowski-Zorn Lemma 0.5.1, $\mathcal{A}$ has a maximal element, the required ultrafilter.

(b) Take $b \in B$ such that $b \neq a$ and $a'b \neq 0$. The last condition means that $b \nleq a$. Consider the filter $G := (a'b)\uparrow = \{c \in B \mid c \geq a'b\}$. By (a), the filter G is contained in an ultrafilter U of B, and clearly $a' \in U$. By Theorem 3.7.2, one has $a \notin u$.

(c) By Exercise 3.7B, the filter F determines a congruence θ of B. Consider the projection $\operatorname{nat}\theta : B \to B^\theta$; $b \mapsto b^\theta$. By (b), there is an ultrafilter U' of B^θ such that $a^\theta \notin U'$. Now $U := (\operatorname{nat}\theta)^{-1}(U')$ is an ultrafilter of B (see Exercise 3.7G), and clearly $F \subseteq U$ and $a \notin U$. □

Let the A_i, for $i \in I$, be Ω-algebras, and let F be a proper filter of the Boolean algebra $\mathcal{P}(I)$, i.e. $\varnothing \notin F$. Define a binary relation θ_F on $\prod_{i \in I} A_i$ by

$$(a,b) \in \theta_F \text{ iff } E_{a,b} := \{i \in I \mid ia = ib\} \in F,$$

i.e. the equalizer $E_{a,b}$ of the parallel pair $I \overset{a}{\underset{b}{\rightrightarrows}} \bigsqcup_{i \in I} A_i$ is an element of F.

Lemma 3.7.4. *The relation θ_F is a congruence relation on $\prod_{i \in I} A_i$.*

Proof. Clearly θ_F is reflexive and symmetric. If $(a,b), (b,c) \in \theta_F$, then $E_{a,b}$, $E_{b,c} \in F$, and hence $E_{a,b} \cap E_{b,c} \in F$. Since $E_{a,b} \cap E_{b,c} \subseteq E_{a,c}$, it follows that $E_{a,c} \in F$, whence $(a,c) \in \theta_F$, and θ_F is an equivalence relation. Now for an n-ary ω in Ω, and $(a_1, b_1), \ldots, (a_n, b_n) \in \theta_F$, one has $E_{a_1,b_1} \cap \cdots \cap E_{a_n,b_n} \subseteq E_{a_1\ldots a_n\omega, b_1\ldots b_n\omega}$, hence $E_{a_1\ldots a_n\omega, b_1,\ldots,b_n\omega} \in F$, and $(a_1 \ldots a_n\omega, b_1 \ldots b_n\omega) \in \theta_F$. □

Definition 3.7.5. Given a non-empty set I indexing Ω-algebras A_i for $i \in I$, and a proper filter F in $\mathcal{P}(I)$, the *reduced product* $\prod_{i \in I} A_i / F$ (of the algebras A_i by the filter F) is defined to be the quotient algebra $(\prod_{i \in I} A_i)^{\theta_F}$. If F is an ultrafilter U, then the reduced product $\prod_{i \in I} A_i / U$ is called an *ultraproduct*. □

For a class $\underline{K}$ of Ω-algebras, $P_r\underline{K}$ denotes the class of isomorphic copies of all reduced products $\prod_{i \in I} A_i / F$ of $\underline{K}$-algebras, and $P_u\underline{K}$ denotes the class of isomorphic copies of all ultraproducts $\prod_{i \in I} A_i / U$ of $\underline{K}$-algebras.

Lemma 3.7.6. *Let $\{A_1, \ldots, A_k\}$ be a finite set of finite Ω-algebras. Let $I = I_1 \sqcup \cdots \sqcup I_k$, the disjoint union of the sets $I_1, \ldots, I_k$, and let $A := A_1^{I_1} \times \cdots \times A_k^{I_k}$.*

If U is an ultrafilter of the Boolean algebra $\mathcal{P}(I)$, then the ultraproduct A/U is isomorphic to one of the algebras $A_1, \ldots, A_k$.

Proof. By Theorem 3.7.2, there is j such that $I_j \in U$. Let $A_j = \{a_1, \ldots, a_l\}$. Choose $b_1, \ldots, b_l \in A$ such that for all $i \in I_j$, one has $ib_1 = a_1, \ldots, ib_l = a_l$. It follows that for any given $a \in A$, one has $E_{a,b_1} \cup \cdots \cup E_{a,b_l} \supseteq I_j$. Hence $E_{a,b_1} \cup \cdots \cup E_{a,b_l} \in U$, and again by Theorem 3.7.2, one has $E_{a,b_1} \in U$ or $\ldots$ or $E_{a,b_l} \in U$. Hence $a \in b_1^{\theta_U}$ or $\ldots$ or $a \in b_l^{\theta_U}$, and any two of these elements are distinct. Consequently $A = \{b_1^{\theta_U}, \ldots, b_l^{\theta_U}\}$. Define

$$h: A/U \to A_j; \ b_i^{\theta_U} \mapsto a_i,$$

where $i = 1, \ldots, l$. The mapping h is an isomorphism. The proof is left as Exercise 3.7H. □

Lemma 3.7.7 [Frayne, Morel, Scott]. *Let $I = \bigcup_{j \in J} I_j$, the disjoint union of sets I_j for $j \in J$. Let F be a filter of the Boolean algebra $\mathcal{P}(J)$, and for each $j \in J$, let F_j be a filter of the Boolean algebra $\mathcal{P}(I_j)$. Set*

$$F^* := \{X \subseteq I \mid \{j \in J \mid X \cap I_j \in F_j\} \in F\}.$$

Then F^* is a filter of $\mathcal{P}(I)$. Moreover, for Ω-algebras A_i with $i \in I$,

$$\prod_{i \in I} A_i / F^* \cong (\prod_{j \in J} (\prod_{i \in I_j} A_i/F_j))/F.$$

Proof. The mapping

$$\alpha: \prod_{i \in I} A_i/F^* \to (\prod_{j \in J} (\prod_{i \in I_j} A_i/F_j))/F; \ a^{\theta_{F^*}} \mapsto b^{\theta_F},$$

where b is defined by $I_j b = (a|_{I_j})^{\theta_{F_j}}$, is the required isomorphism. The proof is left as Exercise 3.7J. □

3. VARIETIES

Lemma 3.7.8. Let $\underline{K}$ be a class of Ω-algebras. Then the following hold:
(a) $P\underline{K} \subseteq P_r\underline{K}$;
(b) $P_rP_r\underline{K} \subseteq P_r\underline{K}$;
(c) $P_r\underline{K} \subseteq SPP_u\underline{K}$;
(d) $SP_r\underline{K} = SPP_u\underline{K}$.

Proof. The proofs of (a) and (b) are left as Exercises 3.7K and 3.7L.

(c) Let F be a filter of $\mathcal{P}(I)$, and let J be the set of ultrafilters of $\mathcal{P}(I)$ containing F. Given Ω-algebras A_i, for $i \in I$, define

$$\alpha_U : \prod_{i \in I} A_i/F \to \prod_{i \in I} A_i/U; \ a^{\theta_F} \mapsto a^{\theta_U}$$

for $U \in J$, and then define

$$\alpha : \prod_{i \in I} A_i/F \to \prod_{U \in J}(\prod_{i \in I} A_i/U); \ a^{\theta_F} \mapsto a \prod_{U \in J} \alpha_U.$$

The mapping α injects. To see this, first note that $F = \bigcap J$ [Exercise 3.7M]. If $a^{\theta_F} \neq b^{\theta_F}$, then $E_{a,b} \notin F$. By Theorem 3.7.3(c), there is an ultrafilter U extending F with $E_{a,b} \notin U$. Thus $a\alpha_u \neq b\alpha_u$, and α is one-to-one. The remaining details are left as Exercise 3.7N.

(d) Applying (c), (a) and (b) one obtains the following inclusions:

$$SP_r\underline{K} \subseteq SPP_u\underline{K} \subseteq SP_rP_r\underline{K} \subseteq SP_r\underline{K},$$

whence

$$SP_r\underline{K} = SPP_u\underline{K}. \ \square$$

Reduced products can be constructed as directed colimits of certain direct products. Let A_i, for $i \in I$, be Ω-algebras and let F be a filter of the Boolean algebra $\mathcal{P}(I)$. Consider the filter F as a semilattice $(F, +)$, where for X and Y in F one has

$$X + Y := X \cap Y,$$

and $X \leq Y$ iff $X \supseteq Y$. Define

$$\varphi_{X,Y} : A_X := \prod_{i \in X} A_i \to A_Y := \prod_{i \in Y} A_i;\ f \mapsto f|_Y.$$

Define a functor $G : (F) \to (\Omega)$ by

$$\begin{array}{ccc} Y & & A_Y \\ \uparrow & \xrightarrow{G} & \uparrow \varphi_{X,Y} \\ X & & A_X \end{array}.$$

One obtains a semilattice ordered system $((A_X)_{X \in F}, (\varphi_{X,Y})_{X \leq Y})$ similar to that of Example 2.3.10. The Ω-algebra structure is defined on the disjoint union of the A_X as in Example 2.3.10, and we know that the directed colimit $\varinjlim G$ is (isomorphic to) the quotient algebra $(\biguplus_{X \in F} A_X, \Omega)^\delta$.

Proposition 3.7.9. *Let A_i, for $i \in I$, be (non-empty) Ω-algebras. For any (proper) filter F of the Boolean algebra $\mathcal{P}(I)$, the reduced product $\prod_{i \in I} A_i/F$ and the directed colimit $\varinjlim G$ are isomorphic, i.e.*

$$\prod_{i \in I} A_i/F \cong \varinjlim_{i \in I} G.$$

Proof. First note that each element of the quotient algebra $(\biguplus_{X \in F} A_X)^\delta$ can be represented as f^δ for f in A_I. Define the mapping

$$h : (\biguplus_{X \in F} A_X)^\delta \to \prod_{i \in I} A_i/F;\ f^\delta \mapsto f^{\theta_F},$$

where $f \in A_I$. The mapping h is an isomorphism. To show that h is bijective, it is enough to note that for f, g in $\prod_{i \in I} A_i$, one has $f^\delta = g^\delta$ iff there is X in F with $f|_X = g|_X$, and $f^{\theta_F} = g^{\theta_F}$ iff $\{i \in I \mid if = ig\} \in F$. The remaining details of the proof are left as Exercise 3.7R. $\square$

3. VARIETIES

Corollary 3.7.10. *Let $\underline{K}$ be a class of Ω-algebras. Then*

$$P_r\underline{K} \subseteq P\underline{K}. \quad \Box$$

Mal'cev's Quasivariety Theorem 3.7.11. *Let $\underline{K}$ be a class of Ω-algebras. Then $\underline{K}$ is a quasi-equational class if and only if $\underline{K}$ is closed under subalgebras and reduced products, i.e. $S\underline{K} \subseteq \underline{K}$ and $P_r\underline{K} \subseteq \underline{K}$.*

Proof. ($\Rightarrow$) If $\underline{K}$ is a quasi-equational class, then $\underline{K}$ is closed under subalgebras, direct products and directed colimits by Theorem 3.6.10. In particular, $S\underline{K} \subseteq \underline{K}$ and $DP\underline{K} \subseteq \underline{K}$. By Corollary 3.7.10, $P_r\underline{K} \subseteq DP\underline{K} \subseteq \underline{K}$. Hence $\underline{K}$ is closed under subalgebras and reduced products.

($\Leftarrow$) Now suppose that $S\underline{K} \subseteq \underline{K}$ and $P_r\underline{K} \subseteq \underline{K}$. Since by Lemma 3.7.8(a), $P\underline{K} \subseteq P_r\underline{K}$, it follows that $P\underline{K} \subseteq P_r\underline{K} \subseteq \underline{K}$. Consequently $\underline{K}$ is a prevariety, and hence is defined by a set $\text{Imp}(\underline{K})$ of implications. Let $\text{Qid}(\underline{K})$ be the set of quasi-identities true in $\underline{K}$. Let $\underline{L}$ be the class of algebras satisfying all the quasi-identities in $\text{Qid}(\underline{K})$. Certainly $\underline{K} \subseteq \underline{L}$. It will be shown that $\underline{K}$ is equal to the quasi-equational class $\underline{L}$.

If the class $\underline{K}$ is properly contained in the class $\underline{L}$, then there is an algebra A lying in the class $\underline{L}$, but not in the class $\underline{K}$. Thus there is then an implication φ in $\text{Imp}(\underline{K})$ that is not satisfied by the algebra A. In other words, there is a mapping $f : \arg(\varphi) \to A$ such that for the unique homomorphic extension $\bar{f} : \arg(\varphi)\Omega \to A$, one has $w_m\bar{f} = w'_m\bar{f}$ for all $m \in M$, but $w\bar{f} \neq w'\bar{f}$. For each finite subset i of M, i.e. for each $i \in \mathcal{P}_{<\infty}(M)$, define the quasi-identity

$$(\varphi_i) \qquad (\&_{m \in I}\ w_m = w'_m) \to (w = w')$$

with the same w_m, w'_m, w, w' as in φ. It is evident that $A \not\models \varphi_i$. Since $A \models \text{Qid}(\underline{K})$, it follows that for no $i \in \mathcal{P}_{<\infty}(M)$ does the quasi-identity φ_i belong to $\text{Qid}(\underline{K})$. Hence for each $i \in \mathcal{P}_{<\infty}(M)$, there is an algebra B_i in $\underline{K}$ such that

$B_i \nvDash \varphi_i$. A filter F will be defined on the Boolean algebra $\mathcal{P}(\mathcal{P}_{<\infty}(M))$ such that the reduced product $\prod_{i \in \mathcal{P}_{<\infty}(M)} B_i/F$ does not satisfy the implication φ. This leads to a contradiction, since a reduced product of $\underline{K}$-algebras B_i must be a $\underline{K}$-algebra.

For each $m \in M$, let

$$m\!\uparrow := \{i \in \mathcal{P}_{<\infty}(M) \mid m \in i\}$$

and

$$M\!\uparrow := \{m\!\uparrow \mid m \in M\}.$$

Note that $m\!\uparrow \in \mathcal{P}(\mathcal{P}_{<\infty}(M))$ and $M\!\uparrow \subseteq \mathcal{P}(\mathcal{P}_{<\infty}(M))$. The filter F is defined to be the filter of $\mathcal{P}(\mathcal{P}_{<\infty}(M))$ generated by the set $M\!\uparrow$. By Lemma 3.7.1,

$$F = \{A \in \mathcal{P}(\mathcal{P}_{<\infty}(M)) \mid \exists m_1, \ldots, m_k \in M.\ A \supseteq \{m_1, \ldots, m_k\}\!\uparrow\},$$

where $\{m_1, \ldots, m_k\}\!\uparrow = \{X \supseteq \mathcal{P}_{<\infty}(M) \mid \{m_1, \ldots, m_k\} \in X\}$. In particular, for each $m \in M$, the set $m\!\uparrow$, identified with $\{m\}\!\uparrow$, is a member of F.

It will be shown that the reduced product $\prod_{i \in \mathcal{P}_{<\infty}(M)} B_i/F$ does not satisfy φ. First, recall that for each $i \in \mathcal{P}_{<\infty}(B)$, one has $B_i \nvDash \varphi_i$. This means that there is a set mapping $f_i : \arg(\varphi_i) \to B_i$ such that, for the unique homomorphic extension $\bar{f}_i : \arg(\varphi_i)\Omega \to B_i$, one has $w_m f_i = w'_m \bar{f}_i$ for all $m \in i$, but $w \bar{f}_i \neq w' \bar{f}_i$. Each mapping f_i may be extended to a new mapping $f'_i : \arg(\varphi) \to B_i$ by assigning any fixed element of B_i to each $x \in \arg(\varphi) - \arg(\varphi_i)$. Let

$$f' := \prod_{i \in \mathcal{P}_{<\infty}(M)} f'_i : \arg(\varphi) \to \prod_{i \in \mathcal{P}_{<\infty}(M)} B_i;$$

$$x \mapsto \left(\mathcal{P}_{<\infty}(M) \to \bigcup_{i \in I} B_i;\ i \mapsto x f_i \right),$$

and let $\bar{f}' : \arg(\varphi)\Omega \to \prod_{i \in \mathcal{P}_{<\infty}(M)} B_i$ be the unique homomorphic extension. It will be shown that for each $m \in M$, one has $w_m \bar{f}' = w'_m \bar{f}'$ but $w \bar{f}' \neq$

3. VARIETIES

$w'\bar{f}'$. First note that $\{i \in \mathcal{P}_{<\infty}(M) \mid w_m \bar{f}'_i = w'_m \bar{f}'_i\} = \{i \in \mathcal{P}_{<\infty}(M) \mid m \in i\} = m\uparrow \in F$. Now recall that for $a, b \in \prod_{i \in I} B_i$, the pair (a, b) is in θ_F precisely if $\{i \in I \mid ia = ib\} \in F$. It follows that for each $m \in M$, one has $(w_m \bar{f}', w'_m \bar{f}') \in \theta_F$. On the other hand, $\{i \in \mathcal{P}_{<\infty}(M) \mid w \bar{f}'_i = w' \bar{f}'_i\} = \varnothing$, and $\varnothing \notin F$. This implies that $(w \bar{f}'_i, w' \bar{f}'_i) \notin \theta_F$, and hence $\prod_{i \in \mathcal{P}_{<\infty}(M)} B_i / F \not\models \varphi$, which contradicts the assumption that $\underline{K} \models \varphi$. Therefore $\underline{K} = \underline{L}$, and $\underline{K}$ is a quasi-equational class. $\square$

Mal'cev's Theorem 3.7.12. *Let $\underline{K}$ be a class of Ω-algebras, and let $Q(\underline{K})$ be the smallest quasi-equational class containing $\underline{K}$. Then $Q(\underline{K}) = SP_r\underline{K}$.*

Proof. Since $Q(\underline{K})$ is quasi-equational and $\underline{K} \subseteq Q(\underline{K})$, it follows that $P_r\underline{K} \subseteq Q(\underline{K})$ and hence also $SP_r\underline{K} \subseteq Q(\underline{K})$. Since $Q(\underline{K})$ is the smallest quasi-equational class containing $\underline{K}$, one only needs to prove that $SP_r\underline{K}$ is quasi-equational. By Theorem 3.7.11, the class $SP_r\underline{K}$ is quasi-equational if $SSP_r\underline{K} \subseteq SP_r\underline{K}$ and $P_r SP_r\underline{K} \subseteq SP_r\underline{K}$. The first inclusion is obvious. To prove the second, suppose that algebras A_j, for $j \in J$, are in $SP_r\underline{K}$, i.e. there are $\underline{K}$-algebras C_i, where $i \in I_j$, and filters F_j of $\mathcal{P}(I_j)$ with

$$A_j \leq \prod_{i \in I_j} C_i / F_j.$$

Take an arbitrary filter F in the Boolean algebra $\mathcal{P}(J)$. Then Lemma 3.7.7 and Exercise 3.7.O imply that

$$\prod_{j \in J} A_j / F \leq (\prod_{j \in J}(\prod_{i \in I_j} C_i / F_j))/F \cong \prod_{i \in I} C_i / F^*$$

for a suitable filter F^*. Consequently, $P_r SP_R \underline{K} \subseteq SP_r\underline{K}$. $\square$

Corollary 3.7.13. *For a class $\underline{K}$ of Ω-algebras,*

$$DSP\underline{K} = SP_r\underline{K}.$$

In particular, a prevariety is closed under directed colimits if and only if it is closed under reduced products.

Proof. By Lemma 3.7.8(a), $P\underline{K} \subseteq P_r\underline{K}$, which implies $DSP\underline{K} \subseteq DSP_r\underline{K}$. By Theorem 3.7.12, $SP_r\underline{K}$ is quasi-equational, and hence is closed under directed colimits, i.e. $DSP_r\underline{K} \subseteq SP_r\underline{K}$. Conversely, since by Corollary 3.7.10, $P_r\underline{K} \subseteq DP\underline{K}$, it follows that $SP_r\underline{K} \subseteq SDP\underline{K}$. On the other hand, by Exercise 2.3F, $SD\underline{K} \subseteq DS\underline{K}$, and hence $SP_r\underline{K} \subseteq SDP\underline{K} \subseteq DSP\underline{K}$. Consequently, $DSP\underline{K} = SP_R\underline{K}$. □

Corollary 3.7.14. *For a class $\underline{K}$ of Ω-algebras*

$$Q\underline{K} = SP_r\underline{K} = SPP_u\underline{K} = DSP\underline{K}.$$

Proof. By Lemma 3.7.8(d), $SP_r\underline{K} = SPP_u\underline{K}$. □

Corollary 3.7.15. *If $\underline{K}$ is a finite set of finite Ω-algebras, then $Q\underline{K} = SP\underline{K}$.*

Proof. Apply Lemma 3.7.6. □

The last corollary of Theorem 3.7.11 characterizes the class of subreducts (i.e. subalgebras of reducts) of algebras in a given quasivariety.

Corollary 3.7.16. *Let $\underline{Q}$ be a quasivariety of Ω_2-algebras of a given type $\tau : \Omega_2 \to \mathbb{N}$. Let $\Omega_1 \subseteq \mathfrak{P}\Omega_2$. Then the class $\underline{K}$ of Ω_1-algebras isomorphic to Ω_1-subreducts of $\underline{Q}$-algebras is also a quasivariety.*

Proof. This follows directly by Theorem 3.7.11, since the class $\underline{K}$ is closed under subalgebras and reduced products [Exercise 3.7S]. □

We conclude this section with two results exhibiting a certain method of creating a new prevariety or quasivariety from two given ones.

Let $\underline{K}$ be a prevariety of idempotent algebras of a given type $\tau : \Omega \to \mathbb{N}$. Let $\underline{R}$ and $\underline{S}$ be *subprevarieties* of the prevariety $\underline{K}$, i.e. subclasses of $\underline{K}$ that are also prevarieties. The *Mal'cev product* $\underline{R} \circ_K \underline{S}$ of $\underline{R}$ and $\underline{S}$ *relative to* $\underline{K}$ consists of $\underline{K}$-algebras A with a congruence θ such that A^θ is in $\underline{S}$, and each θ-class a^θ is in $\underline{R}$. Note that by the idempotency of $\underline{K}$, each θ-class is always a subalgebra of A. If $\underline{K}$ is the class $\underline{\tau}$ of all τ-algebras, then the Mal'cev product of $\underline{R}$ and $\underline{S}$ relative to $\underline{\tau}$ is called simply the *Mal'cev product* of $\underline{R}$ and $\underline{S}$, and is denoted by $\underline{R} \circ \underline{S}$.

Theorem 3.7.17. *The Mal'cev product $\underline{R} \circ_K \underline{S}$ of subprevarieties $\underline{R}$ and $\underline{S}$ of an idempotent prevariety $\underline{K}$ of τ-algebras is again a prevariety.*

Proof. We have to show that the class $\underline{R} \circ_K \underline{S}$ is closed under subalgebras and direct products.

Let A be a $\underline{K}$-algebra in the class $\underline{R} \circ_K \underline{S}$. Then there is a congruence θ on A such that A^θ is in $\underline{S}$ and each θ-class a^θ is in $\underline{R}$. Let B be a subalgebra of A. Then there is a congruence θ' on B such that $B^{\theta'} \leq A^\theta$. Since the quotient algebra A^θ is in the prevariety $\underline{S}$, it follows that the quotient algebra $B^{\theta'}$ is also in $\underline{S}$. Now for any b in B, the θ'-class $b^{\theta'}$ is a subalgebra of the $\underline{R}$-algebra b^θ, and hence is also an $\underline{R}$-algebra. Therefore the algebra B is a member of the Mal'cev product $\underline{R} \circ_K \underline{S}$.

Now suppose that A is in $\underline{K}$ and A is (isomorphic to) a direct product $\prod_{i \in I} A_i$ of algebras A_i, where each A_i is a member of $\underline{R} \circ_K \underline{S}$. Then for each i in I, there is a congruence θ_i on A_i such that the quotient algebra $A_i^{\theta_i}$ is in $\underline{S}$ and each θ_i-class $a_i^{\theta_i}$ is in $\underline{R}$. Consider the homomorphism

$$h : \prod_{i \in I} A_i \xrightarrow{\pi_i} A_i \xrightarrow{\mathrm{nat}\,\theta_i} A_i^{\theta_i},$$

the composite of the projection π_i and the natural homomorphism $\mathrm{nat}\,\theta_i$. Use the universality property (2.3.3) for products to extend h to a corresponding homomorphism

$$\bar{h} : \prod_{i \in I} A_i \to \prod_{i \in I} A_i^{\theta_i}.$$

Let $\theta := \ker \bar{h}$. The First Isomorphism Theorem 1.2.3 shows that

(3.7.1) $$A^\theta \cong \prod_{i \in I} A_i^{\theta_i}.$$

Since for each i in I, the algebra $A_i^{\theta_i}$ is in the prevariety $\underline{S}$, it follows that A^θ is also in $\underline{S}$. Now for each a in A one has

(3.7.2) $$a^\theta \cong \prod_{i \in I}(a\pi_i)^{\theta_i},$$

and each $(a\pi_i)^{\theta_i}$ is a subalgebra of the $\underline{R}$-algebra A_i. Hence the algebra a^θ is also in the prevariety $\underline{R}$.

Since A^θ is in $\underline{S}$ and each θ-class a^θ is in $\underline{R}$, it follows that the direct product A is in the Mal'cev product $\underline{R} \circ_K \underline{S}$. □

As the final result of this chapter, we obtain a similar result for quasivarieties.

Theorem 3.7.18. *The Mal'cev product $\underline{R} \circ_K \underline{S}$ of subquasivarieties $\underline{R}$ and $\underline{S}$ of an idempotent quasivariety $\underline{K}$ is again a quasivariety.*

Proof. In view of Corollary 3.7.13 and Theorem 3.7.17, it is sufficient to show that the prevariety $\underline{R} \circ_K \underline{S}$ is closed under reduced products.

Assume that B is a $\underline{K}$-algebra isomorphic to $\prod_{i \in I} A_i/F$, where all the A_i are $\underline{R} \circ_K \underline{S}$-algebras and F is a filter in $\mathcal{P}(I)$. Let $A \cong \prod_{i \in I} A_i$. Let θ and θ_i be congruences on A and A_i, respectively, defined as in the proof of Theorem 3.7.17. By Theorem 3.7.17, the algebra A is in $\underline{R} \circ_K \underline{S}$, and both (3.7.1) and (3.7.2) hold. Similarly as before, consider the canonical homomorphism

$$g : \prod_{i \in I} A_i/F \to \prod_{i \in I} A_i^{\theta_i}/F; \ (i \mapsto a_i)^{\theta_F} \mapsto (i \mapsto a_i^{\theta_i})^{\theta_F}.$$

Then there is a congruence $\lambda \in Cg\, B$ such that

$$B^\lambda \cong \prod_{i \in I} A_i^{\theta_i}/F.$$

Since all the $A_i^{\theta_i}$ are in the quasivariety $\underline{S}$, it follows that the reduced product $\prod A_i^{\theta_i}/F$, and hence the algebra B^λ, are also in $\underline{S}$.

On the other hand, there is a congruence $\varphi \in Cg\, A$ such that

$$(A^\theta)^\varphi \cong \prod_{i \in I} A_i^{\theta_i}/F \cong B^\lambda.$$

In this isomorphism, each λ-class b^λ of B^λ corresponds to some $(a^\theta)^\varphi$ for a in A. However, by (3.7.2),

$$(a^\theta)^\varphi \cong (\prod_{i \in I}(a\pi_i)^{\theta_i})^{\theta_F} \cong b^\lambda.$$

But since $\prod(a\pi_i)^{\theta_i}$ is in the quasivariety $\underline{R}$, it follows that b^λ is also a member of $\underline{R}$. Consequently, the algebra B is in the class $\underline{R} \circ_K \underline{S}$. □

For more information about quasivarieties, see [Mal'cev 1966, 1967, 1971, 1973], [Gorbunov 1994, 1998], [Wechler 1992].

Exercises

3.7A. Prove Lemma 3.7.1.

3.7B. Let B be a Boolean algebra, and let θ be a binary relation on B. Show that the following conditions are equivalent:

(a) θ is a congruence on the Boolean algebra B;

(b) 1^θ is a filter of B, and for $a, b, \in B$,

$$(a,b) \in \theta \text{ iff } (a'+b)(a+b') \in 1^\theta.$$

Deduce that the filters of a Boolean algebra B determine the congruence relations of B.

3.7C. Let B be a Boolean algebra. Show that the set $Fi(B)$ of filters of a Boolean algebra B is closed under arbitrary intersections. Deduce that the poset $(Fi(B), \subseteq)$ is a complete lattice.

3.7D. Define a subset Y of a set X to be *cofinite* if its complement $X - Y$ is finite. Let $\mathcal{P}_{<\infty}(X)$ and $\mathcal{P}^{<\infty}(X)$ denote the sets of finite and of cofinite subsets of X, respectively. Define $\mathcal{P}_{<\infty}^{<\infty}(X) := \mathcal{P}_{<\infty}(X) \cup \mathcal{P}^{<\infty}(X)$.

(a) Show that under the set-theoretical operations $\cup, \cap$ and $'$, the set $\mathcal{P}_{<\infty}^{<\infty}(X)$ is a Boolean algebra.

(b) Show that in the Boolean algebra $\mathcal{P}_{<\infty}^{<\infty}(X)$, the set of all cofinite subsets of X is the unique non-principal filter.

3.7E. Let F be a proper filter in a Boolean algebra B and let $a \in B$. Let $F_a := \{ac \mid c \in F\}\uparrow = \{b \in B \mid \forall c \in F,\ b \geq ac\}$. Show that F_a is the smallest filter containing both F and a.

3.7F. Let B be a Boolean algebra. Let θ_F be the congruence of B determined by a filter F of B.

(a) Show that if F is an ultrafilter of B, then B^{θ_F} is isomorphic to the two-element Boolean algebra $\underline{2}$.

(b) Show that if $h : B \to \underline{2}$ is a Boolean homomorphism, then $h^{-1}(1)$ is an ultrafilter of B.

3.7G. Let $h : B_1 \to B_2$ be a homomorphism of Boolean algebras. Show that if U is an ultrafilter of B_2, then $h^{-1}(U)$ is an ultrafilter of B_1.

3.7H. Complete the proof of Lemma 3.7.6.

3.7I. Let A_i, for $i \in I$, be Ω-algebras, and let F be a filter of the Boolean algebra $\mathcal{P}(I)$. Show that if F is principal, then the reduced product $\prod_{i \in I} A_i / F$ is isomorphic to one of the algebras A_i.
[Hint: Note that if F is principal, then $F = \{J \subseteq I \mid i \in J\}$ for some i in I.]

3.7J. Complete the proof of Lemma 3.7.7.

3.7K. Prove Lemma 3.7.8(a).

3.7L. Use Lemma 3.7.7 to prove Lemma 3.7.8(b).

3. VARIETIES   179

3.7M. Show that each proper filter in a Boolean algebra $\mathcal{P}(I)$ is an intersection of the ultrafilters containing it.

3.7N. Complete the proof of Lemma 3.7.8(c).

3.7O. Show that for a filter F in the Boolean algebra $\mathcal{P}(I)$, if Ω-algebras $A_i \leq B_i$ for $i \in I$, then $\prod_{i \in I} A_i/F \leq \prod_{i \in I} B_i/F$.

3.7P. Prove that the following hold for a class $\underline{K}$ of Ω-algebras:

    (a) $P_u S \underline{K} \subseteq SP_u \underline{K}$;   (b) $P_r S \underline{K} \subseteq SP_r \underline{K}$.

3.7Q. Complete the proof of Proposition 3.7.9.

3.7R. Complete the proof of Corollary 3.7.16.

3.7S. Let $\underline{K}$ be an idempotent prevariety. Let $\underline{R}$, $\underline{R}'$, $\underline{S}$, and $\underline{S}'$ be subprevarieties of $\underline{K}$. Show the following:

    (a) $\underline{R} \subseteq \underline{R} \circ_K \underline{R}$;

    (b) $\underline{R} \subseteq \underline{R} \circ_K \underline{S}$ and $\underline{S} \subseteq \underline{R} \circ_K \underline{S}$;

    (c) $\underline{R}' \subseteq \underline{R}$ and $\underline{S}' \subseteq \underline{S}$ imply $\underline{R}' \circ_K \underline{S}' \subseteq \underline{R} \circ_K \underline{S}$;

    (d) $\underline{R} \subseteq \underline{S}$ and $\underline{R}' \subseteq \underline{S}$ imply $\underline{R} \circ_S \underline{R}' = (\underline{R} \circ_K \underline{R}') \cap \underline{S}$.

CHAPTER IV

CONSTRUCTING ALGEBRAS and QUASIVARIETIES

The basic algebraic constructions, such as forming subalgebras, different types of products, and more general limits, were discussed in previous sections. However, these constructions are not always sufficient to give a clear description of the structure of algebras in a (pre- or quasi-) variety. Among many other ways of describing the structure of algebras, let us mention two here. One belongs to the realm of embeddability theorems. For example, each integral domain can be embedded into a field, or each commutative cancellative semigroup into a commutative group. There are many more theorems of this kind in algebra. One given algebra is embedded into another, usually one with a better known and richer structure, one in which calculation is easier. We will use such a method in one of the subsequent chapters to describe the structure of certain idempotent and entropic algebras. The second method to be discussed in this chapter shows how to build up an algebra from smaller pieces, usually better known algebras of the same type, and indexed by elements of another algebra. Finally, let us mention that these two methods may be combined to give one more way of characterizing algebras.

If an algebra (A, Ω) is of the same type as the algebras of a prevariety $\underline{P}$, then (A, Ω) has a $\underline{P}$-algebra as a homomorphic image. In fact, (A, Ω) has a largest homomorphic image in $\underline{P}$, the $\underline{P}$-replica of (A, Ω). If the $\underline{P}$-algebras are idempotent, and $h : (A, \Omega) \to (I, \Omega)$ is a homomorphism onto a $\underline{P}$-algebra, then the ker h-classes of (A, Ω) are subalgebras of (A, Ω). This gives a decomposition of (A, Ω) as a disjoint union of subalgebras over its image in $\underline{P}$.

Reversing such a decomposition, we will give some construction methods to obtain an Ω-algebra from a set of Ω-algebras, say (A_i, Ω) for $i \in I$, indexed by elements of another Ω-algebra (I, Ω). The general idea of such constructions depends on converting the algebra (I, Ω) into a small category, assigning to each morphism $i \to j$ of this category a homomorphism $(A_i, \Omega) \to (E_j, \Omega)$ of algebras, where (E_j, Ω) is an extension or superalgebra of (A_j, Ω), and then using these homomorphisms to define the Ω-algebra structure on the disjoint union $A = \bigcup (A_i \mid i \in I)$ of the sets A_i. We will do this in such a way that the algebras (A_i, Ω) will be subalgebras of (A, Ω), while the algebra (I, Ω) will be a quotient of (A_i, Ω). We will investigate properties and special cases of such constructions, and show that algebras in certain classes may be represented as "sums" of the type described above.

4.1. Algebraic quasi-ordering of an algebra

Let (I, Ω) be an algebra of type $\tau : \Omega \to \mathbb{N}$. Define the following relation $\preceq$ on the set I:

(4.1.1)
$$\preceq := \{(i,j) \mid \exists t(x_1 \ldots x_n) \in X\Omega \text{ and} \\ \exists \, i_1, \ldots, i_{k-1}, i_{k+1}, \ldots, i_n \in I \text{ with} \\ j = i_1 \ldots i_{k-1} i i_{k+1} \ldots i_n t\}.$$

It is easy to check that the relation $\preceq$ is a quasi-order. Moreover, it is the smallest reflexive and transitive relation containing the following relation:

(4.1.2)
$$\sigma := \{(i,j) \mid \exists \, \omega \in \Omega \text{ and } i_1, \ldots, i_{k-1}, i_{k+1}, \ldots, i_{\omega\tau} \in I \\ \text{with } j = i_1 \ldots i_{k-1} i \, i_{k+1} \ldots i_{\omega\tau} \omega\}$$

[Exercises 4.1A, 4.1B]. The relation $\preceq$ is called the *algebraic quasi-order* of the algebra (I, Ω).

4. CONSTRUCTING ALGEBRAS

Example 4.1.1. If (I, Ω) is an Ω-semilattice, then the algebraic quasi-order of (I, Ω) coincides with the semilattice order $\leq$ [Exercise 4.1C]. □

Proposition 4.1.2. *A quasi-order ρ on a set A is an algebraic quasi-order of some plural algebra (A, Ω) if and only if ρ is directed.*

Proof. ($\Rightarrow$) If ρ is the algebraic quasi-order of an algebra (A, Ω), and a_i and a_j are in A, then

$$a_i, \ a_j \preceq b_1 \ldots b_{i-1} a_i b_{i+1} \ldots b_{j-1} a_j b_{j+1} \ldots b_n \omega$$

for any $b_1, \ldots, b_n$ in A and n-ary ω in Ω, with $n \geq 2$.

($\Leftarrow$) Let ρ be directed. Then one can define a groupoid operation $\cdot$ on A so that ρ is the algebraic quasi-order $\preceq$ of $(A, \cdot)$. For a and b in A with $a \rho b$ and $b \rho a$, define $ab := b$ and $ba := a$. For a and b in A with $a \rho b$ but (b, a) not in ρ, define $ab := b =: ba$. Now if a and b are not comparable, then choose an element c in A with $a \rho c$ and $b \rho c$. Then define $ab := c =: ba$. The groupoid $(A, \cdot)$ is idempotent, and obviously its algebraic quasi-order $\preceq$ coincides with ρ. □

The algebraic quasi-order $\preceq$ of an algebra (A, Ω) is *full* if $\preceq$ is the universal relation on A, i.e. if $\preceq \ = A \times A$. Contrary to the situation described in Example 4.1.1 and Exercises 4.1D and 4.1E, the algebraic quasi-order $\preceq$ of an algebra is frequently full.

Example 4.1.3. A *left (right) division groupoid* $(G, \cdot)$ is a set G with a binary operation denoted by $\cdot$ such that, for each a in G, the mapping

$$L_a : G \to G; \ g \mapsto ag$$

$$(R_a : G \to G; \ g \mapsto ga)$$

is surjective, i.e. for all a and b in G there is g in G with $ag = b$ (or with $ga = b$). In particular, the class of left (right) division groupoids contains the class of left (right) quasigroups. A *division* groupoid is both a left and a right division groupoid. Clearly, in all three cases, for all a and b in G, one has $(a,b) \in \sigma$ and hence $(a,b) \in \preceq$. The quasi-order $\preceq$ is full. In particular, the algebraic quasi-order of each group and quasigroup is full. □

It is quite common for the algebraic quasi-order of an algebra to be full. A variety of Ω-algebras is called *strongly irregular* if it satisfies an identity $x \cdot y = x$ for a binary Ω-word $x \cdot y$. Note that the majority of known and interesting varieties of algebras are strongly irregular.

Proposition 4.1.4. *Let $\underline{V}$ be a strongly irregular variety of algebras. Then the algebraic quasi-order of each $\underline{V}$-algebra is full.*

Proof. For any $\underline{V}$-algebra (A, Ω) and a, b in A, one has $a \cdot b = a$ and $b \cdot a = b$, whence $b \preceq a$ and $a \preceq b$. □

Example 4.1.5. Lattices are strongly irregular, and hence have full algebraic quasi-orders that do not coincide with the lattice order. □

Example 4.1.6. The algebraic quasi-order of the free idempotent semigroup over x and y cannot be full, since there is no $a \neq x$, y with $a \preceq x$ or $a \preceq y$. □

An algebra (I, Ω) is called *naturally quasi-ordered* if its algebraic quasi-order $\preceq$ satisfies the following condition for all ω in Ω, and $a_1, \ldots, a_{\omega\tau}$, $b_1, \ldots, b_{\omega\tau}$ in I:

(4.1.3)   If $a_i \preceq b_i$ for $i = 1, \ldots, \omega\tau$, then $a_1 \ldots a_{\omega\tau}\omega \preceq b_1 \ldots b_{\omega\tau}\omega$.

Proposition 4.1.7. *Let (I, Ω) be an idempotent algebra with algebraic quasi-order $\preceq$. Then the following conditions are equivalent:*

(a) (I, Ω) *is naturally quasi-ordered;*

(b) *For all ω in Ω, $a_1, \ldots, a_{\omega\tau}$, a in I and $i = 1, \ldots, \omega\tau$,*

(4.1.4)   if $a_i \preceq a$, then $a_1 \ldots a_{\omega\tau}\omega \preceq a$;

4. CONSTRUCTING ALGEBRAS

(c) *The relation α defined by Theorem 0.6.1 is a congruence of (I, Ω), and the quotient (I^α, Ω) is an Ω-semilattice.*

Proof. (a) $\Rightarrow$ (b) If $a_1, \ldots, a_{\omega\tau} \preceq a$, then by (4.1.3), $a_1 \ldots a_{\omega\tau}\omega \preceq a \ldots a\omega = a$.

(b) $\Rightarrow$ (a) Let $a_i \preceq b_i$ for $i = 1, \ldots, \omega\tau$. Then $a_i \preceq b_i \preceq b_1 \ldots b_{\omega\tau}\omega$, and by (4.1.4), $a_1 \ldots a_{\omega\tau}\omega \preceq b_1 \ldots b_{\omega\tau}\omega$.

(a) $\Rightarrow$ (c) Let $(a_i, b_i) \in \alpha$ for $i = 1, \ldots, \omega\tau$. This means that $a_i \preceq b_i$ and $b_i \preceq a_i$. Then condition (4.1.3) implies that $(a_1 \ldots a_{\omega\tau}\omega, b_1 \ldots b_{\omega\tau}\omega) \in \alpha$, i.e. α is a congruence relation. Now let $\{a_1^\alpha, \ldots, a_n^\alpha\} \subseteq I^\alpha$. For each n-ary ω in Ω, one has $a_i \preceq a_1 \ldots a_n\omega$, and hence $a_i^\alpha \leq a_1 \ldots a_n\omega^\alpha$. This means that $a_1 \ldots a_n\omega^\alpha$ is an upper bound for a_i^α. If b^α is an upper bound for all a_i^α, i.e. $a_i^\alpha \leq b^\alpha$, then $a_i \preceq b$. By (b), one has $a_1 \ldots a_n\omega \preceq b$, and hence $a_1^\alpha \ldots a_n^\alpha\omega = a_1 \ldots a_n\omega^\alpha \leq b^\alpha$. It follows that $a_1 \ldots a_n\omega^\alpha$ is the least upper bound of all the a_i^α.

(c) $\Rightarrow$ (a) Let $a_i \preceq b_i$ for $i = 1, \ldots, n$. Then $a_i^\alpha \leq b_i^\alpha$, and for any n-ary ω in Ω, $(a_1 \ldots a_n\omega)^\alpha = a_1^\alpha \ldots a_n^\alpha\omega \leq b_1^\alpha \ldots b_n^\alpha\omega = (b_1 \ldots b_n\omega)^\alpha$. Hence $a_1 \ldots a_n\omega \preceq b_1 \ldots b_n\omega$. $\square$

Note that the algebraic quasi-order of (I^α, Ω) coincides with the semilattice order. Moreover, since α-classes have full algebraic quasi-order, the Ω-semilattice (I^α, Ω) is the semilattice replica of (I, Ω).

Corollary 4.1.8. *If (I, Ω) is an idempotent, naturally quasi-ordered algebra, then (I, Ω) decomposes as the union of subalgebras (i^α, Ω), each with the full algebraic quasi-order, over the semilattice (I^α, Ω).* $\square$

Example 4.1.9. Each normal band [Exercise 1.5E] is naturally quasi-ordered. Indeed, if $(I, \cdot)$ is a normal band, and $a, b, c \in I$ with

$$a, b \preceq c = a_1 \ldots a_{i-1} a\, a_{i+1} \ldots a_{j-1} b a_{j+1} \ldots a_n,$$

then the entropic law implies that $c = a_1 \ldots (ab) \ldots a_n$. By Proposition 4.1.7, $(I, \cdot)$ is naturally quasi-ordered. $\square$

Example 4.1.10. In the free idempotent and entropic groupoid $(G, \cdot)$ (cf. Exercise 1.5E) on two generators x and y, one has $x, y \preceq xy, yx$, but it is not true that $xy \preceq yx$ or that $yx \preceq xy$. It follows that this algebra is not naturally quasi-ordered. □

Exercises

4.1A. Prove that the relation $\preceq$ of (4.1.4) is a quasi-order.

4.1B. Prove that the algebraic quasi-order of an algebra (I, Ω) is the smallest reflexive and transitive relation on I containing the relation σ of (4.1.2).

4.1C. Verify the claim of Example 4.1.1.

4.1D. *Directoids* – algebraic models of upward directed ordered sets – are groupoids satisfying the identities

$$xx = x,$$
$$xy \cdot x = xy,$$
$$y \cdot xy = xy,$$
$$x(xy \cdot z) = xy \cdot z$$

[Ježek, Quackenbush 1990]. The ordering relation of a directoid is defined by $x \leq y$ iff $xy = y$. Show that the algebraic quasi-order $\preceq$ of a directoid coincides with the order $\leq$.

4.1E. Let $\underline{\underline{Q}}$ be the quasivariety of Ω-algebras defined by the quasi-identities of the form

$$(x_1 \ldots x \ldots x_m s = y \text{ and } y_1 \ldots y \ldots y_n t = x) \to (x = y)$$

for all Ω-words s and t. Show that the algebraic quasi-order of each $\underline{\underline{Q}}$-algebra is an order.

4.1F. Show that the relation σ on a groupoid $(G, \cdot)$ defined by (4.1.2) is full if and only if $GL_a \cup GR_a = G$ for each a in G.

4.1G. [Romanowska, Traina 1999] Prove that a quasi-order ρ on a set I is an algebraic quasi-order of some idempotent, naturally quasi-ordered algebra (I, Ω) if and only if the order $\leq$ of the quotient (I^ρ, Ω) is a semilattice order.

4.1H. Let $(A, \cdot)$ be a normal band, and let γ be the relation on A defined by

$$(a, b) \in \gamma :\Leftrightarrow a = aba \text{ and } b = bab.$$

(a) Show that γ is the semilattice replica congruence on $(A, \cdot)$.

(b) Show that each γ-class is a rectangular band.

(c) Show that the relation α defined by Theorem 0.6.1 coincides with γ.

4.1I. Show that each affine space over a field, as defined in Section 1.6.4, is naturally quasi-ordered.

4.2. Functorial sums

Let (I, Ω), and (A_i, Ω) for $i \in I$, be algebras of a (plural) type $\tau : \Omega \to \mathbb{N}$. Let $\preceq$ be the algebraic quasi-order of (I, Ω). Then $(I, \preceq)$ may be considered as a category (I) with elements of I as objects and a unique morphism $i \to j$ if and only if $i \preceq j$. (Cf. Exercise 2.1A.) The objects of the category (Ω) are the Ω-algebras, and the morphisms of (Ω) are the Ω-homomorphisms. Let a covariant functor $F : (I) \to (\Omega)$ be given. Then the *functorial sum* $(A, \Omega) = IF$ of the algebras $(A_i, \Omega) = iF$ over the algebra (I, Ω) by the functor F is the disjoint union $\bigcup (A_i \mid i \in I)$ of the underlying sets with the Ω-algebra structure given by

$$(4.2.1) \quad \omega : A_{i_1} \times \cdots \times A_{i_n} \to A_i \, ; \, (a_{i_1}, \ldots a_{i_n}) \mapsto a_{i_1} \varphi_{i_1, i} \ldots a_{i_n} \varphi_{i_n, i} \omega,$$

for each n-ary ω in Ω, and $i = i_1 \ldots i_n\omega$, where for $k = 1, \ldots, n$, the mapping $\varphi_{i_k,i} : i_k F \to iF$ is the Ω-homomorphism $(i_k \to i)F$. Note that all the iF are subalgebras of the sum IF, and that there is an Ω-homomorphism

$$\pi : (A, \Omega) \to (I, \Omega) \, ; \, a_i \mapsto i,$$

where $a_i \in A_i$, called *projection*. The algebra (I, Ω) is called the *indexing algebra* for the sum. Note that functorial sums have also been called *Agassiz sums*. A functorial sum IF that is isomorphic neither to (I, Ω) nor to any of the iF is called *proper*. If the indexing algebra is an Ω-semilattice, then the functorial sum is called a *Płonka sum*. Note that in the case where (A, Ω) is a Płonka sum, the indexing semilattice is usually interpreted as a join semilattice. Płonka sums will be discussed in next section. If the indexing algebra has a full algebraic quasi-order, then one obtains another common special case.

Proposition 4.2.1. Let $F : (I) \to (\Omega)$ be a functor, and let $(A, \Omega) = IF$ be the functorial sum of $(A_i, \Omega) = iF$ over (I, Ω). If the algebraic quasi-order $\preceq$ of (I, Ω) is full, then for any two i, j in I, the algebras iF and jF are isomorphic. Moreover, the sum IF is isomorphic to the direct product of (I, Ω) and iF, i.e.

$$(A, \Omega) \cong (A_i, \Omega) \times (I, \Omega).$$

Proof. Since the algebraic quasi-order $\preceq$ of (I, Ω) is full, it follows that for any i, j in I, one has $i \preceq j$ and $j \preceq i$, whence there are homomorphisms $\varphi_{i,j} : iF \to jF$ and $\varphi_{j,i} : jF \to iF$. Obviously, $\varphi_{i,j}\varphi_{j,i} = 1_{iF}$ and $\varphi_{j,i}\varphi_{i,j} = 1_{jF}$. Hence $\varphi_{i,j}$ and $\varphi_{j,i}$ are mutually inverse isomorphisms. The remaining part is an obvious consequence of 0.3. □

Note that if a variety $\underline{V}$ of Ω-algebras is an independent join of two subvarieties $\underline{V}_1$ and $\underline{V}_2$, then each $\underline{V}$-algebra can be represented as a functorial sum of $\underline{V}_1$-algebras over a $\underline{V}_2$-algebra, namely as the direct product of a $\underline{V}_1$-algebra and a $\underline{V}_2$-algebra. (Cf. Lemma 3.5.10.)

4. CONSTRUCTING ALGEBRAS

If the indexing algebra (I, Ω) of a functorial sum IF is idempotent and naturally quasi-ordered, then the composition π nat α of the projection π and the natural homomorphism nat α induced by the congruence α of Proposition 4.1.7 is again an Ω-homomorphism. The preimage $B_i := (\pi \text{ nat } \alpha)^{-1}(i^\alpha)$, for $i^\alpha \in I^\alpha$, is a subalgebra of IF. The algebra (B_i, Ω) itself is a functorial sum of the (A_j, Ω) for $j \in i^\alpha$. By Proposition 4.2.1, $(B_i, \Omega) \cong (A_i, \Omega) \times (i^\alpha, \Omega)$. This implies the following.

Proposition 4.2.2. *Let an algebra (A, Ω) be a functorial sum of algebras (A_i, Ω) over an idempotent, naturally quasi-ordered algebra (I, Ω). Then (A, Ω) decomposes as the disjoint union of subalgebras isomorphic to direct products $(A_i, \Omega) \times (i^\alpha, \Omega)$ over the semilattice (I^α, Ω).* □

Lemma 4.2.3. *Let $F : (I) \to (\Omega)$ be a functor, and let $(A, \Omega) = IF$ be the functorial sum of $(A_i, \Omega) = iF$ over (I, Ω). Then for each derived operation $x_1 \ldots x_n w$, and for elements $a_k \in A_{i_k}$, with $k = 1, \ldots, n$, one has*

$$a_1 \ldots a_n w = a_1 \varphi_{i_1, i} \ldots a_n \varphi_{i_n, i} w,$$

where $i = i_1 \ldots i_n w$.

Proof. The proof goes by induction on the number of occurrences of Ω-operations in the corresponding word w. The result holds by (4.2.1) if this number is 1. Otherwise, suppose $a_1 \ldots a_n w = a_1 \ldots a_k (a_{k+1} \ldots a_{k+m} \omega) a_{k+m+1} \ldots a_n v$ for an m-ary operation ω in Ω and $(n - m + 1)$-ary derived operation v. Let $j = i_{k+1} \ldots i_{k+m} \omega$ and $i = i_1 \ldots i_n w = i_1 \ldots j \ldots i_n v$. Then by the induction hypothesis,

$$a_1 \ldots a_n w = a_1 \ldots a_k (a_{k+1} \varphi_{i_{k+1}, j} \ldots a_{k+m} \varphi_{i_{k+m}, j} \omega) a_{k+m+1} \ldots a_n v =$$
$$a_1 \varphi_{i_1, i} \ldots a_k \varphi_{i_k, i} (a_{k+1} \varphi_{i_{k+1}, j} \ldots a_{k+m} \varphi_{i_{k+m}, j} \omega) \varphi_{j, i} a_{k+m+1} \varphi_{i_{k+m+1}, i} \ldots a_n \varphi_{i_n, i} v$$
$$= a_1 \varphi_{i_1, i} \ldots (a_{k+1} \varphi_{i_{k+1}, i} \ldots a_{k+m} \varphi_{i_{k+m}, i} \omega) \ldots a_n \varphi_{i_n, i} v$$
$$= a_1 \varphi_{i_1, i} \ldots a_n \varphi_{i_n, i} w,$$

as required. The penultimate equality holds by the functoriality of the sum. □

Proposition 4.2.4. *The identities satisfied by a proper functorial sum IF of iF over (I, Ω) are precisely the identities satisfied by (I, Ω) and each of the iF.*

Proof. Let $x_1 \ldots x_n w = y_1 \ldots y_m v$ be an identity satisfied by (I, Ω) and each iF. Suppose $a_k \in A_{i_k}$ and $b_l \in A_{j_l}$, where $k = 1, \ldots, n$ and $l = 1, \ldots, m$. Since $w = v$ holds in (I, Ω), one has $i_1 \ldots i_n w = j_1 \ldots j_m v =: i$. By Lemma 4.2.3, one has $a_1 \ldots a_n w = a_1 \varphi_{i_1,i} \ldots a_n \varphi_{i_n,i} w$ and $b_1 \ldots b_m v = b_1 \varphi_{j_1,i} \ldots b_n \varphi_{j_m,i} v$. Since the identity $w = v$ holds in (A_i, Ω), one has $a_1 \varphi_{i_1,i} \ldots a_n \varphi_{i_n,i} w = b_1 \varphi_{j_1,i} \ldots b_m \varphi_{j_m,i} v$. Hence $w = v$ holds in $IF = (A, \Omega)$. Finally, note that an identity satisfied in IF obviously holds in all the $iF = (A_i, \Omega)$, since they are subalgebras, and in the indexing algebra, since it is a homomorphic image of IF. □

Corollary 4.2.5. *Let $\underline{K}$ and $\underline{L}$ be classes of Ω-algebras. Then the set of identities satisfied in the class $s_f(\underline{K}, \underline{L})$ of algebras isomorphic to functorial sums of $\underline{K}$-algebras over $\underline{L}$-algebras coincides with $\mathrm{Id}(\underline{K}) \cap \mathrm{Id}(\underline{L})$, i.e. the equation*

$$\mathrm{Id}(s_f(\underline{K}, \underline{L})) = \mathrm{Id}(\underline{K}) \cap \mathrm{Id}(\underline{L})$$

holds. □

Corollary 4.2.6. *Let $\underline{V}$ and $\underline{W}$ be varieties of Ω-algebras. Then*

$$s_f(\underline{V}, \underline{W}) \subseteq \underline{V} + \underline{W}. \quad \square$$

Functorial sums are closely connected to directed colimits. Let $(A, \Omega) = IF$ be a functorial sum of $(A_i, \Omega) = iF$ over an idempotent, naturally quasi-ordered (I, Ω). Define the following relation δ on the set A. For $a_i \in A_i$ and $b_j \in A_j$,

(4.2.2) $\quad (a_i, b_j) \in \delta :\Leftrightarrow \exists\, k \in I$ with $i, j \preceq k$ and such that $a_i \varphi_{i,k} = b_j \varphi_{j,k}$.

4. CONSTRUCTING ALGEBRAS

It is easy to check that the relation δ is a congruence on IF [Exercise 4.2A]. (Cf. Exercise 2.3H and Example 2.3.10.)

Proposition 4.2.7. *Let $F : (I) \to (\Omega)$ be a functor, and let $(A, \Omega) = IF$ be the functorial sum of $(A_i, \Omega) = iF$ over an idempotent, naturally quasi-ordered algebra (I, Ω). Then*
$$IF^\delta \cong \varinjlim F.$$

Proof. This follows directly by the definition of a functorial sum, and the construction of a directed colimit (Cf. Exercise 2.3H). □

Note that the sums of algebras described in Example 2.3.10, and in the introductions to Lemma 3.6.2 and Proposition 3.7.9 are in fact Płonka sums.

Exercises

4.2A. Verify that the relation δ defined by (4.2.2) is a congruence relation.

4.2B. On a functorial sum IF as in Proposition 4.2.7, define the following relation γ. For $a_i \in iF$ and $b_j \in jF$, set $(a_i, b_j) \in \gamma$ iff $i \preceq j$ and $b_j = a_i \varphi_{i,j}$. Let $\check{\gamma}$ denote the converse of γ, i.e. the relation defined by $(a, b) \in \check{\gamma}$ iff $(b, a) \in \gamma$. Show that $\delta = \gamma \circ \check{\gamma}$.

4.2C. Let $\tau : \Omega \to \mathbb{Z}^+$ be a plural type. Let F be a functor, and let IF be the Płonka sum of the Ω-algebras iF over an Ω-semilattice I as above. Extend I to I_0 by adding a new element $0 \notin I$ and putting $0 < i$ for each i in I. Extend the functor F to $F_0 : (I_0) \to (\Omega)$ by setting $0F$ to be the limit $\varprojlim F$ of the functor F.

(a) Describe the Płonka sum $I_0 F_0$ of the functor F_0. (Cf. Exercise 2.3F.)

(b) Show that $IF \not\cong I_0 F_0$, but $\varinjlim F \cong \varinjlim F_0$.

4.2D. Let $\mathrm{Id}_r(\Omega)$, $\mathrm{Id}_{rr}(\Omega)$, $\mathrm{Id}_{lr}(\Omega)$ denote the sets of all regular, right regular and left regular identities of type $\tau : \Omega \to \mathbb{N}$. (Cf. Exercise 1.5B.) Let $\underline{\underline{V}}$ be a variety of Ω-algebras. Show the following:

(a) $\mathrm{Id}(s_f(\underline{V}, \underline{\mathrm{Sl}})) \subseteq \mathrm{Id}_r(\Omega)$;

(b) $\mathrm{Id}(s_f(\underline{V}, \underline{\mathrm{Lz}})) \subseteq \mathrm{Id}_{lr}(\Omega)$;

(c) $\mathrm{Id}(s_f(\underline{V}, \underline{\mathrm{Rz}})) \subseteq \mathrm{Id}_{rr}(\Omega)$.

4.2E. Let (I, Ω) be an algebra with a directed algebraic quasi-order $\preceq$. Define the functor F and the functorial sum $(A, \Omega) = IF$ as in Proposition 4.2.7. Show that $IF^\delta = \varinjlim F$.

4.3. Płonka sums and regularizations

Płonka sums, i.e. functorial sums over semilattices, provide an excellent tool for representing algebras in regularizations of strongly irregular varieties. (Cf. Section 1.5.) Note that any idempotent irregular variety $\underline{V}$ is always strongly irregular. Indeed, any irregular identity $x_1 \ldots x_n w = y_1 \ldots y_m v$ true in $\underline{V}$ implies the identity $xy \ldots yw = y \ldots yv = y$. In fact, such a variety can always be defined by a set of regular identities together with one strongly irregular identity. In what follows, an identity is called *binary* if it involves precisely two variables.

Lemma 4.3.1. *Let $\underline{V}$ be an idempotent irregular variety of Ω-algebras. Let $\Sigma = \Sigma_r \cup \Sigma_{ir}$ be a basis for the identities true in $\underline{V}$, where Σ_r consists of regular and Σ_{ir} of irregular identities. Then Σ_{ir} includes strongly irregular binary identities, and $\underline{V}$ has a basis consisting of regular identities and any one of the strongly irregular binary identities.*

Proof. Choose any strongly irregular binary identity $x \cdot y = x$ satisfied in $\underline{V}$. If σ is an irregular identity in Σ_{ir} different from $x \cdot y = x$, with variables $x_1, \ldots, x_n$, then form a new identity σ' obtained from σ by replacing each variable x_i by $x_i \cdot x_1 \cdot \ldots \cdot x_n$. Obviously the variety $\underline{V}$ satisfies σ'. Set $\Sigma' := \{\sigma' \mid \sigma \in \Sigma_{ir}, \sigma \neq (x \cdot y = x)\}$. It is easy to see that the set $\Sigma_r \cup \Sigma' \cup \{x \cdot y = x\}$ is a basis for the identities true in $\underline{V}$, as desired. □

4. CONSTRUCTING ALGEBRAS

Płonka's Theorem 4.3.2. *Let $\underline{V}$ be an idempotent irregular variety of Ω-algebras defined by a set Σ of regular identities and an identity of the form $x \cdot y = x$. Then the following classes coincide:*

(a) *the regularization $\underline{\widetilde{V}}$ of $\underline{V}$;*

(b) *the class $S_p(\underline{V}) := S_f(\underline{V}, \underline{Sl})$ of Płonka sums of $\underline{V}$-algebras;*

(c) *the variety $\underline{\overline{V}}$ of Ω-algebras defined by the identities Σ and (for each n-ary ω in Ω) the following:*

(4.3.1) $$\begin{cases} x \cdot x = x, \\ x \cdot yz = xy \cdot z, \\ x \cdot yz = x \cdot zy, \\ x_1 \ldots x_n \omega \cdot y = (x_1 \cdot y) \ldots (x_n \cdot y)\omega, \\ y \cdot x_1 \ldots x_n \omega = y \cdot x_1 \cdot \ldots \cdot x_n. \end{cases}$$

Note that the first three identities define $\cdot$ as a left-normal band operation [Exercise 1.5F], the fourth shows that $\cdot$ distributes from the right over each ω, and the fifth that $\cdot$ "breaks" ω from the left.

Proof. By Corollary 4.2.5 and Example 1.5.4, it follows that each Płonka sum of $\underline{V}$-algebras satisfies the regular identities true in $\underline{V}$. Hence $S_p(\underline{V}) \subseteq \underline{\widetilde{V}}$.

Now note that all the identities defining $\underline{\overline{V}}$ are regular. They are also satisfied in $\underline{V}$, and hence in $\underline{\widetilde{V}}$. This implies $\underline{\widetilde{V}} \subseteq \underline{\overline{V}}$.

Let (A, Ω) be a $\underline{\overline{V}}$-algebra. Define a binary relation β on A by

$$(a, b) \in \beta \text{ iff } a \cdot b = a \text{ and } b \cdot a = b.$$

Clearly β is reflexive and symmetric. Since $(A, \cdot)$ is a left-normal band, it is also transitive. To show that β is a congruence relation on (A, Ω), let ω be an n-ary operation in Ω, and suppose that $(a_1, b_1), \ldots, (a_n, b_n) \in \beta$. Then

$$a_1 \ldots a_n \omega \cdot b_1 \ldots b_n \omega = a_1 \ldots a_n \omega \cdot a_1 \ldots a_n \omega \cdot b_1 \cdot \ldots \cdot b_n$$
$$= a_1 \ldots a_n \omega \cdot a_1 \cdot \ldots \cdot a_n \cdot b_1 \cdot \ldots \cdot b_n = a_1 \ldots a_n \omega \cdot a_1 \cdot \ldots \cdot a_n$$
$$= a_1 \ldots a_n \omega \cdot a_1 \ldots a_n \omega = a_1 \ldots a_n \omega,$$

and similarly $b_1 \ldots b_n\omega \cdot a_1 \ldots a_n\omega = b_1 \ldots b_n\omega$. Thus $(a_1 \ldots a_n\omega, b_1 \ldots b_n\omega) \in \beta$, and β is a congruence on (A, Ω). At the same time, β is a congruence on $(A, \cdot)$, and the corresponding quotient $(A^\beta, \cdot)$ is a semilattice $(I, \cdot)$. As an Ω-semilattice, (I, Ω) is the quotient (A^β, Ω), since $a_1 \ldots a_n\omega \cdot a_1 \cdot \ldots \cdot a_n = a_1 \ldots a_n\omega$ as above, and $a_1 \cdot \ldots \cdot a_n \cdot a_1 \ldots a_n\omega = a_1 \cdot \ldots \cdot a_n$. Hence $(a_1 \ldots a_n\omega, a_1 \cdot \ldots \cdot a_n) \in \beta$. Obviously, the β-classes are $\underline{V}$-subalgebras of (A, Ω). Define the object part of the functor $F : (I) \to (\Omega)$ by $a^\beta F = (a^\beta, \Omega)$. Define the morphism part of the functor by

$$(a^\beta \to b^\beta)F : (a^\beta, \Omega) \to (b^\beta, \Omega) \; ; \; x \mapsto x \cdot b.$$

This is mapping into b^β, since $x^\beta \cdot b^\beta = a^\beta \cdot b^\beta = b^\beta$. It is well defined, since $(b, b') \in \beta$ implies $x \cdot b = x \cdot b \cdot b' = x \cdot b' \cdot b = x \cdot b'$. It is an Ω-homomorphism, since $\cdot$ distributes from the right over ω. Now $(a^\beta \to a^\beta)F$ is the identity on a^β, since $(x, a) \in \beta$ implies $x \cdot a = x$. For $a^\beta \to b^\beta \to c^\beta$ in (I) and $(x, a) \in \beta$, one has $x(a^\beta \to b^\beta)F(b^\beta \to c^\beta)F = x \cdot b \cdot c = x \cdot c = x(a^\beta \to c^\beta)F$. Thus F is indeed a functor. Moreover, (A, Ω) is the Płonka sum IF, since for $a_1, \ldots, a_n$ in A with $a = a_1 \cdot \ldots \cdot a_n$, one has the equation $(a_1(a_1^\beta \to a^\beta)F) \ldots (a_n(a_n^\beta \to a^\beta)F)\omega = (a_1 \cdot a) \ldots (a_n \cdot a)\omega = a_1 \ldots a_n\omega \cdot a = a_1 \ldots a_n\omega$, as required. The containment $\underline{\overline{V}} \subseteq s_p(\underline{V})$ follows. Finally, since $\underline{\widetilde{V}} \subseteq \underline{\overline{V}}$ and $\underline{\overline{V}} \subseteq s_p(\underline{V})$, it follows that $\underline{\widetilde{V}} \subseteq s_p(\underline{V})$, and hence $s_p(\underline{V}) = \underline{\overline{V}} = \underline{\widetilde{V}}$. $\square$

Note that Lemma 4.3.1 and Theorem 4.3.2 remain true if one assumes more generally that the variety $\underline{V}$ is strongly irregular. Note as well that the decomposition of a $\underline{\widetilde{V}}$-algebra as a Płonka sum of $\underline{V}$-algebras does not depend on the choice of the identity $x \cdot y = x$. A binary operation $\cdot$ satisfying the identities (4.3.1) is called a *decomposition* or *partition* operation. In particular, if an idempotent regular variety $\underline{W}$ has a partition operation $\cdot$, then each $\underline{W}$-algebra is a Płonka sum of algebras satisfying the identity $x \cdot y = x$. Note that the semilattice (I, Ω) of the proof of Płonka's Theorem 4.3.2 is the semilattice replica of the Płonka sum (A, Ω).

Example 4.3.3. The word $x \cdot y \cdot x$ determines a partition operation in all normal bands [Exercise 4.3A]. It follows that each normal band is a Płonka sum of rectangular bands. □

The characterization of algebras in regularized varieties as Płonka sums helps to classify subvarieties of the regularization.

Theorem 4.3.4. *The lattice $\mathcal{L}(\widetilde{\underline{V}})$ of subvarieties of the regularization $\widetilde{\underline{V}}$ of an idempotent irregular variety $\underline{V}$ is isomorphic to the direct product $\mathcal{L}(\underline{V}) \times \underline{2}$ of the lattice $\mathcal{L}(\underline{V})$ of subvarieties of $\underline{V}$ and the two-element chain $\underline{2}$.*

Proof. It will first be shown that each (perforce irregular) subvariety $\underline{W}$ of $\underline{V}$ is covered by its regularization $\widetilde{\underline{W}}$ in $\mathcal{L}(\widetilde{\underline{V}})$. Suppose that $\underline{W} \subseteq \underline{U} \subseteq \widetilde{\underline{W}}$ for some variety $\underline{U}$. Then all the regular identities satisfied by all $\underline{U}$-algebras are also satisfied by $\underline{W}$-algebras, whence $\widetilde{\underline{U}} = \widetilde{\underline{W}}$. Now suppose that $\underline{U}$ is irregular, i.e. $\underline{U} \subset \widetilde{\underline{U}}$. By the results of Section 3.5, $\underline{U}$ is the least subvariety of $\underline{V}$ containing the free $\underline{U}$-algebra $\mathfrak{P}U$. By Płonka's Theorem 4.3.2, $\mathfrak{P}U$, as an irregular $\widetilde{\underline{W}}$-algebra, must lie in $\underline{W}$, and so $\underline{U} = \underline{W}$. Thus $\widetilde{\underline{W}}$ covers $\underline{W}$.

Now let $\mathcal{L}_r(\widetilde{\underline{V}})$ be the set of regular subvarieties of $\widetilde{\underline{V}}$ ordered by inclusion. We will show that the regularization mapping

(4.3.2) $$r : \mathcal{L}(\underline{V}) \to \mathcal{L}_r(\widetilde{\underline{V}}); \ \underline{W} \mapsto \widetilde{\underline{W}}$$

is an order-preserving isomorphism. It injects by Płonka'a Theorem 4.3.2. Let $\underline{U}$ be a regular subvariety of $\widetilde{\underline{V}}$. The free $\underline{U}$-algebra $\mathfrak{P}U$ is a member of $\widetilde{\underline{V}}$, and hence is a Płonka sum of $\underline{V}$-algebras, say U_i, for i in the semilattice replica I of $\mathfrak{P}U$. Let $\underline{W}$ be the least subvariety of $\underline{V}$ containing all U_i for $i \in I$. By Płonka's Theorem, the identities satisfied by $\mathfrak{P}U$ are precisely the regular identities satisfied by all the U_i. Thus $\underline{U} = \widetilde{\underline{W}}$, and the mapping (4.3.2) surjects.

Now both $\mathcal{L}(\underline{V})$ and $\mathcal{L}_r(\underline{V})$ are meet semilattices. Then for subvarieties $\underline{U}$ and $\underline{W}$ of $\underline{V}$, one has

$$(\underline{U} \cap \underline{W})r = \widetilde{\underline{U} \cap \underline{W}} = s_p(\underline{U} \cap \underline{W}) = s_p(\underline{U}) \cap s_p(\underline{W})$$
$$= \widetilde{\underline{U}} \cap \widetilde{\underline{W}} = \underline{U}r \cap \underline{W}r.$$

Thus r is a meet semilattice isomorphism, and hence is order-preserving. □

An obvious corollary to Proposition 4.2.7 shows how Płonka sums are connected to directed colimits [Exercise 4.3I]. The next theorem establishes a connection with subdirect products.

Theorem 4.3.5. *Let (A, Ω) be a proper Płonka sum of algebras (A_i, Ω) over a (join) semilattice $(I, \leq)$. Then (A, Ω) is a subdirect product of the algebras (B_i, Ω), where $B_i = A_i$ in the case where $\{i\}$ is a sink of (I, Ω), and otherwise $B_i = A_i \cup \{\infty_i\}$, where $\{\infty_i\}$ is a sink of (B_i, Ω).*

Proof. For j in I, let $P_j := \bigcup(A_i \mid i \leq j)$. Clearly P_j is a subalgebra of (A, Ω), and is itself a Płonka sum. Let δ_j be the congruence on (P_j, Ω) defined as δ in (4.2.2). Then it is easy to see that $(P_j, \Omega)^{\delta_j} \cong (A_j, \Omega)$. Define the relation $\mu_j \subseteq A \times A$ as follows:

$$(a,b) \in \mu_j :\Leftrightarrow (a,b) \in \delta_j \text{ or } \{a,b\} \subseteq A - P_j.$$

Note that $A - P_j$ is a sink in (A, Ω). Denote the μ_j-class $A - P_j$ by ∞_j. It is easy to check that $(A, \Omega)^{\mu_j} \cong (A_j \cup \{\infty_j\}, \Omega)$, with $\{\infty_j\}$ a sink of $(A, \Omega)^{\mu_j}$ in the case $A - P_j \neq \varnothing$. Otherwise, $(A, \Omega)^{\mu_j} \cong (A_j, \Omega)$, and $\{j\}$ is a sink of (I, Ω). By the definition of μ_j, it is clear that $\bigcap (\mu_j \mid j \in I) = \widehat{A}$. Hence, by Lemma 3.1.2, (A, Ω) is a subdirect product of all the $(A, \Omega)^{\mu_j}$. □

Note that in the algebra B_i of Theorem 4.3.5, ∞_i behaves like a "zero" of the algebra, and the algebra (B_i, Ω) can obviously be considered as the Płonka

4. CONSTRUCTING ALGEBRAS

sum of (A_i, Ω) and $(\{\infty_i\}, \Omega)$ over the two-element (join) semilattice $\underset{\sim}{2}$ with $0 < 1$. In such a situation we will sometimes say that (B_i, Ω) is an *extension of* (A_i, Ω) *by zero*.

Only one more step is required to obtain a characterization of subdirectly irreducible Płonka sums.

Lemma 4.3.6. *Let* (A, Ω) *be a subdirectly irreducible extension of an algebra* (A_0, Ω) *by zero. Then* (A, Ω) *is the Płonka sum of the subdirectly irreducible subalgebra* (A_0, Ω) *and the one-element subalgebra* $(A_1, \Omega) = (\{\infty\}, \Omega)$ *over the two-element (join) semilattice* $\underset{\sim}{2}$ *with* $0 < 1$.

Proof. It is enough to note that the congruence relations of (A_0, Ω) are exactly the congruence relations of (A, Ω) restricted to (A_0, Ω). It follows that (A, Ω) is subdirectly irreducible if and only if (A_0, Ω) is. □

Theorem 4.3.7. *Let $\underline{V}$ be an idempotent irregular variety of Ω-algebras. Then a $\underline{\widetilde{V}}$-algebra* (A, Ω) *is subdirectly irreducible if and only if it is one of:*

(a) *a two-element semilattice;*

(b) *a subdirectly irreducible $\underline{V}$-algebra;*

(c) *the Płonka sum of a subdirectly irreducible $\underline{V}$-algebra* (A_0, Ω) *and a one-element $\underline{V}$-algebra* (A_1, Ω) *over the semilattice* $\underset{\sim}{2}$.

Proof. Let (A, Ω) be a subdirectly irreducible $\underline{\widetilde{V}}$-algebra. If (A, Ω) is a semilattice, then by Example 3.1.1 it has precisely two elements. Otherwise, the algebra (A, Ω) is a Płonka sum of $\underline{V}$-algebras (A_i, Ω). By Theorem 4.3.5, either (A, Ω) is isomorphic to some (A_i, Ω), and hence is a $\underline{V}$-algebra, or it is isomorphic to some (B_i, Ω) with $B_i = A_i \cup \{\infty_i\}$, where $\{\infty_i\}$ is a sink of (B_i, Ω). Then by Lemma 4.3.6, (A, Ω) is the Płonka sum of a subdirectly irreducible $\underline{V}$-algebra (A_i, Ω) and a one-element algebra $(\{\infty_i\}, \Omega)$. □

Theorem 4.3.7 remains true if one assumes more generally that the variety $\underline{V}$ is strongly irregular.

Płonka's Theorem 4.3.2 also gives a method for describing free algebras in the regularization $\widetilde{\underline{V}}$ of an idempotent irregular variety $\underline{V}$ as above. By Theorem 4.3.2, the free $\widetilde{\underline{V}}$-algebra $X\widetilde{V}$ on a set X is the Płonka sum of certain $\underline{V}$-subalgebras over a semilattice $(I, \cdot)$. These $\underline{V}$-subalgebras and the semilattice are described in the following theorem. First recall that the free join semilattice on a set X is the set $(\mathcal{P}_{0<\infty}(X), \cup)$ of finite non-empty subsets of X under the union operation.

Theorem 4.3.8. *Let $\underline{V}$ be an idempotent irregular variety of Ω-algebras. Then the free $\widetilde{\underline{V}}$-algebra $X\widetilde{V}$ on a set X is the Płonka sum of the functor $F : (\mathcal{P}_{0<\infty}(X)) \to (\Omega)$ given by*

$$
\begin{array}{ccc}
X_2 & & X_2V \\
i \uparrow & \mapsto & \uparrow iF, \\
X_1 & & X_1V
\end{array}
$$

where the unique arrow $X_1 \to X_2$ for subsets $X_1 \subseteq X_2$ denotes the inclusion mapping i, and iF is the (unique) extension of i to an Ω-homomorphism. □

In other words, the free $\widetilde{\underline{V}}$-algebra $X\widetilde{V}$ is the Płonka sum of the free $\underline{V}$-algebras YV on finite non-empty subsets Y of X over the free semilattice $(\mathcal{P}_{0<\infty}(X), \cup)$. The proof of Theorem 4.3.8 is sketched in Exercises 4.3D and 4.3E.

We conclude this section with a discussion of several notions that are very useful in specifying the cardinalities of free algebras, and their behaviour under regularization.

For a variety $\underline{V}$ of Ω-algebras, let a_n denote the cardinality of the free algebra $\{1, \ldots, n\}V$ for each positive integer n. Recall [Proposition 3.5.2] that a variety $\underline{V}$ is locally finite if each a_n is finite.

Definition 4.3.9. A locally finite variety $\underline{V}$ is *analytic* if the power series

$$V(x) = \sum_{n=0}^{\infty} a_{n+1} x^n$$

4. CONSTRUCTING ALGEBRAS 199

converges in a neighborhood of the origin in the complex plane. In this case, the function $V(x)$ is called the *generating function* of the variety $\underline{V}$. An analytic variety is called *rational* if its generating function is a rational function. □

Example 4.3.10. (a) The variety $\underline{\text{Tr}}$ of trivial bands is rational, with $\text{Tr}(x) = \sum_{n=0}^{\infty} x^n = (1-x)^{-1}$.

(b) The varieties $\underline{\text{Set}}$, $\underline{\text{Lz}}$, and $\underline{\text{Rz}}$ are rational, with $\text{Set}(x) = \text{Lz}(x) = \text{Rz}(x) = \sum_{n=0}^{\infty}(n+1)x^n = \text{Tr}'(x) = (1-x)^{-2}$.

(c) The variety $\underline{\text{Sl}}$ of semilattices is rational, with $\text{Sl}(x) = \sum_{n=0}^{\infty}(2^{n+1}-1)x^n = 2(1-2x)^{-1} - (1-x)^{-1} = (1-2x)^{-1}(1-x)^{-1}$. □

An immediate consequence of Theorem 4.3.8 is that the regularization of a locally finite plural variety of Ω-algebras is again locally finite. An analogous but slightly less immediate consequence is that regularization preserves analyticity and rationality.

Proposition 4.3.11. *Let $\underline{V}$ be an irregular analytic plural variety. Then the regularization $\underline{\widetilde{V}}$ is again analytic. Its generating function is given by*

$$(1-x)^2 \widetilde{V}(x) = V(x/(1-x)).$$

Proof. Let $\widetilde{a}_n$ be the cardinality of the free $\underline{\widetilde{V}}$-algebra $\{1,\ldots,n\}\underline{\widetilde{V}}$. By Theorem 4.3.8, one has $\widetilde{a}_{n+1} = \sum_{p=0}^{n} a_{p+1} \binom{n+1}{p+1}$. Thus

$$x^2 \widetilde{V}(x) = \sum_{n=0}^{\infty} \widetilde{a}_{n+1} x^{n+2} = \sum_{n=0}^{\infty} x^{n+2} \sum_{p=0}^{n} a_{p+1} \binom{n+1}{p+1}$$

$$= \sum_{p=0}^{\infty} \sum_{r=0}^{\infty} x^{p+r+2} a_{p+1} \binom{p+r+1}{p+1}$$

$$= \sum_{p=0}^{\infty} a_{p+1} x^{p+2} \sum_{r=0}^{\infty} \binom{p+r+1}{p+1} x^r$$

$$= \sum_{p=0}^{\infty} a_{p+1} \left(\frac{x}{1-x}\right)^{p+2} = \left(\frac{x}{1-x}\right)^2 V\left(\frac{x}{1-x}\right),$$

the various rearrangements being possible since the latter function is analytic in a neighborhood of the origin if $V(x)$ is. The result follows on multiplying by $(1-x)^{-2} \cdot x^{-2}$. □

Corollary 4.3.12. *Let $\underline{V}$ be an irregular rational plural variety. Then the regularization $\underline{\widetilde{V}}$ is also rational.*

Proof. It follows using the formula of Proposition 4.3.11 for the generating function $\widetilde{V}(x)$ that this is rational if $V(x)$ is. □

Proposition 4.3.11 allows one to determine generating functions for all varieties of normal bands. (See Exercises 4.3C, 1.5E and 1.5F.)

Example 4.3.13. (a) $\mathrm{Sl}(x) = \widetilde{\mathrm{Tr}}(x) = (1-x)^{-2}\mathrm{Tr}(x/(1-x))$
$= (1-x)^{-1}(1-2x)^{-1}$. [Cf. Example 4.3.10 (c).]
(b) $\mathrm{Re}(x) = \sum_{n=0}^{\infty}(n+1)^2 x^n = x\sum_{n=0}^{\infty}(n+1)nx^{n-1} + \sum_{n=0}^{\infty}(n+1)x^n$
$= x\mathrm{Tr}''(x) + \mathrm{Tr}'(x) = (1+x)(1-x)^3$.
(c) For the variety $\underline{\mathrm{NB}}$ of normal bands,
$\mathrm{NB}(x) = \widetilde{\mathrm{Re}}(x) = (1-x)^{-2}\mathrm{Re}(x/(1-x)) = (1-2x)^{-3}$.
(d) For the variety $\underline{\mathrm{Ln}}$ of left normal bands,
$\mathrm{Ln}(x) = \widetilde{\mathrm{Lz}}(x) = (1-x)^{-2}\mathrm{Lz}(x/(1-x)) = (1-2x)^{-2}$. A similar result holds for the variety $\underline{\mathrm{Rn}}$ of right normal bands. □

Exercises

4.3A. Verify the claim of Example 4.3.3.

4.3B. Show that any two partition operations for the regularization of an idempotent irregular variety coincide.

4.3C. Show that $\underline{\mathrm{Tr}}$, $\underline{\mathrm{Lz}}$, $\underline{\mathrm{Rz}}$, $\underline{\mathrm{Re}}$, $\underline{\mathrm{Ln}}$, $\underline{\mathrm{Rn}}$ and $\underline{\mathrm{Sl}}$ are the only proper subvarieties of the variety $\underline{\mathrm{NB}}$ of normal bands. Describe the algebras in $\underline{\mathrm{Ln}}$, $\underline{\mathrm{Rn}}$ and $\underline{\mathrm{Sl}}$ as corresponding Płonka sums. Draw the Hasse diagram of the lattice $\mathcal{L}(\underline{\mathrm{NB}})$.

4. CONSTRUCTING ALGEBRAS

4.3D. Let $\underline{V}$ be an idempotent irregular variety of Ω-algebras. Prove that each $\underline{\widetilde{V}}$-algebra (A, Ω) satisfies:

(a) the last two identities of (4.3.1) for any derived operation;

(b) the identities

$$x_1 \ldots x_m t \cdot y_1 \ldots y_n w = x_1 \ldots x_m t \cdot y_1 \cdot \ldots \cdot y_n,$$

$$x_1 \ldots x_m t = x_1 \ldots x_m t \cdot x_1 \cdot \ldots \cdot x_m,$$

$$(x_1 \cdot \ldots \cdot x_m)(x_2 \cdot x_3 \cdot \ldots \cdot x_m \cdot x_1) \ldots (x_m \cdot x_1 \cdot \ldots \cdot x_{m-1})t = x_1 \ldots x_m t.$$

4.3E. Let $\underline{\widetilde{V}}$ be the regularization of an idempotent irregular variety $\underline{V}$ of Ω-algebras. For an Ω-word w, recall that $\arg(w)$ is the set of variables in w.

(a) Show that the relation γ defined on $X\widetilde{V}$ by

$$(t, w) \in \gamma :\Leftrightarrow \arg(t) = \arg(w)$$

is a congruence relation, and coincides with the congruence β defined in the proof of Theorem 4.3.2.

(b) Show that the quotient $X\widetilde{V}^\gamma$ is isomorphic to the Ω-semilattice $\mathcal{P}_{0<\infty}(X)$.

(c) Let $Y := \{y_1, \ldots, y_r\} \subseteq X$, and let $a_i := y_i \cdot y_{i+1} \cdot \ldots \cdot y_r \cdot y_1 \cdot \ldots \cdot y_{i-1}$. Let $(A, \Omega) = \langle \{a_i\} | i = 1, \ldots, r \rangle$ be the subalgebra of $X\widetilde{V}$ generated by all the a_i. Show that the smallest β-class containing all the a_i coincides with A.

(d) Show that (A, Ω) is the free $\underline{V}$-algebra YV.

4.3F. Let $\underline{V}_1$ and $\underline{V}_2$ be independent (cf. Section 3.5) varieties of Ω-algebras, $\underline{V}_1$ defined by $x \cdot y = x$ and $\underline{V}_2$ defined by $x \cdot y = y$. Let $\underline{V}$ be a variety of the same type satisfying the identity $(x \cdot y) \cdot x = y$. Show the following:

$$s_f(\underline{V}, \widetilde{\underline{V}_1 \vee \underline{V}_2}) = s_p(\underline{V} \vee \underline{V}_1 \vee \underline{V}_2) = \underline{V} \vee \widetilde{\underline{V}_1} \vee \underline{V}_2.$$

4.3G. Consider the variety $\underline{F}$ of affine spaces over a field F. Consider the varieties $\underline{Lz}$ and $\underline{Rz}$ as varieties of the same type as $\underline{F}$. Show that the claim of Exercise 4.3F holds for these three varieties. Find the subvarieties of the variety $\underline{Lz} \vee \widetilde{\underline{Rz} \vee \underline{F}}$, and describe the structure of their members.

[Hint: Consider the word $xy\underline{1}x\underline{0}$.]

4.3H. Find all subdirectly irreducible normal bands.

4.3I. Formulate the corollary to Proposition 4.2.7 that explains how Płonka sums are connected to directed colimits.

4.3J. Let (A, Ω) be an algebra with a (derived) normal band operation · such that · breaks each (n-ary) ω in Ω from the left:

$$y \cdot (x_1 \ldots x_n \omega) = y \cdot x_1 \cdot \ldots \cdot x_n,$$

and · *distributes from both sides* over each ω in Ω:

$$(x_1 \ldots x_n \omega) \cdot y \cdot (x_1 \ldots x_n \omega) = (x_1 \cdot y \cdot x_1) \cdot \ldots \cdot (x_n \cdot y \cdot x_n) \omega.$$

  (a) Let (I, Ω) be the semilattice replica of (A, Ω). Show that (A, Ω) is a Płonka sum of subalgebras (A_i, Ω) for $i \in I$, each satisfying the identity $x \cdot y \cdot x = x$.

  (b) Show that if the operation · commutes with the Ω-operations, then each (A_i, Ω) is the direct product of two algebras, one satisfying $x \cdot y = x$ and the other $x \cdot y = y$.

4.3K. [Płonka 1967d, Padmanabhan 1971, Gierz, Romanowska 1991] A *dissemilattice* is an algebra $(A, +, \cdot)$ with two binary semilattice operations $+$ and · satisfying the distributive law $x(y + z) = xy + xz$. It is *distributive* if $x + yz = (x + y)(x + z)$.

  (a) Show that $x + xy$ defines a partition operation in each distributive dissemilattice. Deduce that each such algebra is a Płonka sum of distributive lattices.

(b) Give examples of dissemilattices that are not distributive.

4.3L. Let $\widetilde{\underline{V}}$ be the regularization of an idempotent irregular variety $\underline{V}$ of Ω-algebras. Show that each $\widetilde{\underline{V}}$-algebra is naturally quasi-ordered.

4.3M. Let I be a semilattice. For each element i of I, let $(M_i, +, R)$ be a module over a commutative, unital ring R. Show that the Płonka sum M of the M_i over I is a semimodule $(M, +, 0, R)$ over the (semi)ring R.

4.4. Płonka sums and quasiregularizations

Let $\underline{V}$ be a strongly irregular variety of plural Ω-algebras, and let $\underline{Sl}$ be the variety of Ω-semilattices, as before. As observed in Section 4.3, joining $\underline{V}$ and $\underline{Sl}$ acts as an operator on the set of identities true in $\underline{V}$ that chooses the regular ones, and gives the identities true in the regularization $\widetilde{\underline{V}}$. On the other hand, the regularization $\widetilde{\underline{V}}$ is the smallest variety containing both $\underline{V}$ and $\underline{Sl}$. Now we can look at $\underline{V}$ and $\underline{Sl}$ as quasivarieties, and ask what is the smallest quasivariety containing both $\underline{V}$ and $\underline{Sl}$. We call this quasivariety the *quasiregularization* of $\underline{V}$, and denote it by $Q(\underline{V}, \underline{Sl})$. We will be interested in properties of quasiregularizations similar to those we investigated for regularizations.

Let $\underline{V}$ be an (idempotent) irregular variety of a fixed plural type $\tau : \Omega \to \mathbb{N}$, and let (A, Ω) be a member of the regularization $\widetilde{\underline{V}}$. Assume that (A, Ω) is a Płonka sum of its subalgebras (A_i, Ω) over a semilattice (I, Ω) via a functor F. Let δ be the congruence on (A, Ω) defined by (4.2.2). Proposition 4.2.8 shows that (A^δ, Ω) is the colimit of the functor F. Note that (A^δ, Ω) is in $\underline{V}$. Now in $\operatorname{Cg} A$, one has $\delta \vee \sigma = A \times A$, where $\sigma = \ker \pi$ [Exercise 4.4A]. For a binary word $x \cdot y$ in $X\Omega$, consider the following quasi-identity:

(q) $\quad (x \cdot y = x \ \& \ y \cdot x = y \ \& \ x \cdot z = z \cdot x = z \ \& \ y \cdot z = z \cdot y = z) \to (x = y)$.

Lemma 4.4.1. *Let $\underline{V}$ be a strongly irregular variety satisfying the identity $x \cdot y = x$. Then*

(a) $\underline{V} \models q$;

(b) $\underline{Sl} \models q$;

(c) *For any non-trivial $\underline{V}$-algebra (A, Ω), the Płonka sum (A^∞, Ω) of (A_0, Ω) and (A_1, Ω), where $A_0 = A$ and $A_1 = \{\infty\}$, does not satisfy q.*

Proof. In $\underline{V}$, the premises of q imply that $x = x \cdot z = z \cdot x = z$, and similarly $y = z$, from which one obtains $x = y$. And in $\underline{Sl}$, one has $x = x \cdot y = y \cdot x = y$.

Now let a and b be distinct elements of (A, Ω). Then $x = a$, $y = b$ and $z = \infty$ satisfy the premises of q, but not the conclusion. □

In particular, Lemma 4.4.1 implies that the quasivariety $Q(\underline{V} \cup \underline{Sl})$ generated by $\underline{V} \cup \underline{Sl}$ satisfies the following inclusions:

$$\underline{V}, \underline{Sl} \subsetneq Q(\underline{V} \cup \underline{Sl}) \subsetneq \widetilde{\underline{V}}.$$

Lemma 4.4.2. *Let (A, Ω) be in the regularization $\widetilde{\underline{V}}$ of a strongly irregular variety $\underline{V}$. Assume that (A, Ω) is a Płonka sum of subalgebras (A_i, Ω) over (I, Ω). Then the following are equivalent:*

(a) $(A, \Omega) \in SP(\underline{V} \cup \underline{Sl})$;

(b) *For every $i \leq j$ in $(I, \leq)$, the homomorphism $\varphi_{i,j}$ is injective;*

(c) *In CgA, $\delta \wedge \sigma = \widehat{A}$, the identity relation on A;*

(d) $(A, \Omega) \models q$.

Proof. $(a) \Rightarrow (b)$ Without loss of generality, one can assume that (A, Ω) is a subalgebra of a product $(B \times J, \Omega)$, where (B, Ω) is in $\underline{V}$ and (J, Ω) is in $\underline{Sl}$. The product $(B \times J, \Omega)$ is a Płonka sum of $(B \times \{j\}, \Omega)$ over (J, Ω), in which the homomorphisms between the summands are isomorphisms. The sum homomorphisms on (A, Ω) are restrictions of those on $(B \times J, \Omega)$, and are therefore injective.

(b) $\Rightarrow$ (c) Let $(a,b) \in \delta \wedge \sigma$. Then there are i, $j \in I$ with $i \leq j$ such that $a\varphi_{i,j} = b\varphi_{i,j}$. Since $\varphi_{i,j}$ is injective, one obtains $a = b$.

(c) $\Rightarrow$ (a) Since $\delta \wedge \sigma = \widehat{A}$, there is a subdirect embedding of (A, Ω) into $(A, \Omega)^\delta \times (A, \Omega)^\sigma$. Now $(A, \Omega)^\delta$ is in $\underline{V}$, and $(A, \Omega)^\sigma$ is in $\underline{Sl}$. Thus (A, Ω) is in $SP(\underline{V} \cup \underline{Sl})$.

(d) $\Rightarrow$ (b) Let (A, Ω) satisfy q. For $i \leq j$ in $(I, \leq)$ and a, b in A, let $a\varphi_{i,j} = b\varphi_{i,j}$. Substituting a, b, $a\varphi_{i,j}$ for x, y, z in q, one concludes that $a = b$.

(a) $\Rightarrow$ (d) By Lemma 4.4.1, all members of $\underline{V} \cup \underline{Sl}$ satisfy q, and hence also their products and subalgebras satisfy q. $\square$

With the class $\underline{V}_{SI}$ of subdirectly irreducibles in $\underline{V}$, (a) may be replaced by

(a') $\qquad\qquad (A, \Omega) \in P_S(\underline{V}_{SI} \cup \{\underset{\sim}{2}\}).$

Let $\widetilde{\underline{V}}_q$ be the subquasivariety of $\widetilde{\underline{V}}$ defined by the quasi-identity q.

"Quasi-Płonka's" Theorem 4.4.3. *Let $\underline{V}$ be an irregular variety of idempotent Ω-algebras defined by a set Σ of regular identities and an identity of the form $x \cdot y = x$. Then the following classes coincide:*

(a) *the quasiregularization $Q(\underline{V}, \underline{Sl})$ of $\underline{V}$,*

(b) *the class $S_{pi}(\underline{V})$ of Płonka sums of $\underline{V}$-algebras with injective sum homomorphisms,*

(c) *the subquasivariety $\widetilde{\underline{V}}_q$ of $\widetilde{\underline{V}}$ defined by the quasi-identity q,*

(d) *the class $SP(\underline{V} \cup \underline{Sl})$,*

(e) *the class $P_S(\underline{V}_{SI} \cup \{\underset{\sim}{2}\})$.*

Proof. Lemmata 4.4.1 and 4.4.2 imply the inclusions:

$$Q(\underline{V}, \underline{Sl}) \subseteq \widetilde{\underline{V}}_q = SP(\underline{V} \cup \underline{Sl}) \subseteq Q(\underline{V}, \underline{Sl}). \quad \square$$

A counterpart of Theorem 4.3.4 for the lattice $\mathcal{L}_q(\widetilde{\underline{V}})$ of subquasivarieties of $\widetilde{\underline{V}}$ is not known in general. However, one can provide some interesting partial results in the case where $\underline{V}$ is locally finite and minimal as a quasivariety.

Lemma 4.4.4. *If a variety $\underline{V}$ of Ω-algebras has a unique subdirectly irreducible algebra (A, Ω) embeddable into every non-trivial $\underline{V}$-algebra, then $\underline{V}$ is a minimal quasivariety. Conversely, any locally finite variety which is minimal as a quasivariety has such an algebra (A, Ω).*

Proof. Obviously $\underline{V} = V(A)$. If $\underline{Q}$ is a non-trivial subquasivariety of $V(A)$, then the assumption implies that A is in $S(\underline{Q}) = \underline{Q}$. Thus Corollary 3.1.14 yields $V(A) = SP(A) \subseteq Q(A) \subseteq \underline{Q}$, whence $\underline{V} = \underline{Q}$.

If $\underline{V}$ is a locally finite minimal variety, then Corollary 3.5.4 to Magari's Theorem 3.5.2 shows that $\underline{V}$ contains a simple finite plain algebra A, namely a non-trivial algebra of minimal cardinality. (Cf. Section 1.2.) Since $\underline{V}$ is minimal as a quasivariety, one has $\underline{V} = V(A) = SP(A)$. If B is a subdirectly irreducible $\underline{V}$-algebra, then $B \leq A$. Since A is plain, the algebra B must be isomorphic to A. Indeed, for each non-trivial finite $\underline{V}$-algebra B, one has $SP(B) = V(A)$. Hence A is in $SP(B)$. Thus $A \leq B$, since A is subdirectly irreducible. Now each infinite $\underline{V}$-algebra C is a directed colimit of finite subalgebras, as described in Exercise 2.3C. Hence $A \leq C$, also. □

Note that the variety $\underline{\underline{Sl}}$ of Ω-semilattices satisfies the conditions of Lemma 4.4.4. The unique subdirectly irreducible member is the two-element Ω-semilattice $\underset{\sim}{2}$, which embeds into each non-trivial Ω-semilattice. Thus $\underline{\underline{Sl}}$ is a minimal quasivariety.

Theorem 4.4.5. *Let $\underline{V}$ be a locally finite irregular variety of idempotent Ω-algebras that is minimal as a quasivariety. Then the lattice $\mathcal{L}_q(\underline{\widetilde{V}})$ of subquasivarieties of the regularization $\underline{\widetilde{V}}$ consists of the following five members:*

$$\begin{array}{ccc} \underline{V} & \longrightarrow \underline{\widetilde{V}}_q \longrightarrow \underline{\widetilde{V}} \\ \uparrow & \uparrow \\ \underline{\underline{Tr}} & \longrightarrow \underline{\underline{Sl}} \end{array}$$

4. CONSTRUCTING ALGEBRAS

Proof. By Lemma 4.4.4, $\underline{V}$ has a unique finite subdirectly irreducible algebra A embeddable into each non-trivial $\underline{V}$-algebra, and $\underline{V} = SP(A)$. Let $\underline{P}$ be a non-trivial subquasivariety of $\underline{\widetilde{V}}_q$ distinct from both $\underline{V}$ and $\underline{Sl}$. Since both $\underline{V}$ and $\underline{Sl}$ are minimal, there is an algebra B that is in $\underline{P}$ but not in $\underline{V} \cup \underline{Sl}$. The algebra B is a non-trivial Płonka sum of $\underline{V}$-algebras B_i over a semilattice I. In particular, there is an element u of I such that B_u is a non-trivial $\underline{V}$-algebra, and therefore A embeds into B_u, and hence into B. It follows that

(4.4.1) $$\underline{V} = SP(A) \subseteq \underline{P}.$$

Now for $i < j$ in $(I, \leq)$ and any $b_i \in B_i$, the set $\{b_i, b_i \varphi_{i,j}\}$ forms a subalgebra of B isomorphic to $\underset{\sim}{2}$. This implies that

(4.4.2) $$\underline{Sl} = SP(\underset{\sim}{2}) \subseteq \underline{P}.$$

Combining (4.4.1) and (4.4.2), one obtains

$$\underline{\widetilde{V}}_q = Q(\underline{V}, \underline{Sl}) \subseteq \underline{P} \subseteq \underline{\widetilde{V}}_q.$$

Now let $\underline{P}$ be a subquasivariety of $\underline{\widetilde{V}}$ not contained in $\underline{\widetilde{V}}_q$. We will show that $\underline{P} = \underline{\widetilde{V}}$. Since by Theorem 4.3.7, $\underline{\widetilde{V}}_{SI} = \{A, A^\infty, \underset{\sim}{2}\}$, and both A and $\underset{\sim}{2}$ embed into A^∞, it follows that $\underline{\widetilde{V}} = Q(A^\infty)$. So it suffices to show that A^∞ is in $\underline{P}$.

Let B be in $\underline{P}$ but not in $\underline{\widetilde{V}}_q$. Then B is a Płonka sum of subalgebras B_i over a semilattice I. By Lemma 4.4.2, there are elements $i \leq j$ in $(I, \leq)$ such that the sum homomorphism $\varphi_{i,j}$ is not injective. Let a, b be in B_i with $c := a\varphi_{i,j} = b\varphi_{i,j}$. Then $B'_i := \varphi_{i,j}^{-1}(c)$ is a non-trivial subalgebra of B_i containing both a and b. Since $\underline{V}$ is a minimal quasivariety, $A \leq B'_i$. Now $B'_i \cup \{c\}$ is a subalgebra of B containing a copy of A^∞. Thus A^∞ is in $\underline{P}$, as desired. □

In more general situations, only the following is known. Let $\underline{M}$ be a variety of idempotent Ω-algebras defined by a single identity of the form $x \cdot y = x$. By Theorem 4.3.4, there is a lattice isomorphism $r : \mathcal{L}(\underline{M}) \to \mathcal{L}_r(\widetilde{\underline{M}})$; $\underline{V} \mapsto \widetilde{\underline{V}}$. Let $\mathcal{L}_{rq}(\widetilde{\underline{M}})$ be the class consisting of $\widetilde{\underline{V}}_q$ for $\underline{V}$ in $\mathcal{L}(\underline{M})$. One has the following relationship.

Theorem 4.4.6. *The poset $\mathcal{L}_{rq}(\widetilde{\underline{M}})$ is a lattice, and the composite mapping $\mathcal{L}(\underline{M}) \to \mathcal{L}_{rq}(\widetilde{\underline{M}}) \to \mathcal{L}_r(\widetilde{\underline{M}})$; $\underline{V} \mapsto \widetilde{\underline{V}}_q \mapsto \widetilde{\underline{V}}$ is an isomorphism.*

Proof. The mapping $\mathcal{L}(\underline{M}) \to \mathcal{L}_{rq}(\widetilde{\underline{M}})$ is certainly surjective and order preserving. For injectivity, it suffices to prove that if $\underline{V} \le \underline{M}$, then $\underline{V} = \underline{M} \cap \widetilde{\underline{V}}_q$. Since $\underline{V}$ is defined relatively to $\widetilde{\underline{V}}$ by the identity $x \cdot y = x$, one has

$$\underline{V} = \underline{M} \cap \widetilde{\underline{V}} \supseteq \underline{M} \cap \widetilde{\underline{V}}_q \supseteq \underline{V}. \quad \square$$

Note that Theorem 4.4.6 says nothing about the subquasivarieties of $\underline{M}$ that fail to be varieties.

It is easy to see that the varieties $\underline{Lz}$ and $\underline{Rz}$ (considered as algebras of an arbitrary but fixed plural type) satisfy the assumptions of Theorem 4.4.5. Hence Theorem 4.4.5 provides a description of the lattice of subquasivarieties of $\widetilde{\underline{Lz}}$ and $\widetilde{\underline{Rz}}$. A more interesting example is given below.

Example 4.4.7. Consider any variety $\underline{F}$ of affine F-spaces over a field F [Section 1.6.4]. Such a variety is idempotent and irregular, taking $x \cdot y := xy\underline{0}$. Moreover it is a minimal variety. The affine space $(F, P, \underline{F})$ is the unique subdirectly irreducible member of $\underline{F}$, and it embeds into each non-trivial affine space. By Lemma 4.4.4, $\underline{F}$ is a minimal quasivariety. The proof of Theorem 4.4.5 goes through for $\underline{F}$, even if it fails to be locally finite. Therefore, the lattice $\mathcal{L}_q(\widetilde{\underline{F}})$ is determined by Theorem 4.4.5. $\square$

The regularization of an irregular variety $\underline{V}$ was originally defined as the variety satisfying the regular identities holding in $\underline{V}$. So it is natural to ask if a similar type of definition would be possible for the quasiregularization $\widetilde{\underline{V}}_q$ of $\underline{V}$. To answer this question, one needs to determine the quasi-identities that hold in the (quasi)variety $\underline{\underline{Sl}}$.

Let

$$(q') \qquad (e_{10} = e_{11} \ \& \ldots \& \ e_{n0} = e_{n1}) \rightarrow (e_{00} = e_{01})$$

be a quasi-identity of a plural type $\tau : \Omega \rightarrow \mathbb{N}$. For $i = 0, 1, \ldots, n$ and $j = 0, 1$, let $\arg(e_{ij})$ be the set of variables in e_{ij}, and let $\arg(q')$ be the set of variables in q'. For $X \subseteq \arg(q')$, define the *span* $\mathrm{sp}(X)$ *of* X to be the smallest subset A of $\arg(q')$ such that for all $i = 1, \ldots, n$,

$$(4.4.3) \qquad \begin{cases} X \subseteq A, \\ \arg(e_{ij}) \subseteq A \Leftrightarrow \arg(e_{i1-j}) \subseteq A. \end{cases}$$

Note that the set $A = \arg(q')$ always satisfies the two conditions above, and that the intersection of any family of sets satisfying these conditions also satisfies them. In Exercise 4.4C, the reader is asked to show that the span defines a closure operator on the set $\arg(q')$.

Define the *span* $\mathrm{sp}(e_{0k})$ of e_{0k} (for $k = 0, 1$) to be $\mathrm{sp}(\arg(e_{0k}))$. We call the quasi-identity q' *regular* if $\mathrm{sp}(e_{00}) = \mathrm{sp}(e_{01})$. For example, the quasi-identity of groupoid type

$$(x \cdot y = x \ \& \ y \cdot x = y \ \& \ x \cdot z = z \cdot x = z \ \& \ y \cdot z = z \cdot y = z) \rightarrow (x = y)$$

is regular since $\mathrm{sp}(x) = \{x, y\} = \mathrm{sp}(y)$, whereas the quasi-identity

$$(x \cdot x' = y' \ \& \ x' \cdot y' = t \ \& \ x' = y' \cdot z \ \& \ x = y \cdot z) \rightarrow (x'y' \cdot xy = x \cdot z)$$

is irregular since $\mathrm{sp}(x'y' \cdot xy) = \{x', y', x, y, z, t\} \neq \{x, y, z\} = \mathrm{sp}(x \cdot z)$.

Lemma 4.4.8. *If a quasi-identity q' is regular, and the conclusion of q' is irregular, then q' is satisfied in the two-element Ω-semilattice $\underset{\sim}{2}$.*

Proof. Suppose q' is not satisfied in $\underset{\sim}{2}$. This means that there is a homomorphism $h : \{x,y\}\mathrm{Sl} \to \underset{\sim}{2}$ from the free semilattice over $\{x,y\}$ onto the semilattice $\underset{\sim}{2}$ such that for each $i = 1, \ldots, n$, $e_{ij}h = e_{e1-j}h$ but $e_{00}h \neq e_{01}h$. Without loss of generality, assume that $e_{00}h = 0$ and $e_{01}h = 1$. A general form of the conclusion of q' is

$$e_{00} = x'_1 \ldots x'_k x_{k+1} \ldots x_l y'_1 \ldots y'_m y_{m+1} \ldots y_n$$
$$= x_{k+1} \ldots x_l z_1 \ldots z_p = e_{01},$$

where for $i = 1, \ldots, k$ and $j = 1, \ldots, m$ one has $x'_i h = 0 = y'_j h$, and for $i = k+1, \ldots l$ and $j = m+1, \ldots, n$ and $r = 1, \ldots, p$ one has $x_i h = y_j h = z_r h = 1$. Let $X' := \{x'_1, \ldots, x'_k\}$, $X := \{x_{k+1}, \ldots, x_l\}$, $Y' := \{y', \ldots, y'_m\}$, $Y := \{y_{m+1}, \ldots, y_n\}$ and $Z := \{z_1, \ldots, z_p\}$.

First note that no primed variable is contained in $\arg(e_{01})$. Now if $\arg(e_{ij}) \subseteq \arg(e_{01})$ for some $i = 1, \ldots n$ and $j = 0, 1$, then

$$\arg(e_{i1-j}) \subseteq \arg(e_{01}) \cup B$$

for some $B \subseteq Y$. It follows that

(4.4.4) $\qquad \arg(e_{01}) \subseteq \mathrm{sp}(e_{01}) \subseteq \arg(e_{01}) \cup Y.$

Obviously

(4.4.5) $\qquad X' \cup Y' \not\subseteq \mathrm{sp}(e_{01}).$

On the other hand, $\arg(e_{ij})$ may contain variables from $X' \cup Y'$. This happens in the case $\arg(e_{ij}) \subseteq \arg(e_{00})$ or in the case where $\arg(e_{ij})$ contains variables both from $X' \cup Y'$ and $X \cup Y \cup Z$. It follows that $\arg(e_{i1-j})$ must also contain at least one variable from $X' \cup Y'$, and maybe some from $X \cup Y \cup Z$. Consequently,

(4.4.6) $\qquad \arg(e_{00}) \subseteq \operatorname{sp}(e_{00}) \subseteq \arg(q').$

Now (4.4.4), (4.4.5) and (4.4.6) show that $\operatorname{sp}(e_{00}) \neq \operatorname{sp}(e_{01})$, providing a contradiction. Hence the semilattice $\underset{\sim}{2}$ satisfies q'. □

Note that if q' is regular and with an irregular conclusion, then at least one of the premises must be irregular, too. In this case, the possible regular premises have no influence on the span of e_{00} and e_{01}.

Lemma 4.4.9. *Let all premises and the conclusion of the quasi-identity q' be irregular. If q' is satisfied in the Ω-semilattice $\underset{\sim}{2}$, then q' is regular.*

Proof. Suppose q' is irregular. It will be shown that the semilattice $\underset{\sim}{2}$ does not satisfy q'. One may assume that the conclusion of q' has the form

$$e_{00} = x_1 \ldots x_k y_1 \ldots y_m = x_1 \ldots x_k z_1 \ldots z_n = e_{01}.$$

Denote $X := \{x_1, \ldots, x_k\}$, $Y := \{y_1, \ldots, y_m\}$, $Z := \{z_1, \ldots, z_n\}$. Any one of the sets X, Y, Z may be empty. Obviously $\operatorname{sp}(e_{00}) = \operatorname{sp}(X) \cup Y \cup A$ and $\operatorname{sp}(e_{01}) = \operatorname{sp}(X) \cup Z \cup B$, where A is built up from those premises of q' that have some elements of Y (and maybe also some of X and Z) on one side, and B is built up from those premises of q' that have some elements of Z (and maybe also some of X and Y) on one side. Since $\operatorname{sp}(e_{00}) \neq \operatorname{sp}(e_{01})$, one has $Y \cup A \neq Z \cup B$. Without loss of generality one may assume that $C := \operatorname{sp}(e_{00}) - \operatorname{sp}(e_{01})$ is non-empty, and that there are elements $c_1, \ldots, c_r \in C$, $d_1, \ldots, d_s \in \operatorname{sp}(e_{00})$ and possibly $a_1, \ldots, a_p \in \operatorname{sp}(e_{00}) - C$ such that for some i, j one has $\arg(e_{ij}) = \{a_1, \ldots, a_p, c_1, \ldots, c_r\}$ and $\arg(e_{i1-j}) = \{d_1, \ldots, d_s\}$. Then $\arg(e_{i1-j}) \cap C \neq \emptyset$. Indeed if not, then $\arg(e_{i1-j}) \subseteq \operatorname{sp}(e_{00}) - C \subseteq \operatorname{sp}(e_{01})$, implying also that $\arg(e_{ij}) \subseteq \operatorname{sp}(e_{01})$, which gives a contradiction. It follows that each of the premises containing elements of C on one side contains elements of C on the other.

Now it is easy to see that the quasi-identity q' is not satisfied in the semilattice $\underset{\sim}{2}$. Just substitute 0 for all elements of C and 1 for all other arguments. □

Theorem 4.4.10. *A quasi-identity q' is satisfied in the variety $\underline{\underline{Sl}}$ of Ω-semilattices if and only if it is regular.*

Proof. This follows by Lemmata 4.4.8 and 4.4.9, since if the conclusion of q' is regular, then q' is regular, and if all the premises of q' are regular, but the conclusion is irregular, then q' is irregular. □

Note that since the variety $\underline{\underline{Sl}}$ of Ω-semilattices is a minimal quasivariety, it follows that $\underline{\underline{Sl}}$ satisfies precisely the regular quasi-identities of type $\tau : \Omega \to \mathbb{N}$.

Corollary 4.4.11. *Let $\underline{\underline{Q}}$ be a quasivariety of Ω-algebras. Let $\widetilde{\underline{\underline{Q}}}$ be its quasiregularization, i.e. the smallest quasivariety containing both $\underline{\underline{Q}}$ and $\underline{\underline{Sl}}$. Then $\widetilde{\underline{\underline{Q}}}$ satisfies precisely the regular quasi-identities holding in the quasivariety $\underline{\underline{Q}}$.* □

Exercises

4.4A. Let (A, Ω) be a Płonka sum of irregular $\underline{\underline{V}}$-algebras (A_i, Ω) over an Ω-semilattice (I, Ω) via a functor F. Show that the colimit $AF \cong (A^\delta, \Omega)$ is a $\underline{\underline{V}}$-algebra. Show that in $\mathrm{Cg}A$, one has $\delta \vee \sigma = A \times A$.

4.4B. Consider the varieties $\underline{\underline{Lz}}$, $\underline{\underline{Rz}}$ and $\underline{\underline{Sl}}$ as varieties of Ω-algebras.

(a) Describe the quasiregularization $\widetilde{\underline{\underline{Lz}}}_q$ of $\underline{\underline{Lz}}$.

(b) Describe all quasi-identities true in $\underline{\underline{Lz}}$.

(c) Describe all regular quasi-identities true in $\underline{\underline{Lz}}$.

(d) Do the same for the variety $\underline{\underline{Rz}}$.

4.4C. Prove that the span defines a closure operator on the set $\arg(q')$. (Cf. Section 3.2.)

4. CONSTRUCTING ALGEBRAS

4.4D. Find all closed subsets of the set of variables of the quasi-identity of groupoid type

$$(x \cdot y = x \,\&\, y \cdot v = u) \to (t = x \cdot v).$$

Exhibit a subset that satisfies the two conditions of (4.4.3), but is not closed.

4.5. Non-functorial sums

In this section we describe the most general construction method for algebras having a homomorphism onto an idempotent, naturally quasi-ordered algebra. As before, a new algebra will be constructed as a "sum" of given algebras over an (idempotent, naturally quasi-ordered) algebra. Such "sums" have extremely broad applicability. However, being so general, they lose some of the elegance and specificity of functorial sums. We have to pay a price for generality! The conditions on a functorial sum, although natural and elegant, are very strong, and do not obtain in general. It is not true that each algebra (A, Ω) projecting onto an idempotent, naturally quasi-ordered algebra (I, Ω) is a functorial sum over that algebra.

Example 4.5.1. Consider the closed unit interval $I = [0, 1]$ of $\mathbb{R}$ as an algebra $(I, \underline{I}^\circ)$. (Cf. Exercise 1.1E, 1.6.2 and Example 1.4.3.) Recall that for each $p \in I^\circ$, one has a binary operation

$$\underline{p} : I \times I \to I; \ (a, b) \mapsto ab\underline{p} := a(1 - p) + bp.$$

The semilattice replica $(S, \underline{I}^\circ)$ of $(I, \underline{I}^\circ)$ has three elements, say 0, 1, s with $0, 1 \leq s$. The corresponding fibres are $I_0 = \{0\}$, $I_1 = \{1\}$ and $I_s = I^\circ$. The algebra $(I, \underline{I}^\circ)$ cannot be reconstructed as a Płonka sum of I_0, I_1 and I_s. Indeed, if it could be, then for $0\varphi_{0,s} = x \in I^\circ$ and y, p in I°, we would have $y p = 0 y \underline{p} = (0 \varphi_{0,s}) y \underline{p} = x y \underline{p} = x(1 - p) + yp$, a contradiction. □

Definition 4.5.2. Let (I, Ω) be an algebra with algebraic quasi-ordering $\preceq$. For each i in I, let an algebra (A_i, Ω) be given together with a superalgebra (E_i, Ω). For $i \preceq j$ in $(I, \preceq)$, let $\varphi_{i,j} : (A_i, \Omega) \to (E_j; \Omega)$ be an Ω-homomorphism such that:

(a) for each (n-ary) ω in Ω and for $i_1, \ldots, i_n$ in I with $i_1 \ldots i_n \omega = i$,

$$(A_{i_1} \varphi_{i_1,i}) \ldots (A_{i_n} \varphi_{i_n,i}) \omega \subseteq A_i \ ;$$

(b) for each $i_1 \ldots i_n \omega = i \preceq j$ in $(I, \preceq)$,

$$a_{i_1} \varphi_{i_1,i} \ldots a_{i_n} \varphi_{i_n,i} \omega \varphi_{i,j} = a_{i_1} \varphi_{i_1,j} \ldots a_{i_n} \varphi_{i_n,j} \omega,$$

where for $k = 1, \ldots, n$, one has $a_{i_k} \in A_{i_k}$.

Then the disjoint union $A = \bigcup (A_i \mid i \in I)$, equipped with the operations

(4.5.1) $\qquad \omega : A_{i_1} \times \cdots \times A_{i_n} \to A_i \ ; \ (a_{i_1}, \ldots, a_{i_n}) \mapsto a_{i_1} \varphi_{i_1,i} \ldots a_{i_n} \varphi_{i_n,i} \omega$

for each ω in Ω, all $(i_1, \ldots, i_n) \in I^n$, and $i = i_1 \ldots i_n \omega$, is called the *sum of the algebras* (A_i, Ω) *over the algebra* (I, Ω) *by the mappings* $\varphi_{i,j}$, or more briefly a *sum* of the (A_i, Ω). □

Note that the left hand side of the equality in (b) is defined, since condition (a) holds. The condition (b) generalizes the functoriality in functorial sums. The algebra (A, Ω) has a *projection* onto (I, Ω), the Ω-homomorphism

$$\pi : (A, \Omega) \to (I, \Omega) \ ; \ a_i \mapsto i$$

for all $a_i \in A_i$. The algebras (A_i, Ω) are called the *sum fibres*, and the homomorphisms $\varphi_{i,j}$ the *sum homomorphisms*.

In what follows, we will be interested in sums of algebras over idempotent naturally quasi-ordered algebras satisfying certain additional conditions. Let (B, Ω) be any algebra, and (P, Ω) a subalgebra of (B, Ω). A congruence θ on

4. CONSTRUCTING ALGEBRAS

(B, Ω) *preserves* the subalgebra (P, Ω) if the restriction of the natural projection $B \to B^\theta$; $b \mapsto b^\theta$ to the subalgebra (P, Ω) injects. The algebra (B, Ω) is said to be an *envelope* of a subalgebra (P, Ω) if equality is the only congruence on (B, Ω) preserving (P, Ω).

A sum (A, Ω) of algebras (A_i, Ω) over an idempotent naturally quasi-ordered algebra (I, Ω) is called a *generalized coherent Lallement sum* if for each $i \in I$, the extension (E_i, Ω) of (A_i, Ω) is an envelope, and moreover

(c) $a_i \varphi_{i,i} = a_i$,

(d) $E_i = \{ a_j \varphi_{j,i} \mid j \preceq i \}$.

Throughout this book, we will simply speak of "Lallement sums" instead of "generalized coherent Lallement sums." However, readers should be careful not to confuse this with the usage of [MT]. (See the notes at the end of the chapter.) Note that the fibres of a Lallement sum are subalgebras of the sum. The significance of Lallement sums resides in the following theorem.

The Sum Theorem 4.5.3. *Let (A, Ω) be an algebra having a homomorphism onto an idempotent, naturally quasi-ordered algebra (I, Ω), with corresponding fibres (A_i, Ω) for $i \in I$. Then (A, Ω) is a Lallement sum of (A_i, Ω) over (I, Ω).*

Proof. First note that for a proper congruence θ of (A, Ω) preserving a subalgebra (S, Ω) of (A, Ω), the quotient (A^θ, Ω) is an envelope of (S, Ω) if and only if θ is a maximal congruence of (A, Ω) preserving (S, Ω). Now let $P_j := \bigcup \{ A_i \mid i \preceq j \}$. Since (I, Ω) is naturally quasi-ordered and idempotent, Proposition 4.1.7 shows that P_j is a subalgebra of (A, Ω). Obviously A_j is a subalgebra of (P_j, Ω). Let Θ be the set of congruences on (P_j, Ω) preserving (A_j, Ω). The set Θ is nonempty, since it contains the equality relation. Each subchain $\{ \theta_k \mid k \in K \}$ of Θ is bounded above by the congruence $\bigcup (\theta_k \mid k \in K)$. By the Kuratowski-Zorn Lemma, Θ has a maximal element $\mu = \mu(j)$. By the observation at the beginning of the proof, $(E_j, \Omega) := (P_j^\mu, \Omega)$ is an envelope of (A_j, Ω). (Here and later we identify the μ-class a_j^μ, for $a_j \in A_j$, with the element a_j.) For $i \preceq j$ in $(I, \preceq)$, the mapping $\varphi_{i,j}$ is defined by

$$\varphi_{i,j} : A_i \to E_j \; ; \; a_i \mapsto a_i^{\mu(j)}.$$

Evidently, each mapping $\varphi_{i,i}$ is an inclusion. For each (n-ary) ω in Ω, $i_1 \ldots i_n \omega = i$ in I, and a_k in A_{i_k} for $k = 1, \ldots, n$, one has

$$a_1 \varphi_{i_1, i} \ldots a_n \varphi_{i_n, i} \omega = a_1^{\mu(i)} \ldots a_n^{\mu(i)} \omega$$
$$= (a_1 \ldots a_n \omega)^{\mu(i)} = a_1 \ldots a_n \omega \in A_i.$$

It follows that the conditions (a) and (c) for Lallement sums are satisfied. Now for each $i_1 \ldots i_n \omega = i \preceq j$, one has

$$a_1 \varphi_{i_1, i} \ldots a_n \varphi_{i_n, i} \omega \; \varphi_{i,j} = a_1^{\mu(i)} \ldots a_n^{\mu(i)} \omega \; \varphi_{i,j}$$
$$= (a_1 \ldots a_n \omega)^{\mu(i)} \; \varphi_{i,j} = a_1 \ldots a_n \omega \; \varphi_{i,j}$$
$$= (a_1 \ldots a_n \omega)^{\mu(j)} = a_1^{\mu(j)} \ldots a_n^{\mu(j)} \omega$$
$$= a_1 \varphi_{i_1, j} \ldots a_n \varphi_{i_n, j} \omega,$$

whence the condition (b) holds as well. The conditions (d) and (4.5.1) are obviously satisfied. In particular, (b) implies that all the $\varphi_{i,j}$ are homomorphisms. □

The representation of (A, Ω) as a Lallement sum obtained in the proof of Theorem 4.5.3 is called *canonical*, and an algebra (A, Ω) represented in this way is called a *canonical Lallement sum*. A sum (A, Ω) of algebras (A_i, Ω) for $i \in I$ is called *strict* if the superalgebras (E_i, Ω) coincide with (A_i, Ω). It is called *proper* if it is isomorphic neither to (I, Ω) nor to any of the (A_i, Ω). It is called a *semilattice sum* if the indexing algebra (I, Ω) is a semilattice.

Corollary 4.5.4. *Let (A, Ω) be an algebra having a homomorphism onto an Ω-semilattice (I, Ω) with corresponding fibres (A_i, Ω) for $i \in I$. Then (A, Ω) is a semilattice Lallement sum of (A_i, Ω) over (I, Ω).* □

4. CONSTRUCTING ALGEBRAS

If the algebraic quasi-order of (I, Ω) in Theorem 4.5.3 is not full, then by Proposition 4.1.7, (I, Ω) itself is a (non-trivial) semilattice Lallement sum of i^α over the Ω-semilattice (I^α, Ω). The mapping π nat α is an Ω-homomorphism onto (I^α, Ω). Each preimage algebra $B_{i^\alpha} = (\pi \text{ nat } \alpha)^{-1}(i^\alpha)$ is a subalgebra of (A, Ω), and is itself a Lallement sum of (A_j, Ω) for $j \in i^\alpha$.

Corollary 4.5.5. Let (A, Ω) be a Lallement sum of (A_i, Ω) over (I, Ω). Then (A, Ω) is a semilattice Lallement sum of the preimages (B_i^α, Ω) of i^α over the Ω-semilattice (I^α, Ω). Each fibre (B_i^α, Ω) is a Lallement sum over (i^α, Ω), and each (i^α, Ω) has a full algebraic quasi-order. □

Note that each functorial sum is a strict Lallement sum. The functoriality of the $\varphi_{i,j}$ implies the conditions (a) and (b). In particular, each Płonka sum is a strict semilattice sum. As we have seen in Section 4.3, Płonka sums may be used to represent algebras in certain varieties. The "natural" classes of algebras representable by Lallement sums are certain prevarieties. Note, however, that unlike the representation of algebras in regularized varieties as Płonka sums of irregular algebras, the representation of algebras in such prevarieties as Lallement sums is not necessarily unique. In addition to the representation given by Theorem 4.5.3, an algebra may have other decompositions as a Lallement sum. Let $\underline{Q}$ be a quasivariety of Ω-algebras of type $\tau : \Omega \to \mathbb{N}$. Let $\underline{R}$ be a quasivariety of Ω'-algebras of type $\tau' : \Omega' \to \mathbb{N}$, such that each basic operation ω of Ω may be interpreted as some derived operation of $\underline{R}$. Let $\underline{R}^\tau$ be the quasivariety of Ω-subreducts of type τ of $\underline{R}$-algebras. By Theorem 3.7.18, the class $\underline{Q} \circ \underline{R}^\tau$ is a quasivariety.

Corollary 4.5.6. Let $\underline{Q}$ and $\underline{R}$ be quasivarieties as above. If all $\underline{R}^\tau$-algebras are idempotent and naturally quasi-ordered, then each Ω-algebra in the quasivariety $\underline{Q} \circ \underline{R}^\tau$ is a Lallement sum of $\underline{Q}$-algebras over an $\underline{R}^\tau$-algebra. □

Example 4.5.7. The relation γ of Exercise 4.1H is the semilattice replica congruence on a normal band $(A, \cdot)$. All γ-classes are rectangular bands. Corollary 4.5.4 shows that each band is a semilattice Lallement sum of rectangular bands. However, it requires more refined methods to describe the detailed structure of bands. □

Example 4.5.8. Let $\underline{V}$ and $\underline{W}$ be idempotent varieties of Ω-algebras. Suppose that the $\underline{W}$-algebras are naturally quasi-ordered. Suppose additionally that there are Ω-words xyv and xyw such that $\underline{W}$ satisfies the identity $xyv = xyw$, and $\underline{V}$ satisfies the identity $xyv = x = yxw$. Then the relation σ defined on any $\underline{V} \vee \underline{W}$-algebra by

$$(a,b) \in \sigma :\Leftrightarrow abv = baw = a \text{ and } bav = abw = b$$

is the $\underline{W}$-replica congruence. The σ-classes are in the variety $\underline{V}$ [Exercise 4.6D]. By Theorem 4.5.3, each $\underline{V} \vee \underline{W}$-algebra is a Lallement sum of $\underline{V}$-algebras over a $\underline{W}$-algebra. □

Similarly as in the case of Płonka sums, Lallement sums may be also presented as certain subdirect products.

Lemma 4.5.9. *Let (A, Ω) be a proper canonical Lallement sum of algebras (A_i, Ω) over an algebra (I, Ω) with a full algebraic quasi-order. Then (A, Ω) is a subdirect product of the envelopes (E_i, Ω) for $i \in I$, i.e.*

$$(4.5.2) \qquad (A, \Omega) \leq_s \prod_{i \in I}(E_i, \Omega).$$

Proof. Let μ_i be a maximal congruence on (A, Ω) preserving (A_i, Ω), as in the proof of Theorem 4.5.3. Set $\mu := \bigcap_{i \in I} \mu_i$. It is clear that for no $i \in I$ and $a_i \neq b_i$ in A_i is (a_i, b_i) in μ. It follows that $\mu = \widehat{A}$. By Lemma 3.1.2, (A, Ω) is a subdirect product of all the $(A, \Omega)^{\mu_i}$. As $(A, \Omega)^{\mu_i} = (E_i, \Omega)$, one readily obtains (4.5.2). □

Theorem 4.5.10. *Let (A, Ω) be a proper canonical Lallement sum of (A_i, Ω) over an idempotent naturally quasi-ordered algebra (I, Ω). Then (A, Ω) is a subdirect product of the algebras (C_i, Ω) for $i \in I$, where $C_i = E_i$ for $i \in j^\alpha$ in the case where $\{j^\alpha\}$ is a sink of (I^α, Ω), and otherwise $C_i = E_i \cup \{\infty_i\}$, where $\{\infty_i\}$ is a sink in (C_i, Ω). Thus*

$$(A, \Omega) \leq_s \prod_{i \in I}(C_i, \Omega).$$

Proof. Let $P_j := \bigcup \{A_i \mid i \preceq j\}$, and let μ'_j be a maximal congruence on (P_j, Ω) preserving (A_j, Ω), as in the proof of Theorem 4.5.3. Define the relation $\mu_j \subseteq A \times A$ as follows:

$$(a, b) \in \mu_j :\Leftrightarrow (a, b) \in \mu'_j \text{ or } \{a, b\} \subseteq A - P_j.$$

Note that for all k in j^α, one has $k \preceq j$, whence $B_{j^\alpha} \subseteq P_j$. More generally, for each $k \preceq j$, one has $B_{k^\alpha} \subseteq P_j$. Moreover, B_{j^α} is a sink in (P_j, Ω). Note also that $A - P_j$ is a sink in (A, Ω) [Exercise 4.5E]. Denote the μ_j-class $A - P_j$ by ∞_j. It is easy to check that μ_j is a congruence of (A, Ω), and that $(A, \Omega)^{\mu_j} = (E_j \cup \{\infty_j\}, \Omega)$, with $\{\infty_j\}$ a sink of $(A, \Omega)^{\mu_j}$, if $A - P_j \neq \varnothing$ [Exercise 4.5E]. Now let $\mu := \bigcap_{j \in I} \mu_j$. Similarly as in Lemma 4.5.10, it is clear that for no $i \in I$ and $a_i \neq b_i$ in A_i is (a_i, b_i) in μ. It follows that $\mu = \widehat{A}$. By Lemma 3.1.2, (A, Ω) is a subdirect product of all the $(A, \Omega)^{\mu_j}$. If j^α is the greatest element of $(I^\alpha, \leq)$, and hence a sink, then (similarly as in Lemma 4.5.10) for $i \in j^\alpha$, one has $(A, \Omega)^{\mu_i} = (E_i, \Omega)$. □

Corollary 4.5.11. *Let (A, Ω) be a proper canonical semilattice Lallement sum of (A_i, Ω) over the Ω-semilattice (I, Ω). Then (A, Ω) is a subdirect product of the algebras (C_i, Ω) (for $i \in I$), where $C_i = E_i$ if $\{i\}$ is a sink of (I, Ω), and otherwise $C_i = E_i \cup \{\infty_i\}$, where $\{\infty_i\}$ is a sink in (C_i, Ω).* □

As a corollary, one obtains a description of Płonka sums as subdirect products. (See Theorem 4.3.5.)

Similarly as in the case of Płonka sums, Theorem 4.5.10 yields a characterization of subdirectly irreducible Lallement sums.

First note the following. Let (A, Ω) be a subdirectly irreducible algebra, and let a and b be elements of A collapsed by the monolithic congruence m. If the m-class a^m is a subalgebra of (A, Ω), then A is an envelope of a^m.

Lemma 4.5.12. *Let $\underline{Q}$ be a prevariety, and let $\underline{R}$ be an idempotent prevariety of the same type as $\underline{Q}$. Then a $\underline{Q} \circ \underline{R}$-algebra (A, Ω) is subdirectly irreducible if and only if it is*

(a) *a subdirectly irreducible $\underline{R}$-algebra, or*

(b) *a subdirectly irreducible $\underline{Q} \circ \underline{R}$-envelope of a $\underline{Q}$-algebra.*

Proof. If (A, Ω) is a subdirectly irreducible $\underline{Q} \circ \underline{R}$-algebra, then there is a congruence θ on (A, Ω) with A^θ in $\underline{R}$ and all θ-classes a^θ in $\underline{Q}$. If θ is the equality relation, then (A, Ω) is in $\underline{R}$. Otherwise, for a and b collapsed by the monolithic congruence m, (A, Ω) is an envelope of the $\underline{Q}$-subalgebra a^θ. $\square$

Note that the extension (B, Ω) of (A, Ω) by zero is always an envelope of (A, Ω). Let us call an envelope (E, Ω) of (A, Ω) *adequate* if it does not coincide with (A, Ω), and is not the extension of (A, Ω) by zero. If the $\underline{Q} \circ \underline{R}$-algebras are Lallement sums of $\underline{Q}$-algebras over $\underline{R}$-algebras, then the subdirectly irreducible members of $\underline{Q} \circ \underline{R}$ admit a more detailed description as follows.

Theorem 4.5.13. *Let $\underline{Q}$ be a prevariety of Ω-algebras. Let $\underline{R}$ be an idempotent prevariety, of the same type as $\underline{Q}$, consisting of naturally quasi-ordered algebras. Then a $\underline{Q} \circ \underline{R}$-algebra (A, Ω) is subdirectly irreducible if and only if it is of one of the following types:*

(a) *a subdirectly irreducible $\underline{R}$-algebra or $\underline{Q}$-algebra;*

(b) *a subdirectly irreducible adequate $\underline{Q} \circ \underline{R}$-envelope of a $\underline{Q}$-algebra;*

(c) *(if $\underline{Q} \circ \underline{R}$ is closed under Płonka sums) a subdirectly irreducible $\underline{Q} \circ \underline{R}$-envelope of a $\underline{Q}$-algebra (A_0, Ω) extended by zero.*

4. CONSTRUCTING ALGEBRAS

Proof. It suffices to prove the "only if" part. Let (A, Ω) be subdirectly irreducible. Then (A, Ω) is a Lallement sum of $\underline{Q}$-algebras (A_i, Ω) over an $\underline{R}$-algebra (I, Ω). Let (A, Ω) be a proper sum. By Theorem 4.5.10, the algebra (A, Ω) is a subdirect product of all the (C_i, Ω). Since (A, Ω) is subdirectly irreducible, it is isomorphic to one of the (C_i, Ω).

We will show that the semilattice (I^α, Ω) has a greatest element. Let $R_j := A - P_j$, where $P_j := \bigcup \{A_i \mid i \npreceq j\}$. Note that for each k with $j \not\preceq k$, one has $B_{k^\alpha} \subseteq R_j$, and R_j is a sink in (A, Ω). For each $j^\alpha \in I^\alpha$, define the relation $\theta_{j^\alpha} \subseteq A \times A$ as follows:

$$(a, b) \in \theta_{j^\alpha} :\Leftrightarrow a = b \text{ or } \{a, b\} \subseteq R_j.$$

It is easy to check that θ_{j^α} is a congruence on (A, Ω) [Exercise 4.5F]. Now if for each j^α in I^α there is a k^α in I^α with $j^\alpha < k^\alpha$ and $j^\alpha \neq k^\alpha$, then $\theta_{k^\alpha} < \theta_{j^\alpha}$ and $\theta_{k^\alpha} \neq \theta_{j^\alpha}$. Consequently, there is no non-trivial congruence μ such that $\mu \leq \theta_{j^\alpha}$ for all $j \in I$. It follows that (I^α, Ω) must have a greatest element, say 1, and that $B_1 = (\pi \text{ nat } \alpha)^{-1}(1)$ is a sink in (A, Ω).

If $i \in 1$, then (A, Ω) is isomorphic to an envelope (E_i, Ω) of (A_i, Ω). Otherwise, if $i \notin 1$, then (A, Ω) is isomorphic to the Płonka sum of an envelope (E_i, Ω) of (A_i, Ω) with a one-element Ω-algebra. Now the congruence relations of (E_i, Ω) are precisely the congruence relations of (C_i, Ω) restricted to (E_i, Ω). It follows that (C_i, Ω) is subdirectly irreducible precisely if (E_i, Ω) is. □

Theorem 4.5.13 reduces the problem of characterizing subdirectly irreducible algebras in the prevariety $\underline{Q} \circ \underline{R}$ to that of characterizing subdirectly irreducible $\underline{R}$-algebras, $\underline{Q}$-algebras and subdirectly irreducible adequate $\underline{Q} \circ \underline{R}$-envelopes of $\underline{Q}$-algebras. Note however that neither are envelopes (E_i, Ω) of fibres (A_i, Ω) of (A, Ω) necessarily $\underline{Q} \circ \underline{R}$-algebras, nor is the quasivariety $\underline{Q} \circ \underline{R}$ necessarily closed under Płonka sums over the semilattice $\underset{\sim}{2}$. On the other hand, if $\underline{V}$ is a variety containing both $\underline{Q}$ and $\underline{R}$ and contained in $\underline{Q} \circ \underline{R}$, then all the canonical envelopes (E_i, Ω) are $\underline{V}$-algebras. If additionally $\underline{V}$ is regular, then $\underline{V}$ is also closed under Płonka sums. In this case subdirectly irreducible $\underline{V}$-algebras may be of all three types mentioned in Theorem 4.5.13.

Note as well that in the case where a subdirectly irreducible $\underline{Q} \circ \underline{R}$-algebra (A, Ω) is an envelope of some (A_i, Ω), and the algebraic quasi-order of (I, Ω) is full, neither (A_i, Ω) nor (I, Ω) need be subdirectly irreducible. This is shown by the following.

Example 4.5.14. Consider the groupoid $(G, \cdot)$ with the following multiplication table

·	0	2	1	3	a	b	x	y
0	0	0	2	2	1	3	0	2
2	2	2	0	0	3	1	2	0
1	3	3	1	1	0	2	3	1
3	1	1	3	3	2	0	1	3
a	b	b	b	b	a	a	b	b
b	a	a	a	a	b	b	a	a
x	x	x	x	x	y	y	x	x
y	y	y	y	y	x	x	y	y

The groupoid is idempotent and entropic. (This will follow easily by the results of Section 5.6.) In fact, G also satisfies the identities $x(x \cdot xy) = x = xy \cdot y$. The groupoid is subdirectly irreducible, and is an envelope of the subgroupoid $G_0 = \{0, 1, 2, 3\}$ that is the reduct $(\mathbb{Z}_4, \underline{2})$ of the group $\mathbb{Z}_4$ [Exercise 4.5H]. The monolithic congruence has two 2-element classes $\{0, 2\}$ and $\{1, 3\}$. All other classes consist of one element. Moreover, $(G, \cdot)$ is a Lallement sum of the subgroupoids $(G_0, \cdot), (G_1, \cdot)$ and $(G_2, \cdot)$, where $G_1 = \{a, b\}$ and $G_2 = \{x, y\}$, over a 3-element left-zero band $(I, \cdot)$. Note that neither $(G_0, \cdot)$ nor $(I, \cdot)$ is subdirectly irreducible. □

Now consider a subdirectly irreducible $\underline{Q} \circ \underline{R}$-algebra (A, Ω) with a non-trivial semilattice replica congruence α on the indexing algebra (I, Ω). Then the semilattice (I^α, Ω) is non-trivial, and the algebra (A, Ω) has a homomorphism onto the two-element semilattice. In this case one can provide some further information on the structure of (A, Ω).

Lemma 4.5.15. *Let (E, Ω) be an envelope of its non-trivial sink (A, Ω). Then (E, Ω) is subdirectly irreducible if and only if (A, Ω) is subdirectly irreducible.*

Proof. First note that since A is a sink in (E, Ω), the relation $\sigma \subseteq E \times E$ defined by $(a, b) \in \sigma$ iff $\{a, b\} \subseteq A$ or $a = b$ is a congruence on (E, Ω). More generally, each congruence $\theta_A \in \mathrm{Cg}A$ can be extended to a congruence $\theta_E \in \mathrm{Cg}E$ by defining $(a, b) \in \theta_E$ iff $(a, b) \in \theta_A$ or $a = b$ for all $a, b \in E$. Since (E, Ω) is an envelope of (A, Ω), each non-trivial congruence $\theta \in \mathrm{Cg}E$ determines a non-trivial congruence θ_A on (A, Ω) which can be extended to the congruence θ_E of (E, Ω). Obviously $\theta_E \leq \theta$. It follows that m_A is the monolithic congruence of (A, Ω) iff m_E is the monolithic congruence of (E, Ω). □

Proposition 4.5.16. *Let (A, Ω) be a subdirectly irreducible algebra having a homomorphism onto the two-element (join) semilattice $\underset{\sim}{2}$ with $0 < 1$. Then (A, Ω) is the Płonka sum of a subdirectly irreducible subalgebra (A_0, Ω) and a one-element subalgebra (A_1, Ω), or (A, Ω) is an envelope of a non-trivial subdirectly irreducible sink (A_1, Ω).*

Proof. Let $f : (A, \Omega) \to (\{0, 1\}, \Omega)$ be a homomorphism onto $\underset{\sim}{2}$. Clearly, both $A_0 = f^{-1}(0)$ and $A_0 = f^{-1}(1)$ are subalgebras of (A, Ω), and A_1 is a sink in (A, Ω). If the only congruence of (A, Ω) preserving (A_1, Ω) is equality, then (A, Ω) is an envelope of the sink (A_1, Ω). Now suppose that there are non-trivial congruences of (A, Ω) preserving (A_1, Ω). We will show that $f^{-1}(1)$ consists of precisely one element. Suppose, on the contrary, that this is not the case. Define two relations φ and ψ on A as follows. Elements x and y of A are φ-related precisely when x and y are in $f^{-1}(1)$ or $x = y$, while ψ is a maximal congruence relation that preserves $f^{-1}(1)$, i.e. for any x and y in $f^{-1}(1)$, the ψ-classes x^ψ and y^ψ differ. It is easy to see that the intersection of ψ and φ is the equality relation, contradicting the fact that (A, Ω) is subdirectly irreducible. It follows that $f^{-1}(1)$ consists of exactly one element, say ∞, and it is clear that (A, Ω) is the Płonka sum of $(f^{-1}(0), \Omega)$ and $(\{\infty\}, \Omega)$. Now the congruence relations of $(f^{-1}(0), \Omega)$ are exactly the congruence relations of (A, Ω) restricted to $f^{-1}(0)$. It follows that (A, Ω) is subdirectly irreducible if and only if $(f^{-1}(0), \Omega)$ is. □

Note that the homomorphism f in the proof induces the homomorphism h of the Ω-semilattice (I^α, Ω) onto $\underset{\sim}{2}$, and that $f = \pi$ nat $\alpha\ h$. Note that $h^{-1}(1)$ is also a sink in (I^α, Ω).

Consider again a subdirectly irreducible $\underline{\underline{Q}} \circ \underline{\underline{R}}$-algebra (A, Ω) with a non-trivial semilattice replica congruence α on the indexing algebra (I, Ω). Assume that (A, Ω) is an envelope of a non-trivial subdirectly irreducible sink (S, Ω). Then (S, Ω) contains $B_1 = (\pi\ \text{nat}\ \alpha)^{-1}(1)$ as a sink, where 1 is the largest element of the semilattice (I^α, Ω). The sink B_1 may or may not be trivial. Assume that it is non-trivial. Now the algebra (A, Ω) can also be considered as a semilattice sum of algebras (B_{i^α}, Ω) over the Ω-semilattice (I^α, Ω). Corollary 4.5.1 and an argument similar to that in the proof of Theorem 4.2 show that (A, Ω) is also an envelope of B_1 as a subdirectly irreducible sink. In particular, the monolithic congruence m of (A, Ω) must be contained in the congruence $\sigma \in CgA$ defined by $(a, b) \in \sigma$ iff $\{a, b\} \subseteq B_1$ or $a = b$.

Now if $\underline{V}$ is a regular variety as described after Theorem 4.5.13, and (A, Ω) is a subdirectly irreducible $\underline{V}$-algebra with a non-trivial semilattice replica, then (A, Ω) has the form described above. In fact, if the corresponding envelopes are non-trivial, one can form new subdirectly irreducibles from given ones by alternating iterations of two constructions: extending by zero and constructing envelopes.

4. CONSTRUCTING ALGEBRAS

The characterization of subdirectly irreducibles becomes simpler in the case of semilattice sums.

Corollary 4.5.17. *Let $\underline{\underline{Q}}$ be a prevariety of Ω-algebras. Then a $\underline{\underline{Q}} \circ \underline{\underline{Sl}}$-algebra is subdirectly irreducible if and only if it is of one of the following types:*

  (a) *a two-element semilattice $\underset{\sim}{2}$;*

  (b) *an adequate $\underline{\underline{Q}} \circ \underline{\underline{Sl}}$-envelope of a subdirectly irreducible $\underline{\underline{Q}}$-algebra;*

  (c) *an adequate $\underline{\underline{Q}} \circ \underline{\underline{Sl}}$-envelope of a subdirectly irreducible $\underline{\underline{Q}}$-algebra (A_0, Ω) extended by zero.* □

Note that Theorem 4.3.7 may be recovered as a corollary of Corollary 4.5.17. Note also that the detailed determination of the subdirectly irreducible envelopes of $\underline{\underline{Q}}$-algebras in Theorem 4.5.13 and Corollary 4.5.17 may still be a difficult problem.

The last theorem of this section will show that, under certain quite natural conditions, Lallement sums embed into functorial sums.

Theorem 4.5.18. *An algebra (A, Ω) is a subalgebra of a functorial sum (B, Ω) of certain fibres (B_j, Ω) over an idempotent, naturally quasi-ordered algebra (J, Ω) if and only if (A, Ω) is a Lallement sum of certain fibres (A_i, Ω) over (I, Ω) by mappings $\varphi_{i,j}$ that satisfy the following condition for all i, i', j, k in I:*

$$(4.5.3) \quad \begin{cases} \forall\, i,\ i' \preceq j \preceq k,\ \forall\, a_i \in A_i,\ \forall b_{i'} \in A_{i'}, \\ a_i \varphi_{i,j} = b_{i'} \varphi_{i',j} \Rightarrow a_i \varphi_{i,k} = b_{i'} \varphi_{i',k}\,. \end{cases}$$

Proof. ($\Rightarrow$) Let (A, Ω) be a subalgebra of a functorial sum (B, Ω) of (B_j, Ω) over (J, Ω) by mappings $\psi_{i,j}$ with sum projection $\pi : B_j \to \{j\}$. Define $I := \pi(A)$, and for each $i \in I$, define $A_i := \pi^{-1}(i) \cap A$. It is easy to see that $(I, \Omega) \leq (J, \Omega)$ and $(A_i, \Omega) \leq (A, \Omega)$. For each $i \in I$, define

$$E_j := \{a_i \psi_{i,j} \mid i,\ j \in I,\ i \preceq j,\ a_i \in A_i\}.$$

Then E_j is a subalgebra of (B_j, Ω). Indeed, since (I, Ω) is idempotent and naturally quasi-ordered, it follows that for $k_1, \ldots, k_n$, j in I with all $k_i \preceq j$, one also has $k = k_1 \ldots k_n \omega \preceq j$ for any (n-ary) ω in Ω. Hence for any a_i in A_{k_i} with $i = 1, \ldots, n$,

$$a_1 \psi_{k_1, j} \ldots a_n \psi_{k_n, j} \omega = a_1 \psi_{k_1, k} \ldots a_n \psi_{k_n, k} \omega \, \psi_{k, j}$$
$$= a_1 \ldots a_n \omega \, \psi_{k, j} \in E_j.$$

Obviously (A_j, Ω) is a subalgebra of (E_j, Ω). Now let $\mu(j)$ be the relation on the union $P_j := \bigcup (A_i \mid i \preceq j)$ defined by

$$(a_k, b_l) \in \mu(j) \;:\Leftrightarrow\; a_k \psi_{k, j} = b_l \psi_{l, j}$$

for $k, l \preceq j$. It is easy to see that P_j is a subalgebra of (A, Ω), that $\mu(j)$ is a congruence on (P_j, Ω) preserving (A_j, Ω), and that $(E_j, \Omega) \cong (P_j, \Omega)^{\mu(j)}$ [Exercise 4.5I]. Moreover (E_j, Ω) is an envelope of (A_j, Ω). To show this, let $\lambda \geq \mu(j)$ be a congruence on (P_j, Ω) preserving (A_j, Ω). For $k, l \preceq j$, let $(a_k, b_l) \in \lambda$. Since $(a_k \psi_{k, j}, a_k) \in \lambda$ and $(b_l \psi_{l, j}, b_l) \in \lambda$, it follows that $(a_k \psi_{k, j}, b_l \psi_{l, j}) \in \lambda$. Since λ preserves (A_j, Ω), it follows that $a_k \psi_{k, j} = b_l \psi_{l, j}$. This implies that $\lambda \leq \mu(j)$, and consequently that $\lambda = \mu(j)$ is a maximal congruence preserving (A_j, Ω). This proves that (E_j, Ω) is an envelope of (A_j, Ω).

For $i \preceq j$ in $(I, \preceq)$, define $\varphi_{i,j} : A_i \to E_j$; $a_i \mapsto a_i \psi_{i,j}$. It is easy to check that the mappings $\varphi_{i,j}$ satisfy all the conditions for a Lallement sum [Exercise 4.5I]. Now for $i, i' \preceq j \preceq k$ in $(I, \preceq)$ and a_i in A_i, $b_{i'}$ in $A_{i'}$, if $a_i \varphi_{i,j} = b_{i'} \varphi_{i',j}$, then

$$a_i \varphi_{i,k} = a_i \psi_{i,k} = a_i \psi_{i,j} \psi_{j,k} = b_{i'} \psi_{i',j} \psi_{j,k} = b_{i'} \psi_{i',k} = b_{i'} \varphi_{i',k}.$$

Consequently, (A, Ω) is a Lallement sum of the (A_i, Ω) over (I, Ω), and satisfies (4.5.3).

($\Leftarrow$) We will define a functor $F : (I) \to (\Omega)$. For each i in I, the object iF is the envelope (E_i, Ω) of (A_i, Ω). For $j \preceq k$, a morphism $\psi_{j,k} := (j \to k)F : (E_j, \Omega) \to (E_k, \Omega)$ is needed. Recall that $E_j = \{a_i \varphi_{i,j} \mid i \preceq j\}$. Define $x\psi_{jk} = a_i \varphi_{ij} \psi_{jk} := a_i \varphi_{i,k}$. The definition is good, since if $a_i \varphi_{i,j} = b_{i'} \varphi_{i',j}$ for $i' \preceq j$, then $a_i \varphi_{i,k} = b_{i'} \varphi_{i',k}$ by condition (4.5.3). To see that $\psi_{j,k}$ is an Ω-homomorphism, consider $x_1 = a_1 \varphi_{i_1,j}, \ldots, x_n = a_n \varphi_{i_n,j}$ in E_j, where $i_1, \ldots, i_n \preceq i = i_1 \ldots i_n \omega \preceq j$. Then for ($n$-ary) ω in Ω,

$$x_1 \psi_{j,k} \ldots x_n \psi_{j,k} \omega = a_1 \varphi_{i_1,k} \ldots a_n \varphi_{i_n,k} \omega$$
$$= a_1 \varphi_{i_1,i} \ldots a_n \varphi_{i_n,i} \omega \; \varphi_{i,k}$$
$$= a_1 \varphi_{i_1,i} \ldots a_n \varphi_{i_n,i} \omega \; \varphi_{i,j} \psi_{j,k}$$
$$= a_1 \varphi_{i_1,j} \ldots a_n \varphi_{i_n,j} \omega \; \psi_{j,k} = x_1 \ldots x_n \omega \; \psi_{j,k},$$

the second and fourth equalities holding by condition (b) in the definition of a Lallement sum. Thus $\psi_{j,k}$ is indeed a homomorphism. Also, $x\psi_{j,j} = a_i \varphi_{i,j} \psi_{j,j} = a_i \varphi_{i,j} = x$, so $\psi_{j,j}$ is the identity on jF. Now suppose $j \preceq k \preceq l$ in $(I, \preceq)$. Then

$$x\psi_{j,k}\psi_{k,l} = a_i \varphi_{i,k} \psi_{k,l} = a_i \varphi_{i,l} = x\psi_{j,l},$$

completing the verification that F is a functor. It remains to check that the Lallement sum (A, Ω) is a subalgebra of the functorial sum (B, Ω) of $(B_i, \Omega) = (E_i, \Omega)$ over (I, Ω). Consider elements $a_1 = a_1 \varphi_{i_1, i_1}, \ldots, a_n = a_n \varphi_{i_n, i_n}$ of A. Then for an n-ary ω in Ω, the operation $a_1 \ldots a_n \omega$ in the sum (B, Ω) is calculated as

$$a_1 \psi_{i_1, i} \ldots a_n \psi_{i_n, i} \omega = a_1 \varphi_{i_1, i} \ldots a_n \varphi_{i_n, i} \omega$$

(where $i = i_1 \ldots i_n \omega$), i.e. just as $a_1 \ldots a_n \omega$ in the Lallement sum. $\square$

The embedding of a Lallement sum into a functorial sum obtained in the proof of Theorem 4.5.18 is called *canonical*.

Corollary 4.5.19. *An algebra (A, Ω) is a subalgebra of a Płonka sum (B, Ω) of fibres (B_j, Ω) over a semilattice (J, Ω) if and only if it is a semilattice Lallement sum of fibres (A_i, Ω) over a semilattice (I, Ω) by mappings $\varphi_{i,j}$ satisfying the condition (4.5.3).* □

Exercises

4.5A. Find all congruences on the convex set $(I, \underline{I}^\circ)$ of Example 4.5.1. Show that each quotient $(I^\theta, \underline{I}^\circ)$, for a non-trivial congruence θ, is an $\underline{I}^\circ$-semilattice.

4.5B. Consider the algebra $(A, \underline{I}^\circ)$ defined on the closed triangle with vertices $(0,0), (1,0)$ and $(0,1)$ in the Euclidean space $\mathbb{R}^2$. Find its $\underline{I}^\circ$-semilattice replica.

4.5C. Let (A, Ω) be a Lallement sum of (A_i, Ω) with envelopes (E_i, Ω) over (I, Ω).

(a) [Ćirić, Petković and Bogdanović 2000] Define the relation θ on A by
$$(a_i, b_j) \in \theta :\Leftrightarrow \exists k \in I.\ \forall l \succeq k,\ a_i \varphi_{i,l} = b_j \varphi_{j,l}.$$
Show that θ is a congruence relation of (A, Ω).

(b) Let θ_j be the restriction of the congruence θ to the subalgebra (P_j, Ω) with $P_j = \bigcup \{A_i \mid i \preceq j\}$. Show how the congruence θ_j is related to the congruence μ defined in the proof of Theorem 4.5.3.

4.5D. Verify the claims of Example 4.5.8.

4.5E. Complete the proof of Theorem 4.5.10 by showing that:

(a) B_{j^α} is a sink of (P_j, Ω);

(b) $A - P_j$ is a sink of (A, Ω);

(c) μ_j is a congruence of (A, Ω) and $(A, \Omega)^{\mu_j} = (E_j \cup \{\infty_j\}, \Omega)$ in the case

$A - P_j \neq \varnothing$, and $(A, \Omega)^{\mu_j} = (E_j, \Omega)$ in the case $j \in 1^\alpha$.

4.5F. Complete the proof of Theorem 4.5.13 by showing that:
- (a) R_j is a sink in (A, Ω);
- (b) θ_{i^α} is a congruence of (A, Ω).

4.5G. Represent each functorial sum over an idempotent naturally quasi-ordered algebra as a subdirect product, and describe all subdirectly irreducible functorial sums.

4.5H. Verify the claims of Example 4.5.14.

4.5I. Verify the claims of Theorem 4.5.18.

4.5J. Show that a Lallement sum (A, Ω) of (A_i, Ω) over (I, Ω) satisfies all the identities true for all the envelopes (E_i, Ω) and for (I, Ω).

4.5K. [Romanowska 1986] Consider the variety $\underline{G}$ of groupoids defined by the identity $x \cdot y = x \cdot z$, and the variety $\overline{\underline{G}}$ of groupoids defined by the identities

$$x \cdot (x \cdot y) = x \cdot y,$$
$$x \cdot (y \cdot z) = x \cdot (z \cdot y),$$
$$(x \cdot y) \cdot z = (x \cdot (y \cdot z))^2,$$
$$x \cdot (y \cdot (x \cdot z)) = x \cdot (y \cdot z).$$

Let $[x_1 \ldots x_n] := (x_1 \cdot (x_2 \cdot (\ldots (x_{n-1} \cdot x) \ldots)))$.

- (a) Show that for each groupoid word $xx_1 \ldots x_n w$ with x as the leftmost variable, the following identity holds in $\overline{\underline{G}}$:

$$xx_1 \ldots x_n w = [xx_1 \ldots x_n]^{2^k},$$

where $y^{2^0} := y$, $y^{2^{n+1}} := y^{2^n} \cdot y^{2^n}$ for $n \geq 0$.

- (b) Prove that the variety $\overline{\underline{G}}$ is the regularization of $\underline{G}$.
- (c) Show that each $\overline{\underline{G}}$-groupoid is a semilattice sum of $\underline{G}$-groupoids, but not necessarily a Płonka sum.

Notes on Chapter 4.

The notion of an algebraic quasi-order was introduced and investigated in [A. Romanowska and S. Traina 1999], see also [M. Ćirić, T. Petković and S. Bogdanović 2001]. Related concepts appeared earlier in semigroup theory [B.M. Schein 1974], [T. Tamura 1975], [M. Ćirić and S. Bogdanović 1996].

Functorial sums (though under different names) were first investigated in semigroup theory. The idea of such a sum over a semilattice goes back at least to [A.H. Clifford 1941]. He showed that certain semigroups (known today as Clifford semigroups) can be characterized as being disjoint unions of groups, or equivalently (in our terminology) as Płonka sums of groups. Clifford's work was accompanied by a number of further papers investigating the structure of semigroups using decomposition over a semilattice and (in some cases) the construction of a "strong semilattice of semigroups", i.e. a Płonka sum in our terminology. See e.g. [D.Mclean 1954], [T. Tamura and N. Kimura 1954, 1955], [N. Kimura 1958], [T. Tamura 1964, 1972], [M. Yamada 1955, 1956], [M. Yamada and T. Kimura 1958], [B.M. Schein 1963], [M. Petrich 1964], [V.N. Saliĭ 1969b, 1970], [I.I. Mel'nik 1970a,b], [M.S. Putcha 1973, 1978], [M. Ćirić and S. Bogdanović 1996], [S. Bogdanović and M. Ćirić 1998].

The concept of a functorial sum of semigroups over a semilattice was then generalized to arbitrary algebras by J. Płonka [1967b], and investigated in subsequent papers by him [1967a,b,c,d,e, 1968a,b, 1969a,b, 1971b, 1974a, 1984a,b, 1985b, 1988, 1989a,b] and others, e.g. [V.N. Saliĭ 1969b], [I.I. Mel'nik 1969, 1971], [A. Mitschke 1973, 1977], [R. Padmanabhan 1971], [B. Jónsson and E. Nelson 1974], [B. John 1975], [H.P. Gumm 1976], [A. Romanowska 1978, 1986]. See also [J. Płonka 1973a, 1985b] and [F. Pastijn and T. Reynaerts 1977] for related results. Semilattices as Ω-semilattices were introduced by G. Grätzer and J. Płonka [1970], and their meaning for regularizations of varieties was pointed out by B. Jónsson and E. Nelson [1974] and [B. John 1975]. The semilattice replicas of algebras were characterized in [A. Romanowska and J.D.H. Smith 1991b]. The name "regular identities" was introduced by J. Płonka. Note however that semigroup theorists call regular identities "homotypical", and irregular ones "heterotypical". In [D. Pigozzi 1979], regular identities were called "variable-uniform", and in [V.N. Saliĭ1969a,b], "normal". Lemma 4.3.1 was proved by [I.I. Mel'nik 1971], see also [A. Romanowska 1986]. Płonka's Theorem 4.3.2 is based mainly on Płonka's results [1967b, 1968b, 1969b, 1974a] with some improvements introduced in [A. Romanowska 1986] and in [MT]. The equivalence of (a) and (c) was shown in [A. Romanowska and J.D.H. Smith 1985b]. Note that the identities of (c) give a basis for the identities of the regularization. In particular, if the variety $\underline{V}$ has a finite basis, then so does its regularization. This fact was proved at first by I.I. Mel'nik [1971], where a different basis was found. In [E. Graczyńska 1983b] a syntactical proof of this fact was presented. See also [E. Graczyńska 1981, 1983a,b, 1987, 1989a,b, 1990] and [E. Graczyńska and F. Pastijn 1982] for some other syntactical considerations.

The lattice of subvarieties of the regularization of a strongly irregular variety was described by J. Dudek and E. Graczyńska [1981], following the earlier description of the semigroup mode varieties. (See e.g. [J.M. Howie 1976]. For more information about semigroup varieties, see [T. Evans 1971] and [A.J. Aizenstat and B.K. Boguta 1979]. For commutative monoid varieties, see [T.J. Head 1968].) The representation of Płonka sums as subdirect products follows by a more general result of Section 5. (See [A. Romanowska 1985] and [J. Płonka 1989b, 1990] for some special cases.) Subdirectly irreducible Płonka sums were characterized by H. Lakser, R. Padmanabhan and C.R. Platt [1970]. Free Płonka sums were investigated in [J. Płonka 1971b], [S.A. Liber 1974], [A. Mitschke 1977], and [A. Romanowska 1978]. "Free" products in varieties $\underline{\widetilde{V}}$ were investigated in [B. Jónsson and E. Nelson 1974]. The concepts of analytic and rational varieties were introduced in the monograph [MT].

Many results concerning regularizations of concrete strongly regular varieties appeared before the full content of Theorem 4.3.2 was known, either by looking for regularizations of known varieties, or by finding that certain varieties are regularized. We refer the reader to the survey [J. Płonka and A. Romanowska 1992] for further information and references. Note that in ring theory there is also a concept of a "semilattice sum of rings" that is related, but not identical, to our semilattice sums. (See e.g. [J. Weissglas 1973], [J. Gardner 1974, 1989], [M.S. Putcha 1981], [M.J. Chick and J. Gardner 1987].)

Much of the beauty of Płonka's theory disappears on regularizing irregular but not strongly irregular varieties. Among the few papers specially investigating such situations, let us mention [J. Płonka 1969a], [V.N. Salii̇̆ 1971], [A. Mitschke 1977], [A. Romanowska 1986], [E. Graczyńska, D. Kelly, P. Winkler 1986]. In particular, V.N. Salii̇̆ has shown that each semigroup in the regularization $\widetilde{\underline{V}}$ of an irregular variety $\underline{V}$ of semigroups embeds into a Płonka sum of $\underline{V}$-algebras. As shown in [E. Graczyńska, D. Kelly and P. Winkler 1986], this is not true in general. (See also [A. Romanowska 1986].)

Functorial sums over arbitrary algebras were first introduced in [G. Grätzer and J. Sichler 1974] under the name "Agassiz sums". They were then investigated by E. Graczyńska and A. Wroński [1978], and by J. Kuras [1984, 1985, 1987]. The approach presented in this book comes from [A. Romanowska and S. Traina 1999]. Predecessors of functorial sums can be observed in semigroup theory. See e.g. [A.H. Clifford 1954], [B.M. Schein 1974, 1996], [M. Petrich 1977].

Quasiregularizations were discussed in [C. Bergman and A. Romanowska 1996]. (Lemma 4.4.4 goes back to [C. Bergman and R. McKenzie 1990].) The concept of a regular quasi-identity resulted from discussions on the topic at the Banach Center in Warsaw during the meeting "Modes, modals, related structures and applications" in March 1997 (C. Bergman, K. Kearnes, A. Romanowska, J.D.H. Smith).

Nonfunctorial sums over semilattices appeared for the first time in semigroup theory. (See [G. Lallement 1967], [I. E. Burmistrovich 1965], and also [M. Petrich 1973, 1974].)

General Lallement sums over (multi-)semilattices were introduced and investigated first in [MT], [A. Romanowska 1985, 1986], and [A. Romanowska and J.D.H. Smith 1989a,b], following earlier papers [A. Romanowska 1982a, 1984] and [A. Romanowska and J.D.H. Smith 1981] where the construction was considered only for certain types of algebras. Semilattice Lallement sums as defined in this chapter were there called "coherent" Lallement sums. These sums were again studied in [A. Romanowska and J.D.H. Smith 1991b], where the first versions of the Sum Theorem 4.5.3 and Theorem 4.5.18 were given. Generalized coherent Lallement sums over arbitrary algebras, and their connection to functorial sums, were investigated in [A. Romanowska and A. Traina 1999], where also the most general version of Theorems 4.5.3 and 4.5.18 was published. A connection with subdirect products and the description of subdirectly irreducible sums was given in [A. Romanowska 2001].

CHAPTER V

INTRODUCTION TO MODES

Fix a type $\tau : \Omega \to \mathbb{N}$ of algebras. A τ-algebra (A,Ω) is said to be *entropic* if each basic operation ω is a homomorphism $\omega : (A^{\omega\tau}, \Omega) \to (A, \Omega)$.

For two operations ω and φ in Ω, say m-ary ω and n-ary φ, what does it mean to say that ω is a homomorphism

(5.1) $$\omega : (A^m, \varphi) \to (A, \varphi) \ ?$$

Take $(a_{11}, \ldots, a_{1m}), \ldots, (a_{n1}, \ldots, a_{nm})$ in A^m. Then (5.1) becomes

$$((a_{11}, \ldots, a_{1m}), \ldots, (a_{n1}, \ldots, a_{nm}))\varphi \, \omega$$
$$= (a_{11} \ldots a_{n1}\varphi) \ldots (a_{1m} \ldots a_{nm}\varphi) \, \omega$$
$$= (a_{11} \ldots a_{1m}\omega) \ldots (a_{n1} \ldots a_{nm}\omega) \, \varphi.$$

Equivalently, it means that the entropic identity

(E) $$(x_{11} \ldots x_{1m}\omega) \ldots (x_{n1} \ldots x_{nm}\omega)\varphi$$
$$= (x_{11} \ldots x_{n1}\varphi) \ldots (x_{1m} \ldots x_{nm}\varphi)\omega$$

is satisfied in (A, Ω), or that the operations ω and φ commute. Note that the identity (E) holds for each pair of operations in Ω, and thus in particular also for $\omega = \varphi$. If ω and φ are equal binary operations denoted by an infix $\cdot$, then (E) becomes

(5.2) $$(x_{11} \cdot x_{12}) \cdot (x_{21} \cdot x_{22}) = (x_{11} \cdot x_{21}) \cdot (x_{12} \cdot x_{22}).$$

For two different infix binary operations $\cdot$ and $+$, (E) becomes

(5.3) $\qquad (x_{11} \cdot x_{12}) + (x_{21} \cdot x_{22}) = (x_{11} + x_{21}) \cdot (x_{12} + x_{22}).$

One may also consider the variables x_{ij} in (E) as elements of the $m \times n$ matrix $[x_{ij}]$. If one denotes by r_i the i-th row and by c_j the j-th column of the matrix, then (E) can be written as

$$r_1 \omega \ldots r_n \omega \varphi = c_1 \varphi \ldots c_m \varphi \omega.$$

If $\omega : (A^0, \Omega) \to (A, \Omega), \varnothing \mapsto a$ is a nullary operation, then entropicity means that $\varnothing \ldots \varnothing \varphi \omega = \varnothing \omega = a = \varnothing \omega \ldots \varnothing \omega \varphi$, i.e. $\varnothing \omega = a$ is a subalgebra of (A, Ω). In what follows we assume that τ is a plural type.

The identities (E), (5.2), and (5.3) have been given various names in the literature. Among others, the following have appeared: entropic, medial, alternation, bi-commutative, bisymmetric, commutative, surcommutative, abelian. The word "entropic", in use in this context for over half a century, refers to the "inner turning" of x_{12} and x_{21} in (5.2). There is also a connection with the information-theoretic concept of entropy, see [Smith 1990]. Entropic algebras are characterized by the following.

Proposition 5.1. *A τ-algebra A is entropic if and only if, for each τ-algebra X, the morphism set $\underline{\tau}(X, A)$ is a subalgebra of the product algebra $\underline{\mathrm{Set}}(X, A) = A^X$.*

Proof. First recall that the product algebra on the set A^X is defined by

$$\omega : (A^X)^m \to A^X; (f_1, \ldots, f_m) \mapsto f_1 \ldots f_m \omega = (f : X \to A),$$

where

$$f : X \to A \,; x \mapsto xf = x(f_1 \ldots f_m \omega) = (xf_1 \ldots xf_m)\omega$$

for each (m-ary) ω in Ω.

5. INTRODUCTION TO MODES

($\Rightarrow$) Suppose first that (A, Ω) is entropic. Let $a_1, \ldots, a_n$ be in X, and let $f_1, \ldots, f_m$ be homomorphisms from (X, Ω) to (A, Ω). Then the entropic law (E) gives

$$(a_1(f_1 \ldots f_m \omega)) \ldots (a_n(f_1 \ldots f_m \omega))\varphi$$
$$= (a_1 f_1 \ldots a_1 f_m \omega) \ldots (a_n f_1 \ldots a_n f_m \omega)\varphi$$
$$= (a_1 f_1 \ldots a_n f_1 \varphi) \ldots (a_1 f_m \ldots a_n f_m \varphi)\omega$$
$$= (a_1 \ldots a_n \varphi f_1) \ldots (a_1 \ldots a_n \varphi f_m)\omega$$
$$= (a_1 \ldots a_n \varphi)(f_1 \ldots f_m \omega),$$

implying that $f = f_1 \ldots f_m \omega$ is a homomorphism from (X, Ω) to (A, Ω).

($\Leftarrow$) Now suppose that the set $\underline{\tau}(X, A)$ is a subalgebra of (A^X, Ω) for each τ-algebra X. Then in particular $\underline{\tau}(A^m, A)$ is a subalgebra of $\underline{\text{Set}}(A^m, A)$. Consider the projections $\pi_i : (A^m, \Omega) \to (A, \Omega); (a_1, \ldots, a_m) \mapsto a_i$. Then $\pi_1 \ldots \pi_m \omega$ is also a homomorphism. Now for a_{ij} in A with $i = 1, \ldots, m$ and $j = 1, \ldots, n$ one has

$$(a_{11} \ldots a_{1n} \varphi) \ldots (a_{m1} \ldots a_{mn} \varphi)\omega$$
$$= (a_{11} \ldots a_{1n} \varphi, \ldots, a_{m1} \ldots a_{mn} \varphi)(\pi_1 \ldots \pi_m \omega)$$
$$= ((a_{11}, \ldots, a_{m1}) \ldots (a_{1n}, \ldots, a_{mn})\varphi)(\pi_1 \ldots \pi_m \omega)$$
$$= ((a_{11}, \ldots, a_{m1})(\pi_1 \ldots \pi_m \omega) \ldots (a_{1n}, \ldots, a_{mn})(\pi_1 \ldots \pi_m \omega))\varphi$$
$$= (a_{11} \ldots a_{m1} \omega) \ldots (a_{1n} \ldots a_{mn} \omega)\varphi.$$

This implies that the algebra (A, Ω) is entropic. $\square$

Corollary 5.2. *If $\underline{K}$ is a prevariety of entropic algebras, then for each pair A, B of $\underline{K}$-algebras, the morphism set $\underline{K}(A, B)$ is again a $\underline{K}$-algebra.* $\square$

Now let $\underline{V}$ be a variety of entropic algebras. Fix a $\underline{V}$-algebra B. Then by

Corollary 5.2, there is a functor

(5.4) $\underline{V}(B,?) : \underline{V} \to \underline{V} ; (f : X \to Y) \mapsto (\underline{V}(B,X) \to \underline{V}(B,Y) ; h \mapsto hf).$

Exactly as in the familiar case where $\underline{V}$ is the variety of modules over a commutative ring, the functor (5.4) has a left adjoint

(5.5) $? \otimes B : \underline{V} \to \underline{V} ; (f : X \to Y) \mapsto (f \otimes B : X \otimes B \to Y \otimes B),$

yielding an adjunction

(5.6) $\underline{V}(A \otimes B, C) \cong \underline{V}(A, \underline{V}(B,C)).$

Writing the unit in the form

(5.7) $\eta_A : A \to \underline{V}(B, A \otimes B) ; a \mapsto (b \mapsto a \otimes b)$

for a $\underline{V}$-algebra A, the counit is just the evaluation

$$\varepsilon_A : \underline{V}(B,A) \otimes B \to A ; f \otimes b \mapsto bf$$

of homomorphisms. The image of a $\underline{V}$-algebra A under (5.5) is called the *tensor product* $A \otimes B$ of A and B. Using the notation (5.7), the algebra $A \otimes B$ is generated by its set $\{a \otimes b \mid a \in A,\ b \in B\}$ of *primitive elements*. Since (5.7) is a $\underline{V}$-morphism, the maps $A \to A \otimes B ; x \mapsto x \otimes b$ and $B \to A \otimes B ; y \mapsto a \otimes y$ are homomorphisms for fixed elements a of A and b of B. See the monograph [PMA, III 3.6 and IV 2.4] for a detailed discussion of tensor products of $\underline{V}$-algebras. The reader is encouraged to work out the details of the relevant proofs.

Recall that a τ-algebra (A, Ω) is *idempotent* if for each basic operation ω, the identity

(I) $x \ldots x\omega = x$

is satisfied. In other words, each singleton subset of A is actually a subalgebra.

5. INTRODUCTION TO MODES

Definition 5.3. A *mode* is an idempotent and entropic algebra. □

As an immediate consequence of the definition, one obtains.

Proposition 5.4. *Products, quotients, and subreducts of modes are modes.* □

Exercise 5A shows that the idempotent and entropic laws for modes are satisfied not only for the basic operations, but also for all derived operations. Identities with such a property are called *hyperidentities*, and a class of varieties of the same type defined by a given set of hyperidentities is called a *hypervariety*.

Corollary 5.5. *The class of mode varieties of type* $\tau : \Omega \to \mathbb{N}$ *is a hypervariety.* □

Another characterization of modes is given by the following observation.

Proposition 5.6. *A τ-algebra (A, Ω) is a mode if and only if each polynomial operation of (A, Ω) is a homomorphism.*

Proof. ($\Rightarrow$) Suppose that A is a mode. Let a be an element of A. Since $\{a\}$ is a subalgebra of (A, Ω), it follows that the constant nullary operation with value a is a homomorphism. Since A is entropic, the basic operations are homomorphisms. Thus each operation derived from $A \cup \Omega$, i.e. each polynomial operation, is a homomorphism.

($\Leftarrow$) Conversely, suppose that each polynomial operation of (A, Ω) is a homomorphism. Since the basic operations are polynomial operations, A is entropic. Since the nullary constant operation with the value a is a homomorphism for each element a of A, its image $\{a\}$ is a subalgebra of the algebra A. Thus (A, Ω) is also idempotent. □

Consider a τ-mode (A, Ω). Suppose that ω is an m-ary postfix operation, and that $\cdot$ is an infix binary operation. Then the idempotent and entropic laws imply

$$x_1\ldots x_m\omega \cdot y = x_1\ldots x_m\omega \cdot y\ldots y\omega = (x_1 \cdot y)\ldots(x_m \cdot y)\omega,$$

$$y \cdot x_1\ldots x_m\omega = y\ldots y\omega \cdot x_1\ldots x_m\omega = (y \cdot x_1)\ldots(y \cdot x_m)\omega,$$

so the following identities of *right-* and *left-distributivity* hold in (A, Ω):

(LD) $\qquad x_1\ldots x_m\omega \cdot y = (x_1 \cdot y)\ldots(x_m \cdot y)\omega,$

(RD) $\qquad y \cdot x_1\ldots x_m\omega = (y \cdot x_1)\ldots(y \cdot x_m)\omega.$

Hence each mode satisfies the identities (LD) and (RD) for any derived operations, one of any arity and the other binary [Exercise 5C].

The aim of this chapter is to present some basic properties and basic examples of modes. We encountered some of them already in previous chapters. Some will be discussed later in a much broader context within subsequent chapters.

Exercises

5A. Show that a mode (A, Ω) satisfies the idempotent and entropic laws for any derived operations.

5B. Show that a mode (A, Ω) satisfies the entropic laws for any polynomial operations.

5C. Show that any binary derived operation in a mode (A, Ω) distributes from the right and from the left over any m-ary derived operation for $m \geq 2$.

5D. [Etherington 1959a] Let $\mathbb{C}^*$ denote the multiplicative group of nonzero complex numbers. Let γ be the identity map, and let β denote complex conjugation. Define the operation $*$ on $\mathbb{C}^*$ by $x*y := (x\gamma) \cdot (y\beta)$. Show that $(\mathbb{C}^*, *)$ is entropic.

5E. [Soublin 1971] Let $(A, +)$ be an abelian group. Let γ and β be two commuting automorphisms of $(A, +)$. Define the operation $*$ on A by $x * y := x\gamma + y\beta$. Show that $(A, *)$ is entropic.

5.1. Modes of submodes

Let (A, Ω) be a τ-algebra, and let $X_1, \ldots, X_n$ be non-empty subsets of the set A. For an n-ary operation ω of Ω, set

(5.1.1) $$X_1 \ldots X_n \omega := \{x_1 \ldots x_n \omega \mid x_i \in X_i\}.$$

This is called the *complex ω-product* of the subsets X_i. One obtains an algebra $(2^A - \{\varnothing\}, \Omega)$ called the *complex set algebra* of the algebra (A, Ω). The definition (5.1.1) extends to linear derived operations w of (A, Ω):

$$X_1 \ldots X_n w := \{x_1 \ldots x_n w \mid x_i \in X_i\}.$$

If a derived operation w has arguments of multiplicity greater than 1, then the situation is more complicated. For example, consider the semigroup $(\mathbb{R}, \cdot)$ of real numbers with the usual multiplication, and let $xw := x \cdot x$. Then, for the subsets $X_1 := \mathbb{R}$ and $X_2 := \mathbb{R}$ of $\mathbb{R}$, the subset $X_1 \cdot X_2 = \{x \cdot y \mid x, y \in \mathbb{R}\} = \mathbb{R}$, while $\{xw \mid x \in \mathbb{R}\} = \{x \cdot x \mid x \in \mathbb{R}\} = \{x \in \mathbb{R} \mid x \geq 0\} \neq X_1 \cdot X_2$.

However, each non-linear word w can be obtained from a linear word w' by identification of some variables. If the set of arguments of w is $\{x_1, \ldots, x_m\}$ and is contained in the set $\{x_1, \ldots, x_n\}$ of arguments of w', then $X_1 \ldots X_m w$ is defined to be the set $\{x_1 \ldots x_n w' \mid x_i \in X_i\}$. Let $x_1 \ldots x_n w = y_1 \ldots y_m t$ be a linear identity satisfied in (A, Ω). Then

$$\begin{aligned} X_1 \ldots X_n w &= \{x_1 \ldots x_n w \mid x_i \in X_i\} \\ &= \{y_1 \ldots y_m t \mid y_j \in Y_j\} \\ &= Y_1 \ldots Y_m t, \end{aligned}$$

so that the identity also holds in the complex set algebra of (A, Ω). Identities of (A, Ω) that are not linear do not necessarily carry over to the complex set algebra.

If (A, Ω) is a mode, let AS denote the set of non-empty subalgebras of (A, Ω). For an n-ary operation ω of Ω, consider the complex ω-product $A_1 \ldots A_n \omega$ of subalgebras $A_1, \ldots, A_n$ of AS. By the entropic law, $A_1 \ldots A_n \omega$ is itself a nonempty subalgebra of (A, Ω), since for each operation φ in Ω of arity m, and for elements a_{ij} of A_i with $1 \leq i \leq n$ and $i \leq j \leq m$, one has

$$(a_{11} \ldots a_{n1}\omega) \ldots (a_{1m} \ldots a_{nm}\omega)\varphi =$$
$$(a_{11} \ldots a_{1m}\varphi) \ldots (a_{n1} \ldots a_{nm}\varphi)\omega \in A_1 \ldots A_n\omega.$$

Thus (5.1.1) makes AS into an algebra (AS, Ω), a subalgebra of the complex set algebra of (A, Ω). We will call (AS, Ω) the *complex algebra of subalgebras* of (A, Ω), or briefly just the *complex algebra*.

Proposition 5.1.1. *If (A, Ω) is a mode, then (AS, Ω) is again a mode satisfying each linear identity satisfied by (A, Ω).*

Proof. For a non-empty subalgebra S of (A, Ω), one has $S \ldots S\omega = \{a_1 \ldots a_n\omega \mid a_i \in S\} \subseteq S$. Conversely, since (A, Ω) is idempotent, $S = \{a \mid a \in S\} = \{a \ldots a\omega \mid a \in S\} \subseteq S \ldots S\omega$, whence $S \ldots S\omega = S$ and (AS, Ω) is idempotent.

Now (AS, Ω), as a subalgebra of the complex set algebra of (A, Ω), satisfies each linear identity satisfied by (A, Ω). In particular, this is true for the linear entropic identities, so that (AS, Ω) is a mode. □

For a mode (A, Ω), let AP denote the set of non-empty finitely generated subalgebras of (A, Ω).

Proposition 5.1.2. *Let ω be an n-ary operation of a mode (A, Ω). If, for each $1 \leq i \leq n$, the subalgebra A_i of A is finitely generated by the set X_i, then the complex ω-product $A_1 \ldots A_n\omega$ of the subalgebras is finitely generated by the complex ω-product $X_1 \ldots X_n\omega$ of the generating sets.*

Proof. Since for each $i = 1, \ldots, n$ one has $A_i = \langle X_i \rangle = \{x_1 \ldots x_n w \mid x_j \in X_i$ for $j = 1, \ldots, n, \ w \in X\Omega\}$ and $X_1 \ldots X_n\omega = \{x_1 \ldots x_n\omega \mid x_i \in X_i\}$, one also has

$$\langle X_1 \ldots X_n \omega \rangle$$
$$= \{y_1 \ldots y_k w \mid y_i \in X_1 \ldots X_n \omega \text{ for } i = 1, \ldots, k \text{ and } w \in X\Omega\}$$
$$= \{(x_{11} \ldots x_{1n}\omega) \ldots (x_{k1} \ldots x_{kn}\omega)w \mid x_{ij} \in X_i \text{ and } w \in X\Omega\}$$
$$= \{(x_{11} \ldots x_{k1}w) \ldots (x_{1n} \ldots x_{kn}w)\omega \mid x_{ij} \in X_i \text{ and } w \in X\Omega\}.$$

On the other hand,

$$A_1 \ldots A_n \omega = \{a_1 \ldots a_n \omega \mid a_i \in A_i\} =$$
$$\{(x_{11} \ldots x_{1k_1}w_1) \ldots (x_{n1} \ldots x_{nk_n}w_n)\omega \mid x_{ij} \in X_i \text{ and } w_i \in X\Omega\}.$$

Hence obviously $\langle X_1 \ldots X_n \omega \rangle \subseteq A_1 \ldots A_n \omega$. To prove the opposite inclusion, it will be shown by downward induction on j that for $1 \leq j \leq n$, the elements

$$x_1 \ldots x_{j-1} a_j a_{j+1} \ldots a_n \omega$$

lie in $\langle X_1 \ldots X_n \omega \rangle$ for each x_i in X_i and $1 \leq i < j$. This is certainly true for $j = n$, and the required result follows if it can be established for $j = 1$. Suppose, then, that the hypothesis is established for $j + 1$. Let m denote the arity of the derived operation w_j. Then by the idempotent and entropic laws

$$x_1 \ldots x_{j-1} a_j a_{j+1} \ldots a_n \omega = x_1 \ldots x_{j-1} (x_{j1} \ldots x_{jm} w_j) a_{j+1} \ldots a_n \omega$$
$$= (x_1 \ldots x_1 w_j) \ldots (x_{j-1} \ldots x_{j-1} w_j)(x_{j1} \ldots x_{jm} w_j)$$
$$(a_{j+1} \ldots a_{j+1} w_j) \ldots (a_n \ldots a_n w_j)\omega$$
$$= (x_1 \ldots x_{j-1} x_{j1} a_{j+1} \ldots a_n \omega) \ldots (x_1 \ldots x_{j-1} x_{jm} a_{j+1} \ldots a_n \omega) w_j.$$

This latter term already lies in $\langle X_1 \ldots X_n \omega \rangle$ by the induction assumption. Thus the first term does, thereby establishing the hypothesis for j and completing the inductive proof. It follows that $A_1 \ldots A_n \omega = \langle X_1 \ldots X_n \omega \rangle$. □

Proposition 5.1.2 shows that the complex ω-products make the set AP into an algebra (AP, Ω). This algebra is a subalgebra of (AS, Ω), and thus itself a mode.

Proposition 5.1.3. *If (A, Ω) is a mode, then (AP, Ω) is again a mode satisfying each linear identity satisfied by (A, Ω).* □

Finally, note that there is an Ω-homomorphism

$$(A, \Omega) \to (AP, \Omega) \; ; \; a \mapsto \langle a \rangle$$

embedding the algebra (A, Ω) as a subalgebra of (AP, Ω). This homomorphism is called the *canonical* embedding.

Exercises

5.1A. Let $(A, \cdot)$ be the free semilattice on $\{0, 1\}$. Construct the semilattice $(AP, \cdot)$.

5.1B. Consider the closed unit interval $I = [0, 1]$ on the real line as an algebra $(I, \underline{I}^\circ)$. (Cf. Exercise 1.1E, Example 1.4.3, Section 1.6.2, Example 4.5.1 and Exercise 4.5A.) Describe the elements of IP and the operations of $\underline{I}^\circ$ on them.

5.1C. [Pilitowska 1996] Consider an abelian group G as a commutative semigroup $(G, \cdot, ^{-1})$ with a unary operation $^{-1}$ satisfying the identities

$$y^{-1}yx = xyy^{-1} \text{ and } xx^{-1} = yy^{-1}.$$

Let H be a subgroup of G.

(a) Show that $H^{-1} = \{h^{-1} \mid h \in H\} = H$ and $HH = H$.

(b) Show that the algebra $(GS, \cdot, ^{-1})$ is a mode.

5.2. Semigroup and diagonal modes

Example 5.2.1 (Semigroup modes). These algebras are precisely the normal bands discussed in Chapters 1 and 4. We know already that there are eight varieties of normal bands, described by the following Hasse diagram:

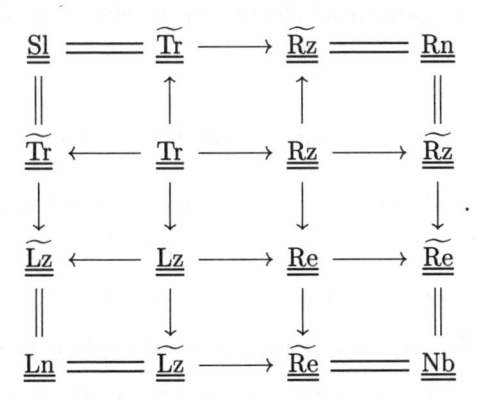

$$
\begin{array}{ccccccc}
\underline{\underline{Sl}} & =\!=\!= & \underline{\underline{\widetilde{Tr}}} & \longrightarrow & \underline{\underline{\widetilde{Rz}}} & =\!=\!= & \underline{\underline{Rn}} \\
\| & & \uparrow & & \uparrow & & \| \\
\underline{\widetilde{Tr}} & \longleftarrow & \underline{Tr} & \longrightarrow & \underline{Rz} & \longrightarrow & \underline{\widetilde{Rz}} \\
\downarrow & & \downarrow & & \downarrow & & \downarrow \\
\underline{\widetilde{Lz}} & \longleftarrow & \underline{Lz} & \longrightarrow & \underline{Re} & \longrightarrow & \underline{\widetilde{Re}} \\
\| & & \downarrow & & \downarrow & & \| \\
\underline{\underline{Ln}} & =\!=\!= & \underline{\underline{\widetilde{Lz}}} & \longrightarrow & \underline{\underline{\widetilde{Re}}} & =\!=\!= & \underline{\underline{Nb}}
\end{array}
$$

Each rectangular band is a direct product of a left-zero band and a right-zero band. Each of the outer varieties $\underline{\widetilde{V}}$ is the regularization of the corresponding inner variety $\underline{V}$, and hence consists precisely of the Płonka sums of $\underline{V}$-algebras. □

The meaning of the presence of a normal band (derived) operation in a mode is clear in view of Theorem 1.3.2 and Płonka's Theorem 4.3.2. (See also Exercises 4.3F and 4.3J.) If a mode (A, Ω) has a derived rectangular band operation $\cdot$, then by Theorem 1.3.2, $(A, \Omega) = (A_1, \Omega) \times (A_2, \Omega)$, where (A_i, Ω) satisfies the identity $x_1 \cdot x_2 = x_i$ for $i = 1$, 2. If (A, Ω) has a left-normal band operation $\cdot$ that distributes and breaks over all Ω-operations from the right-hand side, then (A, Ω) is a Płonka sum of subalgebras satisfying the identity $x \cdot y = x$ [Theorem 4.3.2]. Finally, if (A, Ω) has a normal band operation $\cdot$, then (A, Ω) is a Płonka sum of subalgebras satisfying the identity $xyx = x$ of rectangular bands, and these subalgebras decompose further as direct products, as in the previous case [Exercise 4.3J].

Semigroup modes are easily generalized to diagonal algebras that play a similar role to that played by rectangular bands in the decomposition of algebras into direct products and Płonka sums. (Cf. Exercise 1.5H.)

Example 5.2.2 (Diagonal algebras). Diagonal algebras (A, d) of a type $\tau : d \mapsto n \in \mathbb{Z}^+$ are defined as idempotent algebras satisfying the diagonal identity

(Dl) $\qquad x_{11}\ldots x_{1n}d \ldots x_{n1}\ldots x_{nn}d\, d = x_{11}x_{22}\ldots x_{nn}d.$

For $n = 2$, diagonal algebras are precisely rectangular bands. Exercises 1.3F and 1.5H show that any Ω-mode (A, Ω) having an n-ary diagonal derived operation d is a direct product of n subalgebras (A_i, Ω) satisfying the identity $x_1 \ldots x_i \ldots x_n d = x_i$. Evidently diagonal algebras are modes. Each is a direct product of n *projection* subalgebras (A_i, d) satisfying the identity $x_1 \ldots x_i \ldots x_n d = x_i$. We will sometimes refer to such diagonal algebras as *diagonal modes*. The varieties of diagonal modes may be described similarly as the varieties of rectangular semigroup modes [Exercise 5.2A]. □

Diagonal algebras may be generalized further. Let (A, Ω) be a mode of any finite type $\tau : \Omega \to \mathbb{Z}^+$ such that each (n-ary) ω in Ω is a diagonal operation. First decompose (A, Ω) as a product using the first operation, then decompose each fibre as a product using the second operation, and so on. The final decomposition gives (A, Ω) as a direct product of algebras such that all the Ω-operations are projections.

Exercises

5.2A. Describe the lattice of varieties of n-ary diagonal modes.

5.2B. [Płonka 1966a,b] Show that the following two conditions are equivalent for an idempotent algebra (A, d) with one n-ary operation d:

(a) (A, d) is *associative*, i.e. it satisfies the identities

$$(y_1 x_2 \ldots x_n d) y_2 \ldots y_n d$$
$$= y_1 (x_2 y_2 x_3 \ldots x_n d) y_3 \ldots y_n d$$
$$= y_1 \ldots y_{n-1}(x_2 x_3 \ldots x_n y_n d) d$$

and is *regularly conjugate*, i.e. it satisfies the identities

$$x_i \ldots x_i(x_1 \ldots x_n d)x_i \ldots x_i d = x_i$$

for each $i = 1, \ldots, n$, with $x_1 \ldots x_n d$ in the i-th slot;

(b) (A, d) is diagonal.

5.3. Affine spaces

In Section 1.6.1, we have seen that affine spaces over a field F can be identified with algebras $(A, P, \underline{F})$ having the ternary Mal'cev parallelogram operation P and the binary weighted mean operations $\underline{f}$ for f in F. Moreover, the class $\underline{F}$ of all affine F-spaces is a variety. The notion of an affine space $(A, P, \underline{F})$ over a field F can easily be extended to the case where, instead of a field F, one takes a (unital) ring R. So instead of a vector space over F, consider a unital module $(A, +, R)$ over a unital ring R. As before, define the operation P by

$$P : A \times A \times A \to A; \; (a, b, c) \mapsto abcP := a - b + c,$$

and the operations $\underline{r}$ for r in R by

$$\underline{r} : A \times A \to A; \; (a, b) \mapsto ab\underline{r} := a(1 - r) + br.$$

One obtains an algebra $(A, P, \underline{R})$ that is obviously idempotent. Exercise 5.3A shows that if the ring R is commutative, then the algebra $(A, P, \underline{R})$ is entropic. In what follows we usually assume that the ring R is commutative. (But see Exercise 5.3B for a more general situation.) We will call the algebra $(A, P, \underline{R})$ an *affine space over the ring R* or briefly an *affine R-space*. Note that some authors call such algebras "affine modules" rather than "affine spaces". As in Section 1.6.1, the affine subspaces of an affine R-space are the cosets of R-submodules, and the affine homomorphisms are affine maps [Exercise 5.3C].

Also Proposition 1.6.1.4 holds in this more general case, and the class $\underline{R}$ of all affine R-spaces is a variety [Exercise 5.3D]. As in Lemma 1.6.1.1, the operations P and all $\underline{r}$ for $r \in R$ generate all the derived operations $x_1 a_1 + \cdots + x_n a_n$ of affine R-spaces, and these derived operations are all idempotent derived operations of R-modules.

If the module $(A, +, R)$ is faithful, then the corresponding affine space $(A, P, \underline{R})$ will also be called *faithful*. In a faithful affine R-space any two operations $\underline{r}$ and $\underline{s}$, for distinct r and s in R, are different. On the other hand, in non-faithful affine R-spaces, one may obtain equal operations $\underline{r}$ and $\underline{s}$ from distinct ring elements r and s. As in the case of modules, one can make a faithful affine space from any affine space by dividing the ring R by the corresponding annihilator, and one can extend the type using a ring T having a homomorphism onto the ring R [Exercise 5.3I].

If $\underline{\mathrm{Mod}}_R$ is the variety of R-modules, then the idempotent reducts of R-modules, i.e. the affine R-spaces, form a variety $\underline{R}$. Now if we reintroduce 0 as a constant operation in affine R-spaces, and define module operations as in Exercise 5.3F, we will get back the variety $\underline{\mathrm{Mod}}_R$ of R-modules. Let $\underline{R}^0$ be the variety of *pointed* affine R-spaces $(A, P, \underline{R}, 0)$. (We keep 0 as a constant operation.) Then the varieties $\underline{\mathrm{Mod}}_R$ and $\underline{R}^0$ are obviously equivalent, and hence the free algebras are weakly isomorphic. (Cf. Exercise 3.5L.) If one forgets 0 in affine $\underline{R}^0$-spaces, then one gets the following.

Proposition 5.3.1. Let R be a commutative ring. Let $F_M := (X\mathrm{Mod}_R, +, R)$ be the free R-module over X. Then $F_A := (X\mathrm{Mod}_R, P, \underline{R})$ is the free affine R-space over the set $\{0\} \cup X$. □

Recall that for any ring R, the free R-module over X is the direct sum $\sum_{x \in X} R_x$, where each R_x is isomorphic with the R-module R. This fact, together with Proposition 5.3.1, gives an immediate description of free affine R-spaces.

We will frequently encounter modes that are equivalent to affine spaces. In

5. INTRODUCTION TO MODES

particular, if $(A, \cdot)$ is a binary (or groupoid) mode equivalent to an affine space $(A, P, \underline{R})$, then the operation $\cdot$ is equal to the operation $\underline{r}$ for some r in R. Since all derived operations of $(A, \cdot)$, and hence of $(A, P, \underline{R})$, come from $\underline{r}$, one may assume without loss of generality that r generates the ring R (otherwise, take the subring generated by r). The ring R is a homomorphic image of the free commutative ring $\mathbb{Z}[X]$ on one generator under the homomorphism h extending the mapping $X \mapsto r$. Hence the mode $(A, \cdot)$ is also equivalent to the affine space $(A, P, \underline{\mathbb{Z}[X]})$, where for each t in $\mathbb{Z}[X]$

$$xy\underline{t} = xy\underline{th}.$$

More generally, let (A, Ω) be a mode of any (plural) type $\tau : \Omega \to \mathbb{N} - \{0, 1\}$ equivalent to an affine space $(A, P, \underline{R})$. Consider the free commutative ring $S := \mathbb{Z}[X_{\omega i} \mid \omega \in \Omega, 1 \le i \le \omega\tau]$ on $\sum_{\omega \in \Omega}(\omega\tau)$ commuting indeterminates, and the quotient $T := S/\langle 1 - \sum_{i=1}^{\omega\tau} X_{\omega i} \mid \omega \in \Omega \rangle$ of S by the ideal obtained by setting each sum $\sum_{i=1}^{\omega\tau} X_{\omega i}$ to be 1. Then the algebra (A, Ω) is also equivalent to the affine space $(A, P, \underline{T})$, For $\omega \in \tau^{-1}(n)$, the corresponding operation of the affine space is given by

$$x_1 \ldots x_n \omega = \sum_{i=1}^{n} x_i X_{\omega i}$$

for the indeterminates $X_{\omega 1}, \ldots, X_{\omega n}$ pertaining to ω. Here we do not distinguish between elements of S and their corresponding images in T.

Note that the way we describe the variety $\underline{R}$ in this section does not give an abstract characterization, independently of the variety $\underline{\mathrm{Mod}}_R$ of R-modules. An independent characterization will be given in Chapter 6.

We conclude this section with some comments concerning varieties of affine R-spaces. Similarly to the existence of the 1-1 correspondence between free

R-modules and free affine R-spaces, there is a 1-1 correspondence between the subvarieties of the variety $\underline{\text{Mod}}_R$ and the subvarieties of the variety $\underline{R}$. From module theory, it is well known that the lattice $\mathcal{L}(\underline{\text{Mod}}_R)$ of subvarieties of $\underline{\text{Mod}}_R$ is dually isomorphic to the lattice $I(R)$ of ideals of the ring R. (See e.g. [Ježek, Kepka 1977].) If $\underline{W}$ is a subvariety of the variety $\underline{\text{Mod}}_R$, then the ideal I of R corresponding to the variety $\underline{W}$ is the intersection of the annihilators $\text{Ann}M$ of all R-modules $(M, +, R)$ in $\underline{W}$. Then the variety $\underline{W}$ is the class of all R-modules $(M, +, R)$ that are equivalent to modules $(M, +, R/I)$, i.e. $\underline{W}$ is equivalent to the variety $\underline{\text{Mod}}_{R/I}$ of R/I-modules. Now each ideal I of the ring R determines a subvariety, say $\underline{W}$, of $\underline{\text{Mod}}_R$. Taking all idempotent reducts of modules in $\underline{W}$, together with the empty set, one obtains the corresponding variety of affine spaces over the ring of $\underline{W}$. And by introducing 0 in these affine spaces one recovers the variety (equivalent to) $\underline{W}$.

Exercises

5.3A. Show that if the ring R is commutative and $(A, +, R)$ is an R-module, then the algebra $(A, P, \underline{R})$ is a mode.

5.3B. Let R be a (not necessarily commutative) ring and let $(A, +, R)$ be an R-module. Show that the algebra $(A, P, \underline{R})$ is entropic if and only if the ring $R/\text{Ann}A$ is commutative.

5.3C. Show that

(a) the non-empty affine subspaces of an affine space over a commutative ring R are cosets of R-submodules;

(b) the affine homomorphisms are affine maps.

5.3D. Show that for a commutative ring R, the class $\underline{R}$ of all affine R-spaces is a variety.

5.3E. Let $R = GF(3)$ be the three-element Galois field. Consider the affine R-space $(R \times R, P, \underline{R})$. Find all the affine subspaces. For two lines (1-dimensional affine subspaces) L_1 and L_2, what is the complex product

$L_1L_2\underline{2}$?

5.3F. For a commutative ring R, let $(A, +, R)$ be an R-module and let $(A, P, \underline{R})$ be the corresponding affine space. Show that for a, b in A, one has $a + b = a0bP$, $-x = 0x0P$ and for each r in R, $ar = 0a\underline{r}$. Use these equalities to show that the algebras $(A, +, R)$ and $(A, P, \underline{R}, 0)$ are equivalent, whereas $(A, +, R)$ and (A, P, R) are only polynomially equivalent.

5.3G. Prove Proposition 5.3.1 directly by showing that each identity over $\{0\} \cup X$ true in F_A is satisfied in the variety $\underline{R}$.

5.3H. Let R be a commutative ring.
   (a) Let $\underline{W}$ be a subvariety of the variety $\underline{\mathrm{Mod}}_R$ of R-modules. Describe the identities that are satisfied in $\underline{W}$ in addition to the identities true in $\underline{\mathrm{Mod}}_R$.
   (b) Let $\underline{A}$ be a subvariety of the variety $\underline{R}$ of affine R-spaces. Describe the identities that are satisfied in $\underline{A}$ in addition to the identities true in $\underline{R}$.

5.3I. Given a ring homomorphism $h: T \to S$ and an affine space $(A, P, \underline{S})$, describe the equivalent affine space $(A, P, \underline{T})$. Show that $(A, P, \underline{S})$ and $(A, P, \underline{T})$ are equivalent.

5.3J. Let $(A, +, S)$ be a module over a commutative ring S defined by the action
$$R: S \to \mathrm{End} A$$
of S on the group A.
   (a) Show that for each $s \in S$, the mapping R_s is an affine space homomorphism. (Cf. Exercise 1.6J.)
   (b) The operation $\underline{s}$ (for $s \in S$ with $s \neq 0, 1$) is *cancellative* if it satisfies the quasi-identity
$$(xy\underline{s} = xz\underline{s}) \to (y = z).$$
Show that $\underline{s}$ is cancellative iff R_s is injective. In particular, the

operations $\underline{s}$ for invertible s are all cancellative. (Cf. Exercise 1.6J.)

5.3K. A *combinatorial geometry* is a pair $(X, \mathcal{F})$, where X is a set of "points" and $\mathcal{F}$ is a family of subsets of X called *flats* such that

   (a) $\mathcal{F}$ is closed under intersection;
   (b) there are no infinite chains in the ordered set $\mathcal{F}$;
   (c) $\mathcal{F}$ contains the empty set, all singletons $\{x\}$ for x in X, and the set X;
   (d) for every flat E in $\mathcal{F}, E \neq X$, the flats that cover E in $\mathcal{F}$ partition the remaining points.

(See [van Lint and Wilson 1992].) Consider an n-dimensional vector space V over a field F and the corresponding affine space $(V, P, \underline{F})$. Let $\mathcal{F}$ be the set of all affine subspaces of $(V, P, \underline{F})$. Show that the pair $(V, \mathcal{F})$ is a combinatorial geometry. It is usually called an *affine geometry*, and is denoted by $AG_n(F)$.

5.3L. Describe the affine geometry $AG_2(F)$ for $F = GF(3)$. (Cf. Exercise 5.3E.)

5.4. Quasigroup modes

A *quasigroup mode* is an idempotent and entropic quasigroup. In fact, as Exercise 5.4A shows, it is sufficient to assume the idempotency and entropicity of the multiplication. A rich source of quasigroup modes is provided by affine spaces over commutative rings.

Let R be a commutative ring having a unit p such that $1 - p$ is also a unit of R. Let $(A, P, \underline{R})$ be an affine R-space, and a, b be in A. Then the equations

$$xa\underline{p} = b \text{ and } ay\underline{p} = b$$

both have a unique solution, the first being $x = ab(1 - \underline{p})^{-1}$ and the second $y = ab\underline{p}^{-1}$. This means that the algebra $(A, \cdot) = (A, \underline{p})$ is a quasigroup. The

5. INTRODUCTION TO MODES

corresponding left and right divisions, $/$ and $\backslash$, are defined by

(5.4.1) $\qquad x/y := yx(1-p)^{-1}$ and $x \backslash y := xyp^{-1}$.

In Chapter 6, we will see that in fact each quasigroup mode can be obtained as such a reduct of a suitable affine space, and even more, it is equivalent to this affine space. Note however that if the ring R has no non-identity units, as is the case e.g. for the two-element field $GF(2)$, then affine R-spaces have no quasigroup reducts.

Quasigroup modes appear in combinatorics as algebraizations of certain designs. A typical situation is illustrated by the following example.

Example 5.4.1. A *Steiner triple system* on a set A is a set S of three-element subsets of A, called the *blocks* of the system S, such that each two-element subset of A is contained in exactly one member of S. One defines a binary operation $\cdot$ on A by

$$a \cdot a := a \text{ and } a \cdot b := c \text{ if } \{a,b,c\} \in S.$$

It is easy to see that the groupoid $(A, \cdot)$ defined in this way is idempotent and commutative, and satisfies the identity

(5.4.2) $\qquad x^2 y := x \cdot xy = y.$

It usually fails to be associative or entropic. However, it is a quasigroup. Indeed, if $a \cdot c = b$ then $c = a \cdot ac = ab$. On the other hand, since $a \cdot ab = b$, the equation $a \cdot x = b$, has a unique solution. Similarly, there is a unique d such that $d \cdot a = b$. One can easily see that all the three basic quasigroup operations are equal. Quasigroups with this property are called *totally symmetric*. Quasigroups with commutative multiplication are described as being *commutative*, and idempotent commutative quasigroups satisfying (5.4.2) are called *Steiner quasigroups*.

Proposition 5.4.1.1. Let $(A, \cdot)$ be a Steiner quasigroup. Define S to be the set of three-element subsets $\{a, b, c\}$ of A such that the product of any two elements gives the third. Then S is a Steiner triple system on A. □

The proof of Proposition 5.4.1.1 is left as Exercise 5.4B. It transpires that Steiner triple systems and Steiner quasigroups can indeed be considered as two aspects of the same mathematical object. The following proposition shows the meaning of the entropic law for Steiner quasigroups.

Proposition 5.4.1.2. Let $(A, \cdot)$ be a Steiner quasigroup with corresponding Steiner triple system S. Then the following are equivalent:
(a) $(A, \cdot)$ satisfies the entropic law $xy \cdot zt = xz \cdot yt$;
(b) For two blocks $\{a, b, a \cdot b\}$ and $\{c, d, c \cdot d\}$ of S, the third point of the block determined by $a \cdot b$ and $c \cdot d$ coincides with the third point of the block determined by $a \cdot c$ and $b \cdot d$. □

The proof is left as Exercise 5.4C. In Chapter 6, we will see that entropic Steiner quasigroups are equivalent to affine spaces over the three-element field $GF(3)$. □

Example 5.4.2. If the ring R is a finite field $GF(q)$ for $q \neq 2$, then all elements $p \neq 0, 1$ are invertible, and each defines a quasigroup multiplication on each affine space A over R. In this way one has $q - 2$ quasigroups on the set A. Exercise 5.4D shows that any two of them are *mutually orthogonal*, i.e. for distinct $p, r \neq 0, 1$ in $GF(q)$ and a, b in A the pair of equations

$$a = xy\underline{p},$$
$$b = xy\underline{r}$$

has a unique solution. For a finite algebra A and two Latin squares $(a_{ij}), (b_{ij})$ determined by the tables of $\underline{p}$ and of $\underline{r}$, this means that for each $(a, b) \in A \times A$, there is exactly one index ij such that $(a,b) = (a_{ij}, b_{ij})$. In this way, the finite field $GF(q)$ provides examples of families of mutually orthogonal Latin squares. □

Exercises

5.4A. Show that idempotency and entropicity of a quasigroup $(A, \cdot)$ imply that $(A, \cdot, /, \backslash)$ is a mode.

5.4B. Prove Proposition 5.4.1.1.

5.4C. Prove Proposition 5.4.1.2.

5.4D. Prove that any two distinct quasigroups of Example 5.4.2 are mutually orthogonal.

5.4E. [Evans 1979] Prove that each pair of mutually orthogonal Latin squares on a set A may be equivalently described as an algebra

$$(A, \cdot, /, \backslash, \circ, \phi, \phi, *_1, *_2)$$

such that both $(A, \cdot, /, \backslash)$ and $(A, \circ, \phi, \phi)$ are quasigroups and the following identities hold:

$$(x \cdot y) *_1 (x \circ y) = x,$$
$$(x \cdot y) *_2 (x \circ y) = y.$$

5.4F. Replace the finite field in Example 5.4.2 by a finite commutative ring. Find conditions for p, r in R to define a pair of mutually orthogonal Latin squares on an affine R-space.

5.4G. [Sholander 1949] Let u and v be fixed distinct points of a projective plane Π of order 3. Let L be the line through u and v, and Q the set of points of Π which are not on L. For x and y in Q, define $x * y$ to be the point of intersection of the line through x and u with the line through y and v. Show that $(Q, *)$ is a quasigroup mode.

5.4H. [Mituhisa 1943] Let Q be the set of points of a parabola. Define $*$ on Q by $x*x := x$ and $x*y := z$, where z is the point of Q for which the line through y and z is parallel to the tangent line through x. Show that $(Q, *)$ is a quasigroup mode.

5.4I. [Sholander 1949] Let Q be the set of points in the Euclidean plane, and let A, B and C be three fixed noncollinear elements of Q. Define $*$ on Q by $x*x := x$ and $x*y := z$, where z is the unique point of the plane such that the triangle xyz is similar to the triangle ABC under the correspondence $x \mapsto A$, $y \mapsto B$, $z \mapsto C$. Show that $(Q, *)$ is a quasigroup mode.

5.5. Numbers and binary modes

Reducts of various affine spaces provide further examples of modes. The topic of this section is a set of such examples, relating geometric operations on the real line to finite binary representations of numbers.

Example 5.5.1 (Reflexion and symmetric binary modes). Consider the set $\mathbb{R}$ of real numbers under the binary operation $/ = \underline{-1}$ given by $x/y := xy\underline{-1} = 2x - y$. Geometrically, this operation may be interpreted as a *reflexion*.

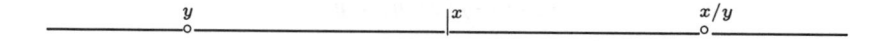

$$\underset{\circ}{\underline{\hspace{2cm} y \hspace{2cm}}} \mid x \underline{\hspace{3cm}} \underset{\circ}{\underline{\hspace{1cm} x/y \hspace{1cm}}}$$

The point x/y on the real line is the image of the point y in a mirror at the point x. We can also say that "x reflects y". Just as the reflexion of a reflexion is the original point, so the identity $x/(x/y) = y$ is satisfied. One is thus led to the following definition.

Definition 5.5.1.1. A *reflexion binary mode* or a *kei mode* is a mode $(A, \cdot)$ with binary operation $\cdot$ satisfying the identity

(R) $\qquad\qquad\qquad x^2 y := x \cdot (x \cdot y) = y. \quad \square$

The variety of all such algebras is denoted Kei.

The mode $(\mathbb{R}, /)$ has the set of integers as the submode $(\mathbb{Z}, /)$ generated by 0 and 1. The aim of this example is to show that the mode $(\mathbb{Z}, /)$ is the free kei mode on the set $\{0, 1\}$, and that its elements may be readily expressed in a standard form corresponding to the binary representations of integers.

First note that for a binary mode $(A, \cdot)$ or $(A, /)$, the identities (I) and (E) defining modes, and their consequences, such as the left and right distributivity (LD) and (RD) of the basic operation, as well as the *flexibility* or *partial associativity* (PA) given below, can be written as:

(I)	$xx = x$	or	$x/x = x$;
(E)	$xy \cdot zt = xz \cdot yt$	or	$(x/y)/(z/t) = (x/z)/(y/t)$;
(LD)	$x \cdot yz = xy \cdot xz$	or	$x/(y/z) = (x/y)/(x/z)$;
(RD)	$xy \cdot z = xz \cdot yz$	or	$(x/y)/z = (x/z)/(y/z)$;
(PA)	$xy \cdot x = x \cdot yx$	or	$(x/y)/x = x/(y/x)$.

Lemma 5.5.1.2. *For elements $a_1, a_2, \ldots, a_m$ in $\{x, y\}$, there are elements $b_1, b_2, \ldots, b_n$ of $\{x, y\}$ such that*

$$(\ldots (a_1 a_2 \cdot a_3) \ldots) a_m \cdot xy = (\ldots (b_1 b_2 \cdot b_3) \ldots) b_n$$

is an identity holding in all Kei-modes.

Proof. The proof is by induction on m. For $m = 1$, this is just reflexion or partial associativity. Let $m > 1$. If $a_m = y$, then by distributivity

$$((\ldots (a_1 a_2 \cdot a_3) \ldots) a_{m-1} \cdot y) \cdot (xy) = ((\cdots (a_1 a_2 \cdot a_3) \ldots) a_{m-1} \cdot x) \cdot y.$$

Let $a_m = x$. By induction, there are elements $c_1, \ldots, c_r$ of $\{x, y\}$ such that

$$(\ldots (a_1 a_2 \cdot a_3) \ldots) a_{m-1} \cdot xy = (\ldots (c_1 c_2 \cdot c_3) \ldots) c_r.$$

Then

$$((\ldots(a_1a_2 \cdot a_3)\ldots)a_{m-1} \cdot x) \cdot xy = ((\ldots(a_1a_2 \cdot a_3)\ldots)a_{m-1} \cdot xy) \cdot (x \cdot xy)$$
$$= (\ldots(c_1c_2 \cdot c_3)\ldots)c_r \cdot y,$$

the first equality holding by distributivity, and the second by the reflexion law. □

Lemma 5.5.1.3. *In the free kei mode* $\{0,1\}$*Kei on the elements 0 and 1, each further element may be expressed in the standard form*

$$(\ldots(x_1x_2 \cdot x_3)\ldots)x_n$$

with $x_1 \neq x_2$ *and* $x_i \in \{0,1\}$.

Proof. The proof is by induction on the length of the element. The shortest element x_1x_2 is already in the standard form. By induction, a longer element not in standard form may be expressed as $[(\ldots(z_1z_2 \cdot z_3)\ldots)z_k \cdot z_{k+1}] \cdot [(\ldots(t_1t_2 \cdot t_3)\ldots)t_l \cdot t_{l+1}]$. There are two cases to consider. If $z_{k+1} = t_{l+1}$, and both are equal to x, say, then

$$[(\ldots(z_1z_2 \cdot z_3)\ldots)z_k \cdot x] \cdot [(\ldots(t_1t_2 \cdot t_3)\ldots)t_l \cdot x]$$
$$= [(\ldots(z_1z_2 \cdot z_3)\ldots)z_k \cdot (\ldots(t_1t_2 \cdot t_3)\ldots)t_l] \cdot x$$
$$= (\ldots(x_1x_2 \cdot x_3)\ldots)x_n \cdot x,$$

the first equality holding by distributivity and the second by induction. If $z_{k+1} = x$ and $t_{l+1} = y$, then

$$[(\ldots(z_1z_2 \cdot z_3)\ldots)z_k \cdot x] \cdot [(\ldots(t_1t_2 \cdot t_3)\ldots)t_l \cdot y]$$
$$= [(\ldots(z_1z_2 \cdot z_3)\ldots)z_k \cdot (\ldots(t_1t_2 \cdot t_3)\ldots)t_l] \cdot xy$$
$$= (\ldots(a_1a_2 \cdot a_3)\ldots)a_m \cdot xy$$
$$= (\ldots(x_1x_2 \cdot x_3)\ldots)x_n,$$

the first equality holding by entropicity, the second by induction, and the third by Lemma 5.5.1.2. □

5. INTRODUCTION TO MODES

Lemma 5.5.1.4. *For elements* $1 = a_0, a_1, \ldots, a_r$ *of* $\{0,1\}$, *the following hold in the* Kei-*mode* $(\mathbb{Z}, /)$:

(a) $\quad (\ldots((0/1)/a_1)\ldots)/a_r = -\sum_{i=0}^{r} a_i 2^{r-i}$,

(b) $\quad (\ldots((1/0)/a_1')\ldots)/a_r' = 1 + \sum_{i=0}^{r} a_i 2^{r-i}$.

Proof. (a) is proved by induction: $0/1 = -1$, while

$$((\ldots((0/1)/a_1)\ldots)/a_r)/a_{r+1} = -\sum_{i=0}^{r} a_i 2^{r-i} - a_{r+1} = -\sum_{i=0}^{r+1} a_i 2^{r+1-i}.$$

(b) First note that the mapping $' : \mathbb{Z} \to \mathbb{Z}$; $n \mapsto 1 - n$ is an automorphism. Indeed $(n/k)' = (2n-k)' = 1-2n+k = 2-2n-1+k = (1-n)/(1-k) = n'/k'$. Then (b) follows from (a) on applying this automorphism. $\square$

Lemmata 5.5.1.2–5.5.1.4 lead to the following theorem.

Theorem 5.5.1.5. *The mode* $(\mathbb{Z}, /)$ *of reflexions on the integers is the free kei mode on the set* $\{0, 1\}$. *A negative integer with binary representation* $-\sum_{i=0}^{r} a_i 2^{r-i}$, *with* $a_0 = 1$, *represents the word* $(\ldots((0/1)/a_1)\ldots)/a_r$, *while an integer* $n > 1$ *for which* $n - 1$ *has the binary representation* $\sum_{i=0}^{r} a_i 2^{r-i}$, *with* $a_0 = 1$, *represents the word* $(\ldots((1/0)/a_1')\ldots)/a_r'$.

Proof. The injection $i : \{0,1\} \to \mathbb{Z}$ extends to a mode homomorphism $i' : (\{0,1\}\mathrm{Kei}, /) \to (\mathbb{Z}, /)$ from the free Kei-mode on $\{0,1\}$ to $\mathbb{Z}$. Since $(\mathbb{Z}, /)$ is generated by $\{0,1\}$, the mapping i' surjects. By Lemma 5.5.1.4, it also injects. Hence it is an isomorphism. $\square$

The groupoid $G^{OP} = (G, *)$ obtained from a groupoid $(G, \cdot)$ by defining $x * y := y \cdot x$ is called the *opposite* of $(G, \cdot)$. The identities satisfied by $(G, *)$ are *opposite* to those true in $(G, \cdot)$. For a variety $\underline{V}$ of groupoids, the *opposite variety* $\underline{V}^{OP}$ consists of opposites of $\underline{V}$-groupoids. For example $\underline{Lz}^{OP} = \underline{Rz}$ and $\underline{Rz}^{OP} = \underline{Lz}$. Then the idempotent and entropic identities (I) and (E) are self-opposite, whereas the opposite form of the reflexion identity (R) is the *symmetric* identity

(S) $$yx^2 := (y \cdot x) \cdot x = y.$$

The variety $\underline{S}$ of *symmetric binary modes* is the opposite $\underline{\text{Kei}}^{OP}$ of the variety $\underline{\text{Kei}}$ of kei modes. One easily obtains the opposite version of Lemmata 5.5.1.2–5.5.1.4 and of Theorem 5.5.1.5. Note that the operation opposite to $/ = \underline{-1}$ in $\mathbb{R}$ and in $\mathbb{Z}$ is the operation $\backslash = \underline{2}$. Indeed, $xy\underline{2} = -x + 2y = yx\underline{(-1)}$. In particular, an integer $n > 1$ for which $n - 1$ has the binary representation $\sum_{i=0}^{r} a_i 2^{r-i}$ with $a_0 = 1$ represents the word $a'_r \backslash (\ldots (a'_1 \backslash (0 \backslash 1))\ldots)$, and a negative integer with binary representation $-\sum_{i=0}^{r} a_i 2^{r-i}$ with $a_0 = 1$ represents the word $a_r \backslash (\ldots (a_1 \backslash (1 \backslash 0))\ldots)$. There is an alternative method of proving the freeness of the symmetric binary mode $(\mathbb{Z}, \backslash)$ based on the following lemma.

Lemma 5.5.1.6. *In the free symmetric binary mode* $\{0,1\}S$ *on the elements 0 and 1, each further element may be expressed in the standard form*

(5.5.1.1) $$(\ldots (x_1 x_2 \cdot x_3) \ldots) x_n$$

with $x_i \in \{0,1\}$ *and* $x_i \neq x_{i+1}$.

Proof. First note that for $a_1, \ldots, a_m$ in $\{x, y\}$ there are $b_1, \ldots, b_n$ of $\{x, y\}$ such that

(5.5.1.2) $$(\ldots (a_1 a_2 \cdot a_3) \ldots) a_m \cdot xy = (\ldots (b_1 b_2 \cdot b_3) \ldots) b_n$$

is an identity true for all symmetric binary modes. Indeed, if $a_m = y$, then $[(\ldots (a_1 a_2 \cdot a_3) \ldots) a_{m-1} \cdot y] \cdot xy = [(\ldots (a_1 a_2 \cdot a_3) \ldots) a_{m-1} \cdot x] y$ by distributivity. And if $a_m = x$, then

$$((\ldots (a_1 a_2 \cdot a_3) \ldots) a_{m-1} \cdot x) \cdot xy$$
$$= [((\ldots (a_1 a_2 \cdot a_3) \ldots) a_{m-1} \cdot x) y \cdot y] \cdot xy$$
$$= [((\ldots (a_1 a_2 \cdot a_3) \ldots) a_{m-1} \cdot x) y \cdot x] y$$

by the symmetric and distributive laws.

The proof that elements of $\{0,1\}S$ may be expressed in the form (5.5.1.1) is by induction, similarly as in the proof of Lemma 5.5.1.3, but using (5.5.1.2) instead of Lemma 5.5.1.2. The proof is left as Exercise 5.5.1B. The symmetric law implies that $x_i \neq x_{i+1}$. □

Theorem 5.5.1.7. *The mode* $(\mathbb{Z}, \backslash)$ *is the free symmetric binary mode on the set* $\{0,1\}$. *For positive k, the following types of integers represent the following words:*

(a) $2k = ((\ldots(x_1 \backslash x_2) \backslash x_3) \ldots) \backslash x_{2k}$, with $x_{2i+1} = 0$ and $x_{2i} = 1$;

(b) $2k+1 = ((\ldots(x_1 \backslash x_2) \backslash x_3) \ldots) \backslash x_{2k+1}$, with $x_{2i+1} = 1$ and $x_{2i} = 0$;

(c) $-2k = ((\ldots(x_1 \backslash x_2) \backslash x_3) \ldots) \backslash x_{2k+1}$, with $x_{2i+1} = 0$ and $x_{2i} = 1$;

(d) $-(2k+1) = ((\ldots(x_1 \backslash x_2) \backslash x_3) \ldots) \backslash x_{2k+2}$, with $x_{2i+1} = 1$ and $x_{2i} = 0$.

Proof. The second part of the proof is similar to the proof of Lemma 5.5.1.4, and is left as Exercise 5.5.1C. The first part is proved similarly as Theorem 5.5.1.5. □

Exercises

5.5.1A. Prove that each groupoid mode $(A, \cdot)$ satisfies the partial associativity (PA).

5.5.1B. Complete the proof of Lemma 5.5.1.6.

5.5.1C. Prove Theorem 5.5.1.7.

5.5.1D. Show that each symmetric binary mode satisfies the identities

$$x \cdot yz = (xz \cdot y)z \quad \text{and} \quad x^2 y = (xy \cdot x)y.$$

5.5.1E. Show that the set $2\mathbb{Z}+1$ of odd integers is a subgroupoid of $(\mathbb{Z}, \backslash)$. In particular, note that in the free symmetric binary mode on two generators, the set of words on the right hand side of formulas (b) and (d) of Theorem 5.5.1.7 is closed under the operation $\backslash$.

5.5.1F. Let $(A, P, \underline{R})$ be an affine space over the ring $R = \mathbb{Z}[X]/\langle X^2 - 1\rangle$ isomorphic to the ring $\mathbb{Z}[k]$ with $k^2 = 1$. Show that the binary mode reduct $(A, \underline{k})$ is a kei mode.

5.5.1G. Let $(A, P, \underline{R})$ be an affine space over the ring $R = \mathbb{Z}[X]/\langle X^n - 1\rangle$ isomorphic to the ring $\mathbb{Z}[k]$ with $k^n = 1$. Show that the binary mode reduct $(A, \cdot) = (A, \underline{k})$ satisfies the identity $x^n y := x(x(\ldots(x \cdot xy)\ldots)) = y$.

5.5.1H. Let $(A, P, \underline{R})$ be an affine space over the ring $R = \mathbb{Z}[X]/\langle X^2 - 2X\rangle$ isomorphic to the ring $\mathbb{Z}[k]$ with $k^2 - 2k = 0$. Show that the binary mode reduct $(A, \underline{k})$ is a symmetric binary mode.

5.5.1I. Let $(A, P, \underline{R})$ be an affine space over the ring $R = \mathbb{Z}[X]/\langle 1 - (1-X)^n\rangle$ isomorphic to the ring $\mathbb{Z}[k]$ with $(1-k)^n = 1$. Show that the binary mode reduct $(A, \cdot) = (A, \underline{k})$ satisfies the identity $xy^n := (\ldots(xy \cdot y)\ldots)y = x$.

Example 5.5.2 (Commutative binary modes). Consider the set $\mathbb{R}$ of real numbers under the binary operation $\cdot = \underline{2^{-1}}$ given by $xy := x \cdot y := xy\underline{2^{-1}} = x2^{-1} + y2^{-1}$. Geometrically, this operation may be interpreted as *mediation*:

```
─────────o──────────────────o──────────────────o─────────
         x                 x · y =             y
                           xy2⁻¹
```

The point $x \cdot y$ is the midpoint of the points x and y on the real line. Just as the midpoint of x and y is the midpoint of y and x, so the commutative law $x \cdot y = y \cdot x$ is satisfied. In analogy with the procedure of the previous example, one is led to the following definition.

Definition 5.5.2.1. A *commutative binary mode* is a mode $(A, \cdot)$ with a binary operation $\cdot$ satisfying the identity

(C) $\qquad\qquad\qquad x \cdot y = y \cdot x. \quad \square$

5. INTRODUCTION TO MODES

The variety of all such modes is denoted $\underline{C}$.

Let $\mathbb{D}$ be the ring of *dyadic* rational numbers, i.e. rationals of the form $m \cdot 2^{-n}$ for integers m and n. Note that the ring $\mathbb{D}$ coincides with $\mathbb{Z}[2^{-1}]$. Set $\mathbb{D}_1 := \mathbb{D} \cap [0,1]$, the *dyadic unit interval*. Then the commutative binary mode $(\mathbb{R}, \cdot)$ has $(\mathbb{D}_1, \cdot)$ as the submode generated by 0 and 1. The aim of this example is to show that $(\mathbb{D}_1, \cdot)$ is the free commutative binary mode on the set $\{0, 1\}$, and that the elements of $(\mathbb{D}_1, \cdot)$ may be readily expressed in a standard form corresponding to the expression of elements of the interior $\mathbb{D}_1^\circ$ of $\mathbb{D}$ as fractions

$$\sum_{i=0}^{r} a_i 2^{i-r-1}$$

for $a_i \in \{0, 1\}$ and $a_0 = 1$.

Lemma 5.5.2.2. *In the free commutative binary mode $\{0,1\}C$ on the elements 0 and 1, each further element may be expressed in the standard form*

$$x_1(\ldots(x_n \cdot 01)\ldots)$$

with n a natural number and each x_i in $\{0,1\}$.

Proof. The proof is by induction on the length of the element. The shortest element is $0 \cdot 1$ or $1 \cdot 0$, and by commutativity, each may be written in the standard form $0 \cdot 1$. By induction, a longer element not in standard form may be expressed as $[z_1(\ldots(z_n \cdot 01)\ldots)] \cdot [t_1(\ldots(t_m \cdot 01)\ldots)]$, a product of standard forms. Then by the entropic law (E) and induction, this may be written as

$$(z_1 \cdot t_1) \cdot [(z_2(\ldots(z_n \cdot 01)\ldots)) \cdot (t_2(\ldots(t_m \cdot 01)\ldots))] = (z_1 \cdot t_1) \cdot u_1(\ldots(u_r \cdot 01)\ldots),$$

where r is a natural number and each u_i is in $\{0,1\}$. If r is 0, then $u_1(\ldots(u_r \cdot 01)\ldots) = 0 \cdot 1$. There are two cases to consider. If $z_1 = t_1$, then $z_1 \cdot t_1 = t_1$ by the idempotent law (I), so that the expression is already in the required standard form $t_1(u_1(\ldots(u_r \cdot 01)\ldots))$. If $z_1 \neq t_1$, then one can assume that $u_1 = z_1$. Otherwise, use commutativity to interchange z_1 and t_1. Then

$$(z_1 \cdot t_1) \cdot u_1(\ldots(u_r \cdot 01)\ldots)$$
$$= (z_1 \cdot u_1) \cdot [t_1(u_2(\ldots(u_r \cdot 01)\ldots))]$$
$$= u_1(t_1(u_2(\ldots(u_r \cdot 01)\ldots))),$$

again in the required standard form. $\square$

Lemma 5.5.2.3. *For elements* $1 = a_0, a_1, \ldots, a_r$ *of* $\{0,1\}$, *the following holds in the commutative binary mode* $(\mathbb{D}_1, \cdot)$:

$$a_r(\ldots(a_1 \cdot 10)\ldots) = \sum_{i=0}^{r} a_i 2^{i-r-1}.$$

Proof. This is proved by induction: $1 \cdot 0 = 0 \cdot 1 = 01\underline{2^{-1}} = 1 \cdot 2^{-1}$, while

$$a_{r+1}(a_r(\ldots(a_1 \cdot 10)\ldots)) = a_{r+1}[a_r(\ldots(a_1 \cdot 10)\ldots)]\underline{2}$$
$$= a_{r+1}2^{-1} + \sum_{i=0}^{r} a_i 2^{i-r-2} = \sum_{i=0}^{r+1} a_i 2^{i-r-2}. \quad \square$$

Theorem 5.5.2.4. *The mode* $(\mathbb{D}_1, \cdot)$ *of mediations on the dyadic unit interval is the free commutative binary mode on the set* $\{0,1\}$. *A proper binary fraction* $\sum_{i=0}^{r} a_i 2^{i-r-1}$ *with* $a_0 = 1$ *represents the word* $a_r(\ldots(a_1 \cdot 10)\ldots)$.

Proof. The injection $i : \{0,1\} \to \mathbb{D}_1$ extends to a mode homomorphism i' : $(\{0,1\}C, \cdot) \to (\mathbb{D}_1, \cdot)$, from the free $\underline{C}$-mode on $\{0,1\}$ to $\mathbb{D}_1$. Since $(\mathbb{D}_1, \cdot)$ is generated by $\{0,1\}$, the mapping i' surjects. Its injectivity follows from Lemma 5.5.2.1. $\square$

There is an alternative method of proving the freeness of the commutative binary mode $(\mathbb{D}_1, \cdot)$. It is based on the expression of elements of the interior

$\mathbb{D}_1^\circ$ as proper fractions $m \cdot 2^{-n}$ with n a positive integer and m an odd integer between 0 and 2^n. For each natural number n, define groupoid words $w_n(x_1, \ldots, x_{2^n})$ in arguments $x_1, \ldots, x_{2^n}$ inductively by

$$w_0(x_1) := x_1,$$
$$w_{n+1}(x_1, \ldots, x_{2^{n+1}}) := w_n(x_1, \ldots, x_{2^n}) \cdot w_n(x_{2^n+1}, \ldots, x_{2^{n+1}}).$$

Lemma 5.5.2.5. *In the free binary mode $\{0,1\}C$ on the elements 0 and 1, each further element may be expressed in the standard form $w_n(x_1, \ldots, x_{2^n})$ with n a positive integer and $x_i = 0$, $x_j = 1$ for $1 \leq i \leq 2^n - m < j \leq 2^n$, m being an odd integer between 0 and 2^n.* □

The proof of Lemma 5.5.2.5 is left as Exercise 5.2.2A. It is easy to see that the word $w_n(x_1, \ldots, x_{2^n})$ as above corresponds to the quotient $m \cdot 2^{-n}$. See also Exercise 5.2.2C.

Exercises

5.5.2A. Prove Lemma 5.5.2.5.

5.5.2B. Deduce from Lemma 5.5.2.5 that $(\mathbb{D}_1, \cdot)$ is the free $\underline{C}$-mode on the set $\{0, 1\}$.

5.5.2C. Below are three tables that show the beginning of the generation process for elements of the free $\underline{C}$-mode on two generators, using three different representations of its elements. Complete the fourth line in each table. The element z in the second table stands for $(xx \cdot xy)(xx \cdot yy) = (xx \cdot xx)(xy \cdot yy)$.

n					
0	0				1
1	$0 \cdot 0$		$0 \cdot 1 = \frac{1}{2}$		$1 \cdot 1$
2		$00 \cdot 01 = \frac{1}{4}$	$00 \cdot 11 = \frac{2}{4}$	$01 \cdot 11 = \frac{3}{4}$	
3		$0(0 \cdot 01) = \frac{1}{8}$	?	?	?

n					
0	x				y
1			$x \cdot y$		
2	$xx \cdot xx$	$xx \cdot xy$	$xx \cdot yy$	$xy \cdot yy$	
3		?	z	?	?

n					
0	0				1
1			0.1		
2		0.01		0.11	
3		0.001	0.011	?	?

5.5.2D. For each $m = 2k+1$, let $l(m)$ be the least integer greater than $\log_2 m$.

(a) Show that $m2^{-l(m)} \in \mathbb{D}_1$ has the binary representation

$$m2^{-l(m)} = a_{l(m)-1}2^{-1} + \cdots + a_1 2^{l(m)-1} + a_0 2^{l(m)}$$
$$= 0, a_{l(m)-1} \ldots a_1 a_0$$

with $a_0 = 1$ and $a_i \in \{0, 1\}$.

(b) Show that the representation above determines the word

$$a_{l(m)-1}(a_{l(m)-2}(\ldots(a_1 \cdot a_0 0)\ldots)). \quad \square$$

Example 5.5.3. Consider again the set $\mathbb{R}$ of real numbers, but this time under the three operations $\underline{-1}$, $\underline{2}$ and $\underline{2^{-1}}$ studied in the previous examples. It is easy to see that $(\mathbb{R}, \underline{2^{-1}}, \underline{-1}, \underline{2})$ is in fact a commutative quasigroup $(\mathbb{R}, \cdot, /, \backslash)$. Since it is obviously a mode, it is a commutative quasigroup mode. It has as a subalgebra the commutative quasigroup mode $(\mathbb{D}, \cdot, /, \backslash) = (\mathbb{D}, \underline{2^{-1}}, \underline{-1}, \underline{2})$ of dyadic rational numbers. The action of the quasigroup operations is illustrated as follows.

x/y	x	$x \cdot y$	y	$x \backslash y$
$= xy\underline{(-1)}$		$= xy\underline{2^{-1}}$		$= xy\underline{2}$
$= yx\underline{2} = y \backslash x$				$= yx\underline{(-1)} = y/x$

It is an immediate consequence of results presented in Section 6.6 that the quasigroup $(\mathbb{D}, \cdot, /, \backslash)$ is the free commutative quasigroup mode on the set $\{0, 1\}$. In the following exercises the reader is asked to find a direct proof of this fact.

Exercises

5.5.3A. Show that the quasigroup $(\mathbb{D}, \cdot, /, \backslash)$ is generated by the elements 0 and 1.

5.5.3B. Describe the standard form for elements of the free commutative quasigroup mode $\{0, 1\}CQM$ on two generators 0 and 1.

5.5.3C. Prove that $(\mathbb{D}, \cdot, /, \backslash)$ is the free commutative quasigroup mode on the set $\{0, 1\}$. $\square$

5.6. Differential groups and groupoids

A *differential group* is defined to be an abelian group A equipped with a group endomorphism $d : A \to A$; $a \mapsto ad$ such that $d^2 = 0$. The endomorphism d is called a *differential* or *boundary operator* of A. Elements of $\operatorname{Ker} d$ are called *cycles* and elements of $\operatorname{Im} d$, *boundaries*. The requirement $d^2 = 0$ means that $\operatorname{Im} d \subseteq \operatorname{Ker} d$. The *homology group* $H(A)$ of the differential group $(A, +, d)$ is defined as the quotient group $\operatorname{Ker} d / \operatorname{Im} d$. Its elements, the cosets $c + \operatorname{Im} d = c + d(A)$ of the group of boundaries, are called *homology classes*. Two cycles c and c' in the same homology class are said to be *homologous*.

For a commutative ring R, the ring $R[d]$ of *dual numbers* over R is the abelian group $R \oplus dR = \{r + ds \mid r, s \in R\}$ with multiplication given by

$$(r + ds)(t + du) = rt + d(st + ru).$$

Note that in particular $d^2 = 0$. The element d is called the *differential*. Elements of the ideal dR are sometimes described as *infinitesimal*, while the projection

$$(5.6.1) \qquad \pi : R[d] \to R \ ; \ r + ds \mapsto r$$

is described as *taking the finite part r* of $r + ds$. Note that the ring $R[d]$ is isomorphic to the ring $R[X]/\langle X^2 \rangle$ of the polynomials over R by the ideal $\langle X^2 \rangle$.

Now for an $R[d]$-module $(A, +, R[d])$, the algebra $(A, +, d)$ is a differential group with the boundary operator d. Under the operation $\underline{d}$ given by $xy\underline{d} = x(1 - d) + yd$, the algebra $(A, \cdot) = (A, \underline{d})$ is a binary mode. To keep the analogy with the notation for dual numbers we use prefix notation for the module operations, and exceptionally write $xy\underline{d} = x - dx + dy$ instead of $x - xd + yd$. The groupoid $(A, \cdot)$ satisfies the identity $x^2 y := x \cdot xy = x$. Indeed, $x^2 y = x - dx + dx - d^2 x + d^2 y = x$. One is thus led to the following definition.

5. INTRODUCTION TO MODES

Definition 5.6.1. A *differential groupoid* or a *differential mode* is a mode $(A, \cdot)$ with a binary operation $\cdot$ satisfying the identity

(D) $$x^2 y := x \cdot (x \cdot y) = x. \quad \square$$

The variety of all differential modes is denoted $\underline{\mathrm{Dm}}$. It can also be defined using the following proposition.

Proposition 5.6.2. *The following sets of identities are equivalent for each groupoid* $(A, \cdot)$:

(a) (I), *the* left-normal *law*

(Ln) $$xy \cdot z = xz \cdot y,$$

and the reduction *law*

(Rd) $$x \cdot yz = xy;$$

(b) (I), (E) *and* (Rd);

(c) (I), (E) *and* (D).

Proof. (a) $\Rightarrow$ (b) By (Rd) and (Lz), it follows that

$$xy \cdot zt = xy \cdot z = xz \cdot y = xz \cdot yt,$$

whence the entropic law holds.

(b) $\Rightarrow$ (c) By (Rd) and (I), it follows that

$$x \cdot xy = xx = x,$$

whence (D) holds.

(c) $\Rightarrow$ (a) By (D) and (E), one obtains

$$x \cdot yz = (x \cdot xy) \cdot (yz) = xy \cdot (xy \cdot z) = xy,$$

whence (Rd) holds. By the distributive law and (Rd),

$$xy \cdot z = xz \cdot yz = xz \cdot y,$$

so the left-normal law is satisfied. □

In particular, for a module $(A, +, R[d])$, the algebra $(A, +, d)$ is a differential group and $(A, \underline{d}))$ is a differential groupoid. We will see later [Exercise 6.4O] that not every differential mode can be obtained as a reduct of some affine $R[d]$-space. However, for those that can be obtained in this way, there are certain very interesting connections between algebraic properties of differential groupoids and homological properties of the corresponding differential groups. The aim of this section is to illustrate such connections, and to show some further "differential" properties of differential groupoids. We start with a certain characterization of the variety $\underline{Dm}$. In what follows, the Mal'cev product $\underline{V} \circ_{\underline{M}} \underline{W}$ of two subvarieties $\underline{V}$ and $\underline{W}$ of a mode variety $\underline{M}$ relative to the class $\underline{M}$ will just be denoted by $\underline{V} \circ \underline{W}$.

Theorem 5.6.3. *The variety $\underline{Dm}$ of differential modes coincides with the Mal'cev power $\underline{Lz} \circ \underline{Lz}$ of the variety $\underline{Lz}$ of left-zero semigroups, relative to the variety $\underline{BM}$ of all binary modes.*

Proof. Let $(A, \cdot)$ be a differential groupoid. The relation γ defined on A by

(5.6.2) \qquad $(a, b) \in \gamma :\Leftrightarrow \forall x \in A,\ xa = xb$

is a congruence of $(A, \cdot)$. This can be shown as follows. Obviously γ is an equivalence relation. Now for $(a_1, b_1) \in \gamma$ and $(a_2, b_2) \in \gamma$, one has $xa_1 = xb_1$ and $xa_2 = xb_2$ for each $x \in A$. By (Rd), this implies that $x \cdot a_1 a_2 = xa_1 = xb_1 = x \cdot b_1 b_2$, showing that $(a_1 a_2, b_1 b_2) \in \gamma$ and consequently that γ is a congruence relation.

To show that $(A, \cdot)$ is in $\underline{Lz} \circ \underline{Lz}$, note first that for each $a \in A$, the congruence class a^γ is in $\underline{Lz}$. Indeed, if a_1 and a_2 are in a^γ, then $(a_1, a_2) \in \gamma$ and for each

$x \in A$ one has $xa_1 = xa_2$. In particular, for $x = a$, one obtains $a_1 = a_1 a_2$, whence a^γ is in Lz. Now the quotient $(A^\gamma, \cdot)$ is also in Lz. Indeed, for any a, b, x in A, the reduction identity implies that $x \cdot ab = xa$, whence $(ab, a) \in \gamma$ and $ab \in a^\gamma$. Consequently $a^\gamma \cdot b^\gamma = a^\gamma$, i.e. $(A^\gamma, \cdot)$ is in Lz.

Now assume that $(A, \cdot)$ is in Lz $\circ$ Lz. There is a congruence θ of $(A, \cdot)$ such that $(A^\theta, \cdot)$ is in Lz and, for each a in A, the subgroupoid $(a^\theta, \cdot)$ is in Lz. Since $(A^\theta, \cdot)$ is in Lz, it follows that $a^\theta b^\theta = (ab)^\theta = a^\theta$. Since $(a^\theta, \cdot)$ is also in Lz, it follows that $a \cdot ab = a$. Hence the identity (D) holds in $(A, \cdot)$. □

Each congruence θ of a differential groupoid $(A, \cdot)$ such that $(A^\theta, \cdot)$ is in Lz and each congruence class $(a^\theta, \cdot)$ is in Lz is called an Lz $\circ$ Lz-*congruence*. The congruence γ defined by (5.6.2) is such a congruence. However, Exercise 5.6A shows that a differential groupoid may have more than one Lz $\circ$ Lz-congruence. Note also that the congruence γ is the kernel of the groupoid homomorphism

(5.6.3) $\qquad R : (A, \cdot) \to (\text{End} A, \cdot), +) : a \mapsto (R_a : A \to A \ ; \ x \mapsto xa)$

from a differential mode $(A, \cdot)$ to the endomorphism monoid $\text{End}(A, \cdot)$ of $(A, \cdot)$ with the operation $+$ defined by

$$x(\varphi + \psi) := x\varphi \cdot x\psi$$

[Exercise 5.6B]. Elements of A related by γ are said to be *cocyclic*.

Another Lz $\circ$ Lz-congruence will play an essential rôle in this section. First, define an *orbit* of x in $(A, \cdot)$ to be the set

$$xR(A) := \{x\varphi \mid \varphi \in R(A)\},$$

where $R(A)$ denotes the submonoid of $(\text{End}(A, \cdot), \cdot)$ generated under map composition by the set $AR := \{R_a \mid a \in A\}$. Two elements a and b of A are said to be *cobordic*, or in the relation β, if their orbits have non-empty intersection:

(5.6.4) $\qquad (a, b) \in \beta :\Leftrightarrow \exists c \in aR(A) \cap bR(A).$

Proposition 5.6.4. *Let $(A, \cdot)$ be a differential mode. Then the following hold.*

(a) *The relation β is a congruence of $(A, \cdot)$.*

(b) *The quotient $(A^\beta, \cdot)$ is the left zero replica of $(A, \cdot)$.*

(c) *The relation β is the least Lz $\circ$ Lz-congruence of $(A, \cdot)$.*

Proof. (a) Obviously, β is reflexive and symmetric. Suppose $(a, b) \in \beta$ and $(b, c) \in \beta$, i.e. there are ν, φ, χ, ψ in $R(A)$ with $a\nu = b\varphi$ and $b\chi = c\psi$. By the left normal law, the monoid $R(A)$ is commutative. Hence $a\nu\chi = b\varphi\chi = c\psi\varphi$, so that $(a, c) \in \beta$ and β is transitive. Now for d in A, $(a \cdot d)\nu = aR_d\nu = a\nu R_d = b\varphi R_d$, so that $(a \cdot d, b) \in \beta$. Similarly, for e in A, $(a, b \cdot e) \in \beta$. Hence $(a \cdot d) \, \beta \, b \, \beta \, a \, \beta \, (b \cdot e)$. By transitivity $(a \cdot d, b \cdot e) \in \beta$, so that β is a congruence of $(A, \cdot)$.

(b) If $(a, b) \in \beta$, then for any c in A, one has $(a \cdot c) \, \beta \, b$. Hence $a^\beta \cdot c^\beta = (a \cdot c)^\beta = b^\beta = a^\beta$, so that $(A^\beta, \cdot)$ is a left zero semigroup. Finally, suppose that $(A^\alpha, \cdot)$ is a left zero semigroup. Set $\nu = R_{a_1} \ldots R_{a_m}$ and $\varphi = R_{b_1} \ldots R_{b_n}$. Then $a^\alpha = a^\alpha R_{a_1^\alpha} \ldots R_{a_m^\alpha} = (a\nu)^\alpha = (b\varphi)^\alpha = b^\alpha R_{b_1^\alpha} \ldots R_{b_n^\alpha} = b^\alpha$, whence $\beta \leq \alpha$.

(c) This is an immediate corollary of (b) and the fact that $(A^\gamma, \cdot)$ is a left zero semigroup. □

Note that since $\beta \leq \gamma$, cobordic elements are cocyclic. For each element a of A, the set

$$a^\gamma \mathrm{nat}\beta = \{b^\beta \mid (a, b) \in \gamma\}$$

is called the *homology set* of $(A, \cdot)$ at a. It is a left zero semigroup under the multiplication it inherits from $(A, \cdot)$. The differential mode $(A, \cdot)$ has *trivial homology* if its homology sets are all singletons, i.e. if each pair of cocyclic elements is cobordic. The opposite extreme is represented by a left zero semigroup, which consists of a single homology set.

5. INTRODUCTION TO MODES

The next theorem describes connections between cycles, boundaries and the homology group of a differential group, and the congruences β and γ of the corresponding differential groupoid.

Theorem 5.6.5. *Let $(A, +, d)$ be a differential group with corresponding differential groupoid $(A, \underline{d})$ defined by $xy\underline{d} := x - dx + dy$. Then the following hold:*

(a) *Elements of A are cocyclic if and only if they differ by a cycle;*

(b) *Elements of A are cobordic if and only if they differ by a boundary;*

(c) *For each cycle $c \in \ker d$, its homology set $c^\gamma \mathrm{nat}\beta$ is equal to the underlying set of the homology group $H(A)$.*

Proof. Let a and b be elements of A.
(a) is a conclusion of the following equivalences:

$$(a, b) \in \gamma \Leftrightarrow R_a = R_b$$
$$\Leftrightarrow \forall x \in A,\ x - dx + da = x - dx + db$$
$$\Leftrightarrow da = db \Leftrightarrow a - b \in \mathrm{Ker}\, d.$$

(b) Assume that $b \in aR(A)$, i.e. $b = a\varphi$ for some $\varphi \in R(A)$. Let $\varphi = R_{a_1} \ldots R_{a_n}$. Then

$$b = a\varphi = aR_{a_1} \ldots R_{a_n} = (\ldots((a \cdot a_1)a_2)\ldots)a_n$$
$$= (\ldots((a - da + da_1)a_2)\ldots)a_n = \ldots$$
$$= a + d(-na + a_1 + \cdots + a_n),$$

whence $b \in a + \mathrm{Im}\, d$, and $aR(A) \subseteq a + \mathrm{Im}\, d$. Now if $b \in a + \mathrm{Im}\, d$, i.e. $b = a + dc$ for some $c \in A$, then $b = a + dc = a - da + da + dc = a(a + c) = aR_{a+c}$, i.e. $b \in aR(A)$, and $a + \mathrm{Im}\, d \subseteq aR(A)$. Consequently $a + \mathrm{Im}\, d = aR(A)$. Now (b) results from the following equivalences:

$$(a,b) \in \beta \Leftrightarrow \exists c \in aR(A) \cap bR(A)$$
$$\Leftrightarrow \exists c \in (a + \operatorname{Im} d) \cap (b + \operatorname{Im} d)$$
$$\Leftrightarrow \exists x, y \in \operatorname{Im} d \text{ with } c = a + x = b + y$$
$$\Leftrightarrow \exists x, y \in \operatorname{Im} d \text{ with } a - b = y - x$$
$$\Leftrightarrow a - b \in \operatorname{Im} d.$$

(c) Let c be a cycle in A, i.e. $dc = 0$. Then for a in A,

$$(c, a) \in \gamma \Leftrightarrow R_c = R_a$$
$$\Leftrightarrow a - c \in \operatorname{Ker} d$$
$$\Leftrightarrow da = dx \Leftrightarrow da = 0,$$

i.e. $c^\gamma = \operatorname{Ker} d$.

Moreover, for b in A,

$$(a, b) \in \beta \Leftrightarrow b - a \in \operatorname{Im} d$$
$$\Leftrightarrow \exists x \in A \text{ with } dx = b - a$$
$$\Leftrightarrow \exists x \in A \text{ with } b = a + dx,$$

i.e. $a^\beta = a + \operatorname{Im} d$. It follows that the homology set $c^\gamma \operatorname{nat} \beta$ at c can be written as

$$c^\gamma \operatorname{nat} \beta = \{a^\beta \mid (a, c) \in \gamma\} = \{a + \operatorname{Im} d \mid a \in \operatorname{Ker} d\} = \operatorname{Ker} d / \operatorname{Im} d = H(A). \quad \square$$

The dual numbers $\mathbb{R}[d]$ over the reals form a differential mode under the multiplication

$$x \cdot y = x - dx + dy.$$

5. INTRODUCTION TO MODES

This differential groupoid provides a convenient framework for certain aspects of real differential calculus. Consider an everywhere-differentiable function $f : \mathbb{R} \to \mathbb{R}$. At each point a of $\mathbb{R}$, the function f has the tangent line approximation $f_a : \mathbb{R} \to \mathbb{R}$ with

(5.6.5) $$f_a(a + x) := f(a) + xf'(a).$$

This tangent line approximation may be used to extend the function $f : \mathbb{R} \to \mathbb{R}$ to a function $f : \mathbb{R}[d] \to \mathbb{R}[d]$ by means of the formula

$$f(a + dx) = f(a) + f'(a)dx$$

for real x, which may be interpreted to mean that the tangent line approximation (5.6.5) is exact for points $a + dx$ infinitesimally close to a. In general, the extended functions are not homomorphisms of the differential groupoid structure on $\mathbb{R}[d]$. However, we will see in the proof of the following theorem that extensions of such functions as (5.6.5) to $\mathbb{R}[d]$ are homomorphisms. This theorem expresses the relationship of a differentiable function to its tangent line approximation by saying that these approximations "repair the failure of the function to be a differential mode homomorphism". The "finite part" map $\pi : \mathbb{R}[d] \to \mathbb{R}$; $x \mapsto x\pi$ is as in (5.6.1).

Theorem 5.6.6. *Let $f : \mathbb{R} \to \mathbb{R}$ be an everywhere-differentiable function. Extend it to*

$$f : \mathbb{R}[d] \to \mathbb{R}[d] \ ; \ x \mapsto f(x\pi) + f'(x\pi)(x - x\pi).$$

Then for all dual numbers x and y,

(5.6.6) $$f(x \cdot y) = f(x) \cdot f_{x\pi}(y).$$

Furthermore, for each real a, any function $g : \mathbb{R}[d] \to \mathbb{R}[d]$ satisfying the functional equation

(5.6.7) $$f(a \cdot x) = f(a) \cdot g(x)$$

for all dual numbers x is infinitesimally close to the tangent line approximation f_a, i.e. for all z in $\mathbb{R}[d]$, $g(a+z)$ and $f_a(a+z)$ are cocyclic.

Proof. Let $a \in \mathbb{R}$ and let $x, y \in \mathbb{R}[d]$. Note that the tangent line approximation of the tangent line approximation f_a is the same line. Then by (5.6.5),

$$\begin{aligned}f_a(y) &= f_a(y\pi + (y - y\pi)) \\ &= f_a(y\pi) + f'_a(y\pi)(y - y\pi) \\ &= [f(a) + f'(a)(y\pi - a)] + f'(a)(y - y\pi) \\ &= f(a) + f'(a)(y - a).\end{aligned}$$

The function $f_a : \mathbb{R} \to \mathbb{R}$ is extended to $f_a : \mathbb{R}[d] \to \mathbb{R}[d]$ by (5.6.5). In particular, for $a = x\pi$ one has $f_{x\pi}(y) = f(x\pi) + f'(x\pi)(y - x\pi)$. Hence

$$\begin{aligned}f(x) \cdot f_{x\pi}(y) &= [f(x\pi) + f'(x\pi)(x - x\pi)] \cdot [f(x\pi) + f'(x\pi)(y - x\pi)] \\ &= f(x\pi) + f'(x\pi)(x - x\pi) - df(x\pi) - f'(x\pi)d(x - x\pi) \\ &\quad + df(x\pi) + f'(x\pi)d(y - x\pi) \\ &= f(x\pi) + f'(x\pi)(x - x\pi - dx + dy) \\ &= f(x - dx + dy) = f(x \cdot y),\end{aligned}$$

verifying (5.6.6). Note similarly that $f_a(x) \cdot f_a(y) = f_a(x \cdot y)$. Finally, if (5.6.7) holds, then for all z in $\mathbb{R}[d]$ one has

$$\begin{aligned}f'(a)dz &= f(a + dz) - f(a) \\ &= f(a \cdot (a + z)) - f(a) \\ &= f(a) \cdot g(a + z) - f(a) \\ &= dg(a + z) - df(a),\end{aligned}$$

whence

$$dg(a + z) = d(f(a) + f'(a)z) = df_a(a + z)$$

i.e. $g(a+z)$ and $f_a(a+z)$ are cocyclic. □

The connections between differentiation of real functions and functional equations in the differential groupoid of real dual numbers given by Theorem 5.6.6 motivate a general definition of differentiation in differential groupoids. It is convenient to revert to postfix notation for functions.

Definition 5.6.7. Let $(A, \cdot)$ be a differential mode. A function $f : A \to A$ is said to be *differentiable* at an element x of A if there is an endomorphism f_x of $(A, \cdot)$, called the *derivative* of f at x, such that

$$(x \cdot y)f = xf \cdot yf_x$$

for all y in A. The function f is *differentiable* (everywhere) if it is differentiable at each element x of A. □

Endomorphisms of $(A, \cdot)$ are of course differentiable, and may be taken as their own derivatives. An immediate consequence of Definition 5.6.7 is the following.

Proposition 5.6.8 (Chain Rule). *If $f : A \to A$ is differentiable at $x \in A$, and $g : A \to A$ is differentiable at xf, then the composite fg is differentiable at x, with*

$$(fg)_x = f_x g_{xf}.$$

Proof. For all y in A, one has $(x \cdot y)fg = (xf \cdot yf_x)g = xfg \cdot yf_x g_{xf}$. Since $f_x g_{xf}$ is an endomorphism of $(A, \cdot)$, the composite fg is differentiable at x, with this endomorphism as a derivative. □

Along with the concept of differentiability given by Definition 5.6.7, there is a corresponding concept of continuity for functions on differential groupoids.

Definition 5.6.9. Let $(A, \cdot)$ be a differential mode, with cobordism relation β. A function $f : A \to A$ is said to be *continuous* at an element x of A if

$$(x, y) \in \beta \implies (xf, yf) \in \beta.$$

The function f is said to be *continuous* (everywhere) if it is continuous at each element of A. □

Note that a continuous function $f : \mathbb{R}[d] \to \mathbb{R}[d]$ on the differential groupoid of real dual numbers yields a well-defined real function f^β such that the diagram

$$\begin{array}{ccc} \mathbb{R}[d] & \xrightarrow{f} & \mathbb{R}[d] \\ \pi \downarrow & & \downarrow \pi \\ \mathbb{R} & \xrightarrow{f^\beta} & \mathbb{R} \end{array}$$

commutes. Continuity of such a function f may be interpreted as saying that it maps infinitesimally close elements to infinitesimally close elements.

Theorem 5.6.10. *Let $(A, \cdot)$ be a differential mode. If a function $f : A \to A$ is differentiable, then it is continuous.*

Proof. Suppose that $x \in A$ and that f is differentiable at each element y of x^β. First, it will be shown that for all ν in $R(A)$,

(5.6.8) $\qquad \exists \nu_y' \in R(A).\ y\nu f = y f \nu_y'.$

This will be proved by induction on the minimal length of ν as a word in the alphabet AR. For the empty word 1, (5.6.8) is true with $1_y' = 1$. Suppose (5.6.8) holds for some $\nu \in R(A)$. Note that f is differentiable at $y\nu$, since $(y\nu, y) \in \beta$ and $(y, x) \in \beta$ imply $(y\nu, x) \in \beta$ by the transitivity of β. Then for z in A one has $y\nu R(z)f = (y\nu \cdot z)f = y\nu f \cdot z f_{y\nu} = y f \nu_y' \cdot z f_{y\nu} = y f \nu_y' R(z f_{y\nu})$, which shows that one can take $\nu_y' R(z f_{y\nu})$ for $[\nu R(z)]_y'$. This verifies that (5.6.8) holds for $\nu R(z)$.

Now recall that $(x, y) \in \beta$. So suppose there are ψ and φ in $R(A)$ with $x\psi = y\varphi$. Then by (5.6.8), $x f \psi_x' = x\psi f = y\varphi f = y f \varphi_y'$, whence $(xf, yf) \in \beta$, showing that f is continuous at x. It follows that f is continuous everywhere. □

Exercises

5.6A. Find all Lz ∘ Lz-congruences of the differential mode $(A, \cdot)$ given by the following table

·	a_1	a_2	b_1	b_2	c
a_1	a_1	a_1	a_1	a_1	a_2
a_2	a_2	a_2	a_2	a_2	a_1
b_1	b_1	b_1	b_1	b_1	b_2
b_2	b_2	b_2	b_2	b_2	b_1
c	c	c	c	c	c

5.6B. Show that the mapping R given by (5.6.3) is a groupoid homomorphism, and that $\gamma = \ker R$.

5.6C. (a) Find an endomorphism d of the group $\mathbb{Z}_8$ such that $d^2 = 0$.

(b) Find the cycles, boundaries and the corresponding homology group of $\mathbb{Z}_8$.

(c) Write the multiplication table of the differential groupoid $(\mathbb{Z}_8, \cdot) := (\mathbb{Z}_8, \underline{d})$.

(d) Find the congruences β and γ of $(\mathbb{Z}_8, \cdot)$.

(e) Illustrate Theorem 5.6.5 for the algebras $(\mathbb{Z}_8, +, d)$ and $(\mathbb{Z}_8, \underline{d})$.

5.6D. Show that each differential mode is a Lallement sum of left zero semigroups over a left zero semigroup.

5.6E. Show that a groupoid $(A, \cdot)$ is idempotent and reductive, i.e. it satisfies (I) and (Rd), iff there is an Lz ∘ Lz-congruence α on $(A, \cdot)$ such that

$$(a, b) \in \alpha \implies \forall x \in A,\ xa = xb.$$

5.6F. [Płonka 1985a] [Romanowska, Roszkowska 1987] Let I be a non-empty set. For each $i \in I$, let a non-empty set A_i be given. For each pair (i, j) in $I \times I$, let $h_{i,j} : A_i \to A_i$ be a mapping satisfying:

(i) $h_{i,i}$ is the identity on A_{ij};

(ii) $h_{i,j}h_{i,k} = h_{i,k}h_{i,j}$.

Define a groupoid structure on the disjoint union A of A_i, for $i \in I$, by

$$a_i \cdot a_j := a_i h_{i,j},$$

where $a_i \in A_i$, and $a_j \in A_j$. The groupoid $(A, \cdot)$ is said to be the Lz-Lz-*sum* of the groupoids $(A_i, \cdot)$.

(a) Show that each Lz-Lz-sum is a differential mode.

(b) Show that each differential mode can be represented as an Lz-Lz-sum.

5.6G. [Romanowska 1988] Let $(A, \cdot)$ be a differential mode and let β be the congruence of $(A, \cdot)$ defined by (5.6.4).

(a) Show that β determines the decomposition of $(A, \cdot)$ as an Lz-Lz-sum of β-classes $(A_i, \cdot) = (a^\beta, \cdot)$ over $(I, \cdot) = (A^\beta, \cdot)$.

(b) Define the directed graph $G(A) = (A, E(A))$ to have elements of A as vertices and triples (a, b, j) such that $a, b \in A_i$ for some $i \in I$ and $ah_{i,j} = b$ as labelled edges. Show that the graph $G(A)$ has the following properties.

(i) $G(A)$ has precisely $|I|$ connected components A_i;

(ii) Each element a of A is an initial point of precisely $|I|$ edges, each labelled by an element of I;

(iii) For each a in A_i and j, k in I, the following diagram commutes:

$$\begin{array}{ccc} a & \xrightarrow{k} & ah_{i,k} \\ {\scriptstyle j}\downarrow & & \downarrow{\scriptstyle j} \\ ah_{i,j} & \xrightarrow{k} & ah_{i,j}h_{i,k} \end{array}$$

(c) Show that each directed graph satisfying the conditions (i)–(iii) uniquely determines a differential mode.

5.6H. Consider an affine space $(A, P, \underline{\mathbb{Z}}_{k^2})$ over the ring $\mathbb{Z}_{k^2}$ for any positive integer k.

(a) Show that the binary mode reduct $(A, \underline{k})$ is a differential mode.

(b) Find the congruences β and γ of the differential mode $(\mathbb{Z}_{k^2}, \underline{k})$.

5.6I. [Krauze 1998] Consider an affine space $(A, P, \underline{\mathbb{Z}}_n)$ over the ring $\mathbb{Z}_n$ for any positive integer n.

(a) Show that there is k in $\mathbb{Z}_n$ such that the binary mode reduct $(A, \underline{k})$ is a differential mode iff n divides k^2.

(b) Describe the corresponding differential group.

5.6J. Find an example of an idempotent groupoid $(A, \cdot)$ that is not a differential groupoid, but does lie in the Mal'cev product $\underline{\mathrm{Lz}} \circ \underline{\mathrm{Lz}}$ (not taken relative to the variety of groupoid modes).

5.6K. Find all continuous functions on a left zero semigroup.

5.7. Multiply differential groups and groupoids

The differential groups and differential groupoids of the previous section can readily be generalized as follows: An *n-differential group* is an abelian group A together with a group endomorphism $d : A \to A$; $a \mapsto ad$ such that $d^n = 0$. The endomorphism d is called an *n-differential operator* of A. The requirement $d^n = 0$ means that there is a sequence of groups and group homomorphisms

(5.7.1) $\quad A \xrightarrow{d_1} A_1 := \operatorname{Im} d \xrightarrow{d_2} A_2 := \operatorname{Im} d^2 \longrightarrow \cdots \xrightarrow{d_n} A_n := \operatorname{Im} d^n = 0,$

with $d_1 = d$ and d_i the restriction of d to A_{i-1} for $i = 2, \ldots, n$, and with $\operatorname{Im} d^i \subseteq \operatorname{Ker} d^{n-i}$. Note that A_{n-2} is a differential group.

Let R be a commutative ring. The ring $R[d]$ (with $d^2 = 0$) of dual numbers, in Section 5.6, will now be replaced by the ring $R[d]$ with $d^n = 0$ [Exercise

5.7A]. Note that the ring $R[d]$ is isomorphic to the ring $R[X]/\langle X^n \rangle$. For an $R[d]$-module $(A, +, R[d])$, the algebra $(A, +, d)$ is an n-differential group. And under the operation $\underline{d}$ given by $xy\underline{d} = x(1-d) + yd$, the algebra $(A, \cdot) := (A, \underline{d})$ is a binary mode. The groupoid $(A, \cdot)$ satisfies the identity

(nD) $\qquad x^n y := x(x \ldots (x \cdot xy) \ldots) = x.$

Indeed,

$$x^n y = (\ldots (x(xy\underline{d})\underline{d}) \ldots)\underline{d}$$
$$= (\ldots (x(x(1-d^2) + yd^2)\underline{d}) \ldots)\underline{d} = \ldots$$
$$= x(1-d^n) + yd^n = x.$$

Definition 5.7.1. A binary mode $(A, \cdot)$ is called an *n-differential mode* if it satisfies the identity (nD). □

The next proposition gives an equivalent definition of n-differential modes.

Lemma 5.7.2. *Let* BM *be the variety of all binary modes* $(A, \cdot)$. *Let n and i be positive integers with $n \geq i$. The following identities are equivalent for the variety* BM:

(a) $\quad x_1(x_2 \ldots (x_{n-1} \cdot x_n y) \ldots) = x_1(x_2 \ldots (x_{n-1} \cdot x_n z) \ldots);$

(b) $\quad x_1(x_2 \ldots (x_{i-1}(x_i^{n-i+1} y)) \ldots) = x_1(x_2 \ldots (x_{i-2} \cdot x_{i-1} x_i) \ldots);$

(nD)
$\qquad x_1^n y = x_1;$

(c) $\quad x_1(x_2 \ldots (x_i(x_{i+1}^{n-i} y)) \ldots) = x_1(x_2 \ldots (x_{i-1} \cdot x_i x_{i+1}) \ldots);$

(d) $\quad x_1(x_2^{n-1} x_3) = x_1 x_2;$

(R_n)
$\qquad x_1(x_2 \ldots (x_{n-1} \cdot x_n y) \ldots) = x_1(x_2 \ldots (x_{n-1} x_n) \ldots).$

Proof. (a) $\Rightarrow$ (b) follows by substituting x_i for $x_{i+1}, \ldots, x_n$ and z in (a). (b) $\Rightarrow$ (nD) follows by substituting x_1 for $x_2, \ldots, x_i$ in (b).

(b) ⇒ (c) As a consequence of the previous implication one obtains $x_1 = x_1^n x_{i+1}$. Then entropicity and (b) imply the following:

$$x_1(x_2\ldots(x_{i+1}^{n-i}y)\ldots) = (x_1^n x_{i+1})(x_2(x_3\ldots(x_i(x_{i+1}^{n-i}y))\ldots))$$
$$= (x_1 x_2)((x_1^{n-1} x_{i+1})(x_3(x_4\ldots(x_{i+1}^{n-i}y))))$$
$$= (x_1 x_2)((x_1 x_3)((x_1^{n-2} x_{i+1})(x_4(\ldots(x_{i+1}^{n-i}y)))))$$
$$= \ldots$$
$$= (x_1 x_2)((x_1 x_3)(\ldots(x_1 x_i)((x_1 x_{i+1})^{n-i+1} y)\ldots))$$
$$= (x_1 x_2)((x_1 x_3)(\ldots((x_1 x_i) \cdot (x_1 x_{i+1}))\ldots))$$
$$= x_1(x_2 \ldots (x_i x_{i+1}) \ldots).$$

(c) ⇒ (b) follows by substituting x_i for x_{i+1} in (c).

(nD) ⇒ (d) follows by the equivalence of (b) and (c) for $i = 1$.

(d) ⇒ (R_n) Applying successively the equivalence of (b) and (c), one obtains the following identities true in $\underline{\text{BM}}$:

$$x_1(x_2(x_3^{n-2} x_4)) = x_1(x_2 x_3),$$
$$x_1(x_2(x_3(x_4^{n-3} x_5))) = x_1(x_2 \cdot x_3 x_4),$$
$$\ldots$$
$$x_1(x_2 \ldots (x_{n-1} \cdot x_n y)\ldots) = x_1(x_2 \ldots (x_{n-2} \cdot x_{n-1} x_n)\ldots).$$

(R_n) ⇒ (d) and (R_n) ⇒ (a) are obvious. □

A binary mode $(A, \cdot)$ satisfying the identity (R_n) is called *n-step left reductive*, or briefly *n-reductive*. The variety of all such modes is denoted $\underline{R}_n$. The class of all n-reductive binary modes for all positive n is the class of *left reductive binary modes*. Lemma 5.7.2 implies the following:

Proposition 5.7.3. *For each positive integer n, the variety of n-differential modes and the variety of n-reductive binary modes coincide.* □

In particular, differential modes are 2-differential, and the variety $\underline{\underline{Dm}}$ of differential modes coincides with the variety of 2-reductive binary modes. [Note that (Rd) coincides with (R_2).]

For each variety $\underline{\underline{R}}_n$, there is also a theorem analogous to Theorem 5.6.3, representing $\underline{\underline{R}}_n$ as a certain Mal'cev power of the variety $\underline{\underline{Lz}}$ of left zero semigroups.

Theorem 5.7.4. *Let k and n be positive integers with $k < n$. Then the variety $\underline{\underline{R}}_n$ of n-reductive binary modes coincides with the Mal'cev product $\underline{\underline{R}}_k \circ \underline{\underline{R}}_{n-k}$ relative to the variety $\underline{\underline{BM}}$ of all binary modes.*

Proof. Let $(A, \cdot)$ be an n-reductive binary mode. Let λ_k be the relation on A defined by

$$(5.7.2) \qquad (a,b) \in \lambda_k :\Leftrightarrow \forall x \in A,\ x^k a = x^k b.$$

It will first be shown that λ_k is a congruence. It is obviously an equivalence relation. Now let $(a_i, b_i) \in \lambda_k$ for $i = 1, 2$, i.e. for each x in A, $x^k a_i = x^k b_i$. Then by the left distributivity (LD), it follows that

$$x^k(a_1 a_2) = x^k a_1 \cdot x^k a_2 = x^k b_1 \cdot x^k b_2 = x^k(b_1 b_2),$$

whence λ_k is a congruence.

If $(a,b) \in \lambda_k$, then for each x in A, one has $x^k a = x^k b$. In particular, for $x = a$, one has $a = a^k a = a^k b$, implying that a^{λ_k} is in $\underline{\underline{R}}_k$.

Now the n-reduction identity (R_n) implies that for a and b in A, and for each x in A,

$$x^k(a^{n-k} b) = x^k(a^{n-k} a) = x^k a.$$

Hence $(a, a^{n-k} b) \in \lambda_k$ and $(a^{\lambda_k})^{n-k} b^{\lambda_k} = a^{\lambda_k}$. Consequently, $(A^{\lambda_k}, \cdot)$ is in $\underline{\underline{R}}_{n-k}$. It follows that each n-reductive mode $(A, \cdot)$ is in $\underline{\underline{R}}_k \circ \underline{\underline{R}}_{n-k}$, whence $\underline{\underline{R}}_n \subseteq \underline{\underline{R}}_k \circ \underline{\underline{R}}_{n-k}$.

Now it will be shown that the opposite inclusion holds as well. Let $(A, \cdot)$ be a mode in $\underline{R}_k \circ \underline{R}_{n-k}$. Then there is a congruence θ on $(A, \cdot)$ such that $(A^\theta, \cdot)$ is in $\underline{R}_{n-k}$ and for each a in A, the subalgebra $(a^\theta, \cdot)$ is in $\underline{R}_k$. It follows that for any a, b in A, one has $(a^{n-k}b)^\theta = (a^\theta)^{n-k}b^\theta = a^\theta$, and hence $(a^{n-k}b, a) \in \theta$. Since each θ-class satisfies (R_n), one has $a = a^k(a^{n-k}b) = a^n b$. It follows that $(A, \cdot)$ satisfies (R_n) and hence belongs to $\underline{R}_n$. □

For a variety $\underline{V}$ of Ω-modes, the *n-iterated Mal'cev power* $\underline{V}^n$ is the Mal'cev product (relative to Ω-modes)

$$\underline{V}^n := (\ldots((\underline{V} \circ \underline{V}) \circ \underline{V}) \ldots) \circ \underline{V}$$

with n occurrences of $\underline{V}$.

Corollary 5.7.5. *Let k and n be positive integers with $k < n$. Then the following Mal'cev products are varieties, and all coincide with the variety $\underline{R}_n$ of n-reductive binary modes:*

$$\underline{R}_1^n = \underline{R}_1 \circ \underline{R}_1^{n-1} = \cdots = \underline{R}_1^{n-1} \circ \underline{R}_1 = \underline{R}_1 \circ \underline{R}_{n-1} = \cdots = \underline{R}_{n-1} \circ \underline{R}_1 = \underline{R}_n.$$ □

Note that $\underline{R}_1$ coincides with $\underline{Lz}$. In particular

(5.7.3) $$\underline{R}_n = (\underline{Lz})^n.$$

The congruences λ_k of an n-reductive mode $(A, \cdot)$ defined by (5.7.2) yield the corresponding decompositions, and form an increasing chain

$$\hat{A} = \lambda_0 \leq \lambda_1 \leq \cdots \leq \lambda_{n-1} \leq \lambda_n = A \times A.$$

Note also that the operator ∘ in (5.7.3) is associative. Corollary 5.7.5 gives a recursive method of constructing n-reductive binary modes from left zero semigroups.

Since each variety $\underline{R}_n$ is strongly irregular, Theorems 4.5.3 and 5.7.4 imply the following:

Theorem 5.7.6. *Each n-reductive binary mode is a Lallement sum of k-reductive modes over an $n - k$-reductive mode, for each $k = 1, \ldots, n - 1$.* □

There are better construction methods for some left reductive binary modes that do not require non-trivial extensions of the fibres. Such constructions for differential groupoids are given in Exercises 5.6F and 5.6G. For a construction of n-reductive binary modes, see Exercise 5.7B. Here we describe one more "Lallement"-like construction that works nicely as a method of building some n-reductive modes.

Let R be a commutative ring, and let I be a left zero semigroup. For each i in I, let A_i be an affine R-space. For arbitrary r in R, consider the groupoids $(A_i, \cdot) := (A_i, \underline{r})$ with the operation $\underline{r}$ given by $xy := xy\underline{r} = x(1-r) + yr$. For each $(i,j) \in I \times I$, let $\varphi_{i,j} : A_i \to A_j$ be a mapping such that

(5.7.4) $$a_i \varphi_{i,i} = a_i r,$$

and

(5.7.5) $$(a_i(1-r) + b_j \varphi_{j,i})\varphi_{i,n} = a_i \varphi_{i,n}(1-r) + b_j \varphi_{j,n} r$$

for a_i in A_i and b_j in A_j. In the disjoint union $A = \biguplus(A_i \mid i \in I)$, define

(5.7.6) $$a_i b_j = a_i b_j \underline{r} := a_i(1-r) + b_j \varphi_{j,i}.$$

Note that the condition (5.7.5) may be written in the form

$$(a_i b_j)\varphi_{i,n} = a_i \varphi_{i,n} \cdot b_j \varphi_{j,n}.$$

Since the conditions of Definition 4.5.2 are satisfied, with the superalgebras E_i equal to A_i, the algebra $(A, \cdot)$ is a strict sum of the algebras $(A_i, \cdot)$ over the left-zero band $(I, \cdot)$. However, it is not a Lallement sum [Exercise 5.7C]. Note also that for $j = i$, one obtains

$$(a_i b_i)\varphi_{i,n} = a_i \varphi_{i,n} \cdot b_i \varphi_{i,n},$$

implying that the $\varphi_{i,j}$ are homomorphisms. The sum $(A, \cdot)$ is also called an $\underline{r}$-*sum of the reducts* $(A_i, \underline{r})$.

Lemma 5.7.7. *The following hold for the proper $\underline{r}$-sum $(A, \cdot) = (A, \underline{r})$ of the reducts $(A_i, \cdot) = (A_i, \underline{r})$ of affine spaces $(A_i, P, \underline{R})$:*

(a) $(A, \cdot)$ *is a mode;*

(b) *For a groupoid word $w(x_1, \cdots, x_n)$, and for $i_1, \ldots, i_n$ in I with $i_1 \ldots i_n w = i_1 =: i$ and for a_k in A_{i_k},*

$$a_1 \ldots a_n w \; \varphi_{i,j} = a_1 \varphi_{i_1,j} \ldots a_n \varphi_{i_n,j} \; w;$$

(c) *If the identity $x_1 \ldots x_n w = y_1 \ldots y_m v$ holds in $(A, \cdot)$, then $x_1 = y_1$;*

(d) *If each groupoid $(A_i, \cdot)$ satisfies an identity $x_1 \ldots x_n w = y_1 \ldots y_m v$, then the sum $(A, \cdot)$ satisfies the identity*

$$x \cdot x_1 \ldots x_n w = x \cdot y_1 \ldots y_m v. \quad \square$$

The proof of Lemma 5.7.7 is left as Exercise 5.7D.

The construction of the sum described above can be used to build $(n+1)$-reductive binary modes as sums of n-reductive modes. For a commutative ring R, consider a quotient R' of the ring $R[d]$ with $d^n = 0$. Let $(A_i, \cdot) = (A_i, \underline{d})$ be (n-reductive) reducts of R'-affine spaces A_i. Let I be a left zero semigroup. With homomorphisms $\varphi_{i,j}$ satisfying (5.7.4) and (5.7.5), and the multiplication on $A = \bigcup (A_i \mid i \in I)$ defined by (5.7.6), the sum $(A, \cdot)$ is a mode. By Lemma 5.7.7 one has the following.

Proposition 5.7.8. *Each $\underline{d}$-sum $(A, \cdot)$ of (n-reductive) reducts $(A_i, \cdot) = (A_i, \underline{d})$ of affine R'-spaces $(A_i, P, \underline{R'})$ is an $(n+1)$-reductive binary mode.* $\quad \square$

In general, a sum $(A, \cdot)$ of $(A_i, \underline{d})$ is not a reduct of an affine space. Consider the groupoid $(A, \cdot)$ given by the table

$\cdot$	0	1	$0'$
0	0	0	1
1	1	1	0
$0'$	$0'$	$0'$	$0'$

It is a quotient of the mode $(\mathbb{Z}_4, \cdot) = (\mathbb{Z}_4, \underline{2})$. On the other hand, it is the $\underline{2}$-sum of two reducts $A_0 = (\{0,1\}, \cdot) = (\{0,1\}, \underline{2})$ and $A_1 = (\{0'\}, \cdot) = (\{0'\}, \underline{2} = \underline{0})$ of affine spaces $\mathbb{Z}_2$ and $\mathbb{Z}_1$ over the ring $\mathbb{Z}_2$, respectively, by the sum homomorphisms $0\varphi_{0,0} = 1\varphi_{0,0} = 0$ and $0'\varphi_{1,0} = 1$, with $\varphi_{1,0}$, $\varphi_{1,1}$ defined in the obvious way. Exercise 6.4P will show that $(A, \cdot)$ cannot be a reduct of an affine space.

Exercises

5.7A. Show that the ring $R[d]$ with $d^n = 0$ is isomorphic to the ring $R[X]/\langle X^n \rangle$. Define the multiplication of the ring $R[d]$, and show that for an $R[d]$-module $(A, +, R[d])$, the algebra $(A, +, d)$ is an n-differential group.

5.7B. [Pilitowska 1996] Let k and n be positive integers, with $k < n$. Let the groupoid $(I, \cdot)$ be an $\underline{\underline{R}}_{n-k}$-mode, and for each i in I, let $(A_i, \cdot)$ be an $\underline{\underline{R}}_k$-mode. For each pair $(i,j) \in I \times I$, let $h_{i,j} : A_i \times A_j \to A_{ij}$; $(a_i, b_j) \mapsto (a_i, b_j)h_{i,j} =: a_i b_j h_{i,j}$ be a mapping such that:

$$a_i a_i h_{i,i} = a_i,$$

$$a_i b_j h_{i,j} \; c_k d_l h_{k,l} \; h_{ij,kl} = a_i c_k h_{i,k} \; b_j d_l h_{j,l} \; h_{ik,jl}.$$

A groupoid structure on the disjoint union $A = \bigcup (A_i \mid i \in I)$ is defined by

$$a_i \cdot a_j := a_i a_j h_{i,j}.$$

Call the groupoid $(A, \cdot)$ an $\underline{\underline{R}}_k$-$\underline{\underline{R}}_{n-k}$-sum of $(A_i, \cdot)$.

(a) Show that $(A, \cdot)$ is n-reductive.

(b) Show that each n-reductive binary mode can be represented in the form of an $\underline{\underline{R}}_k$-$\underline{\underline{R}}_{n-k}$-sum.

5.7C. Show that an $\underline{r}$-sum of reducts $(A, \underline{r})$ of affine R-spaces is not necessarily a Lallement sum.

5.7D. Prove Lemma 5.7.7.

5. INTRODUCTION TO MODES

5.7E. Show that for an n-reductive binary mode $(A, \cdot)$, the complex algebra $(AS, \cdot)$ of submodes of $(A, \cdot)$ is again n-reductive.

5.7F. Consider an affine space $(A, P, \underline{\mathbb{Z}}_{k^n})$ over the ring $\mathbb{Z}_{k^n}$ for given positive integers k and n.

   (a) Show that $(A, \underline{k})$ is an n-reductive mode.

   (b) Determine the congruences $\lambda_1 \leq \cdots \leq \lambda_{n-1}$ of $(A, \underline{k})$.

5.7G. Consider an affine space $(A, P, \underline{\mathbb{Z}}_n)$ over the ring $\mathbb{Z}_n$ for fixed positive n.

   (a) Show that there is an element k of $\mathbb{Z}_n$ such that the binary mode reduct $(A, \underline{k})$ is i-reductive iff n divides k^i.

   (b) Describe the corresponding n-differential group.

5.7H. Let $(A, \cdot) = (A, \underline{d})$ be a reduct of an affine space $(A, P, \underline{R}[d])$ for some commutative ring R with $d^n = 0$. Show that $(A, \cdot)$ is a $\underline{d}$-sum of $\underline{d}$-reducts $(A_i, \underline{d})$ of affine spaces over the ring $R[d]$ with $d^{n-1} = 0$.

5.7I. (a) Show that the directed colimit of n-differential modes $(\mathbb{Z}_{2^n}, \underline{2})$, for $n = 1, 2, \ldots$, is not reductive.

   (b) Show that this directed colimit is a symmetric binary mode.

5.8. Convex sets and barycentric algebras

The class $\underline{C}$ of convex subsets of real vector spaces and the variety $\underline{B}$ generated by the class $\underline{C}$ can be easily generalized by taking any subfield R of the field $\mathbb{R}$ of real numbers instead of the field $\mathbb{R}$. Each such field R inherits the (natural) order of $\mathbb{R}$, so there is an open unit interval $I^\circ =]0, 1[\cap R$ contained in R. Convex sets are defined as in the previous case. The variety $\underline{B}$ of barycentric algebras over the ring R is generated by the class $\underline{Cv}$ of convex subsets of the vector spaces over R.

In what follows we assume that the classes $\underline{Cv}$ and $\underline{B}$ are defined for some fixed subfield R of the field $\mathbb{R}$. Similarly as in the case of barycentric algebras over the field $\mathbb{R}$, the variety $\underline{B}$ contains as a subvariety the variety $\underline{Sl}$ of $\underline{I}°$-semilattices, and also algebras that are neither convex sets nor semilattices. The aim of this section is to give a method for recognizing when a barycentric algebra is a convex set, and to give an axiomatic characterization of the classes $\underline{Cv}$ and $\underline{B}$.

Define the mapping $' : R \to R\ ; r \mapsto 1 - r$. Note that $'$ is involutory, i.e. $r'' = r$, and that $(I°)' = I°$. Let p and q be arbitrary elements of $I°$. Then it is clear that any convex set $(A, \underline{I}°)$ over the ring R satisfies the *idempotent law*

(I) $$xx\underline{p} = x$$

and the *skew commutative* law

(SC) $$xy\underline{p} = yx\underline{p}'.$$

Moreover,

$$\begin{aligned}
xy\underline{p}\ z\ \underline{q} &= x(1-p)(1-q) + yp(1-q) + zq \\
&= xp'q' + ypq' + zq \\
&= xp'q' + \left(y\frac{pq'}{p+q-pq} + z\frac{q}{p+1-pq} \right)(p+q-pq) \\
&= xp'q' + \left(y\frac{pq'}{(p'q')'} + z\frac{q}{(p'q')'} \right)(p'q')'.
\end{aligned}$$

Note that

$$\frac{pq'}{(p'q')'} + \frac{q}{(p'q')'} = 1 \quad \text{and}$$

$$p'q' + (p'q')' = 1.$$

5. INTRODUCTION TO MODES

Hence one sees that $(A, \underline{I}^\circ)$ also satisfies the *skew associative* law

(SA) $\qquad xy\underline{p}\ z\ \underline{q} = x\ yz\underline{(q/(p'q')')}\ \underline{(p'q')'}.$

It is also clear that each barycentric algebra over the ring R satisfies the identities (I), (SC) and (SA). Note that for $p = \frac{1}{2}$, skew commutativity becomes true commutativity, but for no p and q in I° does skew associativity ever become true associativity. Let $\underline{B}'$ be the variety of $\underline{I}^\circ$-algebras $(A, \underline{I}^\circ)$ defined by the identities (I), (SC) and (SA). Obviously $\underline{B} \subseteq \underline{B}'$. We will show that these two varieties coincide. First we describe free $\underline{B}'$-algebras.

Lemma 5.8.1. *Let X be a set. Assume that X is totally ordered by a relation $\leq$. In the free $\underline{B}'$-algebra XB' on the set X, each further element may be expressed in the standard form*

$$(\ldots((x_0 x_1 \underline{p_1}) x_2 \underline{p_2}) \ldots) x_n \underline{p_n}$$

with n a natural number and $x_0 < x_1 < \cdots < x_n$.

Proof. The proof is by induction on the length of the element. The shortest element is $x_0 x_1 \underline{p_1}$ or $x_1 x_0 \underline{p_1}$, and by (SC), each may be written in the standard form. Now consider an element $w_1 w_2 \underline{r}$, and assume by induction that any element of shorter length may be written in the standard form. Suppose w_1 has the standard form $(\ldots((x_0 x_1 \underline{p_1}) x_2 \underline{p_2}) \ldots) x_n \underline{p_n}$ and w_2 has the standard form $(\ldots((y_0 y_1 \underline{q_1}) y_2 \underline{q_2}) \ldots) y_m \underline{q_m}$. Then by induction and skew associativity $w_1 w_2 \underline{r}$ can be written as $w_3 y_m \underline{s}$ for some s in I°, while w_3 can be written in the standard form $(\ldots((z_0 z_1 \underline{s_1}) z_2 \underline{s_2}) \ldots) z_i \underline{s_i}$. If for each $j = 1, \ldots, i$ one has $z_j < y_m$, then $w_3 y_m \underline{s}$ is in the standard form. If there is a j with $z_j < y_m <$

z_{j+1}, then by (SA) and (SC) there are t_k and u_k in $I°$ such that

$$w_3 y_m \underline{s} = ((\ldots(z_0 z_1 \underline{s}_1)\ldots)z_{i-1}\underline{s}_{i-1})(z_i y_m \underline{t}_1)\underline{u}_1$$
$$= ((\ldots(z_0 z_1 \underline{s}_1)\ldots)z_{i-1}\underline{s}_{i-1})(y_m z_i \underline{t}_1')\underline{u}_1$$
$$= ((((\ldots(z_0 z_1 \underline{s}_1)\ldots)z_{i-1}\underline{s}_{i-1})y_m \underline{t}_2)z_i \underline{u}_2$$
$$= \ldots$$
$$= (\ldots(((\ldots(z_0 z_1 \underline{s}_1)\ldots)z_j \underline{s}_j)y_m \underline{t}_l)\ldots)z_i \underline{u}_l,$$

again in the required standard form. □

Now consider the set of functions R^X. Under the usual operations, this is a vector space over R. Under the $\underline{I}°$-operations, R^X is the $\underline{I}°$-reduct of the corresponding affine R-space R^X. For each x in X, define $\delta_x : X \to I$ to be the Kronecker delta function with

$$y \delta_x = \text{if } y = x \text{ then } 1 \text{ else } 0.$$

Let XC be the $\underline{I}°$-subalgebra $\langle \delta_x \mid x \in X \rangle$ of $(R^X, \underline{I}°)$ generated by the functions δ_x. Note that XC is a convex subset of the affine space R^X.

Lemma 5.8.2. *The free $\underline{\underline{B}}'$-algebra XB' is isomorphic to the algebra XC.*

Proof. Let f be a function in XC. By Lemma 5.8.1, the function f can be expressed in the form $f = (\ldots((\delta_{x_0}\delta_{x_1}\underline{p}_1)\delta_{x_2}\underline{p}_2)\ldots)\delta_{x_n}\underline{p}_n$ with $x_0 < x_1 < \cdots < x_n$ and p_i in $I°$. This expression is uniquely defined. Indeed, for $r = n, \ldots, 1$, each $(\ldots(\delta_{x_0}\delta_{x_1}\underline{p}_1)\ldots)\delta_{x_{r-1}}\underline{p}_{r-1}$ is the intersection of the $\underline{I}°$-algebra $\langle \delta_{x_0}, \ldots, \delta_{x_{r-1}} \rangle$ with the line connecting δ_{x_r} and $(\ldots(\delta_{x_0}\delta_{x_1}\underline{p}_1)\ldots)\delta_{x_r}\underline{p}_r$. Now let $g' : (XB', \underline{I}°) \to (XC, \underline{I}°)$ be the homomorphism (uniquely) extending the mapping $g : X \to XC$; $x \mapsto \delta_x$. It is easy to see that the mapping $XC \to XB'; f \mapsto (\ldots((x_0 x_1 \underline{p}_1)x_2 \underline{p}_2)\ldots)x_n \underline{p}_n$ is a well defined two-sided inverse to the homomorphism g'. It follows that $(XB', \underline{I}°)$ and $(XC, \underline{I}°)$ are isomorphic. □

5. INTRODUCTION TO MODES

Corollary 5.8.3. *For any positive integer n and $|X| = n + 1$, the free $\underline{B}'$-algebra XB' is the n-dimensional simplex.* □

Theorem 5.8.4. *The variety $\underline{B}$ of barycentric algebras coincides with the variety $\underline{B}'$ of $\underline{I}°$-algebras defined by the identities* (I), (SC) *and* (SA).

Proof. By Lemma 5.8.2, each $\underline{B}'$-algebra is a homomorphic image of some convex set $XC = XB'$, and hence is a member of the variety $\underline{B}$. It follows that $\underline{B}' \subseteq \underline{B}$, whence $\underline{B}' = \underline{B}$. □

Corollary 5.8.5. *If $(A, \underline{I}°)$ is a barycentric algebra, then so are $(AS, \underline{I}°)$ and $(AP, \underline{I}°)$.*

Proof. This follows by Propositions 5.1.1 and 5.1.3 on noting that skew commutativity and skew associativity are linear identities. □

In the next part of this section, it will be shown that a barycentric algebra $(A, \underline{I}°)$ is a convex set precisely when it is *cancellative*, i.e. it satisfies for each p in $I°$ the following *cancellation* law:

(Cl) $$(xy\underline{p} = xz\underline{p}) \to (y = z).$$

It is easy to see that each convex set is cancellative [Exercise 1.7F]. Conversely, one must show that each cancellative barycentric algebra is a convex set. Let $\underline{Cv}'$ denote the subquasivariety of $\underline{B}$ defined by the cancellation laws (Cl) for all p in $I°$. Obviously $\underline{Cv} \subseteq \underline{Cv}'$.

Theorem 5.8.6. *The quasivariety $\underline{Cv}'$ of cancellative barycentric algebras coincides with the class $\underline{Cv}$ of convex subsets of affine R-spaces.*

Proof. We will show that each cancellative barycentric algebra $(A, \underline{I}°)$ is a convex set. By Lemma 5.8.2 and Theorem 5.8.4, the algebra $(A, \underline{I}°)$ is a homomorphic image of some convex subset $XC = XB$ of the affine R-space

$V = R^X$, say by an $\underline{I}^\circ$-epimorphism $\varphi : XC \to A$. We may also consider the space V as a vector space over R.

Let $S := \{x - y \mid (x,y) \in \ker \varphi\} \subseteq V$. Note that as a congruence of $(A, \underline{I}^\circ)$, the set $\ker \varphi$ is a subalgebra of $(A \times A, \underline{I}^\circ) \le (V \times V, \underline{I}^\circ)$. Hence it is a convex subset of the space $V \times V$. Consider the linear map $\alpha : V \times V \to V$; $(x,y) \mapsto x - y$, a homomorphism of vector spaces. Since $\alpha(\ker \varphi) = S$, it follows that S is a convex subset of V. Since the zero of V is in S, it follows that the vector subspace T of V generated by S is $SR = \{sr \mid s \in S, \ r \in R\}$. Consider the following diagram:

$$\begin{array}{ccc} R^X = V & \xrightarrow{\pi} & V/T = R^X/T \\ {\scriptstyle i}\uparrow & & \uparrow{\scriptstyle j} \\ XC & \xrightarrow{\varphi} & A \end{array}.$$

Here i is the inclusion of XC in V, and π is the natural projection of V onto the quotient vector space V/T. If $(x,y) \in \ker \varphi$, then $x - y \in T$, whence $x\pi = y\pi$ and $(x,y) \in \ker \pi$. Hence $\ker \varphi \subseteq \ker \pi$, so an $\underline{I}^\circ$-homomorphism $j : A \to V/T$ exists making the diagram commute.

We will show that j is injective. It suffices to show that $\ker(i\pi) \subseteq \ker \varphi$. So suppose that $(x,y) \in \ker(i\pi)$. This means that $x - y \in T$. Since $T = SR$, there exists an r in R and (x', y') in $\ker \varphi$ such that $x - y = (x' - y')r$. Without loss of generality, one may assume that $r > 0$. Put $p := r/(r+1)$, so that $0 < p < 1$. Then

$$\begin{aligned} xy'\underline{p} &= x(1-p) + y'p = x\frac{1}{r+1} + y'\frac{r}{r+1} \\ &= (x + y'r)\frac{1}{r+1} \\ &= (y + x'r)\frac{1}{r+1} \\ &= y(1-p) + x'p = yx'\underline{p}, \end{aligned}$$

5. INTRODUCTION TO MODES

the third equality following by $x+y'r = y+x'r$. Since φ is an $\underline{I}^\circ$-homomorphism, one obtains

$$x\varphi \ y'\varphi \ \underline{p} = xy'\underline{p} \ \varphi = yx'\underline{p} \ \varphi = y\varphi \ x'\varphi \ \underline{p}.$$

However $(x', y') \in \ker \varphi$, so that $x'\varphi = y'\varphi$. By the cancellativity of $(A, \underline{I}^\circ)$, it follows that $x\varphi = y\varphi$, so that $(x, y) \in \ker \varphi$, and hence j is injective. Now the injectivity of j implies that A is embeddable in the affine space V/T. Hence A is a convex subset of the space V/T. □

Note that it is possible to relax the cancellation hypothesis in Theorem 5.8.6 as follows.

Proposition 5.8.7. *The class of all barycentric algebras satisfying any one of the cancellation laws* (Cl) *coincides with the quasivariety of cancellative barycentric algebras.*

Proof. It is sufficient to show that if $xy\underline{p} = xz\underline{p}$ holds in a barycentric algebra for some p in I°, then it holds for all such p.

So suppose $ac\underline{p} = bc\underline{p}$ for elements a, b and c of a barycentric algebra $(A, \underline{I}^\circ)$. Then the identity (d) of Exercise 5.8H implies that

$$b \ ac\underline{p} \ (2-p)^{-1} = a \ bc\underline{p} \ (2-p)^{-1}$$

[Exercise 5.8I]. Also by Exercise 5.8H(a) one obtains

$$b \ ac\underline{p} \ (2-p)^{-1} = b \ bc\underline{p} \ (2-p)^{-1} = bc \ \frac{p}{2-p},$$

and similarly $a \ bc\underline{p} \ (2-p)^{-1} = ac\frac{p}{2-p}$. Hence

$$ac\frac{p}{2-p} = bc\frac{p}{2-p}.$$

Now define inductively

$$p_0 := p \quad \text{and} \quad p_{i+1} = \frac{p_i}{2-p_i}.$$

Then $ac\underline{p}_i = bc\underline{p}_i$ for all i. The sequence p_i has limit 0 [Exercise 5.8I], so for arbitrary r in I° one can find an i with $p_i < r$. Let $q := (r-p_i)/(1-p_i)$. Then $0 < q < 1$ and $r = q(1-p_i) + p_i = q - qp_i + p_i = (p'_i q')'$. Now by Exercise 5.8H (b), one obtains the following:

$$ac\underline{r} = ac\underline{(p'_i q')'} = ac\underline{p}_i \, c \, \underline{q}$$
$$= bc\underline{p}_i \, c \, \underline{q} = bc\underline{(p'_i q')'} = bc\underline{r}. \quad \square$$

Exercises

5.8A. Deduce the entropic law for barycentric algebras directly from the skew commutativity and skew associativity.

5.8B. Consider the barycentric algebra $(I, \underline{I}^\circ)$ given by the closed unit interval I of R.

(a) Identify its finitely generated subalgebras. (Cf. Exercise 5.1B.)

(b) Show that $(IP, \underline{I}^\circ)$ is isomorphic to the convex hull of $(0,0), (1,1)$ and $(0,1)$ in R^2, under the $\underline{I}^\circ$-operations.

5.8C. If an affine R-space identity only involves operations $\underline{p}$ with $p \in I^\circ \subseteq R$, then the identity also holds for barycentric algebras. Is it true that skew associativity holds in affine R-spaces for arbitrary p and q in R?

5.8D. Let ABC be a closed triangle in the real plane with vertices A, B and C. Consider the barycentric algebra $(ABC, \underline{I}^\circ)$. (Cf. Exercise 4.5B.) Define the relation θ on ABC as follows:

$(x, y) \in \theta \Leftrightarrow x = y$ or

$\qquad x, y \in AC$ or

$\qquad x, y$ belong to a line segment parallel to AC.

(a) Show that θ is a congruence of $(ABC, \underline{I}^\circ)$.

(b) Show that the quotient $(ABC^\theta, \underline{I}^\circ)$ can be identified with the barycentric algebra $(T, \underline{I}^\circ)$ built as the disjoint union $[0,1] \sqcup]0,1]$,

5. INTRODUCTION TO MODES

where the $\underline{I}^\circ$-operations are defined as follows: For x, y both contained within $[0,1]$ or within $]0,1]$, $xy\underline{p}$ is the usual convex combination. For x in $[0,1]$ and y in $]0,1[$, set $xy\underline{p} := y p \in]0,1]$.

(c) Show that the algebra $(ABC, \underline{I}^\circ)$ has a semilattice as a homomorphic image.

5.8E. [Romanowska, Smith 1990b] Define the set $\mathbb{R}^\infty$ of *extended reals* to be the disjoint union $\mathbb{R} \uplus \{\infty\}$. Show that the algebra $(\mathbb{R}^\infty, \underline{I}^\circ)$, having $(\mathbb{R}, \underline{I}^\circ)$ as a subalgebra and $\infty x \underline{p} = x \infty \underline{p} := \infty$ for any x in $\mathbb{R}^\infty$ and p in I°, is a barycentric algebra. (Such extended reals are very useful in analysis. Cf. Chapter 9.)

5.8F. [Neumann 1970], [Romanowska, Smith 1990b] Let K be a simplicial complex. Let $K^\infty := K \uplus \{\infty\}$. Consider the algebra $(K^\infty, \underline{I}^\circ)$, where $xy\underline{p}$ is the usual convex combination if x and y are contained in a simplex, and otherwise $xy\underline{p} := \infty$. Show that $(K^\infty, \underline{I}^\circ)$ is a barycentric algebra.

5.8G. Let ABC be an open triangle in the real plane. Show that each direction of the plane determines a congruence of the barycentric algebra $(ABC, \underline{I}^\circ)$.

5.8H. Show that the following identities are satisfied in each barycentric algebra $(A, \underline{I}^\circ)$ for any p, q and r in I° with $pq\underline{r} = pr' + qr \neq 0$:

(a) $\qquad x\ xy\underline{p}\ \underline{q} = xy\ \underline{pq}$;

(b) $\qquad xy\underline{p}\ y\ \underline{q} = xy\ \underline{(p'q')'}$;

(c) $\qquad x\ yz\underline{p}\ \underline{q} = xy\ \dfrac{p'q}{(pq)'}\ z\ \underline{pq}$;

(d) $\qquad xy\underline{p}\ xz\underline{q}\ \underline{r} = x\ yz\ \dfrac{qr}{pq\underline{r}}\ \underline{pq\underline{r}}$.

[Hint: Make repeated use of the skew associativity (SA).]

5.8I. Complete the proof of Proposition 5.8.7.

5.8J. [Neumann 1970] Let $(A, \underline{I}^\circ)$ be a barycentric algebra, and let a, b be in A. Show that if $ab\underline{p} = ab\underline{q}$ for some distinct p and q in I°, then $ab\underline{p} = ab\underline{q}$ for all such p and q.

[Hint: The proof is similar to, but easier than, the statement in the proof of Proposition 5.8.7.]

5.8K. Let A be an affine space over a subfield R of the real field $\mathbb{R}$. If A is finite-dimensional, then we consider A with the usual topology induced from $\mathbb{R}$. If A has infinite dimension, a subset X of A is defined to be open if the intersection $X \cap B$ of X with each finite dimensional subspace B of A is open in B. Call this topology τ. Show that, with respect to the topology τ, the following hold for each convex subset C of A:

(a) C is an open subset of A iff for each pair (a, b) of elements of C, there is x in C and r in I° such that $ax\underline{r} = b$.

(b) C is a closed subset of A iff for each pair (a, b) of elements of C, the intersection of the subalgebras $(C, \underline{I}^\circ)$ and $(\langle a, b \rangle, \underline{I}^\circ)$ of $(C, \underline{I}^\circ)$ is finitely generated.

5.9. Free modes on two generators

If $\underline{V}$ is a variety of modes, then the free $\underline{V}$-algebra $\{0, 1\}V$ on the set $\{0, 1\}$ carries a great deal of structure. Of course, it is a $\underline{V}$-algebra. We will show that it also carries two monoid structures.

Example 5.9.1. Consider the free barycentric algebra $\{0, 1\}B = (I, \underline{I}^\circ)$ over a subfield R of the real field $\mathbb{R}$ (cf. Corollary 5.8.3). Besides being a barycentric algebra under the operations $\underline{I}^\circ$, the set I is a subalgebra of the monoid $(R, \cdot, 1)$ with the usual multiplication $\cdot$ of real numbers. The closed unit interval I also carries another monoid structure $(I, \circ, 0)$, where

(5.9.1) $$p \circ q := p + q - pq.$$

5. INTRODUCTION TO MODES

The operation $\circ$ is the so-called *circle composition*. There is also a monoid isomorphism
$$' : (I, \cdot, 1) \to (I, \circ, 0) \; ; \; p \mapsto 1 - p.$$
This isomorphism is *involuntary*, i.e. $p'' = p$ for each p in I. Note, too, that 1 is a *zero* for the circle composition, i.e. $1 \circ p = 1 = p \circ 1$. Then of course 0 is a zero for the usual multiplication. $\square$

Example 5.9.1 leads to the following:

Definition 5.9.2. A *dual monoid* is a set A equipped with two monoid structures $(A, \cdot, 1)$ and $(A, \circ, 0)$ together with an involuntary isomorphism
$$' : (A, \cdot, 1) \to (A, \circ, 0).$$
The *zero element* 0 is a zero for the *multiplication* $\cdot$ and the *identity element* 1 is a zero for the *circle composition* $\circ$. A dual monoid is *commutative* if its monoid structures are commutative. $\square$

Dual monoids form the variety of algebras $(A, \cdot, \circ, ', 0, 1)$ of type $(\{\cdot, \circ\} \times \{2\}) \cup \{(', 1)\} \cup (\{0, 1\} \times \{0\})$ defined by the following identities:

$$x \cdot (y \cdot z) = (x \cdot y) \cdot z,$$
$$x \circ (y \circ z) = (x \circ y) \circ z,$$
$$x \cdot 1 = x = 1 \cdot x, \qquad x \circ 0 = x = 0 \circ x,$$
$$x \cdot 0 = 0 = 0 \cdot x, \qquad x \circ 1 = 1 = 1 \circ x,$$
$$x'' = x, \qquad 0' = 1,$$
$$(x \cdot y)' = x' \circ y',$$
$$(x \circ y)' = x' \cdot y'.$$

The last two identities, the so-called *De Morgan laws*, show that once the multiplication $\cdot$ and the involution $'$ with $0' = 1$ are specified, the circle composition $\circ$ is defined by means of $\cdot$ and $'$.

Example 5.9.3. Boolean algebras $(A, \vee, \wedge, ', 0, 1)$ provide obvious examples of dual monoids. □

Example 5.9.4. Each ring R can be considered as a dual monoid RM. The monoid structures are $(R, \cdot, 1)$ and $(R, \circ, 0)$, the circle composition being given by (5.9.1), while the involution is $' : R \to R \; ; \; r \mapsto 1 - r$. The dual monoid RM is commutative iff the ring R is. □

Now consider the free $\underline{V}$-algebra $\{0,1\}V$. The elements w of $\{0,1\}V$ determine binary derived operations $w : A \times A \to A$ on $\underline{V}$-algebras (A, Ω). It is sometimes convenient to denote elements of $\{0,1\}V$ by the corresponding operations. Then in particular $w_1w_20 = w_1$, while $w_1w_21 = w_2$ and $01w = w$.

Theorem 5.9.5. *Let $\underline{V}$ be a variety of modes. Then the free $\underline{V}$-algebra $\{0,1\}V$ on the set $\{0,1\}$ carries the structure of a commutative dual monoid. The multiplication $\cdot$ is given by*

$$xy(w_1 \cdot w_2) := x \; xyw_1 \; w_2,$$

the zero element is 0, the identity element is 1, and the involution $'$ is given by

$$xy(w') := yxw.$$

Proof. The proof is a direct verification of the requirements of Definition 5.9.2. For the commutativity of the multiplication, one has

$$xy(w_1 \cdot w_2) = x \; xyw_1 \; w_2 = xxw_1 \; xyw_1 \; w_2$$
$$= xxw_2 \; xyw_2 \; w_1 = x \; xyw_2 \; w_1$$
$$= xy(w_2 \cdot w_2).$$

The associativity may be obtained similarly:

$$xy(w_1 \cdot (w_2 \cdot w_3)) = x \; xyw_1 \; (w_2 \cdot w_3)$$
$$= x \; (x \; xyw_1 \; w_2) \; w_3 = x \; (xy(w_1 \cdot w_2)) \; w_3$$
$$= xy((w_1 \cdot w_2) \cdot w_3).$$

Then $xy(1 \cdot w) = x\ xy1\ w = xyw$, so that $(\{0,1\}V, \cdot, 1)$ is a commutative monoid. Further $xy(0 \cdot w) = x\ xy0\ w = xxw = x = xy0$, so that 0 is a zero for the multiplication and $xy0' = yx0 = y = xy1$. □

Example 5.9.6. Consider the free semilattice $\{0,1\}$Sl on 0 and 1, written as a join semilattice. It has three elements 0, 1 and p, where $xyp = x + y$. The monoid operation · of Theorem 5.9.5 gives $xy(p \cdot p) = x\ xyp\ p = xyp$, so the monoid is in fact a meet semilattice corresponding to the chain $0 \to p \to 1$. Note the ordering of the original join semilattice $0 \to p \leftarrow 1$. Note as well that $\{0,1\}$Sl with both semilattice structures $+$ and · is a distributive bisemilattice. (Cf. Exercise 4.3K.) On the other hand, the full commutative dual monoid structure $(\{0,1\}\text{Sl}, \circ, \cdot)$ is the lattice corresponding to the ordering $0 \to p \to 1$. □

Example 5.9.7. Consider the variety <u>Nb</u> of normal bands. The free normal band $\{0,1\}$Nb has elements 0, 1, p with $xyp = xy$, p', b with $xyb = xyx$, and b'. Then the commutative dual monoid structure $(\{0,1\}\text{Nb}, \circ, \cdot)$ is the lattice $(\{0,1\}\text{Nb}, \vee, \wedge)$ corresponding to the ordering

$$\begin{array}{ccccc} p & \longrightarrow & b' & \longrightarrow & 1 \\ \uparrow & & \uparrow & & \\ 0 & \longrightarrow & b & \longrightarrow & p' \end{array} \quad . \quad \square$$

In general, the dual monoid structure on the free <u>V</u>-algebra $\{0,1\}V$ is not a lattice. The multiplication does not need to be idempotent.

The subalgebras of the dual monoid RM of Example 5.9.4 obtained from a ring R are sometimes called *algebraic intervals* [Jamison 1974]. Note that for rings R and S, the assignment M:

$$\begin{array}{ccc} S & & SM \\ \varphi \uparrow & \xmapsto{M} & \uparrow \varphi M \\ R & & RM \end{array},$$

where φ is a ring homomorphism and φM is the corresponding dual monoid homomorphism, defines a functor M from the category $\underline{\text{Ring}}$ of rings to the category $\underline{DM}$ of dual monoids. Given an algebraic interval J in the ring R, and an algebraic interval K in the ring S, one then obtains other algebraic intervals as the images of J under φM and preimages of K. Note that the set $\{0, 1\}$ forms an algebraic interval in any ring.

Example 5.9.8. The unit interval $\mathbb{D}_1$ of the ring $\mathbb{D}$ of rational dyadic numbers (cf. Subsection 5.5.2) forms an algebraic interval. Similar algebraic intervals are obtained from unit intervals of subrings R of the ring $\mathbb{R}$ consisting of numbers m/n with n a power of a prime number p. □

The algebraic intervals of Example 5.9.8 have two more special properties: each such interval J is closed under the operations $\underline{r}$ for r in J, and the additive group of the ring is generated by the interval. More generally, for any commutative ring R, an algebraic interval J different from $\{0, 1\}$ and R is called an *algebraic unit interval* if it is closed under $\underline{r}$ for r in J, and the additive group of R is generated by J. Note that the unit interval I of $\mathbb{R}$ and the unit intervals of Example 5.9.8 are algebraic unit intervals of the corresponding rings $\mathbb{R}$ and R, respectively. Note however that $\mathbb{D}_1$ is also an algebraic interval in $\mathbb{R}$, but is not an algebraic unit interval in this ring.

Exercises

5.9A. Show that the free algebra $\{0, 1\}V$ in the subvariety $\underline{V}$ of $\underline{C}$ defined by the identities $xy^2 = x = y^2x$ consists of three elements $0, 1$ and p with the following multiplication table:

$\cdot$	0	1	p
0	0	p	1
1	p	1	0
p	1	0	p

Find the corresponding dual monoid.

5. INTRODUCTION TO MODES

5.9B. Extend the definition of a barycentric algebra to (ordered) subrings of the real ring $\mathbb{R}$, in particular to the rings of Example 5.9.8.

(a) Are the identities (I), (SC) and (SA) satisfied in such "barycentric" algebras?

(b) By Theorem 5.5.2.3, all the binary operations $\underline{d}$ for d in the open dyadic unit interval $\mathbb{D}_1^\circ$ defined on the closed interval $\mathbb{D}_1$ are generated by one operation $\underline{\frac{1}{2}}$. How many operations are needed to generate all the binary operations $\underline{d}$ for d in the open algebraic intervals of Example 5.9.8?

5.9C. Are there algebraic unit intervals in finite fields?

5.9D. Find an example of a subfield of the real field $\mathbb{R}$ having more than one algebraic unit interval.

5.9E. [Kearnes 1996] Let R be $\mathbb{Q}[\sqrt{2}]$. Show that $J = \mathbb{Q}(\sqrt{2}) \cap [0,1]$ is an algebraic unit interval of $\mathbb{Q}[\sqrt{2}]$. Find other algebraic unit intervals of $\mathbb{Q}[\sqrt{2}]$.

(Hint: Find non-trivial $\mathbb{Q}$-automorphisms of $\mathbb{Q}[\sqrt{2}]$.)

5.9F. [Movsisyan 1998] Let X be a non-empty set, and let $(S, \cdot)$ be a semigroup [or let $(S, \cdot, 1)$ be a monoid]. Define a mapping

$$\varphi : S \to S^{X \times X} \; ; \; s \mapsto (s\varphi : X \times X \to X \; ; \; (x,y) \mapsto xys).$$

Define a multiplication of two functions f and g in $X^{X \times X}$ as follows:

$$xy(f \cdot g) := x \; xyf \; g$$

(and an operation 1 by $xy1 := y$).

(a) Show that

$$\varphi : (S, \cdot, 1) \to (X^{X \times X}, \cdot, 1)$$

is a monoid homomorphism.

The homomorphism φ is called a *binary action of S on the set X*. The action is *faithful* if φ is a monomorphism. To each action φ there corresponds an algebra (X, S) with a semigroup (monoid) of binary operations s, for s in S, satisfying the identities

$$xy(s_1 \cdot s_2) = x\ xys_1\ s_2.$$

(b) Show that for each monoid $(S, \cdot, 1)$, there is a faithful binary action φ on the set S such that the algebra (S, S) satisfies the following identities for each s in S:

$$x\ yzs\ s = xys\ xzs\ s \quad \text{and}$$
$$xys\ uvs\ s = xus\ yvs\ s.$$

(c) Show the same for the semigroup $(S, \cdot)$.

(d) Show that if there is an action of a semigroup $(S, \cdot)$ on S as in (b), then $(S, \cdot)$ is commutative.

Notes on Chapter 5.

The entropic identity was first considered in the case of groupoids and quasigroups in the late nineteen-thirties. Probably the first explicit note is contained in [A.K. Sushkevich 1937, p. 157]. Subsequent papers concerned entropic quasigroups and groupoids. The first paper on the subject was [D.C. Murdoch 1939]. He obtained the entropic law as a generalization of the associative law for quasigroups. For this reason, he called quasigroups satisfying this identity "Abelian". Among other early papers concerning entropic quasigroups are [G.N. Garrison 1940], [K. Toyoda 1940a,b, 1941], [D.C. Murdoch 1941a,b], [A.A. Albert 1943], [R.H. Bruck 1944], [A. Sade 1957, 1959] and [S.K. Stein 1956], and concerning entropic groupoids [I.M.H. Etherington 1949, 1958a,b], [M. Sholander 1949], [D. Frink 1955], [V.D. Belousov 1958].

5. INTRODUCTION TO MODES

The next twenty years brought further development with the papers of e.g. [T. Evans 1961, 1962a,b, 1963], [I.M.H. Etherington 1964], [J. Morgado 1966, 1967], [J. Soublin 1966, 1971], [J. Duplák 1974], [W. Donell 1974], [L. Bénéteau 1975], [D.R. Cecil 1975], [J. Ježek and T. Kepka 1975a,b, 1976, 1977], [P. Das 1976, 1977], [M.A. Taylor 1978], [J. Barnhardt 1978], [J.D.H. Smith 1979], [T. Kepka 1975, 1976, 1978a,b, 1979] and others.

The entropic law appeared in many geometric examples [M. Sholander 1949], [C. Fulton and S. Stein 1957], [I.M.H. Etherington 1964/65], [N.K. Pucharew 1968], [D. Merriel 1970], [E. Doraczyńska 1974], [W. Szmielew 1983], and in the characterization of mean value functions [J. Aczél 1948], [T. Howroyd 1973].

The monograph of J. Ježek and T. Kepka [1983] was the first systematic treatment of entropic groupoids. Note however that the authors used the name "medial" instead of entropic, and reserved the name "entropic" for the class of homomorphic images of "medial" cancellative groupoids.

A more general form of the entropic law was studied in [J. Aczél 1948], [T. Evans 1961, 1962a,b] and [L. Klukovits 1973, 1983]. The equivalence of the entropic law and the homomorphism closure property of Proposition 5.1 was shown in [T. Evans 1961, 1962b] and [L. Klukovits 1975].

A discussion of tensor products in entropic varieties was given in [B.A. Davey and G. Davis 1985]. Their good behaviour may be contrasted with the case of semigroups [P.A. Grillet 1969]. For applications to the study of communicating processes, see [A. Romanowska and J.D.H. Smith 1990a].

Hyperidentities and hypervarieties were first investigated by W. Taylor [1981], and in connection with modes by S. Arworn and K. Denecke [1999a,b]. See also [V.D. Belousov 1965] and [Yu.M. Movsisyan 1998].

The characterization of modes given in Proposition 5.6 was observed by K. Kearnes [1999], and presented in [J.D.H. Smith 1999].

The first papers explicitly devoted to idempotent entropic groupoids were probably those of A. Mitschke [1973], J. Dudek [1974], S. Fajtlowicz and J. Mycielski [1974], G. Grätzer and P. Padmanabhan [1971] and J. Ježek and T. Kepka [1975b, 1976, 1983a].

The first systematic study of modes in general was undertaken in the monograph [MT] of A. Romanowska and J.D.H. Smith [1985c]. Among other things, the monograph contains an extensive study of complex modes of submodes, following the earlier papers of the same authors [1981], [1989a,b]. For complex set-algebras see [A. Shafaat 1974], [G. Grätzer and S. Whitney 1984], [G. Grätzer and H. Lakser 1988], and [C. Brink 1993].

The first results concerning semigroup modes go back to [M. Yamada and N. Kimura 1958]. See also [S.K. Stein 1957], and for entropic semigroups [T. Tamura 1968] and [J.L. Chrislock 1969].

Diagonal algebras were investigated by J. Płonka [1966 a,b], and their generalization by R. Pöschel and M. Reichel [1993].

So far as we know, the first paper considering affine spaces as algebras was [F. Ostermann and J. Schmidt 1966] (see also [W.D. Neumann 1970] and [W. Bos and G. Wolff 1978a,b]), although some ideas may already be found in [M.H. Stone 1949]. It was shown that affine spaces over a certain ring form a variety. Varieties of groupoids equivalent to affine spaces over certain rings were studied by A. Mitschke and H. Werner [1973]. Proposition 5.4.1.2 was first published in that paper. Affine spaces over prime fields were studied in [B. Csákány 1975b], and varieties of quasigroup modes (equivalent to affine spaces) in [B. Csákány and L. Megyesi 1975]. See also [B. Csákány and L. Megyesi 1979] for the generalization to n-ary quasigroup modes, as well as [T. Evans 1976], [V.D. Belousov 1972]. Varieties of "affine modules" over not necessarily commutative rings were investigated in [B. Csákány 1975a,c] and [A. Szendrei 1975].

Idempotent and entropic quasigroups, i.e. quasigroup modes, appeared also

5. INTRODUCTION TO MODES

already in various earlier papers [T. Mituhisa 1943], [J. Aczél 1948], [M. Sholander 1949], [S. Stein 1957], and later in connection with combinatorics, see e.g. [N.S. Mendelsohn 1970], [C.C. Lindner and N.S. Mendelsohn 1973, 1976], [T. Evans 1976, 1982a,b], [M. J. Pelling and D. G. Rogers 1978, 1979]. Many further examples may be found in the survey article [O. Chein 1990]. A bibliography of subsequent papers concerning affine spaces and quasigroup modes will be provided in Chapter 6.

Kei (reflexion) modes appeared for the first time in [T. Mituhisa 1943]. The first description of the free kei modes goes back to [C.C. Lindner and N.S. Mendelsohn 1976]. The approach presented here was given in [MT].

Symmetric binary modes were studied in the doctoral dissertation of B. Roszkowska [1987b]. See also [B. Roszkowska 1987a, 1989, 1999a, 1999b]. Free symmetric binary modes were described in [D. Joyce 1982] and [B. Roszkowska 1987a]. More general idempotent symmetric and (right) distributive groupoids were investigated by R.S. Pierce [1978, 1979]. Motivation for studying such groupoids came from differential geometry and knot theory. See [O. Loos 1968] and [M. Kikkawa 1973, 1974, 1975] for connections with symmetric spaces, [N. Endres 1991, 1993, 1994, 1995, 1996a,b] for connections with algebraic topology, and [D. Joyce 1982a] for connections with knot theory.

Free commutative binary and quasigroup modes were investigated by J. Ježek and T. Kepka [1976]. The approach in this book follows the monograph [MT]. Different mode structures on the set of dyadic rationals were discussed in [Z. Chrapek 1987].

The first examples of differential groupoids appearing in the literature were J. Płonka's "cyclic groupoids" [1971a, 1985a]. The so-called 2-cyclic groupoids form one of four classes of algebras having exactly n n-ary derived operations depending on all their variables for each $n = 1, 2, \cdots$. Differential groupoids in general were studied in [A. Romanowska and B. Roszkowska 1987, 1989] and [A. Romanowska 1988] under the acronymic name "LIR-groupoids" (Left-normal Idempotent Reductive Groupoids). The name "differential mode" appeared in [A. Romanowska and J.D.H. Smith 1991a], investigating homological and differential properties of differential modes. See also [Jung R. Cho 1996]. The more general n-differential modes were investigated under the name of "n-reductive modes" in [A. Pilitowska, A. Romanowska and B. Roszkowska 1996], [A. Pilitowska and A. Romanowska 1998]. (See also [J. Krauze 1998].) The $\underline{r}$-sum construction was introduced by A. Romanowska and B. Roszkowska [1997] in unpublished notes.

Convex sets were described as algebras in [W.D. Neumann 1970]. Neumann provided a characterization of real barycentric algebras, and of convex sets among barycentric algebras as the cancellative ones. For a different proof, and for other characterizations, see [MT] and [A. Romanowska and J.D.H. Smith 1990b]. Barycentric algebras have been studied by several authors under several different names, using different axiomatizations, and concentrating on different aspects of the algebras. See [J. Sobczyńska 1988] for an analysis and comparison of different axiomatizations. We have adopted the set of axioms given by T. Świrszcz [1974, 1975] (who used the name "semiconvex sets"), and who was interested in categorical aspects of convex sets and barycentric algebras. The same name was used by J. Flood [1981], who studied algebraic and geometric aspects. A different axiomatization, and the name "convex prestructures", were given by S.P. Gudder [1973], who applied these algebras to some very interesting topics in physics. (See also [1977] and [1978].) L.A. Skornyakov and his student V.V. Ignatov studied barycentric algebras under the name "convexors". Skornyakov was interested in algebraic aspects [1981, 1982b] and applications in automata theory [19982a, 1985]. Ignatov studied the structure of "convexors" over the reals [1986] and some more general rings [1984]. Some more detailed information will be given in Chapter 7. Finally, one

5. INTRODUCTION TO MODES

should mention a group of category theorists, most notably D. Pumplün and H. Röhrl, working on general "convexity theories" [D. Pumplün 1984, 1985], [H. Röhrl 1991a,b, 1993, 1994] and [D. Pumplün and H. Röhrl 1985a,b, 1990, 1991, 1995]. The "real convexity theory" [D. Pumplün and H. Röhrl 1995] provides algebras they call "convex modules" that are precisely the real barycentric algebras. For general information about convex sets, see [T. Bonnesen and W. Fenchel 1948], [A. Brønsted 1983] and [B. Grünbaum 1967]. For categorical aspects of general modes and entropic algebras, see [F. Linton 1965, 1966] and [A. Kurpiel 1987].

Dual monoids and their rôle for the structure of free algebras on two generators in mode varieties were discussed in [MT]. Algebraic intervals were introduced by R.E. Jamison-Waldner [1974], and algebraic unit intervals were considered by K.A. Kearnes [1996].

CHAPTER VI

MAL'CEV MODES and AFFINE SPACES

The aim of this chapter is to investigate varieties of modes that are equivalent to varieties of affine spaces. We will show that a variety of modes is equivalent to a variety of affine spaces over a commutative ring if and only if it is a variety of Mal'cev modes. Mal'cev modes can be characterized as modes having a ternary operation generalizing the Mal'cev parallelogram operation. The chapter begins with a discussion of general Mal'cev operations and Mal'cev varieties, and then examines connections between the Mal'cev and entropic identities. The central part of the chapter deals with the equivalence theorem mentioned above, and some of its consequences. The latter sections are devoted to affine spaces equivalent to groupoids and quasigroups.

6.1. Mal'cev varieties

A variety $\underline{V}$ of algebras is said to be a *Mal'cev variety* if the congruences on each $\underline{V}$-algebra commute. The name comes from the following theorem of Mal'cev.

Theorem 6.1.1. *Let $\underline{V}$ be a variety of Ω-algebras. Then $\underline{V}$ is a Mal'cev variety if and only if there is a word $xyzP$ in $X\Omega$ such that the identities*

(6.1.1) $$xyyP = x = yyxP$$

are satisfied in $\underline{V}$.

Proof. ($\Leftarrow$) Let A be in $\underline{V}$. Then A satisfies the identities (6.1.1). Suppose that φ and ψ are in CgA, and let $(a,c) \in \varphi \circ \psi$, say $a \, \varphi \, b \, \psi \, c$. Then

$$a = accP \, \psi \, abcP \, \varphi \, aacP = c,$$

whence $a\psi \, abcP \, \varphi \, c$ and $(a,c) \in \psi \circ \varphi$. By symmetry, φ and ψ commute.

($\Rightarrow$) Conversely, let $\underline{V}$ be a Mal'cev variety. Let $F = \{x,y,z\}V$ be the free $\underline{V}$-algebra on the set $\{x,y,z\}$. Consider the congruences $\varphi := cg(x,y)$ and $\psi := cg(y,z)$ on F. Then $x \, \varphi \, y \, \psi \, z$. Since φ and ψ commute, there is an element $xyzP$ in F such that $x \, \psi \, xyzP \, \varphi \, z$. Then $x^\psi = x^\psi y^\psi y^\psi P$ and $z^\varphi = x^\varphi x^\varphi z^\varphi P$. But F^φ and F^ψ are free $\underline{V}$-algebras on the sets $\{x^\varphi, z^\varphi\}$ and $\{x^\psi, y^\psi\}$ respectively. Thus $x = xyyP$ and $z = xxzP$ are identities true in $\underline{V}$, whence (6.1.1) follows. $\square$

The operation P determined in $\underline{V}$-algebras by the word $xyzP$ is called a *Mal'cev operation*, and the identities (6.1.1) are called the *Mal'cev identities*. The Mal'cev parallelogram operation of affine spaces is an obvious example of a Mal'cev operation.

The first consequence of Theorem 6.1.1 is the following very useful property of Mal'cev varieties.

Proposition 6.1.2. *Let $\underline{V}$ be a Mal'cev variety, and let A be a $\underline{V}$-algebra. Then a subalgebra β of the algebra $A \times A$ is a congruence on A if and only if it is a reflexive relation.*

Proof. It is enough to show that a reflexive subalgebra β of $A \times A$ is a symmetric and transitive relation on A. Note that $\widehat{A} \subseteq \beta$. Let (a,b) and (b,c) be in β. Then $b = aabP \, \beta \, abbP = a$, so that β is symmetric. Also $a = abbP \, \beta \, bbcP = c$, so that β is transitive. $\square$

The Mal'cev Theorem 6.1.1 enables one to see that many familiar varieties are Mal'cev varieties. Taking $xyzP := xy^{-1}z$ shows that each variety of groups

6. MAL'CEV MODES

is a Mal'cev variety, as are varieties of rings, modules, etc. Less obviously, quasigroups form a Mal'cev variety with $xyzP := (x/(y \setminus y)) \cdot (y \setminus z)$ [Exercise 6.1A].

Another characteristic feature of Mal'cev algebras is that the so-called centreing congruences are unique.

Definition 6.1.3. Let $\underline{V}$ be a variety of Ω-algebras, and let A be a $\underline{V}$-algebra. Let β and γ be congruences on A. Consider β as a $\underline{V}$-algebra (β, Ω), a subalgebra of the direct square $(A \times A, \Omega)$. (Cf. Section 1.2.) One says that γ *centralizes β by means of a centreing congruence* $(\gamma|\beta) \in Cg\beta$ if and only if the following hold:

(C0) $(a,b) \ (\gamma|\beta) \ (a',b') \Rightarrow a \ \gamma \ a'$;

(C1) $\forall (a,b) \in \beta$, the mapping

$$\pi_1 : (a,b)^{(\gamma|\beta)} \to a^\gamma \ ; \ (a',b') \mapsto a'$$

bijects;

(C2) $a \gamma b \Rightarrow (a,a) \ (\gamma|\beta) \ (b,b)$; (RR)

$(a,b) \ (\gamma|\beta) \ (a',b') \Rightarrow (b,a) \ (\gamma|\beta) \ (b',a')$; (RS)

$(a,b) \ (\gamma|\beta) \ (a',b')$ and $(b,c) \ (\gamma|\beta) \ (b',c')$

$\Rightarrow (a,c) \ (\gamma|\beta) \ (a',c')$. (RT)

The conditions (RR), (RS) and (RT) are described as *respect for the reflexivity, symmetry* and *transitivity of β*, respectively. The situation is best illustrated by the following diagram:

$$\begin{array}{ccc} b & \xleftarrow{\quad \gamma \quad} & b' \\ \| & & \| \\ \beta & & \beta \\ \| & & \| \\ a & \xleftarrow{\quad \gamma \quad} & a' \end{array} \ .$$

Note the following basic properties of the centreing congruences $(\gamma|\beta)$. (For the proofs, see Exercise 6.1B.)

(6.1.2) If γ centralizes β by means of $(\gamma|\beta)$, then β centralizes γ by means of $(\beta|\gamma) \in Cg\gamma$ defined by

$$(a,a')\,(\beta|\gamma)\,(b,b') :\Leftrightarrow (a,b)\,(\gamma|\beta)\,(a',b').$$

(6.1.3) If γ centralizes β by means of $(\gamma|\beta)$, and $\beta' \leq \beta$ for $\beta \in CgA$, then γ centralizes β' by means of $(\gamma|\beta') := (\gamma|\beta) \cap (\beta' \times \beta')$.

(6.1.4) For each $\alpha \in CgA$, α centralizes $\hat{A}$ by means of a unique congruence $(\alpha|\hat{A})$:

$$(a,a)\,(\alpha|\hat{A})\,(b,b) :\Leftrightarrow a\,\alpha\,b.$$

Proposition 6.1.4. Let $\underline{V}$ be a Mal'cev variety of Ω-algebras. Let A be a $\underline{V}$-algebra. Let β and γ be in CgA. If γ centralizes β by means of two congruences $(\gamma|\beta)_1$ and $(\gamma|\beta)_2$ of (β, Ω), then $(\gamma|\beta)_1 = (\gamma|\beta)_2$.

Proof. Let $a\,\beta\,b$ and $a\,\gamma\,a'$. Then for $i = 1, 2$, reflexivity and (RR) imply that $(a,b)\,(\gamma|\beta)_i\,(a,b)$, that $(a,a)\,(\gamma|\beta)_i\,(a,a)$ and that $(a,a)\,(\gamma|\beta)_i\,(a',a')$. It follows that $(a,b) = (aaaP, baaP) = (a,b)(a,a)(a,a)P\,(\gamma|\beta)_i\,(a,b)(a,a)(a',a')P = (aaa'P, baa'P) = (a', baa'P)$. Thus if $(a,b)\,(\gamma|\beta)_1\,(a',b')$, then $(a',b')\,(\gamma|\beta)_1\,(a', baa'P)$, and by (C1), $b' = baa'P$. Similarly, if $(a,b)\,(\gamma|\beta)_2\,(a',b')$, then by (C1) again, $b' = baa'P$. $\square$

Proposition 6.1.5. Let $\underline{V}$ be a Mal'cev variety of Ω-algebras. Let β and γ be congruences on (A, Ω) such that γ centralizes β by means of $(\gamma|\beta)$. Then for $b\,\beta\,a\,\gamma\,a'$ one has

$$(a,b)\,(\gamma|\beta)\,(a',b') \Leftrightarrow b' = baa'P.$$

Similarly, for $a\,\beta\,b\,\gamma\,b'$, one has

$$(a,b)\,(\gamma|\beta)\,(a',b') \Leftrightarrow a' = abb'P.$$

Proof. First note that for $b\ \beta\ a\ \gamma\ a'$, one has

$$(a,b)\ (\gamma|\beta)\ (a,b),$$
$$(a,a)\ (\gamma|\beta)\ (a,a),$$
$$(a,a)\ (\gamma|\beta)\ (a',a') \qquad \text{by (RR)}.$$

On applying the Mal'cev operation P columnwise, one obtains

$$(a,b)\ (\gamma|\beta)\ (a', baa'P).$$

If $(a,b)\ (\gamma|\beta)\ (a',b')$, then $(a',b')\ (\gamma|\beta)\ (a', baa'P)$, so that $b' = baa'P$ follows by property (C1). The other statement is proved similarly. $\square$

A centreing congruence $(A \times A | A \times A)$ will play an essential rôle in the next section. If the algebra $(A \times A, \Omega)$ has a centreing congruence $(A \times A | A \times A)$, then the algebra (A, Ω) is said to be *central*. Note that Proposition 6.1.5 implies that in a central algebra (A, Ω), for all a, a' and b in A, one has the following:

(6.1.5) $\qquad \begin{cases} (a,b)\ (A \times A | A \times A)\ (a',b') \\ \Leftrightarrow b' = baa'P \\ \Leftrightarrow a' = abb'P. \end{cases}$

Proposition 6.1.6. *Let (A, Ω) be an algebra in a Mal'cev variety $\underline{V}$. Let α be a congruence on $(A \times A, \Omega)$. Then α is the centreing congruence $(A \times A | A \times A)$ if and only if for each $a \in A$, one has $(a,a)^\alpha = \widehat{A}$.*

Proof. ($\Rightarrow$) Assume that $\alpha = (A \times A | A \times A)$. Then (6.1.5) holds. Hence for any $a \in A$, one has $(a,a)^\alpha \leq \widehat{A}$, since $(a,a)\ \alpha\ (b,b)$ iff $b = aabP$. Similarly, $\widehat{A} \subseteq (a,a)^\alpha$, since $(a,a)\ \alpha\ (b,c)$ iff $c = aabP = b$.
($\Leftarrow$) Assume that for each $a \in A$ one has $\widehat{A} = (a,a)^\alpha$. Then the conditions (C0) and (RR) are obviously satisfied. Now $(a,b)\ \alpha\ (a',b')$ and $(b,c)\ \alpha\ (b',c')$

together with $(b,b) \; \alpha \; (b',b')$ imply

$$(a,c) = (abbP, bbcP)$$
$$= (a,b)(b,b)(b,c)P$$
$$\alpha \; (a',b')(b',b')(b',c')P$$
$$= (a'b'b'P, b'b'c'P)$$
$$= (a',c'),$$

whence (RT) is satisfied.

To check (RS), let $(a,b) \; \alpha \; (a',b')$. Since $(a,a) \; \alpha \; (a',a')$, $(a,b) \; \alpha \; (a',b')$ and $(b,b) \; \alpha \; (b',b')$, then applying the operation P to both sides again, one obtains $(b,a) \; \alpha \; (b',a')$.

To see the surjectivity of the mapping $\pi_1 : (a,b)^\alpha \to A$; $(a',b') \mapsto a'$, note that $(a,b) \; \alpha \; (a,b)$, $(a,a) \; \alpha \; (a,a)$ and $(a,a) \; \alpha \; (c,c)$. Applying the operation P to both sides as before, one obtains $(a,b) \; \alpha \; (c, bacP)$. Hence for each $c \in A$ there is $d = bacP$ such that $(a,b) \; \alpha \; (c,d)$. The injectivity of π_1 is proved similarly. Let $(c,c') \; \alpha \; (c,c'')$. Then since $(c,c) \; \alpha \; (c,c)$ and $(c',c) \; \alpha \; (c',c)$, one obtains, in a similar way as before, that $(c',c') \; \alpha \; (c',c'')$, which implies $c' = c''$. This shows that (C1) holds. □

Exercises

6.1A. Show that the word $xyzP := (x/(y \setminus y)) \cdot (y \setminus z)$ determines a Mal'cev operation in each quasigroup.

[Hint: Note that $x \cdot (x \setminus x) = x$, and hence $x/(x \setminus x) = x$.]

6.1B. Prove (6.1.2), (6.1.3) and (6.1.4).

6.1C. Show that in groups, $xyzP_1 := xy^{-1}z$ and $xyzP_2 := zy^{-1}x$ define two different Mal'cev operations.

6.1D. In a group, consider the commutator $[x,y] = x^{-1}y^{-1}xy$. Show that $xyzP = xy^{-1}[[x,y],[y,z]]z$ is a Mal'cev operation.

6.2. Mal'cev algebras and entropic laws

Lemma 6.2.1. *Let P be a Mal'cev operation on a non-empty set A. Let $0 \in A$ be any fixed element. Then the following conditions are equivalent:*

(a) *The algebra $(A, P, 0)$ has an abelian group reduct with $xyzP = x-y+z$;*

(b) *The algebra (A, P) is entropic, i.e. P satisfies the entropic identity*

(6.2.1) $$x_{11}x_{12}x_{13}P \ x_{21}x_{22}x_{23}P \ x_{31}x_{32}x_{33}P \ P$$
$$= x_{11}x_{21}x_{31}P \ x_{12}x_{22}x_{32}P \ x_{13}x_{23}x_{33}P \ P;$$

(c) *The algebra (A, P) satisfies the identities*

(6.2.2) $$xyzP = zyxP,$$

(6.2.3) $$(xyzP)zuP = xyuP.$$

Proof. (a) $\Rightarrow$ (b) is obvious.

(b)$\Rightarrow$(c) By the Mal'cev identities (6.1.1) and the entropic identity (6.2.1) one gets the following: $xyzP = yyxP \ yyyP \ zyyP \ P = yyzP \ yyyP \ xyyP \ P = zyxP$, and similarly $xyzP \ zuP = xyzP \ zzzP \ zzuP \ P = xzzP \ yzzP \ zzuP \ P = xyuP$.

(c)$\Rightarrow$(a) Define the operations $+$ and $-$ on A by

$$x + y := x0yP \quad \text{and} \quad -x := 0x0P.$$

Then commutativity of $+$ follows by (6.2.2). The element 0 is obviously the unit for $+$. By (6.2.2) and (6.2.3) one obtains

(6.2.4) $$x + (-y) = (-y) + x = 0y0P0xP = 0yxP = xy0P.$$

In particular, $x + (-x) = xx0P = 0$, whence $-(-x) = 0(-x)0P = (x + (-x))(-x)0P = x0(-x)P(-x)0P = x$. Furthermore, (6.2.4) implies that $(x + (-y)) + z = xy0P0zP = xyzP$. Hence $(x + (-y)) + z = xyzP = zyxP =$

$(z+(-y))+x = x+((-y)+z)$. Since $-(-y) = y$, one gets that $(x+y)+z = x+(y+z)$. Hence $(A,+,-,0)$ is an abelian group, and $xyzP = x-y+z$. □

Note that if P satisfies the entropic law (6.2.1), then for any pair of elements a_0, $a_1 \in A$, there is a mutually inverse pair of P-isomorphisms

(6.2.5) $$f_{i,1-i}: A \to A : c \mapsto a_i a_{1-i} c P$$

with $i = 0, 1$. For by the Mal'cev and entropic laws

$$\begin{aligned} cf_{i,1-i}f_{1-i,i} &= a_{1-i}a_i(a_ia_{1-i}cP)P \\ &= a_{1-i}a_ia_iP \ a_{1-i}a_{1-i}a_iP \ a_ia_{1-i}cP \ P \\ &= a_{1-i}a_{1-i}a_iP \ a_ia_{1-i}a_{1-i}P \ a_ia_icP \ P \\ &= a_ia_icP = c. \end{aligned}$$

Moreover, if A also has an Ω-algebra structure such that (A, P, Ω) is a mode, the mappings $f_{i,1-i}$ are Ω-isomorphisms.

Lemma 6.2.2. *Let $(A, \Omega) = (A, \Omega, P)$ be a Mal'cev algebra. The relation α defined on the direct square $A \times A$ by*

(6.2.6) $$(a,b) \ \alpha \ (c,d) :\Leftrightarrow d = bacP$$

is a congruence relation on $(A, \Omega) \times (A, \Omega)$ if and only if the algebra (A, Ω) satisfies all the entropic laws (E) for $\varphi = P$ and $\omega \in \Omega$, i.e.

(6.2.7) $$\begin{aligned}&(x_{11}\ldots x_{1m}\omega)(x_{21}\ldots x_{2m}\omega)(x_{31}\ldots x_{3m}\omega)P \\ =&(x_{11}x_{21}x_{31}P)\ldots(x_{1m}x_{2m}x_{3m}P) \ \omega \ .\end{aligned}$$

Moreover, for $a \in A$ one has $\widehat{A} = (a,a)^\alpha$.

Proof. For (n-ary) ω in Ω and $i = 1,\ldots,n$, let a_i, b_i, c_i, and d_i be in A and let $(a_i, b_i) \ \alpha \ (c_i, d_i)$. This means that $d_i = b_i a_i c_i P$. The relation α

is a subalgebra of $(A^2, \Omega) \times (A^2, \Omega)$ precisely when $(a_1 \ldots a_n \omega, b_1 \ldots b_n \omega) = (a_1, b_1) \ldots (a_n, b_n) \omega \ \alpha \ (c_1, d_1) \ldots (c_n, d_n) \omega = (c_1 \ldots c_n \omega, d_1 \ldots d_n \omega)$, i.e. when $d_1 \ldots d_n \omega = b_1 \ldots b_n \omega \ a_1 \ldots a_n \omega \ c_1 \ldots c_n \omega \ P$. The last equality holds precisely when (A, Ω) satisfies the entropic laws

$$b_1 a_1 c_1 P \ldots b_n a_n c_n P \ \omega = b_1 \ldots b_n \omega \ a_1 \ldots a_n \omega \ c_1 \ldots c_n \omega \ P$$

for each ω in Ω.

Since $y = yxxP$ holds in A, the relation α is obviously reflexive. Now the first statement of the lemma follows directly by Proposition 6.1.2.

Finally, for any $a \in A$, one has $(a, a)^\alpha \subseteq \widehat{A}$, since $(a, a) \ \alpha \ (b, b)$ if and only if $b = aabP$. Similarly, $\widehat{A} \subseteq (a, a)^\alpha$, since $(a, a) \ \alpha \ (b, c)$ if and only if $c = aabP = b$. □

Note that for $a, b \in A$, one has $(a, b)^\alpha = \{(c, bacP) \mid c \in A\}$.

Corollary 6.2.3. *The relation α defined on $A \times A$ by (6.2.6) is the unique centreing congruence $(A \times A | A \times A)$ on a Mal'cev algebra $(A \times A, \Omega, P)$ if and only if the algebra (A, Ω) satisfies the entropic laws (6.2.7).*

Proof. This follows directly by Propositions 6.1.4 and 6.1.6. □

An algebra (A, Ω) is *diagonally normal* if there is a congruence relation δ on $(A, \Omega) \times (A, \Omega)$ such that the diagonal $\widehat{A}$ is a δ-class, i.e. for each $a \in A$, one has $\widehat{A} = (a, a)^\delta$. In particular, every entropic Mal'cev algebra is diagonally normal. In fact, for Mal'cev algebras the two notions of centrality and diagonal normality coincide.

Proposition 6.2.4. *Let $(A, \Omega) = (A, \Omega, P)$ be a Mal'cev algebra. Then the following conditions are equivalent:*

(a) *The operation P satisfies the entropic laws (6.2.7);*

(b) *The algebra (A, Ω) is diagonally normal, i.e. it is central;*

(c) The algebra (A, Ω) satisfies the quasi-identity

(6.2.8) $\quad (xy_1 \ldots y_n w = xz_1 \ldots z_n w) \to (ty_1 \ldots y_n w = tz_1 \ldots z_n w)$

for each w in $X\Omega$.

Proof. (a)⇔(b) This equivalence follows by Corollary 6.2.3. The congruence α defined by (6.2.6) has the required property.

(b)⇒(c) If (A, Ω) is central, then $(A \times A | A \times A) = \alpha$ from Lemma 6.2.2. For any a, b, $c_1, \ldots, c_n$, $d_1, \ldots, d_n$ in A and $x_1 \ldots x_{n+1} w$ in $X\Omega$, the relations $(a, a) \; \alpha \; (b, b)$ and $(c_i, d_i) \; \alpha \; (c_i, d_i)$ imply that

$$(ac_1 \ldots c_n w, ad_1 \ldots d_n w)$$
$$= (a, a)(c_1, d_1) \ldots (c_n, d_n) w \; \alpha \; (b, b)(c_1, d_1) \ldots (c_n, d_n) w$$
$$= (bc_1 \ldots c_n w, bd_1 \ldots d_n w).$$

Hence if $ac_1 \ldots c_n w = ad_1 \ldots d_n w$, then also $bc_1 \ldots c_n w = bd_1 \ldots d_n w$.

(c)⇒(b) It will be shown that $\widehat{A}$ is a congruence class of the congruence $\varphi := \langle \widehat{A} \rangle_{cg}$ on $(A \times A, \Omega)$. First note that by Proposition 6.1.2, it follows that φ is the subalgebra of $(A^2 \times A^2, \Omega)$ generated by the set $\widehat{A} \times \widehat{A} \cup \{(a, b, a, b) \mid a, b \in A\}$. This implies that the following three conditions are equivalent:

(6.2.9): $\quad (a, a) \; \varphi \; (b, c) \Leftrightarrow b = c$ for all $a, b, c \in A$;

(6.2.10): $\quad \widehat{A}$ is a congruence class of φ;

(6.2.11): $\quad (a, a, b, c) \in \varphi \leq A^2 \times A^2 \Leftrightarrow b = c$ for all $a, b, c \in A$.

So it is enough to show that (c) implies (6.2.11). Let $a, b, c \in A$ be such that $(a, a, b, c) \in \varphi$. By Proposition 1.4.5, it follows that there is an Ω-word w and generators $g_1, \ldots, g_m$ of φ such that

$$(a, a, b, c) = g_1 \ldots g_m w.$$

One may assume that $g_i = (a_i, a_i, b_i, b_i)$ for $i = 1, \ldots, k$ and $g_i = (c_i, d_i, c_i, d_i)$ for $i = k+1, \ldots, m$. Thus

$$(a, a, b, c) = (a_1 \ldots a_k c_{k+1} \ldots c_m w,$$
$$a_1 \ldots a_k d_{k+1} \ldots d_m w, \ b_1 \ldots b_k c_{k+1} \ldots c_m w, \ b_1 \ldots b_k d_{k+1} \ldots d_m w).$$

However, by (6.2.8) and Exercise 6.2G,

$$a_1 \ldots a_k c_{k+1} \ldots c_k w = a = a_1 \ldots a_k d_{k+1} \ldots d_m w$$

implies that

$$b = b_1 \ldots b_k c_{k+1} \ldots c_m w = b_1 \ldots b_k d_{k+1} \ldots d_m w = c.$$

It follows that the condition (6.2.11), and hence also (6.2.10), hold. □

The implication (6.2.8) is called a "term condition" by some model theorists. Note that the equivalence (b)⇔(c) can be proved in a more general setting (see e.g. [McKenzie, McNulty, Taylor 1987]). In particular, any algebra (A, Ω) is diagonally normal if and only if it satisfies the quasi-identity (6.2.8). Some universal algebraists call algebras satisfying (6.2.8) "abelian". However, in the context of this book, the name would be confusing. E.g. by Exercise 6.1A, quasigroups are Mal'cev algebras. But diagonally normal (or central) quasigroups are not necessarily abelian, i.e. commutative and associative. A variety of central and Mal'cev algebras will be called a *central variety*. A variety of diagonally normal algebras will be called *diagonally normal*. Note also that such a variety would be described as "abelian" by those using the name "abelian algebras" for diagonally normal algebras.

As noted before, any variety $\underline{\mathrm{Mod}}_R$ of modules over a fixed ring R is a Mal'cev variety, with $xyzP = x-y+z$. All R-modules $(A, +, R)$ are diagonally normal, i.e. are central, since $\widehat{A}$ is a congruence class of the congruence θ defined by (a, b) θ (a', b') iff $a - b = a' - b'$. Algebras in central varieties are close to modules.

Theorem 6.2.5. Let $(A, \Omega) = (A, \Omega, P)$ be a non-empty Mal'cev algebra. Then (A, Ω) is central if and only if it is polynomially equivalent to a module $(A, +, R)$ over some ring R.

Proof. We only need to prove that a non-empty central Mal'cev algebra (A, Ω) is polynomially equivalent to a module over some ring. First note that since (A, Ω) is a Mal'cev algebra, it has a Mal'cev operation P. By Proposition 6.2.4, the operation P commutes with all operations in Ω, and hence also with itself. Let $0 \in A$ be any fixed element. Then by Lemma 6.2.1, there is an abelian group structure on A with

(6.2.12) $$x + y = x0yP \quad \text{and} \quad -x = 0x0P.$$

The element 0 is the identity of the group, and $xyzP = x - y + z$.

To define the ring of the module we are constructing, take R to be the set of all derived operations of $(A, \Omega, 0)$ of the form $r : A \to A$; $y \mapsto 0yw$ for $xyw \in \{x, y\}\Omega$ such that $0r = 0$. Since P commutes with all the operations of (A, Ω), one has

$$(x+y)r = (x0yP)r = (xr, 0r, yr)P$$
$$= (xr, 0, yr)P = xr + yr.$$

This implies that all the elements r are endomorphisms of the group $(A, +, -, 0)$. Clearly R is closed under composition $\cdot$, since for r and s in R with $xr = 0xw$ and $xs = 0xv$, one has $x(rs) = (xr)s = (0(0xw))v$. Also, the elements $r + s$ and $-r$ defined by

$$x(r+s) := xr + xs = (xr, 0, xs)P \quad \text{and}$$
$$x(-r) := -(xr) = (0, xr, 0)P$$

are in R. The zero and identity elements of R are defined by $x0 := 0$ and $x1 := x$. Thus $(R, +, -, 0, \cdot, 1)$ is a subring of the ring of endomorphisms of $(A, +, -, 0)$. Then clearly $(A, +, R)$ is an R-module, and the set of derived operations of $(A, +, R)$ is contained in the set of derived operations of $(A, \Omega, 0)$.

To finish the proof, it suffices to show that each derived operation $x_1 \ldots x_n w$ of $(A, \Omega, 0)$ is expressible by the polynomial operations of the module $(A, +, R)$. Define $x_1 \ldots x_n w' := x_1 \ldots x_n w - 0 \ldots 0w$. Then w' defines a derived operation of $(A, \Omega, 0)$. Since P commutes with w' and $0 \ldots 0w' = 0$, one obtains

$$(x_1 + y_1, \ldots, x_n + y_n)\, w' = (x_1 0 y_1 P) \ldots (x_n 0 y_n P)\, w'$$
$$= (x_1 \ldots x_n w')\, 0\, (y_1 \ldots y_n w')\, P$$
$$= x_1 \ldots x_n w' + y_1 \ldots y_n w'.$$

Hence by induction

$$x_1 \ldots x_n w' = x_1 0 \ldots 0 w' + 0 x_2 0 \ldots 0 w' + \cdots + 0 \ldots 0 x_n w'.$$

It follows that

$$x_1 \ldots x_n w = x_1 0 \ldots 0 w' + \cdots + 0 \ldots 0 x_n w' + 0 \ldots 0 w$$

(6.2.13)
$$=: x_1 r_1 + \cdots + x_n r_n + 0 \ldots 0w,$$

where $x_i r_i := 0 \ldots 0 x_i 0 \ldots 0 w'$ and each r_i is in R, so that each derived operation of $(A, \Omega, 0)$, and hence each polynomial operation of (A, Ω), is indeed a polynomial operation of the module $(A, +, R)$. □

Corollary 6.2.6. *Let (A, Ω) be an entropic Mal'cev algebra. Then (A, Ω) is central. Moreover, if it is non-empty, then it is polynomially equivalent to a module $(A, +, R)$ over some commutative ring R.*

Proof. For r and s in R with $xr = 0xw$ and $xs = 0xv$, one has

$$x(rs) = (xr)s = 0(0xw)v = (00w)(0xw)v$$
$$= (00v)(0xv)w = 0(0xv)w = (xs)r$$
$$= x(sr),$$

since $0r = 0$ and $0s = 0$. □

Exercises

6.2A. Show that in a Mal'cev variety, the class of diagonally normal algebras forms a subvariety.

6.2B. Let $\underline{W}$ be the diagonally normal subvariety of a Mal'cev variety $\underline{V}$. Let $xyzP$ and $xyzP'$ be two Mal'cev operations for a $\underline{W}$-algebra (A, Ω). Show that
$$xyzP = xyzP'.$$

6.2C. Use the quasi-identity (6.2.8) to show the following:
   (a) Any idempotent and diagonally normal groupoid is entropic, i.e. is a mode;
   (b) Any idempotent and diagonally normal semigroup is a normal band;
   (c) Non-trivial semilattices are not diagonally normal.

6.2D. Let (A, Ω) be any algebra. Define the center of (A, Ω) to be the relation $\zeta(A) \subseteq A \times A$ defined by $(a,b) \in \zeta(A)$ iff (6.2.8) holds for $x = a$ and $t = b$ or $x = b$ and $t = a$, and each w in $X\Omega$, and each choice of elements $y_1, \ldots, y_n$, $z_1, \ldots, z_n$ in A. Show the following:
   (a) The algebra (A, Ω) is diagonally normal iff $\zeta(A) = A \times A$;
   (b) The center $\zeta(A)$ is a congruence on (A, Ω);
   (c) The usual group-theoretic center $Z(G)$ of a group G is the congruence class of $\zeta(G)$ containing the identity of the group.

6.2E. Let $(Q, \cdot, /, \backslash)$ be an idempotent, central quasigroup. Show the following:
   (a) $xyzP = (x/y)(y \backslash z)$ is a Mal'cev operation on $(Q, \cdot, /, \backslash)$;
   (b) $(Q, \cdot, /, \backslash)$ is entropic iff it satisfies the identity
$$x \cdot yx = xy \cdot x$$
   [Hint: Show that for a fixed element e of Q, $x \cdot y = (xe, e, ey)P$];

(c) $(Q, \cdot, /, \backslash)$ is entropic, i.e. is a quasigroup mode.

[Hint: Note that for a fixed $e \in Q$, one has $x + y = (x/e)(e \backslash y)$ and $xy = xe + ey$ and $xe = (x, ex, e)P$, then use (b).]

6.2F. Show that each reduct of an affine space $(A, P, \underline{R})$ over any commutative ring R is diagonally normal.

6.2G. Show that if a Mal'cev algebra (A, Ω) satisfies the quasi-identity (6.2.8), then it also satisfies the quasi-identities

$$(x_1 \ldots x_k y_{k+1} \ldots y_n w = x_1 \ldots x_k z_{k+1} \ldots z_n w) \to$$

$$(t_1 \ldots t_k y_{k+1} \ldots y_n w = t_1 \ldots t_k z_{k+1} \ldots z_n w)$$

for each w in $X\Omega$.

6.3. Mal'cev modes and affine spaces

Corollary 6.2.6 of the previous section shows that each non-empty Mal'cev mode (A, Ω) is polynomially equivalent to a module $(A, +, R)$ over a certain commutative ring R. In this section we will show that each Mal'cev mode (A, Ω) is in fact equivalent to an affine space $(A, P, \underline{R})$.

First note that the idempotent laws allow us to choose the zero of the module $(A, +, R)$ we constructed in the proof of Theorem 6.2.5 for non-empty A in a more natural way. Consider the congruence α on $(A, \Omega) \times (A, \Omega)$ in Lemma 6.2.2, and the quotient $((A \times A)^\alpha, \Omega)$. Note that for each element e of A and (x, y) in $A \times A$ one has $(x, y)^\alpha = (e, yxeP)^\alpha$, and for $a \neq b$, one has $(e, a)^\alpha \neq (e, b)^\alpha$. Hence the mapping

$$\pi_e : (A \times A)^\alpha \to A \; ; \; (e, a)^\alpha \mapsto a$$

has a two-sided inverse, namely

$$\mu_e : A \to (A \times A)^\alpha \; ; \; a \mapsto (e, a)^\alpha.$$

Clearly π_e, and hence also μ_e, is an Ω-homomorphism. Indeed, for (n-ary) ω in Ω and a_i in A

$$(e, a_1)^\alpha \ldots (e, a_n)^\alpha \omega \, \pi_e = (e \ldots e\omega, a, \ldots, a_n\omega)^\alpha \, \pi_e$$
$$= a_1 \ldots a_n \omega = (e, a_1)^\alpha \pi_e \ldots (e, a_n)^\alpha \pi_e \, \omega.$$

Write the quotient algebra $((A \times A)^\alpha, \Omega)$ as $A \times A/\widehat{A}$. Then clearly (A, Ω) is isomorphic with $A \times A/\widehat{A}$, and one can further work with $A \times A/\widehat{A}$ instead of (A, Ω). As a zero 0 of this algebra, choose the diagonal $\widehat{A}$. As in the proof of Theorem 6.2.5, the algebra $A \times A/\widehat{A}$ is an abelian group with $0 = \widehat{A}$. The ring R of $A \times A/\widehat{A}$ is constructed as before with

$$(x, y)^\alpha r = (\widehat{A}, (x, y)^\alpha) w = ((x, x)^\alpha, (x, y)^\alpha) w$$
$$= ((x, x)(x, y)w)^\alpha = (x, xyw)^\alpha$$

and $(x, y)^\alpha 1 := (x, y)^\alpha$. Note however that the assumption $0r = 0$ is no longer necessary, since one obtains this equality by the idempotent law. The ring R is commutative, and $(A \times A/\widehat{A}, +, R)$ is an R-module. Each derived operation of $(A \times A/\widehat{A}, +, R)$ is a derived operation of $(A \times A/\widehat{A}, \Omega, \widehat{A})$. On the other hand, since $(A \times A/\widehat{A}, \Omega)$ is idempotent, each derived operation $x_1 \ldots x_n w$ of $(A \times A/\widehat{A}, \Omega)$ can be expressed as

(6.3.1) $$x_1 \ldots x_n w = x_1 r_1 + \cdots + x_n r_n + 0 \ldots 0w$$
$$= x_1 r_1 + \cdots + x_n r_n$$

with $r_i \in R$ [cf. (6.2.10)], i.e. as a derived operation of the module $(A \times A/\widehat{A}, +, R)$. Summarizing, one has the following:

Proposition 6.3.1. Let (A, Ω) be a non-empty Mal'cev mode, and let e be a fixed element of A. Then (A, Ω, e) is equivalent to the module $(A, +, R)$ over the commutative ring R defined above. □

Corollary 6.3.2. *Each Mal'cev mode (A, Ω) is equivalent to an affine space $(A, P, \underline{R})$ over a commutative ring R.*

Proof. One may assume that A is non-empty, and take R as above. Fix an element e in A. For each r in R, the operation $\underline{r}$ is defined by $xy\underline{r} = x(1-r)+yr$, and $xr = exw$ for some $xyw \in \{x,y\}\Omega$. Then $x(1-r) = x1 + x(-r) = x + (e, xr, e)P = (x, e, (e, xr, e)P)P = (xeeP, e, (e, xr, e)P)P = (x, xr, e)P = (x, exw, e)P = (xxw, exw, eew)P = (xeeP, xxeP)w = xew$. Now $xy\underline{r}$ is easily calculated as

$$xy\underline{r} = x(1-r) + yr = xew + eyw$$
$$= (xew, e, eyw)P = (xeeP, eeyP)w$$
$$= xyw.$$

It follows that each basic, and hence each derived, operation of $(A, P, \underline{R})$ can be expressed as a derived operation of (A, Ω). On the other hand, each derived operation $x_1 \ldots x_n w$ of (A, Ω) is expressed [via (6.3.1)] as $x_1 r_1 + \cdots + x_n r_n$, and since it is idempotent, $r_1 + \cdots + r_n = 1$, so that $x_1 r_1 + \cdots + x_n r_n$ is a derived operation of $(A, P, \underline{R})$. □

Note in particular that the binary derived operations of (A, Ω) are precisely the operations $xy\underline{r} = x(1-r)+yr$ for $r \in R$. And moreover, there is a bijection between the set of binary derived operations of (A, Ω) and the set R.

The last corollary shows the equivalence of a single Mal'cev mode and a corresponding affine space. The next theorem will demonstrate the equivalence of any variety of Mal'cev modes and a corresponding variety of affine spaces. The ring R of Corollary 6.3.2 is constructed for each Mal'cev mode (A, Ω) separately. What we need now is a ring R that will act on all algebras in a given Mal'cev variety $\underline{V}$ simultaneously. The way we constructed a ring for each Mal'cev mode separately suggests taking as the ring R of the variety $\underline{V}$ the ring based on the free $\underline{V}$-algebra $\{0,1\}V$ on two free generators 0 and 1. Recall that by Theorem 5.9.5, the algebra $\{0,1\}V$ has the structure of a commutative dual monoid with the multiplication $\cdot$ given by $xy(w_1 \cdot w_2) = x\ xyw_1\ w_2$, the zero element 0, the identity element 1 and the involution $'$ given by $xyw' = yxw$. On the other hand, the $\underline{V}$-algebra $\{0,1\}V$ is equivalent, as a Mal'cev mode, to the affine space $(\{0,1\}V, P, \underline{R})$. The ring R consists of the derived operations of the form $r_w : \{0,1\}V \to \{0,1\}V;\ 01t \mapsto 0\ 01t\ w = 01(t \cdot w)$ for t and w in $\{0,1\}V$. In particular, $(01t)r_0 = 01(t \cdot 0) = 0$ and $(01t)r_1 = 01(t \cdot 1) = t$. Hence the ring R is isomorphic to the ring $(\{0,1\}V, +, -, 0, \cdot, 1)$ with group operations defined by (6.2.12), the multiplication $xy(w \cdot t) = x\ xyw\ t$, the zero element 0 and identity element 1. It is easy to see that the dual monoid $(\{0,1\}V, \cdot, ', 0, 1)$ coincides with the dual monoid $(\{0,1\}V, +, \cdot)M$ of Example 5.9.4.

All this leads to the following theorem characterizing varieties of Mal'cev modes as varieties of affine spaces.

Theorem 6.3.3. *Let $\underline{V}$ be a Mal'cev variety of modes. Then*

$$(\{0,1\}V, +, -, 0, \cdot, 1)$$

is a commutative ring. Further, the varieties $\underline{V}$ and $\underline{\{0,1\}V}$ are equivalent.

Proof. We only need to prove the second part. First note that each ring R_A constructed for each $\underline{V}$-algebra (A, Ω) separately must be a homomorphic image of the ring $\{0,1\}V$, say by a homomorphism $h_A : \{0,1\}V \to R_A$. It follows that the corresponding affine space $(A, P, \underline{R}_A)$ can be equivalently considered as the affine space $(A, P, \underline{\{0,1\}V})$ by defining $xy\underline{t}$ for $t \in \{0,1\}V$ as $xy\underline{t} = x(1-t) + yt = x(1-t)h_A + yth_A$. (Cf. Sections 1.6.2 and 5.3.) In this way each $\underline{V}$-algebra (A, Ω) becomes the equivalent affine $\{0,1\}V$-space $(A, P, \underline{\{0,1\}V})$. On the other hand, any basic operation ω of Ω is defined on

$(A, P, \underline{\{0,1\}V})$ using (6.3.1) by

$$x_1 \ldots x_n \omega = x_1 r_1 + \cdots + x_n r_n$$
$$= x_1 t_1 h_A + \cdots + x_n t_n h_A$$
$$= x_1 t_1 + \cdots + x_n t_n$$

for all t_i in $\{0,1\}V$ with $t_i h_A = r_i$, and the operation $x_1 t_1 + \cdots + x_n t_n$ is determined by the operation P and the binary operations $\underline{t}$ for t in $\{0,1\}V$. This shows that the two algebras (A, Ω) and $(A, P, \underline{\{0,1\}V})$ are equivalent. Consequently the varieties $\underline{V}$ and $\underline{\{0,1\}V}$ are equivalent. $\square$

Theorem 6.3.4. *For a commutative ring R, the variety $\underline{R}$ of affine R-spaces is the variety of Mal'cev modes $(A, P, \underline{R})$ of type $\{(P, 3)\} \cup (R \times \{2\})$ satisfying the identities*

(a) $\quad xy\underline{0} = x = yx\underline{1},$

(b) $\quad yxyP = xy\underline{2},$

(c) $\quad xy\underline{p} \; xy\underline{q} \; \underline{r} = xy \; \underline{pqr},$

(d) $\quad (xy\underline{p}, xy\underline{q}, xy\underline{r})P = xy \; \underline{pqrP}.$

Proof. Let $\underline{V}$ be the variety of Mal'cev modes satisfying the given identities. Using these identities (including those defining the Mal'cev operation P as well as the idempotent and entropic identities implicit for modes), one obtains the standard form $01\underline{r}$ with r in R, for elements of $\{0,1\}V$. For example

$$(0 \; 01\underline{p} \; \underline{q})(010P)\underline{r} = ((01\underline{0} \; 01\underline{p} \; \underline{q}), 10\underline{2})\underline{r} = ((01 \; \underline{0pq}), 10\underline{2})\underline{r}$$
$$= ((01 \; \underline{0pq}), 01\underline{(-1)})\underline{r} = 01(\underline{0pq})(\underline{-1})\underline{r}$$

with $(0pq)(-1)\underline{r}$ calculated in the ring R. There is then a ring isomorphism

$$\{0,1\}V \to R \; ; \; 01\underline{r} \mapsto r.$$

Thus by Theorem 6.3.3, the variety $\underline{V}$ is the variety $\underline{R}$. □

Theorem 6.3.4 is the result that captures affine spaces algebraically. It identifies the class $\underline{R}$ consisting of the empty algebra and of reducts $(A, P, \underline{R})$ of R-modules as a variety of algebras, with the $\underline{R}$-algebras being precisely the affine R-spaces, and the homomorphisms precisely the affine mappings.

If the element $2 = 1 + 1$ of the ring R is invertible, so that there is 2^{-1} in R with $2 \cdot 2^{-1} = 1$, then the identities for $\underline{R}$ given in Theorem 6.3.4 may be replaced by a simpler set due to Ostermann and Schmidt [1966].

Corollary 6.3.5. *For a commutative ring R in which 2 is invertible, the variety $\underline{R}$ is equivalent to the variety of modes $(A, \underline{R})$ of type $\underline{R} \times \{2\}$ satisfying the identities* (a) *and* (c) *of Theorem 6.3.4.*

Proof. Let $\underline{V}$ be the variety of modes satisfying the given identities. Define a ternary operation P on each $\underline{V}$-algebra by

(6.3.2) $$xyzP := y \ xz\underline{2}^{-1} \ \underline{2}.$$

Then $xxzP = x \ xz\underline{2}^{-1} \ \underline{2} = xz\underline{0} \ xz\underline{2}^{-1} \ \underline{2} = xz \ \underline{0}\underline{2}^{-1}\underline{2} = xz\underline{1} = z$, and similarly $xzzP = x$, showing that P is a Mal'cev operation. The identities (b) and (d) of Theorem 6.3.4 may be checked similarly [Exercise 6.3A]. It follows by Theorem 6.3.4 that $\underline{R}$ coincides with the variety $\underline{V}$. □

Note in particular that for any field F of characteristic different from 2, the variety $\underline{F}$ of affine F-spaces can be defined using only binary operations. On the other hand, since the binary operations $\underline{0}$ and $\underline{1}$ of the affine spaces over the field $GF(2)$ are just projections, the ternary operation P of $GF(2)$-spaces cannot be composed from the binary operations.

The next theorem characterizes all the varieties of affine spaces that can be defined by binary operations. However, it does not provide any equational base explicitly.

Theorem 6.3.6. *Let R be a commutative ring different from $GF(2)$, and let $(A, P, \underline{R})$ be an affine R-space. Then $(A, P, \underline{R})$ is equivalent to the mode $(A, \underline{R})$ if and only if R has no ideal of index 2.*

Proof. ($\Rightarrow$) Suppose on the contrary that there is an ideal I of index 2 in R. Consider two subclones C and D of the clone of $(A, P, \underline{R})$. The clone D consists of all projections and all derived operations $x_1 r_1 + \cdots + x_n r_n$ with $\sum_1^n r_i = 1$, where all r_i except one belong to I. The clone C is the subclone generated by all the operations $\underline{r}$, for r in R [Exercise 6.3C]. It is clear that $C \subseteq D$. However, D does not contain $x - y + z$. Therefore the binary operations do not generate the Mal'cev operation P.

($\Leftarrow$) Assume that R has no ideal of index 2, so in particular R has no homomorphism onto $GF(2)$. Define

$$J := \{r \in R \mid x - yr + zr \in C\}.$$

Since for any s, $t \in J$ and $r \in R$ one has

$$x - y(s - t) + z(s - t) = (x - ys + zs) - zt + yt \quad \text{and}$$
$$x - ysr + zsr = (x - ys + zs)r + x(1 - r),$$

it follows that $s - t$ and sr are in J, and hence that J is an ideal of R. Moreover, since

$$x - y(r - r^2) + z(r - r^2)$$
$$= (x(1 - r) + zr)(1 - r) + (x(2 - r) + y(r - 1))r$$
$$= xz\underline{r}\ xy\ \underline{r - 1}\ \underline{r},$$

we get that $r - r^2 \in J$ for all $r \in R$. Hence the ring R/J satisfies the identity $x^2 = x$. However, R has no homomorphism onto $GF(2)$. This implies that $J = R$. In particular, $1 \in J$, so $x - y + z \in C$. Consequently, the collection of all $\underline{r}$ generates the Mal'cev operation P. □

In particular, if 2 is invertible in the ring R, then R cannot have a homomorphism onto $GF(2)$. Consequently, the variety $\underline{R}$ does not contain a variety equivalent to $\underline{GF(2)}$. Note by Theorem 6.3.6 that if F is a field different from $GF(2)$, then even in the case of characteristic 2, the variety $\underline{F}$ can be defined only by binary operations. On the other hand, since the rings $\mathbb{Z}$ and $\mathbb{Z}_{2n}$ have ideals of index 2, the varieties $\underline{\mathbb{Z}}$ of integral affine spaces and $\underline{\mathbb{Z}}_{2n}$ of affine $\mathbb{Z}_{2n}$-spaces cannot be defined without the Mal'cev operation P. One has even more:

Proposition 6.3.7. *Each integral affine space* $(A, P, \underline{\mathbb{Z}})$ *is equivalent to the Mal'cev mode* (A, P).

Proof. It is sufficient to define all the binary operations $\underline{n}$ for $n \in \mathbb{Z}$ using the operation P. This can be done as follows. First note that since P is a Mal'cev operation, the operations $\underline{0}$ and $\underline{1}$ can be defined as

$$xy\underline{0} := xyyP \quad \text{and} \quad xy\underline{1} := xxyP.$$

Now assume that for all positive integers k not greater than n, the operations $\underline{k}$ are defined using only the operation P. Recall that in the ring $\mathbb{Z}$, one has $xyzP = x - y + z$. Then by Theorem 6.3.4(d),

(6.3.3) $\qquad xy\underline{n} \; xy \; \underline{n-1} \; xy\underline{n} \; P = xy \; \underline{(n, n-1, n)P}$

$\qquad\qquad = xy \; \underline{(n - n + 1 + n)} = xy \; \underline{(n+1)}.$

It follows that all the operations $\underline{n}$ for $n \in \mathbb{N}$ are generated by P. Now for a positive integer n,

$$xy\underline{(-n)} = x(1+n) - yn = y(1 - (n+1)) + x(1+n)$$
$$= yx \; \underline{n+1},$$

showing that P generates all the operations $\underline{n}$ for $n \in \mathbb{Z}$. $\square$

A similar argument shows that Proposition 6.3.7 holds also for each variety $\underline{\mathbb{Z}}_n$ of affine $\mathbb{Z}_n$-spaces [Exercise 6.3D]. However, as we will see later, the operations of each affine $\underline{\mathbb{Z}}_{2n+1}$-space are generated just by one binary operation.

For any commutative ring R and an integer $n \neq 0$, let the sum of n copies of 1 be denoted by n if n is positive, and otherwise let n denote the sum of n copies of -1. Then it is easy to see that the ring $\underline{R}$ of binary operations of $\underline{R}$ contains as a subring the ring $\mathbb{Z}$ or its homomorphic image $\mathbb{Z}_n$. By Proposition 6.3.7, such a ring is generated by the operation P.

Theorem 6.3.8. *The variety $\underline{\mathbb{Z}}$ of integral affine spaces and the variety $\underline{V}$ of Mal'cev modes (A, P) of type $\{P\} \times \{3\}$ are equivalent.*

Proof. By Theorem 6.3.3, the variety $\underline{V}$ is equivalent to the variety $\underline{R}$, where R is the ring defined on the free $\underline{V}$-algebra $F = \{x, y\}V$ on two generators, say x and y. It is enough to show that the ring R coincides with the ring $\mathbb{Z}$. For 0 and 1 of R,

(6.3.4) \qquad $xy\underline{0} = xyyP$ \quad and \quad $xy\underline{1} = xxyP$.

To simplify notation, denote $xyzP$ by (xyz). Then as in Proposition 6.3.7 one has

$$xy\underline{2} = xy\underline{1}\ xy\underline{0}\ xy\underline{1}\ P = yxyP = (yxy),$$
$$xy\underline{3} = xy\underline{2}\ xy\underline{1}\ xy\underline{2}\ P = yxyP\ y\ yxyPP$$
$$= (y(xyx)y),$$

and for any positive integer $n \geq 2$

(6.3.5) \qquad $xy\underline{n} = (y(x(y(\ldots)y)x)y),$

where y is repeated n times. To show this, suppose that (6.3.5) holds for

$n \leq 2k$. Then

$$xy\underline{2k+1} = xy\underline{2k}\ xy\underline{2k-1}\ xy\underline{2k}\ P$$
$$= ((y(x\ldots(yxy)\ldots x)y)(y(x\ldots(xyx)\ldots x)y)(y(x\ldots(yxy)\ldots x)y))$$
$$= (y((x\ldots(yxy)\ldots x)(x\ldots(xyx)\ldots x)(x\ldots(yxy)\ldots x))y)$$
$$\ldots$$
$$= (y(x(\ldots((yxy)y(yxy))\ldots)x)y)$$
$$= (y(x(\ldots(y(xyx)y)\ldots)x)y).$$

Similarly, one shows that if (6.3.5) holds for $n \leq 2k - 1$, then $xy\underline{2k}$ has the desired form. Now for any positive n, one has

(6.3.6) $\qquad xy\underline{(-n)} = yx\underline{n+1} = (x(y(x(\ldots)x)y)x),$

where x is repeated $n + 1$ times. Obviously the ring $\underline{R}$ of binary operations determined by the elements of R contains the subring of all operations $\underline{n}$ for $n \in \mathbb{Z}$. Each operation $\underline{n}$ is defined by (6.3.5) and (6.3.6) using only the operation P. We will now show that the free $\underline{V}$-algebra F consists of precisely these elements. To do so, it is enough to note that by Theorem 6.3.4(d), for any $k, m, n \in \mathbb{Z}$ one has

$$xy\underline{k}\ xy\underline{m}\ xy\underline{n}\ P = xy\ \underline{kmnP}$$

and $kmnP = k - m + n \in \mathbb{Z}$, whence $xy\ \underline{kmnP}$ also has the form of (6.3.5) or (6.3.6). $\square$

Corollary 6.3.9. *For any positive integer n, the variety $\underline{\mathbb{Z}}_n$ of affine $\mathbb{Z}_n$-spaces and the variety $\underline{V}$ of Mal'cev modes (A, P) of type $\{P\} \times \{3\}$, satisfying the identity*

(6.3.7) $\qquad y(x(y\ldots yP)xP)yP = (y(x(y\ldots y)x)y) = x$

6. MAL'CEV MODES

with n appearances of y, are equivalent.

Proof. Exercise 6.3E. □

Note that the identity (6.3.7) means that in each $\mathbb{Z}_n$-space $(A, P, \underline{\mathbb{Z}_n})$, one has

$$xy\underline{n} = xy\underline{0}.$$

Exercises

6.3A. Complete the proof of Corollary 6.3.5.

6.3B. (a) Verify the following consequences of the identities in Theorem 6.3.4:
   (i) $x\ xy\underline{p}\ \underline{q} = xy\underline{(pq)}$;
   (ii) $x\ xy\underline{p}\ \underline{p} = xy\underline{p^2}$;
   (iii) $xy\underline{p}\ xy\underline{(2p)}\ xy\underline{p}\ P = xy\underline{0}$.

   (b) Show that if F is a finite field of characteristic different from 2, and p is a primitive element of F, then each basic operation of affine F-spaces can be obtained by composition from the operation $\underline{p}$.

   (c) Provide examples of finite and infinite rings, different from fields, for which Corollary 6.3.5 cannot be applied.

6.3C. Show that the set D of operations defined in the proof of Theorem 6.3.6 is indeed a clone.

6.3D. [Płonka 1970] Prove Proposition 6.3.7 for affine $\mathbb{Z}_n$-spaces.

6.3E. Prove Corollary 6.3.9.

6.3F. Show that the variety of affine spaces over the field $GF(2)$ is equivalent to the variety of Mal'cev modes (A, P) satisfying the identities

$$xyyP = yxyP = yyxP = x.$$

6.3G. Consider $A = GF(3) \times GF(3)$ as an affine $GF(3)$-space $(A, \underline{GF(3)})$. Show that the complex algebra AS is not an affine space over $GF(3)$. (Cf. Exercise 5.3E.)

6.3H. Is the reduct $(A, \underline{X})$ of an affine space $(A, P, \underline{\mathbb{Z}[X]})$ equivalent to this space? What is the minimal number of operations sufficient to define an algebra equivalent to $(A, P, \underline{\mathbb{Z}[X]})$?

6.3I. Let (A, Ω) be a reduct of an affine space $(A, P, \underline{F})$ over a field F such that $P \notin \mathfrak{P}\Omega$, but (A, Ω, P) and $(A, P, \underline{F})$ are equivalent. Let $J := \{r \in F \mid \underline{r} \in \mathfrak{P}\Omega\}$. Show the following:

(a) The set J is closed under all $\underline{r}$ for r in J.

(b) The additive group of F is generated by J.

(c) $J \neq \{0, 1\}$.

(d) $J \neq F$.

(e) Deduce that J is an algebraic unit interval in F. (Cf. 5.9.)

6.4. Affine spaces equivalent to groupoids

Consider the variety $\underline{GF(3)}$ of affine $GF(3)$-spaces $(A, P, \underline{GF(3)})$. By Corollary 6.3.5, each such space $(A, P, \underline{GF(3)})$ is equivalent to the mode $(A, \underline{GF(3)})$ satisfying the identities (a) and (c) of Theorem 6.3.4. The set $\underline{GF(3)}$ consists of three operations $\underline{0}, \underline{1}$ and $\underline{2}$. Note however that since $xy\underline{0} = xyyP$ and $xy\underline{1} = xxyP$, the operation $\underline{2}$ generates all the other operations, and $(A, P, \underline{GF(3)})$ is equivalent to $(A, \cdot) = (A, \underline{2})$. As we will see, the class of all such $\underline{2}$-reducts of affine $GF(3)$-spaces also forms a variety, equivalent to the variety $\underline{GF(3)}$. This section will discuss such varieties of binary modes that are equivalent to varieties of affine spaces.

Theorem 6.4.1. *Let $(A, P, \underline{R})$ be a faithful affine space over a finite commutative ring R. Let $(A, \cdot) = (A, \underline{r})$ for some fixed r in R. Then the groupoid $(A, \cdot)$ is equivalent to the affine space $(A, P, \underline{R})$ if and only if r generates the ring R, and both r and $1 - r$ are units of R.*

Proof. ($\Rightarrow$) Assume that $(A, \cdot)$ and $(A, P, \underline{R})$ are equivalent. Define the following operations on A:

$$x_1 x_2 f_1 := x_1 \cdot x_2 ;$$

$$x_1 x_2 \ldots x_{2^{n+1}} f_{n+1} := x_1 x_2 \ldots x_{2^n} f_n \cdot x_{2^n+1} x_{2^n+2} \ldots x_{2^{n+1}} f_n.$$

As $x_1 x_2 f_1 = x_1 \cdot x_2 = x_1(1-r) + x_2 r$, then for n one has

$$x_1 x_2 \ldots x_{2^n} f_n$$

(6.4.1)
$$= x_1(1-r)^n + \sum_{k=1}^{n-1}(x_{j_1} + \cdots + x_{j_s})r^k(1-r)^{n-k} + x_{2^n}r^n$$

$$= \sum_{k=0}^{n}(x_{j_1} + \cdots + x_{j_s})r^k(1-r)^{n-k},$$

where $s = \binom{n}{k}$. Note that the sum of coefficients equals

$$\binom{n}{0}(1-r)^n + \binom{n}{1}(1-r)^{n-1}r + \cdots + \binom{n}{n-1}(1-r)r^{n-1} + \binom{n}{n}r^n = 1$$

[Exercise 6.4A]. Now any derived operation of $(A, \cdot)$ is either f_n for some n, or can be obtained from some f_n by identification of some variables. In particular, for any binary derived operation xyq of $(A, \cdot)$, one has

$$xyq = x(1-v) + yv,$$

where $u = 1 - v$ and v are in $\mathbb{Z}[r]$. Since $(A, \cdot)$ and $(A, P, \underline{R})$ are equivalent, it follows that each binary derived operation of $(A, \cdot)$ is $\underline{q}$ for $q \in R$. Hence $q = v$ for some $v \in \mathbb{Z}[r]$, and the ring R is indeed generated by r.

To show that r and $1 - r$ are invertible, note that the Mal'cev operation P is also obtained from some of the f_n, and hence

$$xyzP = x \sum a_i(1-r)^{m_i} r^{n_i} + y \sum b_j(1-r)^{m_j} r^{n_j} + z \sum c_k(1-r)^{m_k} r^{n_k}.$$

In particular,

$$x = xyyP = x \sum a_i(1-r)^{m_i} r^{n_i} + y(\sum b_j(1-r)^{m_j} r^{n_j} + \sum c_k(1-r)^{m_k} r^{n_k}),$$

which implies that for $w = \sum a_i(1-r)^{m_i} r^{n_i - 1}$ one has

$$x = x \sum a_i(1-r)^{m_i} r^{n_i} = x(wr).$$

Since the module $(A, +, R)$ is faithful, it follows that $wr = 1$ and $r^{-1} = w$. Similarly, one shows that there is an inverse $(1-r)^{-1}$ in R.

($\Leftarrow$) Now assume that the ring R is generated by r, and that both r and $1-r$ are invertible. By Section 5.4, the groupoid $(A, \cdot) = (A, \underline{r})$ is a quasigroup. (Note that $xy(1-r) = yx\underline{r}$.) By Exercise 6.2E, the Mal'cev operation P can be calculated as follows:

$$xyzP = (x/y)(y \setminus z).$$

Since the ring R is finite, both divisions $/$ and $\setminus$ can be defined by means of $\underline{r}$. To show this, note that the semigroup $(\{r, r^2, \ldots\}, \cdot)$ is finite, so that there are natural numbers p and n with $r^p = r^{p+n}$. Hence $1 = r^{-p} r^p = r^{-p} r^{p+n} = r^n$ and $r^{-1} = r^{n-1}$. Similarly, there are q and m such that $(1-r)^q = (1-r)^{q+m}$ and $1 = (1-r)^{-q}(1-r)^{q+m} = (1-r)^m$, whence $(1-r)^{-1} = (1-r)^{m-1}$. Now by (5.4.1) one has

$$\begin{aligned} x/y &= yx\underline{(1-r)^{-1}} = yx\underline{(1-r)^{m-1}} \\ &= y(y \ldots (y(yx\underline{(1-r)})\underline{(1-r)}) \ldots)\underline{(1-r)} \\ &= y(y \ldots (y(xy\underline{r})\underline{(1-r)}) \ldots)\underline{(1-r)} = \ldots \\ &= (\ldots ((xy\underline{r})y\underline{r}) \ldots)y\underline{r} = xy^{m-1} \end{aligned}$$

and

$$\begin{aligned} x \setminus y &= xy\underline{r^{-1}} = xy\underline{r^{n-1}} \\ &= x(x \ldots (x(xy\underline{r})\underline{r}) \ldots)\underline{r} = x^{n-1} y. \end{aligned}$$

This implies that

(6.4.2) $$xyzP = (xy^{m-1})(y^{n-1} z).$$

Note also that $xyzP = (x/y)(y \setminus z) = yx(1-r)^{-1}\ yz\underline{r^{-1}}\ \underline{r} = x - y + z$.

It remains to show that all the binary operations of $(A, P, \underline{R})$ are defined purely in terms of $\underline{r}$. First note that since the ring R is generated by r, it follows that any element q in R is an integral polynomial in one variable r, say $q = ar^k + v$, where the degree of v is less than k. If q has degree 0, i.e. $q = n \in \mathbb{Z}$, then the operations $\underline{n}$ are obtained by means of P as in Proposition 6.3.7. If q has positive degree k, then we show by induction on k that the operation $\underline{q}$ can be obtained by means of the operation P and some binary operations $\underline{p}$, where the degree of p is less than the degree k of q.

Indeed

$$\begin{aligned}
(6.4.3)\qquad xy\underline{q} &= x(1-q) + yq \\
&= x(1 - ar^k - v) + y(ar^k + v) \\
&= x(1-v) + yv - xar^k + yar^k \\
&= xy\underline{v} - (y(1-ar^k) + xar^k) + y \\
&= (xy\underline{v}, yx\underline{ar^k}, y)P \\
&= (xy\underline{v}, y\ (yx\underline{ar^{k-1}})\ \underline{r}, y)P.
\end{aligned}$$

Now assume that for all $p \in R$ with degree less than k, the operations $\underline{p}$ are generated by $\underline{r}$. Then by (6.4.2) and (6.4.3), $\underline{q}$ is also generated by $\underline{r}$. It follows that the operation $\underline{r}$ generates all the other derived operations of the affine space $(A, P, \underline{R})$. □

Note that the first part of Theorem 6.4.1 remains true without the finiteness assumption on the ring R.

Now even if a commutative ring R generated by r is infinite, but there are positive integers n and m such that $r^n = 1$ and $(1-r)^m = 1$, then the affine R-space $(A, P, \underline{R})$ is equivalent to the groupoid $(A, \cdot) = (A, \underline{r})$. The proof is the same as the proof of the second part of Theorem 6.4.1. The groupoid $(A, \cdot)$

satisfies the identities

(6.4.4) $$x^n y = y = y x^m.$$

Indeed, if $r^n = 1$ and $(1-r)^m = 1$, then $xy\underline{r^n} = xy\underline{1}$, i.e. $x^n y = y$ and $xy\underline{(1-r)^m} = xy\underline{1}$, i.e. $yx^m = y$. In particular, the groupoid $(A, \cdot)$ of Theorem 6.4.1 satisfies these identities.

Example 6.4.2. Let $\underline{V}_{n,m}$ for $n, m > 1$ be the variety of binary modes $(G, \cdot)$ defined by the identities (6.4.4). This is a Mal'cev variety, with the Mal'cev operation given by

$$xyzP := (xy^{m-1})(y^{n-1}z).$$

So $\underline{V}_{n,m}$ is equivalent to the variety $\underline{R}_{n,m}$ of affine spaces over some commutative ring $R_{n,m}$. The first part of Theorem 6.4.1 shows that the ring $R_{n,m}$ is generated by one element, say r, and both r and $1-r$ are units of $R_{n,m}$. The identities (6.4.4) hold in a groupoid $(G, \cdot) = (G, \underline{r})$ precisely if $r^n = 1$ and $(1-r)^m = 1$. The ring $R_{n,m}$ is isomorphic to the ring $\mathbb{Z}[X]/\langle X^n - 1, (1-X)^m - 1\rangle$. Since both r and $1-r$ are invertible in $R_{n,m}$, each groupoid in $\underline{V}_{n,m}$ is a quasigroup with divisions $/$ and $\backslash$ defined as in the proof of Theorem 6.4.1. It follows that each variety $\underline{V}_{n,m}$ is also equivalent to a variety of quasigroup modes. Note that the ring $R_{n,m}$ does not need to be finite. (See Example 6.4.3 below.)

Corollary 6.4.2.1. *Let $(A, P, \underline{R})$ be a faithful affine R-space over a finite commutative ring R. Let $r \in R$. Then the groupoid $(A, \cdot) = (A, \underline{r})$ is equivalent to the affine space $(A, P, \underline{R})$ if and only if $(A, \cdot)$ is a member of a variety $\underline{V}_{n,m}$ for some positive n and m.* $\square$

Example 6.4.3. Consider the ring $R(n,t) = \mathbb{Z}[X]/\langle X^n - 1, X^t + X - 1\rangle$, where $t \neq n$. Let $r = X + \langle X^n - 1, X^t + X - 1\rangle$. Then obviously $r^n = 1$ and $r^t = 1 - r$. Since $(1-r)^{n/GCD(n,t)} = r^{nt/GCD(n,t)} = 1$, it follows that

6. MAL'CEV MODES

for $m = n/GCD(n,t)$ one has $(1-r)^m = 1$. Hence the groupoid reduct $(A, \cdot) = (A, \underline{r})$ of any affine $R(n,t)$-space $(A, P, \underline{R}(n,t))$ is a member of the variety $\underline{V}_{n,m}$, and hence is equivalent to the affine space $(A, P, \underline{R}_{n,m})$. Since in the ring $R(n,t)$ one has $r^t = 1 - r$, it follows that all the reducts $(A, \cdot) = (A, \underline{r})$ of the affine $R(n,t)$-spaces also satisfy the identity

(6.4.5) $$x^t y = yx.$$

Indeed,

$$x^t y = (\ldots(x(x\ xy\underline{r})\underline{r})\underline{r}\ldots)\underline{r} = xy\underline{r}^t$$
$$= xy(1-r) = yx\underline{r} = yx.$$

Exercise 6.4C shows that if a binary mode $(A, \cdot)$ satisfies the identities $x^n y = y$ and $x^t y = yx$, then there is an m' such that $(A, \cdot)$ satisfies the identity $yx^{m'} = y$ as well.

Now consider the subvariety $\underline{G}(n,t)$ of the variety $\underline{V}_{n,m}$ defined by the identity (6.4.5), i.e. $\underline{G}(n,t)$ is the variety of binary modes $(G, \cdot)$ defined by

(6.4.6) $$\begin{cases} x^n y = y, \\ x^t y = yx \end{cases}$$

for positive integers n and t, with $t \neq n$. We will show that the variety $\underline{G}(n,t)$ is equivalent to the variety $\underline{R}(n,t)$ of affine $R(n,t)$-spaces. As a subvariety of some $\underline{V}_{n,m}$, the variety $\underline{G}(n,t)$ is equivalent to the variety $\underline{R}$ of affine spaces over some commutative ring R. The ring R is a quotient $R_{n,m}/I$ of the ring $R_{n,m}$. It is also generated by $r = r + I$, and it satisfies $r^n = 1$. The groupoid multiplication is given by $\cdot = \underline{r}$. We have to show that R is isomorphic to $R(n,t)$. First note that the identity (6.4.5) implies that $r^t = 1 - r$. Hence the ring R is a quotient of $R(n,t)$. However, since the reducts $(A, \cdot) = (A, \underline{r})$ of all the affine $R(n,t)$-spaces $(A, P, \underline{R}(n,t))$ [and in particular the reduct $(R(n,t), \underline{r})$ of $(R(n,t), P, \underline{R}(n,t))$] are in the variety of $\underline{G}(n,t)$, it follows that $R \cong R(n,t)$.

The following lemma shows that one may always assume that $t < n$.

Lemma 6.4.3.1. *Let $t < n < s$ and let $s \equiv t \pmod{n}$. Then $R(n,s) = R(n,t)$.*

Proof. Let $s = an + t$. Then

$$\frac{X^{s-t} - 1}{X^n - 1} = \frac{(X^n)^a - 1^a}{X^n - 1} = X^{n(a-1)} + X^{n(a-2)} + \cdots + X^n + 1 \in \mathbb{Z}[X].$$

Hence $X^t + X - 1 = (X^s + X - 1) - X^t \frac{X^{s-t} - 1}{X^n - 1}(X^n - 1)$ is in $\langle X^n - 1, X^s + X - 1 \rangle$. Similarly, $X^s + X - 1$ is in $\langle X^n - 1, X^t + X - 1 \rangle$, whence $\langle X^n - 1, X^s + X - 1 \rangle = \langle X^n - 1, X^t + X - 1 \rangle$. □

Surprisingly, the ring $R(n,t)$ may be infinite.

Theorem 6.4.3.2. *The ring $R(n,t)$ is infinite if and only if $n \equiv 0 \pmod{6}$ and $t \equiv 5 \pmod{6}$.*

Proof. First recall that $R(n,t) = \mathbb{Z}[X]/I$, where $I = \langle X^n - 1, X^t + X - 1 \rangle$. Note that $R(n,t)$ is finite if and only if the ideal I contains an element d of $\mathbb{Z}$. Indeed, $d \in \mathbb{Z}$ implies that $d\mathbb{Z} \subseteq d\mathbb{Z}[X] \subseteq I$, and hence $\mathbb{Z}[X]/I$ is isomorphic to the finite ring $\mathbb{Z}_d[X]/I$. On the other hand, if $\mathbb{Z}[X]/I$ is finite, then it has a finite characteristic χ. Since for each $p \in R$ one has $\chi p = 0$, it follows in particular that for each $m \in \mathbb{Z}$, one has $\chi m = 0$, whence $d = \chi 1 \in I$.

Now $d \in \mathbb{Z} \cap I$ means that there are f and g in $\mathbb{Z}[X]$ such that $d = f(X^n - 1) + g(X^t + X - 1)$, or equivalently $1 = f/d(X^n - 1) + g/d(X^t + X - 1)$ in $\mathbb{Q}[X]$. But this means that $X^n - 1$ and $X^t + X - 1$ are mutually prime in $\mathbb{Q}[X]$.

Now note that $X^n - 1$ and $X^t + X - 1$ are not mutually prime if and only if they have a common solution in the field of complex numbers. Since $X^n = 1$, the solution has to be a root of 1. And since $X^t + X = 1$, it has to be a primitive 6th root of 1. Hence $n \equiv 0 \pmod{6}$ and $k \equiv 5 \pmod{6}$. □

The considerations above may be summarized as follows.

6. MAL'CEV MODES

Theorem 6.4.3.3. *Let n and t be positive integers such that $t < n$. Then the variety $\underline{R}(n,t)$ of affine spaces over the ring $R(n,t)$ is equivalent to the variety $\underline{G}(n,t)$ of binary modes.* □

We conclude this example by showing that the identity (6.4.5) is in fact equivalent to a certain linear identity. This gives another axiomatization for the variety $\underline{G}(n,t)$. A different way of finding this axiomatization is given in Exercise 6.4B.

Lemma 6.4.3.4. *Let $(A, \cdot)$ be a binary mode. Then $(A, \cdot)$ satisfies the identity*

$$(6.4.5) \qquad x^t y = yx$$

if and only if it satisfies the identity

$$x_0(x_1(\ldots(x_{t-1}x_t)\ldots)) = x_t(x_1(\ldots(x_{t-1}x_0)\ldots)).$$

Proof. The implication ($\Leftarrow$) is obvious. We only have to show that the reverse implication ($\Rightarrow$) holds. This can be done by repeated use of the distributive

344   MODES

and entropic laws, as well as (6.4.5):

$$x_0(x_1(\ldots(x_{t-1}x_t)\ldots))$$
$$= x_0 x_1 \cdot (x_0 x_2 \cdot (\ldots (x_0 x_{t-1} \cdot x_0 x_t) \ldots)) \qquad \text{by (LD)}$$
$$= x_0 x_1 \cdot (x_0 x_2 \cdot (\ldots (x_0 x_{t-1} \cdot x_t^t x_0) \ldots)) \qquad \text{by (6.4.5)}$$
$$= x_0 x_1 \cdot (\ldots \quad (x_0 x_t \cdot (x_{t-1} \cdot x_t^{t-1} x_0)) \ldots) \qquad \text{by (E)}$$
$$= (x_0 x_1 \cdot (\ldots \quad (x_t^t x_0 \cdot (x_{t-1} \cdot x_t^{t-1} x_0)) \ldots) \qquad \text{by (6.4.5)}$$
$$= x_0 x_1 \cdot (\ldots \quad (x_t x_{t-1} \cdot x_t^{t-1} x_0) \ldots) \qquad \text{by (RD)}$$
$$= x_0 x_1 \cdot (\ldots \quad (x_0 x_{t-2} \cdot (x_t(x_{t-1} \cdot x_t^{t-2} x_0))) \ldots) \qquad \text{by (LD)}$$
$$= x_0 x_1 \cdot (\ldots \quad (x_0 x_t \cdot (x_{t-2}(x_{t-1} \cdot t_t^{t-2} x_0))) \ldots) \qquad \text{by (E)}$$

$$\ldots\ldots$$

$$= x_0 x_1 \cdot x_t(x_2(\ldots \quad (x_{t-2}(x_{t-1} \cdot x_t x_0)) \ldots))$$
$$= x_0 x_t \cdot x_1(x_2(\ldots \quad (x_{t-2}(x_{t-1} \cdot x_t x_0)) \ldots)) \qquad \text{by (E)}$$
$$= x_t^t x_0 \cdot x_1(x_2(\ldots \quad (x_{t-1} \cdot x_t x_0)) \ldots)) \qquad \text{by (6.4.5)}$$
$$= x_t x_1 \cdot (x_t^{t-1} x_0 \cdot (x_2(\ldots \quad (x_{t-1} \cdot x_t x_0) \ldots))) \qquad \text{by (E)}$$
$$= x_t x_1 \cdot (x_t x_2 \cdot (x_t^{t-2} x_0 \cdot (\ldots (x_{t-1} \cdot x_t x_0) \ldots))) \qquad \text{by (E)}$$

$$\ldots\ldots$$

$$= x_t x_1 \cdot (x_t x_2 \cdot (\ldots \quad (x_t x_{t-2} \cdot (x_t x_{t-1} \cdot x_t x_0)) \ldots))$$
$$= x_t(x_1(x_2(\ldots(x_{t-1} x_0) \ldots))). \qquad \square$$

Example 6.4.4. The varieties $\underline{G}(n,t)$ have interesting subvarieties. One series is given by the varieties $\underline{G}(q)$ of binary modes $(A, \cdot)$ equivalent to the varieties $\underline{\underline{GF(q)}}$ for $q \neq 2$. Let r be a primitive element of $GF(q)$. Theorem 6.4.1 shows that each affine space $(A, \underline{\underline{GF(q)}})$ is equivalent to the groupoid $(A, \underline{r})$. Evidently $r^{q-1} = 1$, and there is a power $t \leq q - 2$ such that $1 - r = r^t$. It follows that the reducts $(A, \cdot) = (A, \underline{r})$ of all the affine $GF(q)$-spaces satisfy the

6. MAL'CEV MODES

identities (6.4.6) for $n = q - 1$, whence they belong to the variety $\underline{G}(q-1,t)$. Exercise 6.4D shows that they also satisfy the identities

(6.4.7) $$x^u y = y^w x$$

for all $1 \leq u, w \leq q - 2$ such that $r^u + r^w = 1$. Note in particular that the identity (6.4.5) is also contained in this set. (Take $u = t$ and $w = 1$.)

Now for a finite field $GF(q)$ with $q \neq 2$ and a primitive element r of $GF(q)$, define $\underline{G}(q)$ to be the variety of binary modes $(A, \cdot)$ satisfying the identities

(6.4.8) $$x^{q-1} y = y,$$
$$x^u y = y^w x$$

for all positive integers u and w such that $1 \leq u, w \leq q-2$ and $r^u + r^w = 1$. As a subvariety of $\underline{G}(q-1,t)$, the variety $\underline{G}(q)$ is equivalent to the variety $\underline{R}$ of affine spaces over some ring R that is a quotient of the ring $R(n,t)$. In particular, the affine space $(R, P, \underline{R})$ is the free $\underline{R}$-algebra on two generators, and is equivalent to the free $\underline{G}(q)$-groupoid $(R, \underline{r})$ on two generators.

Lemma 6.4.4.1. *The free $\underline{G}(q)$-algebra on two generators x and y consists of elements x, xy, $x^2 y$, ..., $x^{q-1} y$, and is isomorphic to $(GF(q), \underline{r})$.*

Proof. One has to show that the set $\{x, xy, \ldots, x^{q-1} y\}$ is closed under groupoid multiplication. First note that for $1 \leq u, w \leq q-2$ with $r^u + r^w = 1$, one has

$$y(x^u y) = y(y^w x) = y^{w+1} x$$
$$= \text{if } w + 1 = q + 1 \text{ then } x \text{ else } x^v y,$$

where v is such that $r^{w+1} + r^v = 1$. Now for any $x^u y$ and $x^w y$ with $0 < u \leq w \leq q - 1$ and with $r^{w-u} + r^s = 1$ one has

$$x^u y \cdot x^w y = x^u y \cdot x^u (x^{w-u} y) = x^u (y \cdot x^{w-u} y)$$
$$= x^u (y^{s+1} x)$$
$$= \text{if } s + 1 = q - 1 \text{ then } x^u y \text{ else } x^{u+v} y = x^{(u+v) \bmod (q-1)} y$$

where $r^{s+1} + r^v = 1$. If $0 < w \leq u \leq q-1$, then the calculation of $x^u y \cdot x^w y$ reduces to the above, since

$$x^u y \cdot x^w y = x^w(x^{u-w} y) \cdot x^w y$$
$$= x^w(x^{u-w} y \cdot y)$$
$$= x^w(x^{u-w} y \cdot x^{q-1} y).$$

For the second part of the lemma, it is sufficient to note that all the non-zero elements of $GF(q)$ are powers of r, and can be generated as $01\underline{r} = 0 \cdot 1$, $01\underline{r^2} = 0 \cdot (0 \cdot 1)$, ..., $01\underline{r^{q-1}} = 0^{q-1} 1$. $\square$

Proposition 6.4.4.2. *The variety $\underline{\underline{GF}}(q)$ of affine spaces over the finite field $GF(q)$ for $q \neq 2$ is equivalent to the variety $\underline{G}(q)$ of binary modes.* $\square$

For further examples, see Exercises 6.4E-J. $\square$

Example 6.4.5. Consider the varieties $\underline{\underline{\mathbb{Z}}}_{2k+1}$ of affine spaces over the rings $\mathbb{Z}_{2k+1}$. Note that each such ring is generated both by $k+1$ and by 2, and that $k+1$ and 2 are mutually inverse in $\mathbb{Z}_{2k+1}$. Note also that $1 - (k+1) \equiv 2k + 2 - (k+1) = k+1$ and $1 - 2 = -1 \equiv 2k$. By Theorem 6.4.1, each faithful affine $\mathbb{Z}_{2k+1}$-space $(A, P, \underline{\underline{\mathbb{Z}}}_{2k+1})$ is equivalent to the groupoid $(A, \cdot) = (A, \underline{r})$, where $r = k+1$ or $r = 2$. The groupoids $(A, \underline{k+1})$ and $(A, \underline{2})$ are also equivalent to quasigroups $(A, \cdot, /, \backslash)$, where $\cdot = \underline{k+1}$ and $/ = \underline{-1}$ and $\backslash = \underline{2}$. In this example, we are interested in groupoid reducts $(A, \cdot) = (A, \underline{2})$ of the affine $\mathbb{Z}_{2k+1}$-spaces. First note that in each such groupoid, $xy^2 = xy\underline{2}\,y\underline{2} = -(-x+2y)+2y = x$. It follows that all the $(A, \underline{2})$ are symmetric binary modes. Now since 2 and $2k+1$ are relatively prime, Euler's Theorem shows that there is an $n = \varphi(2k+1)$ such that $2^n \equiv 1 \mod (2k+1)$. Since $x^n y = x(1 - 2^n) + y 2^n$, it follows that each groupoid $(A, \cdot) = (A, \underline{2})$ also satisfies the identity $x^n y = y$, and is a member of the variety $\underline{\underline{V}}_{n,2}$. Exercise 6.4K shows that the reducts $(A, \underline{2})$ of the spaces $(A, \mathbb{Z}_{2k+1})$ also satisfy the identity

(s_{2k+1}) $\qquad\qquad$ $(\ldots(yx \cdot y)x \ldots)y = x$, where the groupoid word on the left has length $2k+1$.

Now let $\underline{S}_{2k+1}$ be the variety of symmetric binary modes satisfying the identity (s_{2k+1}). It is evident that the class of $\underline{2}$-reducts of $\mathbb{Z}_{2k+1}$-spaces is contained in $\underline{S}_{2k+1}$. We will show that these two classes coincide. The proof comes in a series of lemmata. First, define the following groupoid words $w_i = w_i(x, y)$, and let w'_i be $w_i(y, x)$:

$$w_0 := x, \qquad w_1 := y,$$
$$w_2 := xy, \qquad w_3 := yx \cdot y, \quad \ldots,$$
$$w_{2k} := w'_{2k-1} \cdot y \qquad w_{2k+1} := w'_{2k} \cdot y.$$

Lemma 6.4.5.1. *Each symmetric binary mode satisfies the identities*

(6.4.9) $$\begin{cases} y \cdot w_{k+1} = w_{2k+1}, \\ y \cdot w'_{k+1} = w'_{2k+2} \end{cases}$$

for each positive integer k.

Proof. The proof is by induction on k. For $k = 1$ one has $y \cdot w_2 = y \cdot xy = w_3$ and $y \cdot w'_2 = y \cdot yx = (yx \cdot y)x = w'_4$. Now suppose (6.4.9) holds. Then

$$y \cdot w_{k+2} = (y \cdot w'_{k+1})y \qquad \text{(by distributivity)}$$
$$= w'_{2k+2} \cdot y \qquad \text{(by induction)}$$
$$= w_{2k+3}.$$

Similarly

$$y \cdot w'_{k+2} = (y \cdot w_{k+1}) \cdot yx \qquad \text{(by entropicity)}$$
$$= w_{2k+1} \cdot yx \qquad \text{(by induction)}$$
$$= (w_{2k+1}x \cdot x) \cdot yx \qquad \text{(by symmetry)}$$
$$= (w_{2k+1}x \cdot y)x \qquad \text{(by distributivity)}$$
$$= w'_{2k+4}. \quad \square$$

Lemma 6.4.5.2. *The variety $\underline{S}_{2k+1}$ is a Mal'cev variety.*

Proof. We will show that the word

$$P = (x \cdot y) \cdot w_{k+1}(z, y)$$

defines a Mal'cev operation in each $\underline{S}_{2k+1}$-mode. We have to show that in each $\underline{S}_{2k+1}$-mode $(A, \cdot)$, one has $xyyP = x = yyxP$. First note that by the idempotent and symmetric laws $xyyP = xy \cdot y = x$. Then again by idempotency and Lemma 6.4.5.1, $yyxP = y \cdot w_{k+1} = w_{2k+1}$. Since $\underline{S}_{2k+1}$ is defined by $w_{2k+1} = x$, it follows that $yyxP = x$. □

As a Mal'cev variety of modes, the variety $\underline{S}_{2k+1}$ is equivalent to some variety $\underline{R}$ for a commutative ring R. We will show that the ring R is $\mathbb{Z}_{2k+1}$. To simplify notation, write:

$$(6.4.10) \quad \begin{cases} w_1 := 1, \ w_2 := 1 \setminus 0 = 10\underline{2}, \quad w_3 := (1 \setminus 0) \setminus 1 = w_2 1\underline{2}, \\ \ldots, w_{2i} := w_{2i-1} \setminus 0 = w_{2i-1} 0\underline{2}, \quad w_{2i+1} := w_{2i} \setminus 1 = w_{2i} 1\underline{2}. \end{cases}$$

Lemma 6.4.5.3. *The mode $(\mathbb{Z}_{2k+1}, \setminus) = (\mathbb{Z}_{2k+1}, \underline{2})$ is the free $\underline{S}_{2k+1}$-mode on the set $\{0, 1\}$. The integer $2i + 1$ for $i = 1, \ldots, k$ represents the word w_{2i+1}, while the integer $2i$ represents the word $w_{2k+1-(2i-1)}$.*

Proof. First note that by Exercise 5.5.1E, the subset $W := \{w_i \mid i \in \mathbb{N}\}$ of $\{0, 1\}S$ is closed under $\setminus$. Now the identity (s_{2k+1}) implies that in the free $\underline{S}_{2k+1}$-mode $(\{0, 1\}S_{2k+1}, \setminus)$, W reduces to $\{w_1, \ldots, w_{2k+1} = 0\}$. It follows that in $\{0, 1\}S_{2k+1}$ each element may be expressed as one of the w_i.

The injection $i : \{0, 1\} \to \mathbb{Z}_{2k+1}$ extends to a homomorphism

$$i' : (\{0, 1\}S_{2k+1}, \setminus) \to (\mathbb{Z}_{2k+1}, \setminus).$$

Since $\{0, 1\}$ generates $(\mathbb{Z}_{2k+1}, \setminus)$, the mapping i' surjects. To show that it also injects, it is sufficient to note that the elements of $\mathbb{Z}_{2k+1}$ different from 1 can be calculated as:

$$w_2 = 1 \setminus 0 = 10\underline{2} = -1 \equiv 2k,$$
$$w_3 = (1 \setminus 0) \setminus 1 = (-1)1\underline{2} = 2 + 1 = 3,$$
$$\cdots\cdots$$
$$w_{2i} = w_{2i-1}0\underline{2} = -2i + 1 \equiv 2k - (2i - 2),$$
$$w_{2i+1} = w_{2i}1\underline{2} = 2 - (-2i + 1) = 2i + 1,$$
$$\cdots\cdots$$
$$w_{2k} = 2k - (2k - 2) = 2,$$
$$w_{2k+1} = 2k + 1 \equiv 0. \quad \square$$

Proposition 6.4.5.4. *The variety $\underline{S}_{2k+1}$ of symmetric binary modes satisfying the identity (s_{2k+1}) is equivalent to the variety $\underline{\mathbb{Z}}_{2k+1}$ of affine $\mathbb{Z}_{2k+1}$-spaces.* $\square$

Exercises

6.4A. Show that the sum of coefficients in (6.4.1) equals 1.

6.4B. (a) Show that the reducts $(A, \underline{r})$ of all affine $R(n,t)$-spaces satisfy the identity

$$x_1(x_2(\ldots(x_t x_{t+1})\ldots)) = x_{t+1}(x_2(\ldots(x_t x_1)\ldots)).$$

(b) Deduce that the variety $\underline{G}(n,t)$ may also be defined by the idempotent law, the identity $x^n y = y$, and the identity of (a).

6.4C. Show that if a binary mode $(A, \cdot)$ satisfies the identities $x^n y = y$ and $x^t y = yx$, then there is an m' such that $(A, \cdot)$ satisfies the identity $xy^{m'} = y$ as well.

6.4D. Show that the reducts $(A, \underline{r})$ of all affine $GF(q)$-spaces satisfy the identities (6.4.7).

6.4E. (a) Show that each faithful affine $\mathbb{Z}_7$-space $(A, \underline{\mathbb{Z}_7})$ is equivalent to groupoids

$(A, \underline{2}), (A, \underline{3})$ and $(A, \underline{4})$.

(b) Show that $(A, \underline{2})$ is symmetric and $(A, \underline{4})$ is commutative.

(c) Show that 3 is a primitive element of $\mathbb{Z}_7$, but 2 and 4 are not. Give the axiomatization of the variety $\underline{G}(7)$.

6.4F. [Mitschke, Werner 1973] Show that the following conditions are equivalent for a groupoid $(G, \cdot)$:

(a) $(G, \cdot)$ satisfies the identities $x^2 = x$, $x \cdot yx = y$, $xy \cdot z = zy \cdot x$;

(b) $(G, \cdot)$ satisfies the identities $x^2 = x$, $xy^3 = x$, $xy \cdot z = zy \cdot x$;

(c) $(G, \cdot)$ is a member of the variety $\underline{G}(4)$, equivalent to $\underline{GF(4)}$.

6.4G. [Mitschke, Werner 1973] Show that $R(3, 2) = \mathbb{Z}_2[X]/(X^2 + X - 1) = GF(4)$.

[Hint: In $\mathbb{Z}_2[X]$, $X^3 - 1 = (X + 1)(X^2 + X - 1)$.]

6.4H. [Mitschke, Werner 1973] (a) Show that for the ring $\mathbb{Z}_{25}$ and $r = 3$ one has $r^{12} = 23 = 1 - r$ and $r^{20} = 1$.

(b) Show that for the ring $\mathbb{Z}_9$ and $r = 2$ one has $r^3 = 8 = 1 - r$ and $r^6 = 1$.

6.4I. (a) Show that each reduct $(A, \underline{k+1})$ of an affine $\mathbb{Z}_{2k+1}$-space $(A, \underline{\mathbb{Z}_{2k+1}})$ is commutative. (For $r = k + 1$, one has $r = 1 - r$.)

(b) Show that there is n such that $(k + 1)^n = 1$. Deduce that each $(A, \underline{k+1})$ is a member of the variety $\underline{G}(n, n)$.

6.4J. [Mitschke, Werner 1973] Show that $R(n, 1) = \mathbb{Z}_{2^n - 1}$.

[Hint: In $R(n, 1)$, one has $2^n X^n = 2^n$ and $(2X)^n = 1$. Hence $2^n - 1 = 0$.]

Deduce that $R(2, 1) = GF(3)$, $R(3, 1) = GF(7)$, $R(5, 1) = GF(31)$.

6.4K. Let $(A, \cdot)$ be the reduct $(A, \underline{2})$ of a space $(A, \underline{\mathbb{Z}_{2k+1}})$. Let

$$w_m := (\ldots ((yx \cdot y)x) \ldots)y,$$

where the right hand side is a groupoid word of length m. Show that

$$w_m = xy\underline{m}.$$

Conclude that the reducts $(A, \cdot) = (A, \underline{2})$ of $\mathbb{Z}_{2k+1}$-spaces satisfy the identity (s_{2k+1}).

6.4L. Show that for each prime number p and $r = 2, \ldots, p$, the groupoid $(\mathbb{Z}_p, \underline{r})$ is equivalent to the affine space $(\mathbb{Z}_p, \underline{\mathbb{Z}_p})$.

6.4M. Let R be a commutative ring, and let $r \in R$. Under what condition is the binary operation $\underline{r}$:
   (a) commutative;
   (b) associative;
   (c) a left (or right) zero semigroup operation?

6.4N. Show that a non-trivial semilattice is never equivalent to an affine space.

6.4O. Show that a non-trivial left (or right) zero semigroup is never equivalent to an affine space.

6.4P. Find an example of a differential groupoid that is not a reduct of an affine space.

   [Hint: Note that each reduct of an affine space is diagonally normal (Exercise 6.2F). Find a diagonally abnormal homomorphic image of a differential groupoid that is a reduct of an affine space.]

6.4Q. Let $(A, \cdot)$ be an $\underline{S}_{2k+1}$-mode. Show that for a, b and c in A:
   (a) $a \neq b$ implies $a \neq a \cdot b \neq b$;
   (b) $a \cdot b = c$ if and only if $c \cdot b = a$.
Show that the condition (b) is equivalent to the symmetric identity (S).

6.4R. Define the groupoid words $t_i = t_i(x, y)$ as follows:

$$t_0 := x, \qquad t_1 := y$$
$$t_i := t_{i-2} \cdot t_{i-1} \qquad \text{for } i \geq 2.$$

Show that the identities

$$t_i = w_i$$

hold in the variety $\underline{\underline{S}}_{2k+1}$ for each natural number i, where the w_i are the groupoid words defined in Example 6.4.5.

6.4S. [Kotzig 1970], [Dénes and Keedwell 1972], [Roszkowska 1989] Let $(A, \cdot)$ be a finite $\underline{\underline{S}}_{2k+1}$-mode with r elements. Let $\mathcal{K}_r$ be a complete undirected graph on r vertices labeled by the elements of A. Use Exercises 6.4P and 6.4R to show the following.

(a) The edges (a, b) and (b, c) of $\mathcal{K}_r$ belong to the same closed path of $\mathcal{K}_r$ iff $ab = c$.

(b) The graph $\mathcal{K}_r$ can be partitioned into disjoint simple closed paths whose lengths are divisions of $2k + 1$.

(c) If $2k + 1$ is prime, then the partition comprises $r(r - 1)/2(2k + 1)$ disjoint simple closed paths of lengths $2k + 1$.

6.5. Affine spaces equivalent to quasigroups

Recall that a quasigroup $(A, \cdot, /, \backslash)$ is an algebra with three binary operations of multiplication $\cdot$, right division $/$, and left division $\backslash$ satisfying the identities

(6.5.1)
$$\begin{cases} (x \cdot y)/y = x = (x/y) \cdot y, \\ y \backslash (yx) = x = y(y \backslash x). \end{cases}$$

By Exercise 6.1A, each variety of quasigroups is a Mal'cev variety. By Exercise 6.2E, the Mal'cev operation for quasigroup modes is given by

$$xyzP = (x/y)(y \backslash z).$$

This yields the following:

6. MAL'CEV MODES

Corollary 6.5.1. *Each variety of quasigroup modes is a variety of Mal'cev modes.* □

As a variety of Mal'cev modes, the variety $\underline{QM}$ of quasigroup modes is equivalent to the variety $\underline{R}$ of affine spaces over some ring R. The following theorem determines the ring R.

Theorem 6.5.2. *The variety $\underline{QM}$ of quasigroup modes is equivalent to the variety $\underline{\mathbb{Z}}[p,q,r]$ of affine spaces over the ring $\mathbb{Z}[p,q,r]$, where $p+q = pq$ and $pr = 1$. This ring is isomorphic to the localization $\mathbb{Z}[X]_M$ of the ring $\mathbb{Z}[X]$ at the monoid $(M, \cdot, 1) = (\{X^k(1-X)^l \mid k, l \in \mathbb{N}\}, \cdot, 1)$.*

Proof. First note that the ring R of the variety $\underline{QM}$ is a quotient of the integral polynomial ring $\mathbb{Z}[X, Y, Z]$ with three (commuting) indeterminates X, Y and Z. The quotient is determined by the identities (6.5.1). If we put $xy := xy\underline{X}$, $x/y = xy\underline{Y}$ and $x \setminus y = xy\underline{Z}$, then (6.5.1) implies that

$$(xy)/y = xy\underline{X}\, y\underline{Y} = (x(1-X) + yX)(1-Y) + yY$$
$$= x(1 - X - Y - XY) + y(X - XY + Y) = x$$

and

$$y \setminus (yx) = y\, yx\underline{X}\, \underline{Z} = y(1-Z) + (y(1-X) + xX)Z$$
$$= y(1 - XZ) + xXZ = x,$$

whence $XY - X - Y = 0$ and $XZ - 1 = 0$. It follows that the ring R is a quotient of the ring $\mathbb{Z}[X, Y, Z]/\langle XY - X - Y, XZ - 1\rangle$, isomorphic to the ring $\mathbb{Z}[p, q, r]$ with $p + q = pq$ and $pr = 1$. On the other hand, affine spaces over the ring $\mathbb{Z}[p, q, r]$ are quasigroup modes under $x \cdot y = xy\underline{p}$, $x/y = xy\underline{q}$ and $x \setminus y = xy\underline{r}$. Hence R is isomorphic to $\mathbb{Z}[p, q, r]$.

Now Exercise 6.5A shows that each quasigroup satisfies the identity $y/(x \setminus$

$y) = x$. For the ring $\mathbb{Z}[p,q,r]$, this means that

$$y/(x \setminus y) = y(1-q) + (x(1-r) + yr)q$$
$$= y(1 - q + rq) + x(q - qr) = x.$$

It follows that $q(1-r) = 1$. Together with the equality $pr = 1$, this means that p, q, r and $1-r$ are all invertible. And since $1 - p = 1 - r^{-1} = (r-1)r^{-1}$, it follows that $1 - p$ is invertible, too. Hence $r = p^{-1}$ and $q = p(p-1)^{-1} = 1 - (1-p)^{-1}$. Note also that $xy\underline{q} = xy\underline{(1 - (1-p)^{-1})} = yx\underline{(1-p)^{-1}}$. Hence each element of the ring $\mathbb{Z}[p,q,r]$ can be expressed as an (integer) polynomial over p, p^{-1} and $(1-p)^{-1}$. Now it is easy to see that the ring $\mathbb{Z}[p,q,r]$ and the ring of rationals $f(X)X^{-k}(1-X)^{-l}$ [for $f(X) \in \mathbb{Z}[X]$ and k, $l \in \mathbb{N}$] are isomorphic. □

Since the localization of a Noetherian ring is Noetherian, the ring $\mathbb{Z}[X]_M$ must be Noetherian, too. Moreover, any maximal ideal of $\mathbb{Z}[X]_M$ has a finite index (Exercise 6.5C). Hence the quotient of $\mathbb{Z}[X]_M$ by any maximal ideal is a finite field. This implies the following:

Corollary 6.5.3. *There are countably many varieties of quasigroup modes.* □

Corollary 6.5.4. *The equationally complete varieties of quasigroup modes are equivalent to the varieties of affine spaces over the finite fields $GF(q)$ for $q \neq 2$. Consequently, they are equivalent to varieties of groupoid modes.* □

Exercises

6.5A. Show that each quasigroup satisfies the identities $y/(x \setminus y) = x = (y/x) \setminus y$.

6.5B. [Curtis 1979] An *equihoop* is a binary mode $(A, \cdot)$ satisfying the identity $x \cdot yx = y$.

   (a) For points P and Q in the Euclidean plane $\mathbb{R}^2$, define $P \cdot Q$ to be the point R such that PQR is an equilateral triangle and P, Q, R

are arranged around the triangle in an anticlockwise sense. Prove that $(\mathbb{R}^2, \cdot)$ is an equihoop.

(b) Show that equihoops are quasigroups.

(c) Identify the variety of equihoops as $\underline{\mathbb{Z}}[X]/\langle X^2 - X + 1\rangle$.

6.5C. Show that for any maximal ideal I of $\mathbb{Z}[X]_M$, the quotient $\mathbb{Z}[X]_M/I$ is a finite field.

6.6. Commutative quasigroup modes

Recall that a quasigroup is described as *commutative* if its multiplication is commutative. This section is concerned with commutative quasigroup modes. A basic example of such modes was presented in Example 5.5.3 of Chapter 5. This is the quasigroup $(\mathbb{D}, \cdot, /, \backslash) = (\mathbb{D}, \underline{2^{-1}}, \underline{-1}, \underline{2})$ of dyadic rational numbers. We will show in the current section that this algebra plays a very special role for the variety $\underline{CQM}$ of commutative quasigroup modes. As a Mal'cev variety, $\underline{CQM}$ is equivalent to some variety $\underline{R}$ of affine spaces. We will show that the ring R is $\mathbb{D}$. We will also show that subvarieties of $\underline{CQM}$ are equivalent to certain varieties of symmetric binary modes and of commutative binary modes.

Theorem 6.6.1. *The variety $\underline{CQM}$ of commutative quasigroup modes is equivalent to the variety $\underline{\mathbb{D}}$ of affine $\mathbb{D}$-spaces.*

Proof. By Theorem 6.3.3, the Mal'cev variety $\underline{CQM}$ is equivalent to the variety $\underline{R}$ for a ring R. Thus the problem is to identify R as the ring $\mathbb{D}$ of dyadic rationals. First note that the injection $i : \{0.1\} \to \mathbb{D}$ induces a surjective homomorphism $i' : R \to \mathbb{D}$ from the free $\underline{R}$-algebra on $\{0, 1\}$. On the other hand, there are elements m, r and l of R such that $(R, \cdot, /, \backslash) = (R, \underline{m}, \underline{r}, \underline{l})$. The ring R may be identified with some ring $\mathbb{Z}[m, r, l]$. Now $x \cdot y = y \cdot x$ in $(R, \cdot)$ implies $xy\underline{m} = yx\underline{m}$, i.e. $2m = 1$. Then $(xy)/y = x$ implies $xy\underline{m}\, y\, \underline{r} = x2^{-1}(1-r) + y2^{-1}(1+r) = x$, whence $r = -1$, and $y \backslash (yx) = x$ implies

$y\ yx\underline{m}\ \underline{l} = y(1-l) + (y+x)2^{-1}l = y(1-2^{-1}l) + x2^{-1}l = x$, whence $l = 2$. Thus $\mathbb{Z}[m, r, l] = \mathbb{Z}[2^{-1}] = \mathbb{D}$. □

Corollary 6.6.2. *For a commutative quasigroup mode* $(Q, \cdot, /, \backslash)$, *the reduct* $(Q, \cdot)$ *is a commutative binary mode, and the reduct* $(Q, \backslash)$ *is a symmetric binary mode.* □

The following theorem describes the subvarieties of the variety $\underline{\underline{CQM}}$.

Theorem 6.6.3. *Each non-trivial subvariety of the variety* $\underline{\underline{CQM}}$ *is equivalent to some variety* $\underline{\underline{\mathbb{Z}}}_{2n+1}$ *of affine* $\mathbb{Z}_{2n+1}$*-spaces.*

Proof. First note that the ring $\mathbb{D}$ is the localization of the ring $\mathbb{Z}$ at the monoid $(M, \cdot, 1)$, where $M = \{2^n \mid n \in \mathbb{N}\}$. Since the localization of a principal ideal ring is a principal ideal ring, it follows that $\mathbb{D}$ also has this property. Now for any two integers k and l, the ideal $(k2^{-l})$ generated by $k2^{-l}$ coincides with $k\mathbb{D}$. Indeed $(k2^{-l}) = \{k2^{-l} \cdot m2^n \mid m, n \in \mathbb{Z}\} = \{km2^{-n} \mid m, n \in \mathbb{Z}\} = k\mathbb{D}$. If k is even, say $k = 2^l(2k' + 1)$, then $k\mathbb{D} = (2k' + 1)\mathbb{D}$. So all ideals of $\mathbb{D}$ have the form $k\mathbb{D}$ for an odd positive k.

Since the subvarieties of $\underline{\underline{\mathbb{D}}}$ are determined by the homomorphic images of the ring $\mathbb{D}$, we need to find all of them. Let $\alpha : \mathbb{D} \to R$ be any surjective ring homomorphism. Let $\lambda : \mathbb{Z} \to \mathbb{D}$; $z \mapsto z/1$ be the canonical embedding. Then the image $R = \beta(\mathbb{Z})$ of the composition $\beta = \lambda\alpha : \mathbb{Z} \to R$ must be $\mathbb{Z}_n$ for some positive n. Since all ideals of $\mathbb{D}$ have the form $(2k+1)\mathbb{D}$, it follows that $\alpha^{-1}(0) = (2k+1)\mathbb{D}$ for some natural number k. Now $\mathbb{Z} \cap (2k+1)\mathbb{D} = \{(2k+1)m \mid m \in \mathbb{Z}\}$. Hence $\beta^{-1}(0) = (2k+1)\mathbb{Z}$. Thus $n = 2k+1$. Consequently, the rings $\mathbb{Z}_{2k+1}$ are all the proper quotients of the ring $\mathbb{D}$. □

Note that for the quasigroup reduct $(A, \cdot, /, \backslash) = (A, \underline{2^{-1}}, \underline{-1}, \underline{2})$ of an affine $\mathbb{Z}_{2n+1}$-space, one has $2^{-1} = n + 1$ and $-1 = 2n$.

Lemma 6.6.4. *Each non-trivial quasigroup congruence on* $(\mathbb{D}, \cdot, /, \backslash)$ *is of the*

form
$$\sigma_m := \{(x,y) \in \mathbb{D} \times \mathbb{D} \mid x - y \in m\mathbb{D}\}$$

for an odd natural number m.

Proof. By Theorem 6.6.1, the congruence classes of a quasigroup congruence on $\mathbb{D}$ are precisely the cosets of an ideal of the ring $\mathbb{D}$. Each non-trivial ideal of $\mathbb{D}$ is of the form $m\mathbb{D}$ for an odd natural number m. □

Note that the σ_m are precisely the fully invariant congruences on $(\mathbb{D}, \cdot, /, \backslash)$, and that they determine all the subvarieties of the variety $\underline{CQM}$. In fact, the congruence class 0^{σ_m} of $\mathbb{D}$ coincides with $m\mathbb{D}$, and $\mathbb{D}/m\mathbb{D} = \mathbb{D}^{\sigma_m} \cong \mathbb{Z}_m$. For each positive odd m, let $\underline{CQ}_m$ denote the subvariety of $\underline{CQM}$ equivalent to $\underline{\mathbb{Z}}_m$.

Corollary 6.6.5. *The lattice of subvarieties of the variety $\underline{CQM}$ is isomorphic to the lattice of odd natural numbers with division. The isomorphism is given by $0 \mapsto \underline{CQM}$; $2k+1 \mapsto \underline{CQ}_{2k+1}$.* □

Lemma 6.6.4 will help in finding equational bases for each $\underline{CQ}_m$. For each $m = 2k + 1$, let $l(m)$ be the least integer greater than $\log_2 m$, e.g. $l(3) = 2$, $l(5) = 3$, $l(7) = 3$, $l(9) = 4$. Obviously $m 2^{-l(m)} \in \mathbb{D}_1$, the unit interval of dyadic rationals. Recall (Exercise 5.5.2D) that $m 2^{-l(m)}$ represents the word $w_m(0, 1)$ obtained from 0 and 1 using only the operation $\cdot = \underline{2^{-1}}$. The quotient $m 2^{-l(m)}$ is first represented in the binary form

$$m 2^{-l(m)} = a_{l(m)-1} 2^{-1} + \cdots + a_1 2^{l(m)-1} + a_0 2^{l(m)}$$
$$= 0, a_{l(m)-1} \ldots a_1 a_0$$

with $a_0 = 1$ and $a_i \in \{0, 1\}$. This representation determines the word

$$w_m(0, 1) = a_{l(m)-1}(a_{l(m)-2}(\ldots (a_1 \cdot a_0 0) \ldots)).$$

Corollary 6.6.6. *Each variety $\underline{\underline{CQ}}_{2k+1}$ of commutative quasigroup modes is defined by one additional identity*

(6.6.1) $$xyw_{2k+1} = x.$$

Proof. First note that for $m = 2k+1$, one has $\sigma_m = cg(0, m)$, the smallest congruence containing $(0, m)$. More generally, $\sigma_m = cg(0, mz2^{-n})$ for each integer $z \in \mathbb{Z}$ and natural number $n \in \mathbb{N}$. It is obvious that $\alpha := cg(0, mz2^{-n}) \leq \sigma_m$. By Lemma 6.6.4, the congruence class 0^α is an ideal of the ring $\mathbb{D}$ generated by $mz2^{-n}$. But $\langle mz2^{-n} \rangle = \{mz2^{-n} \cdot k2^{-l} \mid k \in \mathbb{Z}, l \in \mathbb{N}\} = \{mk2^{-l} \mid k \in \mathbb{Z}, l \in \mathbb{N}\} = m\mathbb{D}$. Hence $\sigma_m = cg(0, mz2^{-n})$, and in particular $\sigma_m = cg(0, m2^{-l(m)})$. This implies that in the quasigroup $\mathbb{D}/m\mathbb{D} = \mathbb{D}^{\sigma_m}$, one has $0 = m2^{-l(m)}$. Now 0 corresponds to the word x and $m2^{-l(m)}$ corresponds to the word $w_m(x, y)$ in the free $\underline{\underline{CQM}}$-quasigroup on generators x and y (in place of 0 and 1). Since σ_m is the fully invariant congruence determining the variety $\underline{\underline{CQ}}_m$, it follows that the identity $xyw_m = x$ defines the variety $\underline{\underline{CQ}}_m$. □

The results of the current and previous sections (Example 6.4.5) yield the equivalence of the following varieties:

(6.6.2) $$\underline{\underline{\mathbb{Z}}}_{2k+1} \simeq \underline{\underline{S}}_{2k+1} \simeq \underline{\underline{CQ}}_{2k+1}.$$

We also know (Example 6.4.5) that the affine $\mathbb{Z}_{2k+1}$-spaces $(A, \underline{\underline{\mathbb{Z}}}_{2k+1})$ are eqivalent to the reducts $(A, \underline{k+1})$, and that these reducts are commutative binary modes satisfying the identity (6.6.1). Now by Exercise 6.4I, there is a natural number n such that all $(A, \underline{k+1})$ are members of the variety $\underline{G}(n, n)$. We will show that the varieties of (6.6.1) are also equivalent to varieties of commutative binary modes.

First we need the following lemma.

Lemma 6.6.7. If $(A, \cdot)$ is a sink-free commutative binary mode, then $(A, \cdot)$ is cancellative, i.e. it satisfies the quasi-identity

(6.6.3) $$(xy = xz) \to (y = z).$$

Proof. For each a in A, define the following relation ρ_a on A:

$$(b, c) \in \rho_a :\Leftrightarrow ba = ca.$$

Obviously ρ_a is an equivalence relation. Now let $(b, c), (b', c') \in \rho_a$. By the distributive law $xy \cdot z = xz \cdot yz$ that holds in each groupoid mode, one has $bb' \cdot a = ba \cdot b'a = ca \cdot c'a = cc' \cdot a$. Hence ρ_a is a congruence. The intersection $\rho := \bigcap_{a \in A} \rho_a$ is the equality relation $\widehat{A}$. Indeed,

$$(b, c) \in \rho \Leftrightarrow \forall a \in A, \ ba = ca$$
$$\Rightarrow b = bb = cb = bc = cc = c.$$

For a, b in A, define

$$S(a, b) := \{x \in A \mid (a, b) \in \rho_x\}.$$

The set $S(a, b)$ is a sink in $(A, \cdot)$. Indeed for $x \in S(a, b)$ and $y \in A$, one has $ax = bx$, whence

$$a \cdot bx = ab \cdot ax = ab \cdot bx = b \cdot ax,$$
$$\Rightarrow a \cdot ax = a \cdot bx = b \cdot ax = b \cdot bx,$$
$$\Rightarrow a \cdot yx = ay \cdot ax = (a \cdot ax)(y \cdot ax)$$
$$= (b \cdot bx)(y \cdot bx) = by \cdot bx = b \cdot yx.$$

Consequently $xy \in S(a, b)$ and $S(a, b)$ is a sink. However, since $(A, \cdot)$ has no proper non-empty sinks, it follows that $S(a, b) = A$.

Now if $(b, c) \in \rho_a$, then $a \in S(b, c) = A$. Hence $(b, c) \in \rho_a$ for each a in A, and $(b, c) \in \rho$. But since ρ is the equality relation, it follows that $b = c$. This proves that the cancellation law (6.6.3) holds in $(A, \cdot)$. $\square$

Theorem 6.6.8. *The variety $\underline{C}_{2k+1}$ of commutative binary modes satisfying the identity (6.6.1) is equivalent to the quasigroup variety $\underline{CQ}_{2k+1}$.*

Proof. It is sufficient to show that each $\underline{C}_{2k+1}$-mode $(A, \cdot)$ is equivalent to a $\underline{CQ}_{2k+1}$-quasigroup $(A, \cdot)$. Let $(A, \cdot)$ be a $\underline{C}_{2k+1}$-mode. Since $(A, \cdot)$ is in a strongly irregular variety, $(A, \cdot)$ cannot have an epimorphism $(A, \cdot) \to (\{0, 1\}, \cdot)$ onto the two-element semilattice. By Exercise 6.6A, this means that $(A, \cdot)$ has no proper non-empty sinks. By Lemma 6.6.7, $(A, \cdot)$ is cancellative.

Now the groupoid $(A, \cdot)$ satisfies the identity (6.6.1) that can be written as

$$xyw_{2k+1} = z_{l-1}(z_{l-2}(\ldots(z_1 \cdot yx)\ldots)) = x$$

with $z_i \in \{x, y\}$. In fact Exercise 6.6C shows that $z_{l-1} = y$. Let $u_{2k+1}(x, y) := z_{l-2}(z_{l-3}(\ldots(z_1 \cdot yx)\ldots))$. For $a, b \in A$, the equation $a \cdot x = b$ has a unique solution. Indeed, since $b = a \cdot (abu_{2k+1}) = a \cdot x$, the cancellation law for $(A, \cdot)$, implies that $x = abu_{2k+1}$. It follows that

$$a \setminus b = abu_{2k+1} \qquad \text{and} \qquad b/a = abu_{2k+1}.$$

Hence each $\underline{C}_{2k+1}$-groupoid $(A, \cdot)$ is equivalent to the $\underline{CQ}_{2k+1}$-quasigroup $(A, \cdot, /, \setminus)$. □

Corollary 6.6.9. *The following four varieties are equivalent for each natural number k:*

$$\underline{\mathbb{Z}}_{2k+1} \simeq \underline{S}_{2k+1} \simeq \underline{CQ}_{2k+1} \simeq \underline{C}_{2k+1}. \quad \square$$

Note that if $2k + 1$ is a prime power, Example 6.4.4 provides yet more varieties equivalent to those of Corollary 6.6.9.

Exercises

6.6A. [Romanowska, Smith 1991b] Let $(A, \cdot)$ be a groupoid mode. Let $S_i = (a_i)$ be the principal sink in $(A, \cdot)$ generated by a_i for $i = 1, 2$, and define the complex product $S_1 \cdot S_2 = \{s_1 \cdot s_2 \mid s_1 \in S_1,\ s_2 \in S_2\}$. Show that:

(a) $S_1 \cdot S_2$ is a sink;

(b) $S_1 \cdot S_2 = (a_1 a_2)$;

(c) $S_1 \cap S_2 = (a_1 a_2)$;

(d) The mapping

$$A \to S\ ;\ a \mapsto (a)$$

is a homomorphism onto the semilattice of principal sinks;

(e) $(A, \cdot)$ has no proper non-empty sinks if and only if there is no epimorphism of $(A, \cdot)$ onto the 2-element semilattice. (Cf. Section 7.5.)

6.6B. [Mitschke, Werner 1973] Show that the following conditions are equivalent for a Steiner quasigroup $(A, \cdot)$ (cf. Examples 5.4.1 and 6.4.4):

(a) $(A, \cdot)$ is equivalent to an affine space over $GF(3)$;

(b) $(A, \cdot)$ satisfies the identity

$$x(y \cdot zt) = t(y \cdot zx);$$

(c) $(A, \cdot)$ is entropic;

(d) In the affine space $(A, \underline{GF(3)})$, any tetrahedron has a unique centre, i.e. the third point on the line determined by the third points on two non-adjacent edges.

6.6C. Show that for a $\underline{C}_{2k+1}$-mode $(A, \cdot)$, the identity (6.6.1) can be written as

$$xyw_{2k+1} = y(z_{l-2}(\ldots(z_1 \cdot yx)\ldots)) = x$$

for $z_i \in \{x, y\}$.

Notes on Chapter 6.

The basic Theorem 6.1.1 of this chapter was discovered by A.I. Mal'cev [1954], and assumed fundamental importance in the theory of Mal'cev varieties [J.D.H. Smith 1976], [H.P. Gumm 1979], [J. Hagemann and C. Hermann 1979]. (See also later standard texts for universal algebra such as [S. Burris and H.P. Sankappanavar 1981] and [R. McKenzie, G. McNulty and W. Taylor 1987], where also some further conditions characterizing properties of congruences by satisfaction of certain identities, so-called Mal'cev conditions, can be found.) The basic properties of Mal'cev algebras and varieties were investigated in [J.D.H. Smith 1976]. In particular, centreing congruences were introduced there. See also Chapter III (Centrality) in [O. Chein, H.O. Pflugfelder and J.D.H. Smith 1990]. The papers [J.D.H. Smith 1990b, 1997] discuss the relationships between the concepts of centrality and entropicity of quasigroups, including their connections with homotopies and semisymmetric quasigroups (cf. also [F. Zirilli 1968]). The problem of the uniqueness of Mal'cev operations on an algebra was discussed in [G. Czédli and J.D.H. Smith 1981]. The formulation of Lemma 6.2.1 comes from [A. Szendrei 1986]. The main Theorem 6.2.5 has many predecessors and successors, theorems giving different characterizations of algebras polynomially equivalent to modules. Probably the first such theorem was published by D. C. Murdoch [1941a,b], who showed that each (non-empty) idempotent entropic quasigroup is polynomially equivalent to a module, although he attributed some of the methods he used to A.K. Sushkewitsch. K. Urbanik [1959/1960] proved similar results for so-called Marczewski algebras. Related ideas can be found in [J.P. Soublin 1966] and [F. Ostermann and J. Schmidt 1966]. In general, algebras polynomially equivalent to modules have been characterized by B. Csákány [1964] in terms of the associated system of congruences. Then J.D.H. Smith [1976], and later H.P. Gumm [1979], each proved a version of Theorem 6.2.5 for Mal'cev varieties. Their results were extended to congruence modular varieties by C. Herrmann [1979]. (Alternative proofs were given by H.P. Gumm [1980] and W. Taylor [1982].) For related categorical results, see Mitchell's Embedding Theorem in [B. Mitchell 1965]. The condition (6.2.8) appeared for the first time in [H. Werner 1974].

Affine spaces as algebras were described first by F. Ostermann and J. Schmidt [1966]. They showed that affine spaces over a fixed ring form a variety. Another proof, and a different characterization of such varieties, were provided by B. Csákány [1975a]. (See also [1975b,c].) The proof of the equivalence of a Mal'cev variety of modes and the corresponding variety of affine spaces is modelled on the proof of Theorem 254 in [MT]. The axiomatization for varieties of affine spaces over a commutative ring of Theorem 6.3.4 also comes from [MT]. Theorem 6.3.6 follows from [A. Szendrei 1975]. For discussion of the problem of reducing the number and arity of the basic operations necessary for affine spaces, see [J. Płonka 1970, 1973b, 1974b] and [A. Szendrei 1975, 1977, 1982a,b].

Probably the first paper investigating groupoids equivalent to affine spaces was [A. Mitschke and H. Werner 1973], followed by [B. Ganter and H. Werner 1975]. A first version of Theorem 6.4.1 was communicated to the authors by A. Szendrei. The varieties $\underline{V}_{n,m}$ were considered in [A. Pilitowska, A. Romanowska and J.D.H. Smith 1995]. Varieties of affine spaces over the rings $\underline{R}(n,t)$, and their equivalence with the varieties $\underline{G}(n,t)$, were investigated in [A. Mitschke and H. Werner 1973]. The varieties $\underline{G}(q)$, their equivalence with corresponding varieties of affine spaces, and their relation to combinatorial Steiner systems were discussed in [B. Ganter and H. Werner 1975]. The varieties $\underline{S}_{2k+1}$ of symmetric binary modes were studied by B. Roszkowska [1987a,b, 1989]. For related results, see also [R. Padmanabhan and J. Płonka 1980].

The equivalence of the variety of quasigroup modes with the variety $\underline{\mathbb{Z}}[p,q,r]$ was discovered by B. Csákány and L. Megyesi [1975]. They generalized their result to varieties of n-ary quasigroup modes [1979]. On the other hand, J. Ježek and T. Kepka [1977] thoroughly investigated and characterized entropic quasigroups as polynomially equivalent to modules over certain rings. See also [J.D.H. Smith 1979, 1990].

The varieties of commutative quasigroup modes were described in [J. Ježek and T. Kepka 1975a]. For related results, see also [1976, 1983a], and [J. Gatial 1969, 1971].

Finally note a series of papers by J. Dudek, who characterized varieties of groupoids equivalent to affine spaces over $GF(q)$ for small numbers q by the number of binary operations [1981, 1986a,b, 1991, 1992, 1994a,b]. See also [J. Dudek 1989, 1990], [Jung R. Cho and J. Dudek 2000], and [J. Dudek and J. Tomasik 1996].

CHAPTER VII

SUBREDUCTS OF AFFINE SPACES

We have seen in Chapter 6 that Mal'cev modes may be identified with affine spaces over suitable commutative rings. Reducts and subreducts of affine spaces form another important class of modes close to affine spaces. The next step in generalizing affine spaces is made by considering different types of (functorial and non-functorial) sums of subreducts of affine spaces. The aim of this chapter is to provide general machinery for recognizing such modes as members of certain subprevarieties of varieties of modes, and to give some other characterizations of them. The first task is to assign to each variety $\underline{V}$ of modes its affinization, a certain variety $\underline{R(V)}$ of affine spaces, and to discuss the rôle of the so-called minimal cogenerator for $\underline{R(V)}$, the unique single algebra generating this variety. These concepts are then used to give some characterizations of quasi-affine algebras, i.e. subreducts of affine spaces, among modes in a given variety. As one application, one obtains characterizations of quasi-affine commutative binary modes, barycentric algebras (these are precisely the convex sets), and some complex algebras of convex sets. Another application concerns subdirectly irreducible quasi-affine modes. Subsequent sections deal with subreducts of functorial sums of affine spaces or their subreducts. Section 3 gives a characterization of subreducts of Płonka sums of affine spaces, and Section 4 a characterization of subreducts of functorial sums of cancellative modes. These results are then used in Section 5 to give several descriptions of the structure of barycentric algebras and of commutative binary modes. In Section 6, all varieties and quasivarieties of barycentric algebras are described. Section 7 goes back to discuss the rôle of cancellative modes for the structure of modes in general. It is shown that each cancellative mode embeds as a subreduct into an affine space. This result is then used in Section 8 to show that any Lallement sum of cancellative modes over a naturally quasi-ordered mode embeds as a subreduct into a functorial sum of affine spaces.

7.1. Tensor products of varieties and affinization

This section gives a method for constructing new varieties of modes out of given ones. It is a rich source of examples of varieties of modes. It also gives a very useful method for identifying subclasses of mode varieties consisting of reducts of affine spaces.

Definition 7.1.1. Let $\underline{V}_i$ be varieties of algebras of type $\tau_i : \Omega_i \to \mathbb{N}$, for $i = 1, 2$. Then the *tensor* or *Kronecker product* $\underline{V}_1 \otimes \underline{V}_2$ is the variety of algebras $(A, \Omega_1 \bigcup \Omega_2)$ of type $\tau_1 \bigcup \tau_2$ whose τ_1-reducts lie in $\underline{V}_1$, whose τ_2-reducts lie in $\underline{V}_2$, and which satisfy the identities saying that each basic operation from Ω_1 is a homomorphism of Ω_2-algebras, or equivalently that each operation from Ω_1 commutes with each operation from Ω_2. □

Note that this construction works very nicely in the case of mode varieties. Indeed, one has the following:

Proposition 7.1.2. *If $\underline{V}_1$ and $\underline{V}_2$ are varieties of modes, then so is $\underline{V}_1 \otimes \underline{V}_2$.* □

The symmetry of the entropic laws implies the commutativity

$$\underline{V}_1 \otimes \underline{V}_2 = \underline{V}_2 \otimes \underline{V}_1$$

of the tensor product. The variety $\underline{\text{Set}}$ is an identity for the tensor product

7. SUBREDUCTS OF AFFINE SPACES

operation on varieties of modes, i.e. for a variety $\underline{\underline{V}}$ of modes one has

$$\underline{\underline{\mathrm{Set}}} \otimes \underline{\underline{V}} = \underline{\underline{V}} \otimes \underline{\underline{\mathrm{Set}}} = \underline{\underline{V}}.$$

There are reductions $\pi_i : \underline{\underline{V}}_1 \otimes \underline{\underline{V}}_2 \to \underline{\underline{V}}_i$, by means of which $\underline{\underline{V}}_1 \otimes \underline{\underline{V}}_2$-algebras may be considered just as $\underline{\underline{V}}_i$-algebras, namely $\underline{\underline{V}}_i$-reducts of $\underline{\underline{V}}_1 \otimes \underline{\underline{V}}_2$-algebras. If $\underline{\underline{V}}_1$ and $\underline{\underline{V}}_2$ are varieties of affine R_i-spaces, i.e. $\underline{\underline{V}}_1 = \underline{\underline{R}}_1$ and $\underline{\underline{V}}_2 = \underline{\underline{R}}_2$, then

(7.1.1) $$\underline{\underline{R}}_1 \otimes \underline{\underline{R}}_2 = \underline{\underline{R_1 \otimes R_2}},$$

where the ring $R_1 \otimes R_2$ is the tensor (or Kronecker) product of the rings R_1 and R_2. (See [Jacobson 1964, Definition V.2.1] and Exercise 7.1A.)

For a variety $\underline{\underline{V}}$ of modes, the *affinization* of $\underline{\underline{V}}$ is defined to be the variety $\underline{\underline{\mathbb{Z}}} \otimes \underline{\underline{V}}$. Now $\underline{\underline{\mathbb{Z}}} \otimes \underline{\underline{V}}$ is also a variety of modes, but equipped with the Mal'cev operation P. There is then a ring $R(V)$ such that

$$\underline{\underline{\mathbb{Z}}} \otimes \underline{\underline{V}} = \underline{\underline{R(V)}}.$$

(Cf. Theorem 6.3.3.) The $\underline{\underline{V}}$-reducts of $\underline{\underline{R(V)}}$-algebras are often very useful and informative models of $\underline{\underline{V}}$-algebras.

Example 7.1.3. If S is a commutative ring, then $\underline{\underline{\mathbb{Z}}} \otimes \underline{\underline{S}} = \underline{\underline{\mathbb{Z} \otimes S}} = \underline{\underline{S}}$, so that $R(S) = S$ [Exercise 7.1B]. In other words, varieties of affine spaces over a ring are their own affinizations. □

Example 7.1.4. Let $\underline{\underline{M\tau}}$ denote the variety of all modes of a given (plural) type $\tau : \Omega \to (\mathbb{N} - \{0,1\})$. Let S be the integral polynomial ring $\mathbb{Z}[X_{\omega i} \mid \omega \in \Omega, 1 \leq i \leq \omega\tau]$ over a set $\{X_{\omega i} \mid \omega \in \Omega, 1 \leq i \leq \omega\tau\}$ of $\sum_{\omega \in \Omega} \omega\tau$ commuting indeterminates. Then $R(M\tau)$ is the quotient ring $S/\langle 1 - \sum_{i=1}^{\omega\tau} X_{\omega i} \mid \omega \in \Omega \rangle$ of S by the ideal obtained by setting each sum $\sum_{i=1}^{\omega\tau} X_{\omega i}$ to be 1. For $\omega \in \tau^{-1}(n)$, the corresponding operation on an affine space over $R(M\tau)$ is

(7.1.2) $$x_1 \ldots x_n \omega = \sum_{i=1}^n x_i X_{\omega i}$$

for the indeterminates $X_{\omega 1}, \ldots, X_{\omega n}$ pertaining to ω (cf. 5.3). □

Example 7.1.5. Consider the variety $\underline{\underline{Dm}}$ of differential modes (cf. 5.6). In the notation of Example 7.1.4, let $X_1 = X$ and $X_2 = 1 - X$ be the indeterminates pertaining to groupoid multiplication. Then $R(\mathrm{Dm})$ is a quotient of $\mathbb{Z}[X]$. Now the reduction law $x \cdot yz = xy$ holds for $x \cdot y = x(1 - X) + yX$ in $\underline{R(\mathrm{Dm})}$. Equating coefficients of z in

$$x(1 - X) + (y(1 - X) + zX)X$$
$$= x \cdot yz = xy$$
$$= x(1 - X) + yX$$

shows that $zX^2 = 0$, so that $R(\mathrm{Dm})$ is a quotient of $\mathbb{Z}[X]/\langle X^2 \rangle \cong \mathbb{Z}[d]$, the ring of integral dual numbers with $d^2 = 0$. Conversely, affine spaces over $\mathbb{Z}[d]$ are differential groupoids under $x \cdot y = x(1 - d) + yd$. Thus $R(\mathrm{Dm}) = \mathbb{Z}[d]$. Now $\underline{d}$-subreducts of affine $\mathbb{Z}[d]$-spaces are differential modes. Exercise 7.1D involves the affinizations of certain varieties of differential modes. In particular, all these affinizations reduce to affinizations $\underline{\underline{\mathbb{Z}[d]}}$ with $d^2 = 0 = pd$ of the varieties $\underline{\underline{D}}_{0,p}$ defined by

$$x = xy^p$$

for positive integers p. We will see in Chapter 8 that the variety $\underline{\underline{Dm}}$ is the join of the chain consisting of all the $\underline{\underline{D}}_{0,p}$. In particular, this shows that the quasivariety of $\underline{d}$-subreducts of affine $\mathbb{Z}[d]$-spaces generates the variety $\underline{\underline{Dm}}$ of differential modes. Note however that $\underline{\underline{Dm}}$ contains modes that are not subreducts of affine spaces. (Cf. Example 6.4O.) □

More generally, Exercise 7.1C shows that the affinization of the variety $\underline{\underline{R}}_n$ of n-reductive or n-differential groupoids is the variety of affine spaces over the ring $R(R_n) = \mathbb{Z}[X]/\langle X^n \rangle \cong \mathbb{Z}[d]$ with $d^n = 0$. (Cf. 5.7.)

Example 7.1.6. Consider the variety $\underline{C}$ of commutative binary modes (cf. 5.5 and 6.6). As in Example 7.1.5, let X be the indeterminate pertaining to groupoid multiplication. Now in $\underline{\underline{\mathbb{Z}}} \otimes \underline{\underline{C}}$, equating coefficients of x (or of y) in $x \cdot y = y \cdot x$ gives

$$x(1 - X) + yX = x \cdot y = y \cdot x = y(1 - X) + xX,$$

or $X = \frac{1}{2}$. A similar argument as in Example 7.1.5 shows that $R(C) = \mathbb{Z}[\frac{1}{2}] = \mathbb{D}$. Thus the affinization of $\underline{\underline{C}}$ is (equivalent to) the variety $\underline{\underline{CQM}}$ of commutative quasigroup modes. Now $\frac{1}{2}$-reducts of affine $\mathbb{D}$-spaces are cancellative commutative binary modes. On the other hand, we will see in Chapter 8 that each cancellative $\underline{\underline{C}}$-mode is a $\frac{1}{2}$-subreduct of an affine $\mathbb{D}$-space. This identifies the subclass of $\underline{\underline{C}}$ consisting of subreducts of affine spaces as the subquasivariety of cancellative $\underline{\underline{C}}$-modes. Note however that there are $\underline{\underline{C}}$-modes that are not cancellative [Exercise 7.1F].

Example 7.1.7. Consider the variety $\underline{\underline{B}}$ of barycentric algebras over a subfield R of the real field $\mathbb{R}$. The ring $R(B)$ is in fact the free $\underline{\underline{R(B)}}$-algebra on two generators. Each $\underline{\underline{R(B)}}$-algebra satisfies the identities

(7.1.3)
$$xy(\underline{p+r}) = (xy\underline{p}, x, xy\underline{r})P,$$
$$xy(\underline{-r}) = (x, xy\underline{r}, x)P,$$
$$xy(\underline{pq}) = x \ xy\underline{p} \ \underline{q}.$$

Indeed, this follows directly by Theorem 6.3.4 and (6.2.12), since $(xy\underline{p}, x, xy\underline{r})P = xy \ \underline{p0rP} = xy(\underline{p+r})$ and $(x, xy\underline{r}, x)P = xy \ \underline{0r0P} = xy \ (\underline{-r})$. Now the free $\underline{\underline{R(B)}}$-algebra $R(B)$ consists of all binary operations. In particular, it contains all the operations $\underline{r}$ for all r in $I°$, and by the identities (7.1.3) also for all r in R. Since the ring $R(B)$ is generated as a ring by the set $I°$, it follows that $R(B)$ consists exactly of these operations $\underline{r}$. They are pairwise distinct because the $\underline{\underline{R(B)}}$-algebra R is obviously a member of $\underline{\underline{R(B)}}$. It follows that

$R(B) = R$. Now the class of $\underline{I}°$-subreducts of affine R-spaces is precisely the class $\underline{Cv}$ of convex subsets of R-spaces. □

Example 7.1.8. Consider the variety $\underline{\text{Ln}}$ of left normal bands. Let X be the indeterminate pertaining to groupoid multiplication. Now in $\underline{\mathbb{Z}} \otimes \underline{\text{Ln}}$, equating coefficients of y in $x \cdot y = x(y \cdot y) = xy \cdot y$ gives $yX = yX(1 - X) + yX$, or $yX = yX^2$. Then

$$x \cdot yz = x(1 - X) + y(1 - X)X + zX^2$$
$$= x(1 - X) + zX = xz,$$

whence $x \cdot y = x \cdot zy = x \cdot yz = x \cdot z$ using the left normal law. In particular, $x \cdot y = x \cdot x = x$. Now equating coefficients of y gives $yX = 0$. This means that $R(\text{Ln})$ is a quotient of $\mathbb{Z}[X]/\langle X \rangle \cong \mathbb{Z}$. On the other hand, defining $x \cdot y := xy \underline{0} = x$ in an integral affine space gives an $\underline{\text{Ln}}$-algebra, so that $R(\text{Ln}) = \mathbb{Z}$.

From the argument above, it is easy to see that the subvariety $\underline{\text{Lz}}$ of the variety $\underline{\text{Ln}}$ of left zero semigroups has the same affinization as $\underline{\text{Ln}}$. Indeed, one has

$$\underline{\mathbb{Z}} \subseteq \underline{\mathbb{Z}} \otimes \underline{\text{Lz}} \subseteq \underline{\mathbb{Z}} \otimes \underline{\text{Ln}} = \underline{\mathbb{Z}},$$

so that $R(\text{Lz}) = \mathbb{Z}$. The variety of left normal bands is the regularization of the variety of left zero bands, so one may paraphrase the above as

(7.1.4) $$R(\widetilde{\text{Lz}}) = R(\text{Lz}). \quad \Box$$

The equality (7.1.4) underlies and illustrates a more general equality between the affinization of a variety and its regularization.

Theorem 7.1.9. *Let $\underline{V}$ be an irregular (plural) variety of modes, with regularization $\underline{\widetilde{V}}$. Then $R(V) = R(\widetilde{V})$.*

Proof. By Płonka's Theorem 4.3.2, there is a derived partition operation $\cdot$ in $\underline{\widetilde{V}}$-algebras satisfying the identities true for left normal bands. The corresponding reduction, namely $\underline{\widetilde{V}} \to \underline{\mathrm{Ln}}$; $(A, \Omega) \to (A, \cdot)$, commutes with the affinization to give $\underline{\underline{\mathbb{Z}}} \otimes \underline{\widetilde{V}} \to \underline{\underline{\mathbb{Z}}} \otimes \underline{\mathrm{Ln}}$. But by Example 7.1.8, one has $\underline{\underline{\mathbb{Z}}} \otimes \underline{\mathrm{Ln}} = \underline{\underline{\mathbb{Z}}} \otimes \underline{\mathrm{Lz}}$, so that the partition operation on $\underline{\underline{\mathbb{Z}}} \otimes \underline{\widetilde{V}}$-algebras is a left zero band operation. By Płonka's Theorem again, this means that the $\underline{\widetilde{V}}$-reducts of $\underline{\underline{\mathbb{Z}}} \otimes \underline{\widetilde{V}}$-algebras are $\underline{V}$-algebras, whence $\underline{\underline{\mathbb{Z}}} \otimes \underline{\widetilde{V}} \subseteq \underline{\underline{\mathbb{Z}}} \otimes \underline{V}$. Conversely, $\underline{V} \subseteq \underline{\widetilde{V}}$ implies $\underline{\underline{\mathbb{Z}}} \otimes \underline{V} \subseteq \underline{\underline{\mathbb{Z}}} \otimes \underline{\widetilde{V}}$. Thus $\underline{R(V)} = \underline{\underline{\mathbb{Z}}} \otimes \underline{V} = \underline{\underline{\mathbb{Z}}} \otimes \underline{\widetilde{V}} = \underline{R(\widetilde{V})}$, from which $R(V) = R(R(V)) = R(R(\widetilde{V})) = R(\widetilde{V})$ by Example 7.1.3. □

Exercises

7.1A. [Freyd 1966] Show that, for commutative rings R_1 and R_2, one has

$$\underline{\underline{R}}_1 \otimes \underline{\underline{R}}_2 = \underline{\underline{R_1 \otimes R_2}} \ .$$

7.1B. Prove the equality in Example 7.1.3.

7.1C. Show that
$$R(R_n) = \mathbb{Z}[X]/\langle X^n \rangle.$$

7.1D. [Krauze 1998] Let $\underline{\underline{D}}_{j,j+p}$ be the subvariety of the variety $\underline{\underline{Dm}}$ of differential groupoids defined by the identity

$$xy^j = xy^{j+p}$$

for some integers j and p, with j natural and p positive. Show that $R(D_{j,j+p}) = \mathbb{Z}[X]/\langle X^2, pX \rangle \cong \mathbb{Z}[d]$ with $d^2 = 0 = pd$.

7.1E. Find the affinization of the variety of n-cyclic binary modes $(A, \cdot)$ defined by the identity

$$xy^n = x.$$

7.1F. Find an example of non-cancellative $\underline{\underline{C}}$-modes.

7.2. Subreducts of Mal'cev modes

A mode is called *quasi-affine* if it is a subreduct of an affine space. For a variety $\underline{V}$ of modes, the subclass $Q_a(\underline{V})$ of quasi-affine $\underline{V}$-modes obviously forms a quasivariety. (Cf. Corollary 3.7.16.) By Exercise 7.2B, it is not necessarily a variety. Examples 7.1.6 and 7.1.7 show that the subclass $Q_a(\underline{C})$ of quasi-affine modes contained in the variety $\underline{C}$ of commutative binary modes, and similarly the subclass $Q_a(\underline{B})$ in the variety $\underline{B}$ of barycentric algebras over a subfield R of $\mathbb{R}$, form the quasivarieties of cancellative algebras. On the other hand, Example 7.1.5 shows that quasi-affine algebras that are members of a variety $\underline{V}$ of Ω-modes cannot always be characterized as cancellative $\underline{V}$-algebras. (Cf. Exercise 6.4O.) Since such a variety $\underline{V}$ may contain modes that are not subreducts of affine spaces, one again has the problem of recognizing such subreducts among all $\underline{V}$-algebras. Recall that the quasi-affine modes belonging to the variety $\underline{V}$ embed into affine $\underline{\mathbb{Z}} \otimes \underline{V}$-spaces. The solution to the problem that will be given in this section relates the class $\underline{V}$ and the class $Q_a(\underline{V}) = s_\Omega(\underline{\mathbb{Z}} \otimes \underline{V})$ of Ω-subreducts of affine $\underline{\mathbb{Z}} \otimes \underline{V}$-spaces with the affinization $\underline{R(V)} = \underline{\mathbb{Z}} \otimes \underline{V}$ of the variety $\underline{V}$.

The key concept that will be used in the solution is the concept of a minimal cogenerator of the variety $\underline{\mathrm{Mod}}_R$ of modules over a commutative ring R. It is well known from the theory of modules (see e.g. [Anderson and Fuller 1992]) that there is a *cogenerator* F in $\underline{\mathrm{Mod}}_R$, i.e. for each module A in $\underline{\mathrm{Mod}}_R$ there is a set I such that the module A embeds into the power F^I:

(7.2.1) $$\iota : (A, +, R) \to (F^I, +, R).$$

Moreover, one can always find a *minimal cogenerator* F in $\underline{\mathrm{Mod}}_R$, i.e. a cogenerator F such that if F' is another cogenerator in $\underline{\mathrm{Mod}}_R$, then F embeds into F'. The minimal cogenerator F is uniquely specified up to isomorphism.

The concept of a minimal cogenerator of the variety $\underline{\mathrm{Mod}}_R$ carries over to the corresponding variety $\underline{R}$ of affine spaces. By Proposition 6.3.1, if $(A, P, \underline{R})$

7. SUBREDUCTS OF AFFINE SPACES

is an affine R-space, and e is a fixed element of A, then $(A, P, \underline{R}, e)$ is equivalent to the module $(A, +, R)$. More generally, results of Section 6.3 show that if $\underline{V}$ is a variety of Mal'cev Ω-modes equivalent to the variety $\underline{R} = \{0, 1\}V$, then each "pointed" $\underline{V}$-algebra (A, Ω, e) is equivalent to the "pointed" affine space $(A, P, \underline{R}, e)$ and to the R-module $(A, +, R)$. Then the embedding (7.2.1) has a restriction to the $\underline{R}$-homomorphism

$$\iota : (A, P, \underline{R}) \to (F^I, P, \underline{R}),$$

or equivalently to the Ω-homomorphism

$$\iota : (A, \Omega) \to (F^I, \Omega).$$

Conversely, each subalgebra of each power $(F^I, P, \underline{R})$ of the affine space $(F, P, \underline{R})$ is of course an affine R-space. The algebra (F, Ω) equivalent to the space $(F, P, \underline{R})$ will be called the *minimal cogenerator of the variety* $\underline{V}$.

Example 7.2.1. If the commutative ring R of a variety $\underline{V}$ of Mal'cev modes is a field F, then F is the minimal cogenerator of the variety $\underline{V}$ equivalent to $\underline{F}$. □

Example 7.2.2. (Cf. [Anderson and Fuller 1992], Exercise 13, p. 215.) Let R be a principal ideal domain. The group R^* of units of R acts on R by (right) multiplication. Elements common to an R^*-orbit are described as *associates*. The *primes* in R are non-zero elements p such that pR is a prime ideal in R. In particular, prime elements of a principal ideal ring are precisely its irreducible elements. Let $\mathbb{P}$ denote a set consisting of one representative from each class of associated primes in R. If Q is the field of quotients of R, then $Q/R = \bigoplus \{R_{p^\infty} \mid p \in \mathbb{P}\}$, the direct sum of all $R_{p^\infty} = \{(a/p^n) + R \mid a \in R, n \in \mathbb{Z}\}$, is the minimal cogenerator in the module variety $\underline{\mathrm{Mod}}_R$. And of course, the affine space reduct of Q/R is the minimal cogenerator in the affine space variety $\underline{R}$. In particular, $\mathbb{Q}/\mathbb{Z} \cong \bigoplus\{\mathbb{Z}_{p^\infty} \mid p \in \mathbb{P}\}$ is the minimal cogenerator in the variety $\underline{\mathbb{Z}}$ of integral affine spaces. Similarly, $\mathbb{Q}/\mathbb{D} \cong \bigoplus\{\mathbb{D}_{p^\infty} \mid p \in \mathbb{P}\}$, where $\mathbb{D}_{p^\infty} = \{(m2^k/p^n) + \mathbb{D} \mid k, m, n \in \mathbb{Z}\}$, is the minimal cogenerator in the variety $\underline{\mathbb{D}}$ of affine $\mathbb{D}$-spaces, or equivalently, in the variety $\underline{CQM}$ of commutative quasigroup modes. □

Now let $\underline{V}$ be any variety of Ω-modes, and let $\underline{R(V)} = \underline{\mathbb{Z}} \otimes \underline{V}$ be its affinization. As explained above, the variety $\underline{R(V)}$ has a minimal cogenerator F_V, specified uniquely up to isomorphism. A $\underline{V}$-algebra (A, Ω) is said to be *finitely separable in the variety $\underline{V}$* if and only if for $a \neq b$ in A there is $f_{ab} \in \underline{V}(A, F_V)$ with $af_{ab} \neq bf_{ab}$. (Cf. Exercise 3.1C.)

Theorem 7.2.3. *Let $\underline{V}$ be a variety of Ω-modes, and let $\underline{R(V)}$ be its affinization, with a minimal cogenerator $F = F_V$. Then the following conditions are equivalent for each $\underline{V}$-algebra (A, Ω):*

(a) (A, Ω) *is a subreduct of some $\underline{R(V)}$-algebra;*

(b) (A, Ω) *is finitely separable in $\underline{V}$;*

(c) (A, Ω) *is a subdirect product of Ω-subreducts of F, more precisely*

$$(A, \Omega) \leq_s \prod\{(Af_{ab}, \Omega) \mid (a, b) \in A \times A\}.$$

Proof. (a)⇒(b) Suppose that (A, Ω) is a subreduct of an $\underline{R(V)}$-algebra $(B, P, \underline{R(V)})$. Then there is an embedding

$$\iota : (B, P, \underline{R(V)}) \to (F^I, P, \underline{R(V)}).$$

Hence by Exercise 3.1C, there is an affine space homomorphism

$$h : (B, P, \underline{R(V)}) \to (F, P, \underline{R(V)})$$

such that if $a \neq b$ in $A \subseteq B$, then $ah \neq bh$. Of course h is also an Ω-homomorphism, and h restricted to A is the homomorphism f_{ab} of (A, Ω)

separating the points a and b.

(b)$\Rightarrow$(a) Assume that for any elements a and b of A there is a homomorphism $(A, \Omega) \to (F, \Omega)$ separating these elements. Consider the affine space $(F^A, P, \underline{R(V)})$. With respect to the mapping $0 : A \to F$; $a \mapsto 0$, the affine space F^A becomes a module $M = (F^A, +, R(V))$. Let M^* be the dual module to M. For each a in A, there is a^* in M^* given by

$$a^* : M \to F \ ; \ m \mapsto am,$$

i.e. $ma^* = am$ for $m : A \to F$. Note that $\Omega(A, F) \subseteq F^A$ and $(\Omega(A, F), \Omega) \leq (F^A, \Omega)$. Moreover, (F^A, Ω) is a reduct of $M = (F^A, +, \underline{R(V)})$. The assignment $a \mapsto a^*$ embeds A in M^*. Indeed, for a distinct pair $b \neq c$ of elements of A, there is an Ω-homomorphism $f : (A, \Omega) \to (F, \Omega)$ in M with $bf \neq cf$, and hence $fb^* = bf \neq cf = fc^*$, i.e. $b^* \neq c^*$. In particular, $a \mapsto a^*|_{\Omega(A,F)}$ is an embedding. Further, this embedding is an Ω-homomorphism. Indeed, for (n-ary) ω in Ω and m in $\Omega(A, F)$, one has

$$m(a_1 \ldots a_n \omega)^* = a_1 \ldots a_n \omega \ m \qquad \text{and}$$
$$m(a_1^* \ldots a_n^*)\omega = ma_1^* \ldots ma_n^* \ \omega$$
$$= a_1 m \ldots a_n m \ \omega$$
$$= a_1 \ldots a_n \omega \ m,$$

since m is an Ω-homomorphism. Thus (A, Ω) is a subreduct of M^*.

(b)$\Leftrightarrow$(c) This equivalence follows directly by Corollary 3.1.3 and Exercise 3.1C. $\square$

Now consider the variety $\underline{\underline{B}}$ of barycentric algebras over a subfield R of $\mathbb{R}$. Recall that R is the minimal cogenerator of $\underline{\underline{B}}$. As a corollary of Theorem 7.2.3, one obtains a second characterization of convex sets amongst barycentric algebras in $\underline{\underline{B}}$.

Corollary 7.2.4. *A barycentric algebra B in the variety $\underline{B}$ is a convex set if and only if B is finitely separable, i.e. for $a \neq b$ in B there is $f_{ab} \in \underline{B}(B,R)$ with $af_{ab} \neq bf_{ab}$.* □

One of the most interesting applications of Corollary 7.2.4 is to the barycentric algebra $(AP, \underline{I}^\circ)$ of non-empty polytopes contained in a convex subset $(A, \underline{I}^\circ)$ of a finite-dimensional Euclidean space $\mathbb{R}^d$. There is a similar application to the subalgebra $(AK, \underline{I}^\circ)$ of $(AS, \underline{I}^\circ)$ consisting of compact convex subsets of A.

Theorem 7.2.5. *Let A be a convex subset of a finite-dimensional Euclidean space $\mathbb{R}^d$. Then the sets AP of non-empty polytopes in A and AK of non-empty compact convex subsets of A are convex subsets of real affine spaces. More precisely, the barycentric algebra $(AK, \underline{I}^\circ)$ and its subalgebra $(AP, \underline{I}^\circ)$ lie in the class $\underline{Cv}$.*

Proof. It will be shown that the algebras $(AP, \underline{I}^\circ)$ and $(AK, \underline{I}^\circ)$ are finitely separable, so that Corollary 7.2.4 may be applied to give the desired result.

Denote the inner product in $\mathbb{R}^d$ by $\mathbb{R}^d \times \mathbb{R}^d \to \mathbb{R}$; $(x, y) \mapsto (x|y)$. Define the *support function*

$$H : AK \times \mathbb{R}^d \to \mathbb{R} \; ; \; (C, x) \mapsto \sup\{(c, x) \mid c \in C\}.$$

For a fixed x in $\mathbb{R}^d$, the function $H_x : AK \to \mathbb{R}$; $C \mapsto H(C, x)$ is a $\underline{B}$-homomorphism, since for C, D in AK and p in I° one has $H_x(CD\underline{p}) =$

$$H(CD\underline{p}, x) = \sup\{(y|x) \mid y \in CD\underline{p}\}$$
$$= \sup\{(cd\underline{p}|x) \mid c \in C, \; d \in D\}$$
$$= \sup\{(c|x)(d|x)\underline{p} \mid c \in C, \; d \in D\}$$
$$= \sup\{(c|x)(1-p) + (d|x)p \mid c \in C, \; d \in D\}$$
$$= (1-p)\sup\{(c|x) \mid c \in C\} + p\sup\{(d|x) \mid d \in D\}$$
$$= H(C,x)H(D,x)\underline{p} = H_x(C)H_x(D)\underline{p}.$$

The penultimate equality holds because the real numbers p and $1-p$ are nonnegative.

Now if two compact non-empty convex sets C and D are distinct, then there is a vector x in $\mathbb{R}^d$ for which $H(C,x) \neq H(D,x)$. (See [Bonnesen and Fenchel 1948] p.24, and [Grünbaum 1967], 2.2 Ex. 8(iv).) Thus both AP and AK are finitely separable, as required. $\square$

Another application of Theorem 7.2.3 is to subdirectly irreducible quasi-affine algebras.

Corollary 7.2.6. *For a variety $\underline{V}$ of Ω-modes, the subdirecly irreducible quasi-affine $\underline{V}$-algebras are Ω-subreducts of the cogenerator F_V.* $\square$

Lemma 7.2.7. *Let $\underline{V}$ be a variety of Ω-modes. If a quasi-affine $\underline{V}$-algebra (A, Ω) is a proper semilattice Lallement sum of subalgebras (A_i, Ω) over (I, Ω), then for all $i \leq j$ in $(I, \leq)$:*

(a) $A_i \varphi_{i,j} \cap A_j = \varnothing$;
(b) $\varphi_{i,j} : A_i \to E_j$ *is an injection.*

Proof. First note that the quasi-identity (6.2.7) of Proposition 6.2.4, characterizing diagonally normal algebras, is inherited by subreducts. Let ω be an n-ary operator in Ω. For $i < j$, let a_i, b_i be in A_i and let c_j be in A_j. Suppose, on the contrary, that $a_i \varphi_{i,j} \in A_j$. Then $a_i \ldots a_i c_j \omega = a_i \varphi_{i,j} \ldots a_i \varphi_{i,j} c_j \omega$ is in A_j. But $a_i \ldots a_i b_i \omega$ lies in A_i, and hence is not equal to $a_i \varphi_{i,j} \ldots a_i \varphi_{i,j} b_i \omega$ in A_j. This violates (6.2.7). It follows that (a) holds.

Now for $a_i \neq b_i$, let $a_i \varphi_{i,j} = b_i \varphi_{i,j}$. Assume that $a_i \ldots a_i b_i \omega \neq b_i$. Then $a_i \ldots a_i c_j \omega = a_i \varphi_{i,j} \ldots a_i \varphi_{i,j} c_j \omega = b_i \varphi_{i,j} \ldots b_i \varphi_{i,j} c_j \omega = b_i \ldots b_i c_j \omega$. But $a_i \ldots a_i b_i \omega \neq b_i \ldots b_i \omega = b_i$, again violating the condition (6.2.7). $\square$

In particular, Lemma 7.2.7 shows that if a subdirectly irreducible quasi-affine $\underline{V}$-algebra is a canonical semilattice sum of algebras (A_i, Ω), then for each

$i \in I$, the envelope E_i must be the disjoint union of all the A_j for $j \leq i$. Such sums are in a sense opposite to Płonka sums, where all envelopes coincide with the fibres they contain.

Corollary 7.2.8. *Let $\underline{V}$ be a regular variety of Ω-modes not equivalent to the variety of Ω-semilattices. Let $\underline{R(V)}$ be its affinization, with a minimal cogenerator F_V. If (A, Ω) is a subdirectly irreducible quasi-affine $\underline{V}$-algebra, then (A, Ω) is a disjoint union of subalgebras of (F_V, Ω). Moreover, either (A, Ω) does not contain a non-trivial sink, or (A, Ω) is an envelope of a subalgebra with this property.* □

Note that in the case where the variety $\underline{V}$ is irregular, no $\underline{V}$-algebra has a non-trivial homomorphism onto a non-trivial semilattice. Consequently, all subdirectly irreducible quasi-affine $\underline{V}$-algebras are "sink-free" subalgebras of (F_V, Ω).

Example 7.2.9. Consider the variety $\underline{B}$ of barycentric algebras over a field R as in Corollary 7.2.4. Then R is a minimal cogenerator of the affinization $\underline{R}$ of $\underline{B}$, and the quasi-affine $\underline{B}$-algebras are just convex subsets of affine R-spaces. By Corollary 7.2.6, subdirectly irreducible convex $\underline{B}$-algebras are $\underline{I}^\circ$-subalgebras of $(R, \underline{I}^\circ)$. However, the only subalgebras of $(R, \underline{I}^\circ)$ are proper or improper intervals. Open proper intervals are simple, and any two of them are isomorphic barycentric algebras. By Corollary 7.2.8, envelopes of proper open intervals are closed or half-closed intervals. They provide three other isomorphic types of subdirectly irreducible convex sets. The algebra $(R, \underline{I}^\circ)$ is simple, and is not isomorphic to any proper interval. Indeed, it satisfies the formula $(\forall\, x \in R,\ \forall\, y \in R,\ \forall\, p \in I^\circ,\ \exists z \in R.\ xz\underline{p} = y)$ that is not satisfied in any proper interval.

Proposition 7.2.9.1. *The bounded intervals $([0, 1], \underline{I}^\circ)$, $((0, 1], \underline{I}^\circ)$, $([0, 1), \underline{I}^\circ)$, $((0, 1), \underline{I}^\circ)$ and the closed unbounded intervals $([0, \infty), \underline{I}^\circ)$ and $(R, \underline{I}^\circ)$ of R are,*

7. SUBREDUCTS OF AFFINE SPACES

up to isomorphism, the only subdirectly irreducible quasi-affine barycentric algebras over R. □

Example 7.2.10. By Example 7.1.6, the affinization of the variety $\underline{C}$ of commutative binary modes is the variety $\underline{D}$ of affine $\mathbb{D}$-spaces with the minimal cogenerator $F_\mathbb{D} = \mathbb{Q}/\mathbb{D} \cong \bigoplus \{\mathbb{D}_{p^\infty} \mid p \in \mathbb{P}\}$ of Example 7.2.2. By Corollary 7.2.6, subdirectly irreducible $\underline{C}$-algebras are $\underline{2^{-1}}$-subalgebras of the reduct $(\mathbb{Q}/\mathbb{D}, \underline{2^{-1}})$. Note that $\mathbb{D}_{p^\infty} = \{z/(2^n p^k) + \mathbb{D} \mid z \in \mathbb{Z},\ n,\ k \in \mathbb{N}\} = \{z/p^k + \mathbb{D} \mid z \in \mathbb{Z},\ 0 \le z < p^k,\ n,\ k \in \mathbb{N}\}$. Indeed, since 2^n and p^k are relatively prime, there are integers u and v such that $\frac{z}{2^n p^k} = \frac{u}{2^n} + \frac{v}{p^k}$, whence $\frac{z}{2^n p^k} + \mathbb{D} = \frac{v}{p^k} + \mathbb{D}$. Note that for a prime number $p \in \mathbb{P}$ with $p > 2$ and $r \in \mathbb{Z}^+$, the module $\mathbb{Z}_{p^r} \cong \mathbb{Z}/p^r\mathbb{Z}$ embeds into $\mathbb{D}_{p^\infty}$ via the homomorphism extending the mapping $1 + p^r\mathbb{Z} \mapsto \frac{1}{p^r} + \mathbb{D}$, so that $(\mathbb{Z}_{p^r}, +, \mathbb{D}) \cong (\mathbb{D}_{p^r}, +, \mathbb{D})$ with $\mathbb{D}_{p^r} = \{\frac{z}{p^r} + \mathbb{D} \mid z \in \mathbb{Z}_{p^r}\}$. The simple $\mathbb{D}$-modules are $\mathbb{Z}_p$ for $p > 2$. The $\mathbb{D}$-modules $\mathbb{Z}_{p^r}$ for $r \in \mathbb{N}$ are obviously subdirectly irreducible submodules of $\mathbb{D}_{p^\infty}$. Now $\mathbb{Z}_{p^r}$ can be equivalently considered as a $\mathbb{Z}_{p^r}$-module. Indeed, since p^r and 2^n are relatively prime, there are integers u and v such that $\left(\frac{1}{p^r} + \mathbb{D}\right)\frac{c}{2^n} = \frac{u}{p^r} + \frac{v}{2^n} + \mathbb{D} = \frac{u}{p^r} + \mathbb{D}$. Hence the annihilator $\mathrm{Ann}\mathbb{D}_{p^r}$ of $\mathbb{D}_{p^r}$ equals $\{\frac{c}{2^n} p^r \mid c \in \mathbb{Z},\ n \in \mathbb{N}\}$, and so the ring $\mathbb{D}/\mathrm{Ann}\mathbb{D}_{p^r}$ is isomorphic to $\mathbb{Z}_{p^r}$. It follows that the modules $(\mathbb{Z}_{p^r}, +, \mathbb{D})$ and $(\mathbb{Z}_{p^r}, +, \mathbb{Z}_{p^r})$ are equivalent. The affine spaces $(\mathbb{Z}_{p^r}, P, \mathbb{Z}_{p^r})$ are equivalent to their $\underline{2^{-1}}$-reducts, where $\underline{2^{-1}} = (p^r + 1)/2$. (Cf. 6.6 and Theorem 6.4.1.) It follows that the groupoids $(\mathbb{Z}_{p^r}, \cdot)$ with $x \cdot y := xy(p^r + 1)/2$ are subdirectly irreducible $\underline{C}$-modes. By Theorem 6.6.7 and Corollary 6.6.8, the subvarieties $\underline{C}_{2k+1}$ of $\underline{C}$ contain only cancellative modes, and are equivalent to the varieties $\underline{Z}_{2k+1}$ of affine $\mathbb{Z}_{2k+1}$-spaces and to the varieties $\underline{CQ}_{2k+1}$ of commutative quasigroup modes. It follows that $(\mathbb{Z}_{p^r}, \cdot)$ are the only subdirectly irreducible modes in the varieties $\underline{C}_{2k+1}$. We are left with the following question: Are there subdirectly irreducible quasi-affine $\underline{C}$-modes not contained in varieties $\underline{C}_{2k+1}$? First note that since the ring $\mathbb{D}$ does not have a unique minimal ideal, the groupoid $(\mathbb{D}, \cdot)$ is not subdirectly irreducible. In fact, no proper interval of $\mathbb{D}$ can be a subdirectly irreducible $\underline{C}$-mode [Ježek and Kepka 1975a]. The groupoids $(\mathbb{D}_{p^r}, \underline{2^{-1}})$ are generated by $0 = 0 + \mathbb{D}$ and $1 = p^{-r} + \mathbb{D}$. The $\underline{2^{-1}}$-reducts of $F_{\mathbb{D}}$ generated by finitely many non-zero elements are generated by two elements, and are isomorphic to the unit interval $(\mathbb{D}_1, \underline{2^{-1}})$, hence are not subdirectly irreducible. However, there exist non-finitely generated subdirectly irreducible quasi-affine subgroupoids of $(F_{\mathbb{D}}, \underline{2^{-1}})$. These are the $(\mathbb{D}_{p^\infty}, \cdot) = (\mathbb{D}_{p^\infty}, \underline{2^{-1}})$. To prove this, we need to show that the congruence lattice of $(\mathbb{D}_{p^\infty}, \cdot)$ has a monolith μ. So let θ be a congruence of $(\mathbb{D}_{p^\infty}, \cdot)$. Let $\frac{u}{2^n p^k}$ and $\frac{v}{2^m p^l}$ be in $\mathbb{D}_{p^\infty}$ and be related by θ. Then $\left(0 = \left(\frac{-u}{2^n p^k}\right) \cdot \frac{u}{2^n p^k}, \left(\frac{-u}{2^n p^k}\right) \cdot \frac{v}{2^m p^l}\right) \in \theta$, too. It follows that there is an element a in $\mathbb{D}_{p^\infty}$ with $(0, a) \in \theta$. Without loss of generality one can assume that $a = \frac{c}{p^r} + \mathbb{D}$ for some $0 < c < p^r$ and $r \in \mathbb{N}$. Let θ' be the congruence θ restricted to the subdirectly irreducible subgroupoid $(\mathbb{D}_{p^r}, \cdot)$ of $(\mathbb{D}_{p^\infty}, \cdot)$. The monolith μ' of the congruence lattice of $(\mathbb{D}_{p^r}, \cdot)$ is obviously contained in θ', and it collapses all elements $\mathbb{D}$, $p^{-1} + \mathbb{D}, \ldots, (p-1)p^{-1} + \mathbb{D}$ of the simple subgroupoid $(\mathbb{D}_p, \cdot)$ of $(\mathbb{D}_{p^r}, \cdot)$. Now the monolith μ is the smallest congruence of $(\mathbb{D}_{p^\infty}, \cdot)$ generated by the set $\mathbb{D}_p$. This gives the following:

Proposition 7.2.10.1. *The groupoids $(\mathbb{D}_{p^r}, \underline{2^{-1}})$ and $(\mathbb{D}_{p^\infty}, \underline{2^{-1}})$ for odd prime numbers p and positive integers r are, up to isomorphism, all the subdirectly irreducible quasi-affine commutative binary modes.* □

Exercises

7.2A. Let A be a convex subset of a finite-dimensional Euclidean space $\mathbb{R}^d$. Show that the barycentric algebra $(AS, \underline{I}^\circ)$ of all non-empty convex subsets of A is not a convex set. Hence Theorem 7.2.5 has no analogue for the complex algebras of subalgebras of convex sets.

7.2B. Use the quasi-identity (6.2.8) to show that a homomorphic image of a reduct of an affine space is not necessarily a reduct of an affine space. (Cf. Exercises 6.2F and 6.4O.)

7.3. Subreducts of Płonka sums of Mal'cev modes

The separability condition of the previous section characterizes the subreducts of affine spaces, i.e. the quasi-affine modes, amongst all the members of a given variety $\underline{V}$ of modes. In this section, we will formulate a similar separability condition for characterizing subreducts of Płonka sums of affine spaces amongst all the members of a given variety $\underline{V}$ of modes.

Let $\underline{V}$ be a variety of Ω-modes, and let $\underline{R(V)}$ be its affinization, with minimal cogenerator $F = F_V$. The $\underline{V}$-reduct (F, Ω) of the cogenerator F lies in $\underline{R(V)}$. Let F^∞ be the Płonka sum of (F, Ω) and the one-element algebra $(\{\infty\}, \Omega)$ over the 2-element semilattice $\underset{\sim}{2}$ with $0 < 1$ by the functor that sends the unique nonidentity arrow to the $\underline{V}$-morphism $F \to \{\infty\}$. Note that F^∞ lies in the regularization $\underline{\widetilde{V}}$. A $\underline{V}$-algebra (A, Ω) is said to be *separable in the variety* $\underline{V}$ if and only if, for all $a \neq b$ in A, there is an Ω-homomorphism $f_{ab} \in \widetilde{V}(A, F^\infty)$ with $af_{ab} \neq bf_{ab}$. Note that $\underline{\widetilde{V}}$-algebras are finitely separable if distinct points can be separated by homomorphisms into F, i.e. by "finite homomorphisms". On the other hand, $\underline{\widetilde{V}}$-algebras are "separable" if distinct points can be separated by homomorphisms into F^∞, i.e. by homomorphisms which "may take infinite values". If $\underline{V}$ is an irregular variety of modes, then no $\underline{V}$-algebra can have a non-trivial semilattice quotient. Thus each f_{ab} must map to F. In other words, separable algebras in irregular varieties are finitely separable. The following characterization then becomes especially interesting for regular varieties $\underline{V}$ of modes. Note that in this case $\underline{\widetilde{V}} = \underline{V}$.

Theorem 7.3.1. *Let $\underline{V}$ be a variety of Ω-modes, and let $\underline{R(V)}$ be its affinization, with minimal cogenerator F. Then the following conditions are equivalent* for each $\underline{V}$-algebra (A, Ω):

(a) (A, Ω) is separable in $\underline{V}$;
(b) (A, Ω) is an Ω-subreduct of an $\widetilde{R(V)}$-algebra;
(c) (A, Ω) is a subreduct of a Płonka sum of affine spaces over $R(V)$;
(d) (A, Ω) is a subdirect product of subalgebras of (F^∞, Ω), or more precisely

$$(A, \Omega) \leq_s \prod \{(Af_{ab}, \Omega) \mid (a, b) \in A \times A\}.$$

Proof. (a)$\Rightarrow$(b) Let (A, Ω) be a separable $\underline{V}$-algebra. Define $A^* := \underline{\widetilde{V}}(A, F^\infty)$. By Corollary 5.3, A^* can also be considered as an Ω-algebra contained in the regularization $\underline{\widetilde{V}}$. Define $A^{**} = \underline{\widetilde{V}}(A^*, F^\infty)$. The *evaluation map*

$$e : A \to A^{**}; \ a \mapsto (a^{**} : A^* \to F^\infty; \ f \mapsto af)$$

is clearly well-defined, since for each (n-ary) ω in Ω and $f_1, \ldots, f_n$ in A^* one has

$$(f_1 \ldots f_n \omega) a^{**} = a(f_1 \ldots f_n \omega) = af_1 \ldots af_n \ \omega = f_1 a^{**} \ldots f_n a^{**} \omega.$$

Now e is a $\underline{\widetilde{V}}$-homomorphism, since for each (n-ary) ω in Ω, for $a_1, \ldots, a_n$ in A, and for f in A^*, one has

$$f(a_1^{**} \ldots a_n^{**} \omega) = (fa_1^{**}) \ldots (fa_n^{**})\omega = a_1 f \ldots a_n f \ \omega$$
$$= (a_1 \ldots a_n \omega) f = f(a_1 \ldots a_n \omega)^{**}.$$

Further, e embeds A into A^{**}, since for $a \neq b$ in A, the element f of A^* separating a and b satisfies $fa^{**} = af \neq bf = fb^{**}$. Thus the $\underline{V}$-algebra A is a subalgebra of the $\underline{\widetilde{V}}$-algebra $A^{**} = \underline{\widetilde{V}}(A^*, F^\infty)$, which in turn is a reduct of the $\widetilde{R(V)}$-algebra A^{**}.

(c)$\Rightarrow$(a) Now suppose that a $\underline{V}$-algebra (A, Ω) is a subreduct of the Płonka sum $IG = (B, P, \underline{R(V)})$ of affine $R(V)$-spaces $(B_i, P, \underline{R(V)})$ over a semilattice

7. SUBREDUCTS OF AFFINE SPACES

$(I, +)$ by the functor $G : (I) \to \underline{\underline{R(V)}}$, with projection $\pi : IG \to I$. To show that (A, Ω) is separable in $\underline{\underline{V}}$, it suffices to verify that the $\underline{\underline{\widetilde{V}}}$-algebra (B, Ω) is separable in $\underline{\underline{\widetilde{V}}}$. Fix an element i in I. Then a $\underline{\underline{V}}$-homomorphism $f : (B_i, \Omega) \to (F, \Omega)$ may be extended to a $\underline{\underline{\widetilde{V}}}$-homomorphism $\widetilde{f} : (B, \Omega) \to (F^\infty, \Omega)$ by

(7.3.1) $\qquad a\widetilde{f} := \text{if } a\pi \leq i \text{ then } a\varphi_{a\pi, i}f \text{ else } \infty.$

For each (n-ary) ω in Ω and $a_1, \ldots, a_n$ in B, with $a_1\pi + \cdots + a_n\pi = j$ in I, one must check the homomorphism condition

(7.3.2) $\qquad a_1\widetilde{f} \ldots a_n\widetilde{f}\, \omega = a_1 \ldots a_n \omega\, \widetilde{f}.$

There are two cases to consider. If for each $1 \leq k \leq n$ one has $a_k\pi \leq i$, then

$$a_1 \ldots a_n \omega\, \pi = a_1\pi \ldots a_n\pi\, \omega = a_1\pi + \cdots + a_n\pi$$
$$\leq i + \cdots + i = i.$$

The map $\widetilde{f}$ in (7.3.2) may then be replaced by $\varphi_{a\pi, i}f$. Since $\varphi_{a\pi, i}f$ is a $\underline{\underline{\widetilde{V}}}$-homomorphism, equality holds in (7.3.2) in this case. Otherwise, there is $1 \leq k \leq n$ such that $a_k\pi \not\leq i$, so that $a_k\pi \leq a_1\pi + \cdots + a_k\pi + \cdots + a_n\pi = j \not\leq i$. The right hand side of (7.3.2) is

$$a_1\varphi_{a_1\pi, j} \ldots a_n\varphi_{a_n\pi, j}\omega\, \widetilde{f} = \infty,$$

while the left hand side is

$$a_1\widetilde{f} \ldots a_k\widetilde{f} \ldots a_n\widetilde{f}\, \omega = a_1\widetilde{f} \ldots \infty \ldots a_n\widetilde{f}\, \omega = \infty,$$

so that equality obtains again.

Now consider a pair $a \neq b$ of distinct elements of B. If $a\pi = b\pi = i$, then $af \neq bf$ for a $\underline{\underline{V}}$-homomorphism $f : (B_i, \Omega) \to (F, \Omega)$, whence $a\widetilde{f} \neq b\widetilde{f}$ for the extended $\underline{\underline{\widetilde{V}}}$-homomorphism $\widetilde{f} : (B, \Omega) \to (F^\infty, \Omega)$ given by (7.3.1). If

$a\pi \neq b\pi$, then without loss of generality one can assume that $a\pi \not\leq b\pi = i$. Take $f : B_i \to F$ to be any $\underline{V}$-homomorphism, e.g. a constant map. Then for the extended $\underline{\widetilde{V}}$-homomorphism $\widetilde{f} : (B, \Omega) \to (F^\infty, \Omega)$ given by (7.3.1), one has $b\widetilde{f} \in F$ and $a\widetilde{f} = \infty$, so $a\widetilde{f} \neq b\widetilde{f}$. Thus IG is separable in $\underline{\widetilde{V}}$.

(b)⇔(c) This equivalence follows directly by Płonka's Theorem 4.3.2.

(a)⇔(d) This equivalence follows directly by Exercise 3.1C. □

Now, for a variety $\underline{V}$ of Ω-modes, let $S_e(\underline{V})$ be the subclass of all separable $\underline{V}$-algebras. Theorem 7.3.1(b) and Corollary 3.7.16 show that $S_e(\underline{V})$ is a quasi-variety. As each $S_e(\underline{V})$-algebra decomposes as a sum of quasi-affine $\underline{V}$-algebras over an Ω-semilattice, it is obvious that

$$S_e(\underline{V}) = Q_a(\underline{V}) \circ \underline{Sl},$$

where the Mal'cev product of the quasi-varieties $Q_a(\underline{V})$ and $\underline{Sl}$ is taken relative to $S_e(\underline{V})$.

Theorem 7.3.1 allows one to describe subdirectly irreducible separable $\underline{V}$-algebras. Note that a subalgebra of the algebra (F^∞, Ω) is either a subalgebra of (F, Ω) or a Płonka sum of a subalgebra of (F, Ω) and the singleton Ω-algebra.

Corollary 7.3.2. *If a subdirectly irreducible $S_e(\underline{V})$-algebra is not a semilattice, then it is either a subdirectly irreducible subalgebra of (F, Ω) or a subdirectly irreducible subalgebra of (F, Ω) extended by zero.* □

Example 7.3.3. Consider the variety $\underline{B}$ of barycentric algebra as in Example 7.2.9. Note that by Exercise 4.2D, any Płonka sum of barycentric algebras in $\underline{B}$ is again a barycentric algebra. In particular, Płonka sums of convex sets in $\underline{B}$ lie again in $\underline{B}$. On the other hand, $\underline{I}^\circ$-subreducts of affine R-spaces are obviously barycentric algebras. By Corollary 7.3.2, the subdirectly irreducible $S_e(\underline{B})$-algebras are precisely those listed in Proposition 7.2.9.1, the Płonka sums of these algebras with the singleton $\underline{I}^\circ$-algebra, and the 2-element $\underline{I}^\circ$-semilattice.

We will see in Section 7.5 that these are all the subdirectly irreducible $\underline{B}$-algebras. □

Example 7.3.4. Commutative binary modes in the variety $\underline{C}$ of Example 7.2.10 behave much like the barycentric algebras over the field R. Płonka sums of cancellative $\underline{C}$-modes are again in $\underline{C}$. And 2^{-1}-subreducts of affine $\mathbb{D}$-spaces are members of $\underline{C}$. By Corollary 7.3.2, the subdirectly irreducible $S_e(\underline{C})$-modes are those listed in Proposition 7.2.10.1, the Płonka sums of these algebras with the singleton $\underline{C}$-mode, and the 2-element semilattice. As in the case of barycentric algebras, we will show later that these are all the subdirectly irreducible $\underline{C}$-modes. □

7.4. Subreducts of functorial sums of cancellative modes

In the previous section, we used the separability condition to characterize $\underline{V}$-subreducts of functorial sums (over semilattices) of affine $R(V)$-spaces. One can also consider such subreducts as semilattice sums of $\underline{V}$-subreducts of affine $R(V)$-spaces. (Cf. Section 4.5.) As in the case of barycentric algebras and commutative binary modes, it can happen that such $\underline{V}$-subreducts are precisely the cancellative $\underline{V}$-algebras, and the corresponding sums are semilattice sums of cancellative $\underline{V}$-algebras. In this section we are interested in Lallement sums of cancellative Ω-modes over arbitrary naturally quasi-ordered modes. Cancellativity for Ω-modes is defined similarly as in the case of binary modes. An Ω-mode is said to be *cancellative* if it satisfies the quasi-identity

$$(x_1 \ldots x_{i-1} y x_{i+1} \ldots x_n \omega = x_1 \ldots x_{i-1} z x_{i+1} \ldots x_n \omega) \to (y = z)$$

for each (n-ary) ω in Ω and each $i = 1, \ldots, n$. We will use an approach different from that of Section 7.3 to characterize Lallement sums of cancellative Ω-modes. We will show that they are precisely the subalgebras of functorial sums of cancellative Ω-modes. This gives a good description of algebras in certain quasivarieties of modes. We then apply these results to give certain representation theorems for barycentric algebras and for commutative binary modes in the next section, and show that these algebras are separable.

Let $C_l(\Omega)$ be the quasivariety of cancellative Ω-modes of type $\tau : \Omega \to \mathbb{Z}^+$, and let $\underline{\mathbb{I}}$ be a quasivariety of naturally quasi-ordered Ω-modes. By Corollary 4.5.6, each algebra (A, Ω) in the quasivariety $C_l(\Omega) \circ \underline{\mathbb{I}}$ is a Lallement sum of some $C_l(\Omega)$-subalgebras (A_i, Ω) over an $\underline{\mathbb{I}}$-algebra (I, Ω). As in the proof of Theorem 4.5.3, the envelopes (E_i, Ω) in the Lallement sum may be obtained as quotients (P_j^μ, Ω) of the subalgebras $P_j = \bigcup\{A_i \mid i \leq j\}$ by maximal congruences $\mu(j)$. To make good use of the cancellativity in the characterization theorem below, we will define certain special maximal congruences $\mu(j)$ of (P_j, Ω) and show that the envelopes $E_j = P_j^{\mu(j)}$ are also cancellative.

A mapping $\sigma : \Omega \to \mathbb{Z}^+$ with $\omega\sigma \leq \omega\tau$ for each ω in Ω is called a *subtype* of τ. Define a relation $\mu = \mu(j, \sigma)$ on P_j by:

$$(b, c) \in \mu :\Leftrightarrow \forall\, \omega \in \Omega,\ \forall\, a_1, \ldots, a_{\omega\sigma-1}, a_{\omega\sigma+1}, \ldots, a_{\omega\tau} \in A_j,$$

$$a_1 \ldots a_{\omega\sigma-1} b a_{\omega\sigma+1} \ldots a_{\omega\tau}\omega = a_1 \ldots a_{\omega\sigma-1} c a_{\omega\sigma+1} \ldots a_{\omega\tau}\omega.$$

Lemma 7.4.1. *For each subtype σ of τ, the relation $\mu(j, \sigma)$ is the largest congruence on (P_j, Ω) preserving (A_j, Ω). Moreover $\mu := \mu(j) = \mu(j, \sigma)$ does not depend on the choice of the subtype σ, and the envelope $(E_j, \Omega) = (P_j^\mu, \Omega)$ of (A_j, Ω) is cancellative.*

Proof. It is obvious that $\mu(j, \sigma)$ is an equivalence relation. Now for $i = 1, \ldots, m$, let $k_i \preceq j$ and $l_i \preceq j$. Let $b_i \in A_{k_i}$ and $c_i \in A_{l_i}$. Suppose $(b_i, c_i) \in \mu$, i.e. for each n-ary ω in Ω and $a_1, \ldots, a_n$ in A_j one has $a_1 \ldots a_{\omega\sigma-1} b_i a_{\omega\sigma+1} \ldots a_n\omega = a_1 \ldots a_{\omega\sigma-1} c_i a_{\omega\sigma+1} \ldots a\omega$. Then by the idempotent and entropic laws, the fol-

7. SUBREDUCTS OF AFFINE SPACES 387

lowing holds for each n-ary ω' in Ω:

$$a_1 \ldots a_{\omega\sigma-1}(b_1 \ldots b_m\omega')a_{\omega\sigma+1} \ldots a_n\omega$$
$$= (a_1 \ldots a_1\omega') \ldots (a_{\omega\sigma-1} \ldots a_{\omega\sigma-1}\omega')(b_1 \ldots b_m\omega')(a_{\omega\sigma+1} \ldots a_{\omega\sigma+1}\omega')$$
$$\ldots (a_n \ldots a_n\omega')\omega$$
$$= (a_1 \ldots a_{\omega\sigma-1}b_1 a_{\omega\sigma+1} \ldots a_n\omega) \ldots (a_1 \ldots a_{\omega\sigma-1}b_m a_{\omega\sigma+1} \ldots a_n\omega)\omega'$$
$$= (a_1 \ldots a_{\omega\sigma-1}c_1 a_{\omega\sigma+1} \ldots a_n\omega) \ldots (a_1 \ldots a_{\omega\sigma-1}c_m a_{\omega\sigma+1} \ldots a_n\omega)\omega'$$
$$= a_1 \ldots a_{\omega\sigma-1}(c_1 \ldots c_m\omega')a_{\omega\sigma+1} \ldots a_n\omega.$$

Thus μ is a congruence on (P_j, Ω). By the cancellativity of (A_j, Ω), it is immediate that μ preserves (A_j, Ω).

Let λ be another congruence on (P_j, Ω) preserving (A_j, Ω). Let $(b, c) \in \lambda$ for b in A_i and c in A_k, where $i, k \preceq j$. Then for $a_1, \ldots, a_n$ in A_j and n-ary ω in Ω one has $(a_1 \ldots a_{\omega\sigma-1}b a_{\omega\sigma+1} \ldots a_n\omega, a_1 \ldots a_{\omega\sigma-1}c a_{\omega\sigma+1} \ldots a_n\omega) \in \lambda$. But since both these elements are in A_j, and λ preserves (A_j, Ω), it follows that $a_1 \ldots a_{\omega\sigma-1}b a_{\omega\sigma+1} \ldots a_n\omega = a_1 \ldots a_{\omega\sigma-1}c a_{\omega\sigma+1} \ldots a_n\omega$. Consequently $(b, c) \in \mu$, and μ is the largest congruence on (P_j, Ω) preserving (A_j, Ω). It is clear that μ does not depend on the choice of the subtype σ.

To show that $(E_j, \Omega) = (P_j^\mu, \Omega)$ is cancellative, let $a_1, \ldots, a_n$ be in A_{j_i} for $i = 1, \ldots, n$, and let $j_i \preceq j$. For $k, l \preceq j$ let $b \in A_k$ and $c \in A_l$. For n-ary ω in Ω, let $(a_1 \ldots a_{i-1}b a_{i+1} \ldots a_n\omega)^\mu = (a_1 \ldots a_{i-1}c a_{i+1} \ldots a_n\omega)^\mu$, i.e. for each m-ary ω' in Ω and $d_1, \ldots, d_{p-1}, d_{p+1}, \ldots, d_m$ in A_j and $1 \leq p \leq m$ one has:

$$d_1 \ldots d_{p-1}(a_1 \ldots a_{i-1}b a_{i+1} \ldots a_n\omega)d_{p+1} \ldots d_m\omega'$$
$$= d_1 \ldots d_{p-1}(a_1 \ldots a_{i-1}c a_{i+1} \ldots a_n\omega)d_{p+1} \ldots d_m\omega'.$$

By the idempotent and entropic laws one obtains

$$d_1 \ldots d_{p-1}(a_1 \ldots a_{i-1} x a_{i+1} \ldots a_n \omega) d_{p+1} \ldots d_m \omega'$$
$$= (d_1 \ldots d_{p-1} a_1 d_{p+1} \ldots d_m \omega') \ldots (d_1 \ldots d_{p-1} a_{i-1} d_{p+1} \ldots d_m \omega')$$
$$(d_1 \ldots d_{p-1} x d_{p+1} \ldots d_m \omega')(d_1 \ldots d_{p-1} a_{i+1} d_{p+1} \ldots d_m \omega')$$
$$\ldots (d_1 \ldots d_{p-1} a_n d_{p+1} \ldots d_m \omega') \omega,$$

where $x = b$ or $x = c$. Moreover for $r = 1, \ldots, n$ the elements

$$d_1 \ldots d_{p-1} a_r d_{p+1} \ldots d_m \omega'$$

and the element $d_1 \ldots d_{p-1} x d_{p+1} \ldots d_m \omega'$ belong to A_j. Since (A_j, Ω) is cancellative, it follows that $d_1 \ldots d_{p-1} b d_{p+1} \ldots d_m \omega' = d_1 \ldots d_{p-1} c d_{p+1} \ldots d_m \omega$, whence $b^\mu = c^\mu$. Consequently, (E_j, Ω) is cancellative. $\square$

Theorem 7.4.2. *Let (A, Ω) be a mode in the quasivariety $C_l(\Omega) \circ \underline{\mathbb{I}}$, with projection π onto an $\underline{\mathbb{I}}$-algebra (I, Ω) and corresponding $C_l(\Omega)$-fibres (A_i, Ω). Then (A, Ω) is a subalgebra of a functorial sum of cancellative envelopes (E_i, Ω) over (I, Ω).*

Proof. By Theorem 4.5.3 and Lemma 7.4.1, the algebra (A, Ω) is a Lallement sum of (A_i, Ω) over (I, Ω), and for each i in I, the cancellative envelope of (A_i, Ω) is $(E_i, \Omega) = (P_i^\mu, \Omega)$. The sum homomorphisms $\varphi_{i,j}$ are $\varphi_{i,j} : A_i \to E_j$; $a \mapsto a^{\mu(j)}$. To prove that (A, Ω) embeds into a functorial sum (E, Ω) of (E_i, Ω) over (I, Ω), it is enough to show that the $\varphi_{i,j}$ satisfy the condition (4.5.3) of Theorem 4.5.18. Let i, j, k, l be in I, and let i, $j \preceq k \preceq l$. Let e be in A_i and f in A_j. We will show that $e^{\mu(k)} = f^{\mu(k)}$ implies $e^{\mu(l)} = f^{\mu(l)}$. Let $a_1, \ldots, a_n$ be in A_k and $b_1, \ldots, b_m$ in A_l. For each n-ary ω in Ω and m-ary ω' in Ω, let

$$a_1 \ldots a_{p-1} e a_{p+1} \ldots a_n \omega = a_1 \ldots a_{p-1} f a_{p+1} \ldots a_n \omega.$$

7. SUBREDUCTS OF AFFINE SPACES

Then by the idempotent and entropic laws

$$b_1 \ldots b_{r-1}(a_1 \ldots a_{p-1} x a_{p+1} \ldots a_n \omega) b_{r+1} \ldots b_m \omega'$$
$$= (b_1 \ldots b_{r-1} a_1 b_{r+1} \ldots b_m \omega') \ldots (b_1 \ldots b_{r-1} a_{p-1} b_{r+1} \ldots b_m \omega')$$
$$(b_1 \ldots b_{r-1} x b_{r+1} \ldots b_m \omega')(b_1 \ldots b_{r-1} a_{p+1} b_{r+1} \ldots b_m \omega')$$
$$\ldots (b_1 \ldots b_{r-1} a_n b_{r+1} \ldots b_m \omega')\omega,$$

where $x = e$ or $x = f$. Moreover, for $s = 1, \ldots, n$, the elements

$$b_1 \ldots b_{r-1} a_s b_{r+1} \ldots b_m \omega'$$

and the element $b_1 \ldots b_{r-1} x b_{r+1} \ldots b_m \omega'$ belong to A_l. Since (A_l, Ω) is cancellative, it follows that

$$b_1 \ldots b_{r-1} e\ b_{r+1} \ldots b_m \omega' = b_1 \ldots b_{r-1} f b_{r+1} \ldots b_m \omega',$$

whence $e^{\mu(l)} = f^{\mu(l)}$. □

The embedding of a $C_l(\Omega) \circ \underline{\mathbb{I}}$-mode (A, Ω) into the functorial sum of cancellative envelopes (E_i, Ω) obtained in Theorem 7.4.2 will be called *canonical*.

Note that obviously a subalgebra of a functorial sum of $C_l(\Omega)$-algebras over an $\underline{\mathbb{I}}$-algebra is a member of the quasivariety $C_l(\Omega) \circ \underline{\mathbb{I}}$.

Corollary 7.4.3. *The class $SS_f(C_l(\Omega), \underline{\mathbb{I}})$ of subalgebras of functorial sums of $C_l(\Omega)$-algebras over an $\underline{\mathbb{I}}$-algebra coincides with the quasivariety $C_l(\Omega) \circ \underline{\mathbb{I}}$, i.e.*

$$SS_f(C_l(\Omega), \underline{\mathbb{I}}) = C_l(\Omega) \circ \underline{\mathbb{I}}. \quad \square$$

Note also the following corollary to Theorem 4.5.13.

Corollary 7.4.4. *A $C_l(\Omega) \circ \underline{\mathbb{I}}$-algebra (A, Ω) is subdirectly irreducible if and only if it is of one of the following types:*

(a) *a subdirectly irreducible $\underline{\mathbb{I}}$-algebra;*

(b) *a subdirectly irreducible cancellative Ω-mode;*

(c) *(if $C_l(\Omega) \circ \underline{\mathbb{I}}$ is closed under Płonka sums) a subdirectly irreducible cancellative Ω-mode extended by zero.* □

Now for a variety $\underline{V}$ of Ω-modes, consider the quasivariety $Q_a(\underline{V})$ of $\underline{V}$-subreducts of $\underline{R(V)}$-algebras. Let $C_l(Q_a(\underline{V}))$ be the subquasivariety of cancellative $Q_a(\underline{V})$-algebras. The following corollary to Theorem 7.4.2 characterizes algebras in the quasivariety $C_l(Q_a(\underline{V})) \circ \underline{\mathbb{I}}$.

Corollary 7.4.5. *The following condtions are equivalent for a variety $\underline{V}$ of Ω-modes:*

(a) (A, Ω) *is in the quasivariety* $C_l(Q_a(\underline{V})) \circ \underline{\mathbb{I}}$;

(b) (A, Ω) *is a Lallement sum of cancellative quasi-affine $\underline{V}$-algebras over an $\underline{\mathbb{I}}$-algebra;*

(c) (A, Ω) *is a subalgebra of a functorial sum of cancellative $\underline{V}$-algebras over an $\underline{\mathbb{I}}$-algebra.* $\square$

In particular, Corollary 7.4.5 gives another characterization of the subreducts of Płonka sums of affine $R(V)$-spaces considered in Theorem 7.3.1.

We conclude this section with a version of Theorem 7.4.2 concerning the case where the $\underline{\mathbb{I}}$-algebras have a full algebraic quasi-order $\preceq$, e.g. where $\underline{\mathbb{I}}$ generates an irregular variety. Recall that by Proposition 4.2.1, in such a case a functorial sum is just a direct product.

Corollary 7.4.6. *Assume that $\underline{\mathbb{I}}$-algebras have a full algebraic quasi-order. Then each $C_l(\Omega) \circ \underline{\mathbb{I}}$-algebra (A, Ω) embeds into the direct product of a $C_l(\Omega)$-algebra (E, Ω) and an $\underline{\mathbb{I}}$-algebra (I, Ω), i.e.*

$$(A, \Omega) \leq (E, \Omega) \times (I, \Omega).$$

Proof. By Theorem 7.4.2, each $C_l(\Omega) \circ \underline{\mathbb{I}}$-algebra (A, Ω) embeds into a functorial sum of cancellative envelopes (E_i, Ω) over (I, Ω). By Proposition 4.2.1, all the envelopes (E_i, Ω) are isomorphic, and the functorial sum is isomorphic to any direct product $(E_i, \Omega) \times (I, \Omega)$. $\square$

Example 7.4.7. Consider the class $\underline{Cv}$ of convex subsets of affine R-spaces, i.e. the class of cancellative barycentric algebras over R. Let $\underline{\mathbb{I}}$ be the variety of $\underline{I}^\circ$-left zero bands, i.e. for each $r \in I^\circ$, one has $x \cdot y = xy\underline{r} = x$. Since left zero bands have a full algebraic quasi-order, Corollary 7.4.6 implies that each $\underline{Cv} \circ \underline{\mathbb{I}}$-algebra embeds into the direct product of a $\underline{Cv}$-algebra and an $\underline{I}^\circ$-left zero band. □

Example 7.4.8. Consider the quasivariety $\underline{Cv} \circ \underline{Cv}$ with $\underline{Cv}$ defined as in Example 7.4.7. Let $(A, \underline{I}^\circ)$ be a $\underline{Cv} \circ \underline{Cv}$-algebra with (cancellative) fibres $(A_j, \underline{I}^\circ)$ and an open convex set $(J, \underline{I}^\circ)$ as a $\underline{Cv}$-quotient. Now each open convex set has a full algebraic quasi-order. Indeed, for any i, j in J with $i \neq j$ there is $r \in I^\circ$ and x in J such that $ax\underline{r} = b$, whence $i \preceq j$ [Exercise 5.8K(a)]. As in Example 7.4.7, Corollary 7.4.6 implies that $(A, \underline{I}^\circ)$ embeds into the direct product of the convex set $(E_j, \underline{I}^\circ)$ and the open convex set $(J, \underline{I}^\circ)$. □

7.5. Barycentric algebras and commutative binary modes

The results of the two previous sections will now be used to describe the structure of barycentric algebras and commutative binary modes. We will see that both these types of algebras behave very similarly. It will be shown that they are subreducts of Płonka sums of corresponding affine spaces, and hence are separable. In fact, commutative binary modes may be considered as "barycentric" algebras over the ring $\mathbb{D}$ of dyadic rationals, although of course the ring $\mathbb{D}$ is not a field.

The first aim is to describe the semilattice decomposition of modes in general, and barycentric algebras in particular. In Section 3.3, the semilattice decomposition of Ω-algebras was described using the concept of a wall. We will show that for modes, the concept of a wall in the Decomposition Theorem 3.3.1 may be replaced by the concept of a sink. We start with a lemma, an analogue of Lemma 3.3.9 for walls.

Lemma 7.5.1. For any $a_1, \ldots, a_n$ in a mode (A, Ω), and n-ary ω in Ω, one has

$$(a_1) \cap \cdots \cap (a_n) = (a_1 \ldots a_n \omega).$$

Proof. First note that for sinks $S_1, \ldots, S_n$ of (A, Ω), one has

(7.5.1) $$S_1 \ldots S_n \omega = S_1 \cap \cdots \cap S_n.$$

Indeed, clearly $X_1 \ldots S_n \omega \subseteq S_1 \cap \cdots \cap S_n$. And conversely, if $s \in S_1 \cap \cdots \cap S_n$, then $s = s \ldots s\omega \in S_1 \ldots S_n \omega$, whence the equality (7.5.1) holds. In particular, one obtains

(7.5.2) $$(a_1) \ldots (a_n)\omega = (a_1) \cap \cdots \cap (a_n).$$

Next we will prove that

(7.5.3) $$(a_1) \ldots (a_n)\omega = (a_1 \ldots a_n \omega).$$

First note that $a_i \in (a_i)$ implies $a_1 \ldots a_n \omega \in (a_i)$. Hence the element $a_1 \ldots a_n \omega$ lies in $(a_1) \ldots (a_n)\omega$, and consequently $(a_1 \ldots a_n \omega) \subseteq (a_1) \ldots (a_n)\omega$. To show the reverse inclusion, let $s_1 \ldots s_n \omega$ be an element of $(a_1) \ldots (a_n)\omega$. Note that the principal sink (a) of (A, Ω) is the set of all elements $ax_2 \ldots x_n w$ for derived operations w and elements x_i of A [Exercise 7.5B]. Hence each $s_i \in (a_i)$ has the form

$$s_i = a_i x_{i2} \ldots w_i$$

for some derived operation w_i and elements $x_{i2}, \ldots$ of A. It will be shown by induction on i that there are derived operations v_i and elements $y_{2i}, \ldots$ of A such that

(7.5.4) $$s_1 \ldots s_n \omega = (a_1 \ldots a_{i-1}(a_i x_{i2} \ldots w_i) \ldots (a_n x_{n2} \ldots w_n)\omega) y_{2i} \ldots v_i.$$

7. SUBREDUCTS OF AFFINE SPACES

For $i = 1$, (7.5.4) is just

$$s_1 \ldots s_n \omega = (a_1 x_{12} \ldots w_1)(a_2 x_{22} \ldots w_2) \ldots (a_n x_{n2} \ldots w_n)\omega,$$

with v_1 being the identical unary operation defined by $xv_1 := x$. Then by idempotency and entropicity one has

$$s_1 \ldots s_n \omega = (a_1 x_{12} \ldots w_1)(s_2 \ldots s_2 w_1) \ldots (s_n \ldots s_n w_1)\omega$$

$$= (a_1 s_2 \ldots s_n \omega)(x_{12} s_2 \ldots s_n \omega) \ldots w_1$$

$$=: (a_1 s_2 \ldots s_n \omega) y_{21} \ldots v_1,$$

the formula (7.5.4) for $i = 2$. Suppose (7.5.4) holds for a given i, and write it as

$$s_1 \ldots s_n \omega = (a_1 \ldots a_{i-1}(a_i x_{i2} \ldots w_i) s_{i+1} \ldots s_n \omega) y_{2i} \ldots v_i.$$

Then again idempotency and entropicity give the following:

$$s_1 \ldots s_n \omega = [(a_1 a_1 \ldots w_i) \ldots (a_{i-1} a_{i-1} \ldots w_i)$$

$$(a_i x_{2i} \ldots w_i)(s_{i+1} s_{i+1} \ldots w_i) \ldots (s_n s_n \ldots w_i)\omega] y_{2i} \ldots v_i$$

$$= [(a_1 \ldots a_{i-1} a_i s_{i+1} \ldots s_n \omega)(a_1 \ldots a_{i-1} x_{2i} s_{i+1} \ldots s_n \omega) \ldots w_i] y_{2i} \ldots v_i$$

$$=: (a_1 \ldots a_i(a_{i+1} x_{i+1,2} \ldots w_{i+1}) s_{i+2} \ldots s_n \omega) y_{2,i+1} \ldots v_{i+1}.$$

Thus the formula (7.5.4) is established by induction. Taking $i = n$, it shows that $s_1 \ldots s_n \omega$ lies in $(a_1 \ldots a_n \omega)$, as required. Indeed, for $i = n$, one obtains

$$s_1 \ldots s_n \omega = (a_1 \ldots a_{n-1}(a_n x_{n2} \ldots w_n)\omega) y_{2n} \ldots v_n$$

$$= ((a_1 \ldots a_n \omega)(a_1 \ldots a_{n-1} x_{n-2} \omega) \ldots w_n) y_{2n} \ldots v_n$$

$$\in (a_1 \ldots a_n \omega). \quad \square$$

In particular, it follows that for a and b in A and n-ary ω in Ω, one has $(a) \cap (b) = (a) \cap (b) \cap \cdots \cap (b) = (ab \ldots b\omega)$. This shows that the principal sinks of the mode (A, Ω) form a subsemilattice $(\mathcal{S}, \cap)$ of the meet semilattice of all sinks. Moreover, it shows that the mapping

(7.5.5) $$h : A \to \mathcal{S} : A \mapsto (a)$$

is an Ω-homomorphism from the mode (A, Ω) onto the Ω-semilattice $(\mathcal{S}, \Omega)$ of principal sinks.

If (A, Ω) is a mode, then Proposition 3.3.10 can be rewritten as follows.

Proposition 7.5.2. *Let (A, Ω) be a mode. Then the following conditions are equivalent:*

(a) (A, Ω) *is algebraically open;*

(b) (A, Ω) *has no proper non-empty walls;*

(c) (A, Ω) *has no proper non-empty sinks;*

(d) *there is no epimorphism $(A, \Omega) \to \underset{\sim}{2}$ onto the two-element Ω-semilattice.*

Proof. We only need to prove that (c) and (d) are equivalent. Now (d) implies (c), since a proper non-empty sink S contains a proper sink (s) for s in S, whence the homomorphism (7.5.5) maps (A, Ω) onto a non-trivial semilattice. And such a semilattice always has a homomorphism onto $\underset{\sim}{2}$. Conversely (c) implies (d), since the preimage of 1 under an Ω-epimorphism $(A, \Omega) \to \underset{\sim}{2}$ is a proper non-empty sink of (A, Ω). □

In particular, it follows that for a mode (A, Ω) the semilattice of principal walls and the semilattice of principal sinks are isomorphic. Now the Decomposition Theorem 3.3.11 can be reformulated as follows.

Decomposition Theorem 7.5.3. *The semilattice replica of a mode (A, Ω) is isomorphic to the semilattice of principal walls and to the semilattice of principal sinks. The corresponding ρ-classes are algebraically open subalgebras of (A, Ω).* □

Corollary 7.5.4. *Each mode* (A, Ω) *is a semilattice sum of algebraically open subalgebras.* □

Now it will be shown that if (A, Ω) is a barycentric algebra or a commutative binary mode, then algebraically open subalgebras are always cancellative. This will allow us to use the results of Section 7.4 to represent both types of algebras.

In the case of commutative binary modes, the cancellativity of algebraically open algebras was in fact already proved in Lemma 6.6.7. Then Theorem 7.4.2 and Corollaries 7.4.3, 7.4.5 give the following:

Structure Theorem 7.5.5. *Each commutative binary mode* $(A, \cdot)$ *is*

(a) *a semilattice Lallement sum of cancellative submodes over its semilattice replica and*

(b) *a submode of a Płonka sum of cancellative $\underline{C}$-modes over its semilattice replica.* □

Let $c_l(\underline{C})$ be the class of cancellative $\underline{C}$-modes. Since obviously each $\underline{C}$-mode is in the quasivariety $c_l(\underline{C}) \circ \underline{Sl}$, Corollary 7.4.4 shows that the algebras described in Example 7.3.4 are indeed all subdirectly irreducible $\underline{C}$-modes. Note that they all embed into the Płonka sum $F_{\mathbb{D}}^\infty$ of the reduct $(F_{\mathbb{D}}, \cdot)$ of the minimal cogenerator $F_{\mathbb{D}}$ and the singleton $\{\infty\}$. Then Theorem 7.3.1 implies the following:

Theorem 7.5.6. *Each commutative binary mode* $(A, \cdot)$

(a) *is a $\underline{2^{-1}}$-subreduct of a Płonka sum of affine spaces over $F_{\mathbb{D}}$ and*

(b) *is separable in $\underline{C}$ and*

(c) *embeds into a product of copies of* $(F_{\mathbb{D}}^\infty, \cdot) = (F_{\mathbb{D}}^\infty, \underline{2^{-1}})$. □

Now let us turn our attention to barycentric algebras. We will show that algebraically open barycentric algebras are in fact just open convex sets. We will thus obtain similar structure theorems.

Lemma 7.5.7. *If a barycentric algebra* $(A, \underline{I}^\circ)$ *is algebraically open, then it is a convex set.*

Proof. Let $(A, \underline{I}^\circ)$ be an algebraically open barycentric algebra. If $(A, \underline{I}^\circ)$ has less than two elements, then it is certainly a convex set. So assume for the rest of the proof that A has at least two elements. Suppose that $(A, \underline{I}^\circ)$ is not a convex set. By Theorem 5.8.6, the cancellation laws (Cl) of Section 5.8 break down, i.e. there are elements $a \neq b$ and x in A such that

(7.5.6) $$\exists\, p \in I^\circ.\ xa\underline{p} = xb\underline{p}.$$

By Proposition 5.8.7, (7.5.6) is equivalent to:

(7.5.7) $$\forall\, p \in I^\circ,\ xa\underline{p} = xb\underline{p}.$$

Now for each q in I°, define

$$S := \{x \in A \mid xa\underline{q} = xb\underline{q}\}.$$

The truth and equivalence of (7.5.6) and (7.5.7) show that S is a non-empty subset of A independent of the choice of q. It is a subalgebra of $(A, \underline{I}^\circ)$, since for x and y in S, one has

$$xy\underline{p}\ a\underline{q} = x\ ya\frac{q}{(p'q')'}\ (p'q')'$$
$$= x\ yb\frac{q}{(p'q')'}\ (p'q')' = xy\underline{p}\ b\ \underline{q}$$

by skew associativity. It is proper, since $a, b \in S$ implies

$$a = aa\underline{q} = ab\underline{q} = ba\underline{q}' = bb\underline{q}' = b,$$

a contradiction. Finally, S is a sink of $(A, \underline{I}^\circ)$, since for s in S and y in A and q in I°, one has

$$ys\underline{p}\ a\ \underline{q} = y\ sa\frac{q}{(p'q')'}\ (p'q')'$$
$$= y\ sb\frac{q}{(p'q')'}\ (p'q')' = ys\underline{p}\ b\ \underline{q},$$

7. SUBREDUCTS OF AFFINE SPACES

whence $y s \underline{p} \in S$. By Proposition 7.5.2, the existence of the proper non-empty sink S contradicts the algebraic openness of $(A, \underline{I}^\circ)$. □

The name "algebraic openness" suggests that under an appropriate topology on its affine hull, an algebraically open barycentric algebra should be an open convex set. And indeed, this is the case, when we define a topology on affine spaces over a subfield R of $\mathbb{R}$ as was done in Example 5.8K. If A is a finite dimensional R-space, then the topology is induced from $\mathbb{R}$. If A has infinite dimension, then a subset X of A is defined to be open if the intersection $X \cap B$ of X with each finite-dimensional subspace B of A is open in B.

Proposition 7.5.8. *A barycentric algebra $(A, \underline{I}^\circ)$ is an open convex subset of an affine R-space if and only if it is algebraically open.*

Proof. ($\Rightarrow$) Let A be an open convex subset of an affine R-space V, and let a and b be in A. Define the mapping

$$\varphi : R \to V \ ; \ x \mapsto a + x(b - a).$$

Clearly, φ is an affine mapping, and is continuous. It follows that the set $\varphi^{-1}(A)$ is an open and convex subset of R, i.e. it is an open interval in R, and hence it is simple. Consequently, by Proposition 7.5.2, it has no proper non-empty sinks. It follows that the convex set A also has this property.

($\Leftarrow$) Let A be non-empty and algebraically open. By Lemma 7.5.7, the set A is convex. We will show that A is also open. As a convex set, A embeds into its affine hull $V = \langle A \rangle$. Fix a point $0 \in A$ and consider V as a vector space with 0 as a zero.

If V has finite dimension, then the interior A° of A is non-empty. For an arbitrary a in A and p in I°, the set $aA^\circ \underline{p} = a(1 - p) + A^\circ p$ is an open subset of A, and hence it is contained in A°. It follows that A° is a sink of $(A, \underline{I}^\circ)$. Since $(A, \underline{I}^\circ)$ is algebraically open, $A = A^\circ$, and consequently A is open.

Now assume that V has infinite dimension. Let W be a finite dimensional subspace of V. Since $(A, \underline{I}^\circ)$ is algebraically open, it follows that for any a and b in $A \cap W$, one has $a = ub\underline{p}$ and $b = va\underline{q}$ for some p and q in I° and u and v in A [Exercise 7.5I]. Obviously, u and v are in $A \cap W$. Hence also $A \cap W$ is algebraically open. It will be shown that the space W is generated by the set $A \cap W$. Assume that $0 \neq w \in W$. Since $V = \langle A \rangle$, it follows that $w = \sum a_i r_i$ for a_i in A and $0 \neq r_i \in R$. We may assume that all the r_i are positive. To show this, suppose $r_k < 0$. Then since $(A, \underline{I}^\circ)$ is algebraically open and $0 \in A$, one has $(0) = (a_k)$. Hence there are p in I° and $\bar{a}_k$ in A such that $0 = \bar{a}_k a_k \underline{p}$. Thus
$$a_k = \frac{p-1}{p}\bar{a}_k \text{ and } a_k r_k = \bar{a}_k r_k \frac{p-1}{p} = \bar{a}_k \bar{r}_k,$$
where $\bar{r}_k > 0$. Now let $r = \sum r_i$. Then
$$w = \left(\sum a_i(r_i/r)\right)r =: ar.$$
Since all the a_i are in A, and all the numbers $r_i/r > 0$ and $\sum(r_i/r) = 1$, it follows that $a \in A$, whence $a \in A \cap W$ and $w \in A \cap W$. This shows that $W \subseteq \langle A \cap W \rangle$. Since the opposite inclusion is obvious, it follows that $W = \langle A \cap W \rangle$.

Recall that the space $W = \langle A \cap W \rangle$ has finite dimension, and that $A \cap W$ is algebraically open. As in the first part of the proof, one obtains that $A \cap W$ is an open set in W. By the definition of the topology on V, it follows that A is open. □

As a corollary to Corollary 7.5.4, Lemma 7.5.7 and Proposition 7.5.8, one obtains the following:

Corollary 7.5.9. *Each barycentric algebra is a semilattice Lallement sum of open convex sets.* □

Now Theorem 7.4.2, Corollary 7.4.3, Corollary 7.4.5 and Corollary 7.5.9 give the following:

7. SUBREDUCTS OF AFFINE SPACES

Structure Theorem 7.5.10. *Each barycentric algebra $(A, \underline{I}^\circ)$ over a subfield R of the real field $\mathbb{R}$ is*

(a) *a semilattice Lallement sum of open convex sets over its semilattice replica and*

(b) *a subalgebra of a Płonka sum of convex sets over its semilattice replica.* □

Since obviously each barycentric algebra is in the quasivariety $\underline{Cv} \circ \underline{Sl}$, Corollary 7.4.4 shows that the algebras described in Example 7.3.3 are indeed all the subdirectly irreducible barycentric algebras. Now let R^∞ be the Płonka sum of R and $\{\infty\}$ as in Section 7.3. It is easy to see that each of the subdirectly irreducible barycentric algebras embeds into the minimal cogenerator R^∞. Then Theorem 7.3.1 implies the following:

Theorem 7.5.11. *Each barycentric algebra $(A, \underline{I}^\circ)$ over a subfield R of the real field $\mathbb{R}$*

(a) *is an $\underline{I}^\circ$-subreduct of a Płonka sum of affine spaces over R and*

(b) *is separable in $\underline{B}$ and*

(c) *embeds into a product of copies of $(R^\infty, \underline{I}^\circ)$.* □

Example 7.5.12. Let us go back to Example 7.4.7, and once more consider modes in the quasivariety $\underline{Cv} \circ \underline{Cv}$. Let $(A, \underline{I}^\circ)$ be a $\underline{Cv} \circ \underline{Cv}$-algebra with convex sets $(A_j, \underline{I}^\circ)$ as fibres and a convex set $(J, \underline{I}^\circ)$ as an indexing algebra. Theorem 7.5.11 implies that each barycentric algebra is naturally quasi-ordered [Exercise 7.5K]. Theorem 7.4.2 and Corollary 7.4.5 imply that $(A, \underline{I}^\circ)$ is a Lallement sum of convex sets $(A_j, \underline{I}^\circ)$ over the convex set $(J, \underline{I}^\circ)$, and at the same time a subalgebra of a functorial sum $(E, \underline{I}^\circ)$ of the $(E_j, \underline{I}^\circ)$, the envelopes of $(A_j, \underline{I}^\circ)$, over $(J, \underline{I}^\circ)$. Let $(S, \underline{I}^\circ)$ be the semilattice replica of $(J, \underline{I}^\circ)$ with $\pi : J \to S$ as the projection, and all $\pi^{-1}(s)$ being open convex sets. Example 7.4.8 shows that for any j in $\pi^{-1}(s)$, the subalgebra of $(E, \underline{I}^\circ)$ consisting of all $(E_j, \underline{I}^\circ)$ with j in $\pi^{-1}(s)$ is the direct product of an $(E_j, \underline{I}^\circ)$ with $(\pi^{-1}(s), \underline{I}^\circ)$. □

Exercises

7.5A. [Romanowska and Smith 1991] (a) Show that the complement of a wall in an algebra (A, Ω) is a sink.

(b) Show that the complement of a sink is a wall if it is a subalgebra. Sinks satisfying the latter property are called *prime*.

(c) Show that a prime sink may be described equivalently as a sink S satisfying the following property for each (n-ary) ω in Ω and all $a_1, \ldots, a_n$ in A;

$$(a_1 \ldots a_n \omega \in S) \Rightarrow (\exists 1 \leq i \leq n.\ a_i \in S).$$

7.5B. [Ignatov 1986] Show that the principal sink (a) of a mode (A, Ω) is the set of elements $ax_2 \ldots x_n w$ for derived operations w and elements $x_2, \ldots, x_n$ of A.

7.5C. [MT] Let (A, Ω) be a mode. Let AN denote the set of finitely generated non-empty sinks of (A, Ω). Let AZ denote the set of non-empty sinks of (A, Ω).

(a) Show that AZ forms a submode of (AS, Ω).

(b) Show that AN forms a submode of (AS, Ω).

(c) Show that $(AN, \cup, \Omega)$ and $(AZ, \cup, \Omega)$ are equivalent to distributive lattices, sublattices of $(2^A, \cup, \cap)$.

7.5D. [MT] Let ω be an n-ary operation of Ω. For $i = 1, \ldots, n$, let $S_i = \bigcup_{a_i \in M_i} (a_i)$ (with M_i finite) be a finitely generated sink of (A, Ω). Show that

$$S_1 \ldots S_n \omega = \bigcup \{(a_1 \ldots a_n \omega) \mid (a_1, \ldots, a_n) \in M_1 \times \cdots \times M_m\}.$$

7.5E. [MT] Let (A, Ω) be a mode, and $(L, \cup, \Omega)$ a distributive Ω-lattice. Show that each mode homomorphism $f : (A, \Omega) \to (L, \Omega)$ can be extended to a unique homomorphism

$$f' : (AN, \cup, \Omega) \to (L, \cup, \Omega)$$

7. SUBREDUCTS OF AFFINE SPACES

with $\psi f' = f$, where ψ is an Ω-homomorphism given by

$$\psi : (A, \Omega) \to (AN, \Omega); \ a \mapsto (a)$$

[Hint: Define $f' : AN \to L$ by $\left(\sum_{a \in M} a\psi\right) f' = \sum_{a \in M} af$ for finite subsets M of A. Show that f' is well-defined. To see that it is an Ω-homomorphism, use Exercise 7.5D.]

7.5F. [MT] Use Exercise 7.5E to show that the free distributive lattice on a set X is the algebra $(XS\text{l}N, \cup, \cdot)$ of finitely generated non-empty sinks of the free semilattice $XS\text{l}$ on X.

7.5G. [Romanowska 1985] Let $\underline{V}$ be an irregular variety of Ω-modes. Let $h : (A, \Omega) \to (S, \Omega)$ be a homomorphism of (A, Ω) onto its Ω-semilattice replica (S, Ω), with corresponding fibres $A_s = h^{-1}(s)$, for s in S, lying in the variety $\underline{V}$. Show that a subset B of A is a sink in (A, Ω) iff $B = \bigcup (A_t \mid t \in T)$, where T is a sink in (S, Ω).

7.5H. Find all sinks, walls and the semilattice replica of the barycentric algebra $(T, \underline{I}^\circ)$ of Exercise 5.8D.

7.5I. [Ignatov 1986] Let $\sigma := \ker h$, the kernel of the homomorphism h defined by (7.5.5).

(a) Show that $(a, b) \in \sigma$ iff there are u and v in A, and p and q in I°, such that $ua\underline{p} = b$ and $vb\underline{q} = a$.

(b) Show that for $a \in A$, the principal sink (a) is the set $\{ua\underline{r} \mid u \in A, r \in I^\circ\}$.

(c) Show that A is algebraically open iff for all a, b in A, $(a, b) \in \sigma$.

7.5J. Find all the representations of the barycentric algebra $(T, \underline{I}^\circ)$ of Exercise 7.5H given by Theorems 7.5.10 and 7.5.11.

7.5K. [Romanowska and Traina 1999] Use Structure Theorem 7.5.11 to show that each barycentric algebra is naturally quasi-ordered.

7.5L. Show that for a barycentric algebra $(A, \underline{I}^\circ)$, the congruences $\mu(j, \sigma)$ of Lemma 7.4.1 coincide with the congruences θ_j of Exercise 4.5L and with the congruences γ_j defined on (P_j, Ω) by

$$(b_i, c_k) \in \gamma_j :\Leftrightarrow \exists\, a_j \in A_j \,.\, \exists\, p \in I^\circ.\; a_j b_i \underline{p} = a_j c_k \underline{p}.$$

7.5M. Show that each convex set is an envelope of an open convex sink.

7.5N. Use Lemma 7.2.7 to describe the detailed structure of a convex set $(A, \underline{I}^\circ)$ presented as a Lallement sum of open convex sets $(A_j, \underline{I}^\circ)$ over an $\underline{I}^\circ$-semilattice $(J, \underline{I}^\circ)$. In particular, show that:

(a) $(A, \underline{I}^\circ)$ satisfies the conditions (a) and (b) of Lemma 7.2.7;

(b) all the congruences $\mu(j, \sigma)$ of Lemma 7.4.1 are equality relations, and all the envelopes $(E_j, \underline{I}^\circ)$ are subalgebras of (A, Ω);

(c) the semilattice $(J, \underline{I}^\circ)$ has an upper bound.

7.5O. Show that if a barycentric algebra satisfies the conditions (a), (b) and (c) of Exercise 7.5N, then it is a convex set.

7.5P. [Ignatov 1985] Show that the relation γ defined on a barycentric algebra $(A, \underline{I}^\circ)$ by

$$(a, b) \in \gamma :\Leftrightarrow \exists\, c \in A \,.\, \exists\, p \in I^\circ.\; ca\underline{p} = cb\underline{p}$$

is a congruence relation, and that the algebra $(A^\gamma, \underline{I}^\circ)$ is a convex set.

7.5Q. Let $(A, \underline{I}^\circ)$ be a barycentric algebra and a Lallement sum of open convex sets $(A_j, \underline{I}^\circ)$ over a semilattice $(J, \underline{I}^\circ)$. Show that the relation δ defined on A by

$$(a, b) \in \delta :\Leftrightarrow a \text{ and } b \text{ belong to a subalgebra of}$$
$$(A, \Omega) \text{ that is a convex set}$$

is a congruence relation. Show that it gives a decomposition of $(A, \underline{I}^\circ)$ as a disjoint union of maximal convex sets M_j contained in A over a semilattice K that is a homomorphic image of $(J, \underline{I}^\circ)$. Deduce that $(A, \underline{I}^\circ)$ is a Lallement sum of convex sets $(M_j, \underline{I}^\circ)$ with convex envelopes

$(D_j, \underline{I}^\circ)$ over the semilattice K, and a subalgebra of a functorial sum of $(D_j, \underline{I}^\circ)$ over K.

7.6. Varieties and quasivarieties of barycentric algebras

In this section we describe all the subvarieties and subquasivarieties of the variety $\underline{B}$ of barycentric algebras over a fixed subfield R of the real field $\mathbb{R}$.

We first describe the subvarieties of $\underline{B}$. As noted before, $\underline{B}$ contains the subvariety $\underline{Sl}$ of $\underline{I}^\circ$-semilattices.

Theorem 7.6.1. *The variety $\underline{Sl}$ is the only non-trivial proper subvariety of $\underline{B}$.*

Proof. Let $(A, \underline{I}^\circ)$ be a barycentric algebra which is not in $\underline{Sl}$. By Exercise 5.8J, there exist a and b in A with $ab\underline{p} \neq ab\underline{q}$ for all p and q in I°. Hence a and b generate a subalgebra isomorphic to the unit interval $(I, \underline{I}^\circ)$. Since the free $\underline{B}$-algebras are subalgebras of powers of $(I, \underline{I}^\circ)$ (see Section 5.8), the algebra $(A, \underline{I}^\circ)$ generates the whole variety $\underline{B}$. Thus a proper subvariety of $\underline{B}$ must be contained in $\underline{Sl}$. Since the only nontrivial subvariety of $\underline{Sl}$ is $\underline{Sl}$ itself, the theorem holds. $\square$

As is to be expected, the lattice of quasivarieties of barycentric algebras is much richer than the lattice of varieties. However, it is still easily described.

For each $n = 0, 1, 2, \ldots$, define a barycentric algebra T_n as follows. The algebra $T_0 = I_0$ is the trivial one-element algebra $(\{0_0 = 1_0\}, \underline{I}^\circ)$. For each $n = 1, 2, \ldots$, pick a copy of the interval $(]0, 1], \underline{I}^\circ)$ and its envelope $([0, 1], \underline{I}^\circ)$. The n-th copy of $(]0, 1], \underline{I}^\circ)$ is denoted by I_n. The n-th envelope $([0, 1], \underline{I}^\circ)$ is denoted by E_n, and its elements by i_n. Let T_n be the disjoint union $I_0 \bigcup I_1 \bigcup \ldots \bigcup I_n$ of the sets I_i. The barycentric algebra structure on the set T_n is introduced by defining $(T_n, \underline{I}^\circ)$ to be the Lallement sum of the algebras $(I_0, \underline{I}^\circ), (I_1, \underline{I}^\circ), \ldots, (I_n, \underline{I}^\circ)$ with the envelopes $(E_0 = I_0, \underline{I}^\circ), (E_1, \underline{I}^\circ),$

..., $(E_n, \underline{I}^\circ)$ by the sum homomorphisms $\varphi_{j,k}$ defined for $j < k$ as

$$\varphi_{j,k} : I_j \to E_k; \ i_j \mapsto 0_k.$$

In particular, for $j \leq k$ and r in I^0, one has

$$i_j i_k \underline{r} = i_j \varphi_{j,k} i_k \underline{r} = 0_k i_k \underline{r}.$$

Obviously, T_n is a subalgebra of the Płonka sum of the envelopes $E_0, E_1, \ldots, E_n$. The algebra T_n can be illustrated diagramatically as follows.

(7.6.1)

$$0_n \ \square \underline{} \cdots \underline{} \blacksquare \ 1_n$$

$$\cdots$$

$$\uparrow \ \varphi_{2,3}$$

$$0_2 \ \square \underline{} \cdots \underline{} \blacksquare \ 1_2$$

$$\uparrow \ \varphi_{1,2}$$

$$0_1 \ \square \underline{} \cdots \underline{} \blacksquare \ 1_1$$

$$\uparrow \ \varphi_{0,1}$$

$$\blacksquare$$
$$0_0 = 1_0$$

Note that T_1 is just the unit interval $(I, \underline{I}^\circ)$, and T_2 is the algebra T of Exercise 5.8D. Note also that each T_k is a subalgebra of T_{k+1}. Let $\underline{\underline{B}}_n$ be the quasivariety $Q(T_n)$ generated by T_n, and let $\underline{\underline{\widetilde{B}}}_n$ be its quasiregularization, i.e. $\underline{\underline{\widetilde{B}}}_n = Q(T_n, \underset{\sim}{2})$. Finally, denote by $\underline{\underline{B}}_\omega$ the quasivariety generated by all T_n, and by $\underline{\underline{\widetilde{B}}}_\omega$ its quasiregularization, i.e. the quasivariety generated by all T_n and the $\underline{I}^\circ$-semilattice $\underset{\sim}{2}$. It is clear that the quasivarieties $\underline{\underline{B}}_n$ and $\underline{\underline{\widetilde{B}}}_n$ together with the varieties $\underline{\underline{Tr}}$, $\underline{\underline{Sl}}$ and $\underline{\underline{B}}$ form the following ordered set (7.6.2):

$$
\begin{array}{ccccccccc}
\underline{\underline{Sl}} = \underline{\underline{\tilde{B}}}_0 & \longrightarrow & \underline{\underline{\tilde{B}}}_1 \cdots & \longrightarrow & \underline{\underline{\tilde{B}}}_n & \longrightarrow & \cdots \longrightarrow & \underline{\underline{\tilde{B}}}_\omega & \longrightarrow & \underline{\underline{B}} \\
\uparrow & & \uparrow & & \uparrow & & & \uparrow & & \\
\underline{\underline{Tr}} = \underline{\underline{B}}_0 & \longrightarrow & \underline{\underline{B}}_1 \cdots & \longrightarrow & \underline{\underline{B}}_n & \longrightarrow & \cdots \longrightarrow & \underline{\underline{B}}_\omega. & &
\end{array}
$$

The aim of the remaining part of this section is to prove the following theorem.

Theorem 7.6.2. *The ordered set (7.6.2) is the lattice of all subquasivarieties of the variety $\underline{\underline{B}}$ of barycentric algebras.*

The proof will consist of a series of lemmata and propositions. In particular, several characterizations of the quasivarieties $\underline{\underline{B}}_n$ and $\underline{\underline{\tilde{B}}}_n$ will also be given.

First recall that by Lemma 4.4.4, the variety $\underline{\underline{Sl}}$ of $\underline{\underline{I}}^\circ$-semilattices is a minimal quasivariety. The same is true of the quasivariety $\underline{\underline{B}}_1$. In fact, the following holds.

Lemma 7.6.3. *The quasivariety $\underline{\underline{B}}_1$ coincides with the quasivariety $\underline{\underline{Cv}}$ of convex sets, is generated by any of the unit intervals, and is minimal.*

Proof. Since the closed unit interval $T_1 = I$ is obviously a convex set, and is a subalgebra of any non-trivial convex set, it follows that $\underline{\underline{B}}_1 = Q(T_1) \subseteq \underline{\underline{Cv}}$. To prove the converse inclusion, first note that by Corollary 7.2.4, each convex set A embeds into a power R^k of R, so that $\underline{\underline{Cv}} = SP(R) = Q(R)$. Now the convex set R is a directed colimit of its finitely generated subalgebras. But finitely generated non-trivial subalgebras of R are two-generated, and are all isomorphic to the closed unit interval T_1. Recall that by Corollary 3.7.14, $Q(T_1) = DSP(T_1)$. It follows that $\underline{\underline{Cv}} = Q(R) \subseteq Q(T_1)$, and hence $\underline{\underline{Cv}} = Q(T_1) = \underline{\underline{B}}_1$. Note that also $Q(I) = Q(I^\circ)$.

Since each finitely generated convex set contains the interval I, and is contained in some power of I, it follows that $\underline{\underline{Cv}} = Q(T_1)$ is a minimal quasivariety of barycentric algebras. □

Lemma 7.6.4. *The variety $\underline{B}$ of barycentric algebras is generated by the extension I^∞ of the unit interval I by zero, i.e.*

$$\underline{B} = Q(I^\infty).$$

Proof. The proof is similar to the proof of Lemma 7.6.3. By Theorem 7.5.11, each barycentric algebra embeds into a power $(R^\infty)^k$ of R^∞, so that $\underline{B} = SP(R^\infty) = Q(R^\infty)$. Now R^∞ is a directed colimit of finitely generated nontrivial subalgebras that are isomorphic to I or to I^∞. Hence $\underline{B} = Q(R^\infty) \subseteq Q(I^\infty) \subseteq Q(R^\infty) = \underline{B}$. □

Now consider the following quasi-identities for some p, r and s in I° with $r \neq s$, and $n = 1, 2, \ldots$:

(α_n) $\qquad (\&_{\substack{2 \leq k \leq n \\ i,j < k}} x_k x_i \underline{p} = x_k x_j \underline{p}) \to (x_0 x_1 \underline{r} = x_0 x_1 \underline{s})$;

(α) $\qquad (xy\underline{p} = y) \to (x = y)$;

(β) $\qquad (xz\underline{p} = z = yz\underline{p}) \to (xy\underline{r} = xy\underline{s})$.

Note that, if for elements a, b and c of a barycentric algebra, one has $ab\underline{p} = ac\underline{p}$ for some p in I°, then $ab\underline{p} = ac\underline{p}$ for all p in I°. (Cf. the proof of Proposition 5.8.7.) Similarly, if $ab\underline{r} = ab\underline{s}$ for some distinct r and s, then $ab\underline{r} = ab\underline{s}$ for all such r and s [Exercise 5.8J]. Note also that (α_n) implies (β) and (α) implies (β).

We will show that the satisfaction of any of these quasi-identities in any quasivariety of barycentric algebras is equivalent to the exclusion of certain algebras from these quasivarieties.

First note that the quasi-identity (α_n) is not satisfied in a barycentric algebra $(A, \underline{I}^\circ)$ if and only if there are elements $x_0, x_1, \ldots, x_n$ in A such that

(ψ_n) $\qquad \&_{\substack{2 \leq k \leq n \\ i,j < k}} x_k x_i \underline{p} = x_k x_j \underline{p}$ and $x_0 x_1 \underline{r} \neq x_0 x_1 \underline{s}$.

Similarly, the quasi-identity (α) is not satisfied in $(A, \underline{I}^{\circ})$ if and only if there are x and y in A such that

(ψ) $\qquad\qquad xy\underline{p} = y \text{ and } x \neq y.$

And finally, (β) is not satisfied in $(A, \underline{I}^{\circ})$ if and only if there are x, y and z in A such that

(φ) $\qquad\qquad xz\underline{p} = z = yz\underline{p} \text{ and } xy\underline{r} \neq xy\underline{s}.$

Lemma 7.6.5. Let $(A, \underline{I}^{\circ})$ be a barycentric algebra. Let $a_0, a_1, \ldots, a_n$ be elements of A.

(a) If a_0, a_1 and a_2 satisfy the formula (φ), then they generate a subalgebra isomorphic to I^{∞}.

(b) If I^{∞} is not a subalgebra of A, and a_0 and a_1 satisfy the formula (ψ), then they generate a subalgebra isomorphic to $\underset{\sim}{2}$.

(c) If I^{∞} is not a subalgebra of A, and $a_0, a_1, \ldots, a_n$ satisfy the formula (ψ_n), then they generate a subalgebra isomorphic to T_n.

Proof. It is enough to note that I^{∞} can be defined as an algebra generated by three generators satisfying (φ). Similarly, the semilattice $\underset{\sim}{2}$ can be defined as a barycentric algebra generated by two generators satisfying (ψ). And T_n can be defined as an algebra generated by $n+1$ generators satisfying (ψ_n). $\square$

Denote by $Q(\gamma_1, \gamma_2, \ldots)$ the quasivariety of barycentric algebras defined by quasi-identities $\gamma_1, \gamma_2, \cdots$. Denote by $N(A_1, A_2, \ldots)$ the class of barycentric algebras not having subalgebras isomorphic to $A_1, A_2, \cdots$. The following proposition clarifies the relation between the quasiidentities (α_n), (α) and (β) and the exclusion of certain algebras.

Proposition 7.6.6. *The following hold for any quasivariety $\underline{\underline{Q}}$ of barycentric* algebras:

(a) $\underline{Q}$ is defined by (β) if and only if it coincides with the class $N(I^\infty)$, i.e.

$$Q(\beta) = N(I^\infty);$$

(b) $\underline{Q}$ is defined by (α) if and only if it coincides with the class $N(\underaccent{\tilde}{2})$, i.e.

$$Q(\alpha) = N(\underaccent{\tilde}{2});$$

(c) $\underline{Q}$ is defined by (α_{n+1}) if and only if it coincides with the class $N(T_{n+1}, I^\infty)$ i.e.

$$Q(\alpha_{n+1}) = N(T_{n+1}, I^\infty);$$

(d) $\underline{Q}$ is defined by (α_{n+1}) and (α) if and only if it coincides with the class $N(T_{n+1}, \underaccent{\tilde}{2})$, i.e.

$$Q(\alpha_{n+1}, \alpha) = N(T_{n+1}, \underaccent{\tilde}{2});$$

(e) $\underline{Q}$ is the variety $\underline{\underline{Sl}}$ of semilattices if and only if it coincides with the class $N(T_1)$, i.e.

$$Q(\alpha_1) = \underline{\underline{Sl}} = N(T_1).$$

Proof. (a) We will show that for any barycentric algebra A, one has $I^\infty \leq A$ if and only if A does not satisfy (β), i.e. there are a, b and c in A such that $ac\underline{p} = c = bc\underline{p}$ and $ab\underline{r} \neq ab\underline{s}$. Indeed, if such elements exist, then Lemma 7.6.5(a) shows that a, b and c generate I^∞. On the other hand, if $I^\infty \leq A$, then $0\infty\underline{p} = \infty = 1\infty\underline{p}$ and $01\underline{r} \neq 01\underline{s}$.

(b) We will show that for any barycentric algebra A, one has $\underaccent{\tilde}{2} \leq A$ if and only if A does not satisfy (α), i.e. there are a and b in A such that $ab\underline{p} = b$ and $a \neq b$. But if such elements exist, then Lemma 7.6.5(b) shows that a and b generate a two-element semilattice. On the other hand, if $\underaccent{\tilde}{2} \leq A$, then $01\underline{p} = 1$

and $0 \neq 1$.

(c) We will show that for any barycentric algebra A, one has: $[T_{n+1} \leq A$ or $I^\infty \leq A]$ if and only if A does not satisfy (α_{n+1}), i.e. there are elements $a_0, a_1, \ldots, a_{n+1}$ in A such that (ψ_{n+1}) holds. If such elements exist, then **either** they generate I^∞, **or** by Lemma 7.6.5 they generate T_{n+1}. On the other hand, the generators of I^∞ and of T_{n+1} satisfy (ψ_{n+1}).

(d) follows by (b) and (c) since $\underset{\sim}{2}$ is a subalgebra of I^∞.

(e) is obvious. □

The next step is to clarify the relation between the quasivarieties $\underline{\underline{B}}_n$ and $\underline{\underline{\tilde{B}}}_n$, and the quasivarieties of Proposition 7.6.6. We start with the following.

Proposition 7.6.7. *A barycentric algebra A is a convex set if and only if it does not contain a subalgebra isomorphic to $\underset{\sim}{2}$ or to T_2, i.e. $\underline{\underline{Cv}} = N(T_2, \underset{\sim}{2})$, and hence*

$$\underline{\underline{B}}_1 = Q(T_1) = \underline{\underline{Cv}} = N(T_2, \underset{\sim}{2}) = Q(\alpha_2, \alpha).$$

Proof. The second equality follows by Lemma 7.6.3. Since the algebras T_2 and $\underset{\sim}{2}$ do not satisfy any of the cancellation laws (Cl), it follows that $\underline{\underline{Cv}} \subseteq N(T_2, \underset{\sim}{2})$. Now assume that A is not convex, and that a, b and c are elements of A such that $ab\underline{p} = ac\underline{p}$ and $b \neq c$. If for all r, s one has $bc\underline{r} = bc\underline{s}$, then b and c generate a semilattice. So assume $bc\underline{r} \neq bc\underline{s}$ for some r and s. Then since $N(\underset{\sim}{2}) \subseteq N(I^\infty)$, it follows by Lemma 7.6.5 that a, b and c generate T_2. Consequently, $N(T_2, \underset{\sim}{2}) \subseteq \underline{\underline{Cv}}$, and hence $N(T_2, \underset{\sim}{2}) = \underline{\underline{Cv}}$. The last equality follows by Proposition 7.6.6(d). □

To show how the quasivarieties $\underline{\underline{B}}_n$ and $\underline{\underline{\tilde{B}}}_n$ are related to the quasivarieties of Proposition 7.6.6, one needs more detailed information about the structure of barycentric algebras in the quasivariety $Q(\beta)$. We obtain this by describing the finitely generated relatively subdirectly irreducibles. First note that all the quasivarieties $\underline{\underline{B}}_1, \underline{\underline{B}}_2, \ldots, \underline{\underline{B}}_\omega$ are contained in $Q(\alpha)$, since all the algebras

$T_1, T_2, \ldots$ satisfy the quasi-identity (α). Note also that by Lemma 7.6.5, a $Q(\alpha)$-algebra A always contains a subalgebra isomorphic to T_n for some n. For any barycentric algebra A, let $h(A)$ be the maximal number n such that T_n embeds into A. Call $h(A)$ the *height* of A.

Lemma 7.6.8. *Let A be a finitely generated algebra in $Q(\alpha)$.*

(a) *A is a relatively subdirectly irreducible member of $Q(\alpha)$ if and only if A is isomorphic to some T_n.*

(b) *A is a relatively subdirectly irreducible member of $Q(\alpha_{n+1}, \alpha)$ if and only if A is isomorphic to some T_i for $i = 1, 2, \ldots, n$.*

Proof. The proof that each algebra T_i with $i \leq n$ is relatively subdirectly irreducible in $Q(\alpha_{n+1}, \alpha)$ and in $Q(\alpha)$ is relegated to Exercise 7.6B. We will show that all subdirectly irreducibles have the desired form. So let A be a non-trivial finitely generated $Q(\alpha)$-algebra. Assume that A is a semilattice Lallement sum of open convex sets A_j over a semilattice J in the canonical way. First note that the semilattice J must be finite, and hence it has an upper bound, say n. Then there is no $i < j$ in $(J, \leq)$ and a_i in A_i such that $a_i \varphi_{i,j} \in A_j$. Indeed, otherwise a_i and $a_i \varphi_{i,j}$ would form a two-element subsemilattice of A. It follows that

(7.6.3) $$A_i \varphi_{i,j} \cap A_j = \varnothing$$

for any two elements i and j with $i < j$. Next, if $|A_i| = 1$, then i must be a minimal element of the semilattice J. Otherwise, A would have a two-element subsemilattice. In particular, $A_n = \pi^{-1}(n)$ is not trivial. Moreover A_n is a sink in A. In fact A_n is the smallest sink in A.

Let $\underline{\underline{Q}}$ be any of the quasivarieties $Q(\alpha_{n+1}, \alpha)$ and $Q(\alpha)$. Note that all the $\underline{\underline{Q}}$-congruences of an open convex set are $\underline{\underline{B}}_1$-congruences [Exercise 7.6G]. Then any non-trivial $\underline{\underline{Q}}$-congruence θ on A_n can be extended to the congruence θ_A

on A defined as follows:

$$(a,b) \in \theta_A :\Leftrightarrow (a,b) \in \theta \text{ or } a = b.$$

Now assume that A is relatively subdirectly irreducible in $\underline{\underline{Q}}$. Then A has a non-trivial (open convex) sink A_n. The sink A_n must be isomorphic to I°. Indeed, otherwise there would be a set of $\underline{\underline{Q}}$-congruences θ_k of A_n, for k in K, having as an intersection the equality relation $\widehat{A_n}$ on A_n. Now the intersection of their extensions θ_{kA} to A is the equality relation $\widehat{A}$ on A. It follows that A_n is I°, and hence the smallest sink of A must be an open unit interval.

Now the (canonical) envelope E_n of A_n is a convex set. Since A is finitely generated, it must be the closed unit interval. Call the ends of this interval 0_n and 1_n, so that $E_n = [0_n, 1_n]$. By (7.6.3), for each $i < n$ and b_i in A_i, one has $b_i \varphi_{i,n} = 1_n$ or $b_i \varphi_{i,n} = 0_n$. Set $G_n := \{b_i \in A \mid b_i \varphi_{i,n} = 1_n\}$ and $G'_n := \{b_i \in A \mid b_i \varphi_{i,n} = 0_n\}$. Obviously both G_n and G'_n are (disjoint) subalgebras of A. Moreover, since no $\underline{\underline{Q}}$-congruence on the non-trivial G_n or G'_n collapses any elements of A_n, it follows that the intersection of any two $\underline{\underline{Q}}$-congruences, one generated by some elements of G_n and the other by some elements of G'_n, must be the equality on A_n. Hence at least one of G_n and G'_n must be a singleton, say $G_n = \{1_n\}$ with $A_n \cup \{1_n\} =: I_n$, a half-closed interval. Since A is finitely generated, G'_n is a non-empty subalgebra of A. If $|G'_n| = 1$, then $G'_n = \{0_n\}$ and A is the closed unit interval isomorphic to T_1. So suppose that $|G'_n| > 1$. Then G'_n has a smallest non-trivial sink A_{n-1}. A similar argument as above shows that A_{n-1} is isomorphic to I°, that the (canonical) envelope $E_{n-1} = [0_{n-1}, 1_{n-1}]$ is isomorphic to I, and $G_{n-1} = \{1_{n-1}\}$ with $A_{n-1} \cup \{1_{n-1}\} = I_{n-1}$. If $G'_{n-1} = \{0_{n-1}\}$, then A is isomorphic to T_2 and we are done. If not, then we continue as before, obtaining algebras isomorphic to $T_3, \ldots, T_n$. Since the height of finitely generated algebras in $Q(\alpha_{n+1}, \alpha)$ is at most n, it follows that T_i with $i \leq n$ are all the relatively subdirectly irreducibles in $Q(\alpha_{n+1}, \alpha)$. □

The proof of the following lemma is similar, and is left as Exercise 7.6D.

Lemma 7.6.9. *Let A be a finitely generated algebra in $Q(\beta)$. Then*

(a) *A is a relatively subdirectly irreducible member of $Q(\beta)$ if and only if A is isomorphic to some T_n or to $\underset{\sim}{2}$.*

(b) *A is a relatively subdirectly irreducible member of $Q(\alpha_{n+1})$ if and only if A is isomorphic to $\underset{\sim}{2}$, or to some T_i for $i = 1, 2, \ldots, n$.* □

Corollary 7.6.10. *Let A be a finitely generated barycentric algebra.*

(a) *If A is in the quasivariety $Q(\alpha)$ and $h(A) = n$, then A embeds into some power T_n^j of T_n.*

(b) *If A is in the quasivariety $Q(\beta)$ and $h(A) = n$, then A embeds into a direct product $T_n^j \times \underset{\sim}{2}^i$ for some j and i.* □

Proposition 7.6.11. *The following hold for quasivarieties of barycentric algebras:*

(a) $\underline{\underline{B}}_n = N(T_{n+1}, \underset{\sim}{2})$;

(b) $\underline{\underline{\widetilde{B}}}_n = N(T_{n+1}, I^\infty)$;

(c) $\underline{\underline{B}}_\omega = N(\underset{\sim}{2})$;

(d) $\underline{\underline{\widetilde{B}}}_\omega = N(I^\infty)$.

Proof. (a) Since T_{n+1} is not a subalgebra of T_n, it follows that T_n is a member of $N(T_{n+1}, \underset{\sim}{2})$. Hence $\underline{\underline{B}}_n = Q(T_n) \subseteq N(T_{n+1}, \underset{\sim}{2})$. Now if A is a finitely generated algebra in the class $N(T_{n+1}, \underset{\sim}{2})$, then $h(A) \leq n$, and by Corollary 7.6.10, A embeds into some power T_n^j. Hence A is in $\underline{\underline{B}}_n$. Since each algebra in $N(T_{n+1}, \underset{\sim}{2})$ is a directed colimit of finitely generated algebras in the same class, it follows that $N(T_{n+1}, \underset{\sim}{2}) \subseteq \underline{\underline{B}}_n$.

The proofs of the equalities (b), (c) and (d) are similar, so we will only sketch them.

(b) Since T_n and $\underset{\sim}{2}$ are in $N(T_{n+1}, I^\infty)$, it follows that $\underline{\underline{\widetilde{B}}}_n \subseteq N(T_{n+1}, I^\infty)$. Now

7. SUBREDUCTS OF AFFINE SPACES

if a finitely generated algebra A is in the class $N(T_{n+1}, I^\infty)$, then $h(A) \leq n$, and by Corollary 7.6.10, A is in $\underline{\underline{B}}_n$. Hence $N(T_{n+1}, I^\infty) \subseteq \underline{\underline{B}}_n$.

(c) Since all the T_n satisfy the quasi-identity (α), it follows by Proposition 7.6.6(b) that $\underline{\underline{B}}_\omega \subseteq Q(\alpha) = N(2)$. If A is a finitely generated algebra in $N(2)$, then it has a (finite) height, say k, and by Corollary 7.6.10, it embeds into some T_k^j, and hence is in $\underline{\underline{B}}_\omega$. It follows that $N(2) \subseteq \underline{\underline{B}}_\omega$.

(d) Since 2 and all the T_n are in $N(I^\infty)$, one has $\widetilde{\underline{\underline{B}}}_\omega \subseteq N(I^\infty)$. Since each finitely generated algebra A in $N(I^\infty)$ has finite height, Corollary 7.6.10 again shows that $N(I^\infty) \subseteq \widetilde{\underline{\underline{B}}}_\omega$. □

Corollary 7.6.12. *The quasivariety $\underline{\underline{B}}$ and the quasivarieties $\underline{\underline{B}}_n, \widetilde{\underline{\underline{B}}}_n$ for $n = 0, 1, \ldots, \omega$ are pairwise distinct, and satisfy the following:*

(a) $\underline{\underline{B}}_n = N(T_{n+1}, 2) = Q(\alpha_{n+1}, \alpha)$;
(b) $\widetilde{\underline{\underline{B}}}_n = N(T_{n+1}, I^\infty) = Q(\alpha_{n+1})$;
(c) $\underline{\underline{B}}_\omega = N(2) = Q(\alpha)$;
(d) $\widetilde{\underline{\underline{B}}}_\omega = N(I^\infty) = Q(\beta)$. □

Proposition 7.6.13. *The quasivarieties $\underline{\underline{B}}_n, \widetilde{\underline{\underline{B}}}_n$ for $n = 0, 1, \ldots, \omega$, together with $\underline{\underline{B}}$, are all the quasivarieties of barycentric algebras.*

Proof. Let $\underline{\underline{Q}}$ be a quasivariety of barycentric algebras containing any of the two minimal quasivarieties $\underline{\underline{B}}_1 = \underline{\underline{Cv}}$ and $\widetilde{\underline{\underline{B}}}_0 = \underline{\underline{Sl}}$.

First assume that $\underline{\underline{B}}_1 \leq \underline{\underline{Q}} \leq \underline{\underline{B}}_\omega$. Let n be the maximal height for all the algebras in $\underline{\underline{Q}}$. Assume first that $1 \leq n = k < \infty$. Then T_k is in $\underline{\underline{Q}}$, and hence $\underline{\underline{B}}_k = Q(T_k) \leq \underline{\underline{Q}}$. On the other hand, by Proposition 7.7.11, $\underline{\underline{Q}} \subseteq N(T_{k+1}, 2) = \underline{\underline{B}}_k$, whence $\underline{\underline{Q}} = \underline{\underline{B}}_k$. Now suppose $n = \infty$. Then all the T_k are in $\underline{\underline{Q}}$, and hence $\underline{\underline{B}}_\omega = Q(T_1, T_2, \ldots) \leq \underline{\underline{Q}}$. Now by Proposition 7.6.11, one has $\underline{\underline{Q}} \subseteq N(2) = \underline{\underline{B}}_\omega$. Hence $\underline{\underline{Q}} = \underline{\underline{B}}_\omega$.

The proof is similar in the case $\widetilde{\underline{\underline{B}}}_0 \leq \underline{\underline{Q}} \leq \widetilde{\underline{\underline{B}}}_\omega$. If $1 \leq n = k < \infty$, then T_k and 2 are in $\underline{\underline{Q}}$, and hence, by Proposition 7.6.11, one has $\widetilde{\underline{\underline{B}}}_k = Q(T_k, 2) \leq$

$\underline{\underline{Q}} \subseteq N(T_{k+1}, I^\infty) = \underline{\underline{\widetilde{B}}}_k$. And if $n = \infty$, then $\underline{\underline{\widetilde{B}}}_\omega = Q(2, T_1, T_2, \ldots) \leq \underline{\underline{Q}} \subseteq N(I^\infty) = \underline{\underline{\widetilde{B}}}_\omega$.

Finally, if $\underline{\underline{\widetilde{B}}}_\omega < \underline{\underline{Q}} \leq \underline{\underline{B}}$ and A is an algebra in $\underline{\underline{Q}}$ but not in $\underline{\underline{\widetilde{B}}}_\omega$, then A has I^∞ as a subalgebra. But by Lemma 7.6.4, the algebra I^∞ generates the variety $\underline{\underline{B}}$. Hence $\underline{\underline{Q}} = \underline{\underline{B}}$. □

The proof of Theorem 7.6.2 follows by Propositions 7.6.6, 7.6.7, 7.6.11 and 7.6.13.

Exercises

7.6A. (a) Find all the congruences of the algebras T_2 and T_3.

(b) Find all the $\underline{\underline{B}}_2$-congruences of T_2 and the $\underline{\underline{B}}_3$-congruences of T_3.

(c) Find all the $\underline{\underline{\widetilde{B}}}_2$-congruences of T_2 and the $\underline{\underline{\widetilde{B}}}_3$-congruences of T_3.

7.6B. Show that each algebra T_i, for $i \leq n$, is a relatively subdirectly irreducible member of the quasivariety $\underline{\underline{B}}_n$, and of the quasivariety $\underline{\underline{B}}_\omega$.

7.6C. Show that each algebra T_i, for $i \leq n$, is a relatively subdirectly irreducible member of the quasivariety $\underline{\underline{\widetilde{B}}}_n$ and of the quasivariety $\underline{\underline{\widetilde{B}}}_\omega$. Compare the relatively monolithic congruence of T_n with the respective monolithic congruences obtained in Exercise 7.6B.

7.6D. Prove Lemma 7.6.9.

[Hint: Let A be a semilattice Lallement sum as in the proof of Lemma 7.6.8. Show that if, for some a_i in A_i and $i < j$ in $(J, \leq)$, one has $a_i \varphi_{i,j} = a_j \in A_j$, then $A_i \varphi_{i,j} = \{a_j\}$.]

7.6E. Show that the quasi-identities (α_n) and (β) are quasiregular, whereas the quasiidentity (α) is not.

7.6F. Let A be a barycentric algebra that is a proper Płonka sum of convex sets, and is not a non-trivial direct product. Show that such a Płonka sum is not a member of the quasivariety $\underline{\underline{\widetilde{B}}}_\omega$.

7.6G. Show that for any subquasivariety $\underline{\underline{Q}}$ of $Q(\alpha)$, each $\underline{\underline{Q}}$-congruence of an open convex set is a $\underline{\underline{B}}_1$-congruence.

7.6H. [Ignatov 1985] Show that the relation σ defined on a barycentric algebra $(A, \underline{I}^\circ)$ by

$$(a, b) \in S :\Leftrightarrow \exists\, c \in A \,.\, \exists\, p \in I^\circ \,.\, ca\underline{p} = cb\underline{p} = c$$

is a congruence relation, and that the algebra $(A^\sigma, \underline{I}^\circ)$ is in the quasi-variety $\underline{\underline{B}}_\omega$.

7.7. Embedding cancellative modes into affine spaces

As we have seen in previous sections, cancellative modes, and in particular cancellative quasi-affine modes, play an important rôle in the structure theory of modes. Not all quasi-affine modes are cancellative. However, we will prove in this section that every cancellative mode embeds as a subreduct into an appropriate affine space, i.e. that it is quasi-affine.

The first main result is the following.

Theorem 7.7.1. *Each cancellative mode (C, Ω) of a fixed type $\tau : \Omega \to \mathbb{Z}^+$ embeds as an Ω-subreduct into an affine space.*

The proof of Theorem 7.7.1 is divided into a sequence of lemmata. First we need the following definitions. For any element e of C and each operation ω in Ω, define the i-th *translation by e* to be the mapping

$$(7.7.1) \qquad e_i(\omega) : C \to C;\; x \mapsto e \ldots exe \ldots e\omega$$

with x as the i-th argument of ω. An element e is called an *identity* if, for any ω in Ω, all the i-translations (with $i = 1, \ldots, \omega\tau$) are identity mappings. In what follows, we assume that (C, Ω) is an arbitrarily fixed non-empty cancellative mode, ω is an arbitrary n-ary operation in Ω, and φ is an arbitrary n-ary operation in Ω. Assume additionally that $m \leq n$.

Lemma 7.7.2. *If* (C, Ω) *has an identity element* e, *then for all* $x_1, \ldots, x_m$ *in* C:

(7.7.2) $$x_1 \ldots x_m \omega = x_{1\pi} \ldots x_{m\pi} e \ldots e\varphi$$

for any permutation π *of the set* $\{1, \ldots, m\}$.

Note that in the case $m = n$, there is no e on the right hand side of the equality (7.7.2).

Proof. For any permutation π on $\{1, \ldots, m\}$, and $x_1, \ldots, x_m$ in C, one has

$$x_1 \ldots x_m \omega$$
$$= (e \ldots e x_1 e \ldots e\varphi) \ldots (e \ldots e x_m e \ldots e\varphi)\omega$$
$$= (e \ldots e x_{1\pi} e \ldots e\omega) \ldots (e \ldots e x_{m\pi} e \ldots e\omega)\varphi$$
$$= x_{1\pi} \ldots x_{m\pi} e \ldots e\varphi,$$

with $x_{i\pi}$ as the i-th argument. The second equality follows by the entropic law. □

Lemma 7.7.3. *If* (C, Ω) *has an identity element* e, *then* (C, Ω) *is a reduct of a commutative monoid* $(C, +, e)$ *such that for each* ω *in* Ω,

(7.7.3) $$x_1 \ldots x_m \omega = x_1 + \cdots + x_m.$$

Proof. Define a binary operation $+$ on C by

$$x + y := xye \ldots e\omega.$$

Note that by Lemma 7.7.2, the choice of ω in Ω is irrelevant, and for any φ in

Ω, one has $x + y = xye\ldots e\varphi$. Now for any a, b, c and d in C,

$$(a + b) + (c + d)$$
$$= ((abe\ldots e\omega)(cde\ldots e\omega)e\ldots e)\omega$$
$$= ((abe\ldots e\omega)(cde\ldots e\omega)(e\ldots e\omega)\ldots(e\ldots e\omega))\omega$$
$$= ((ace\ldots e\omega)(bde\ldots e\omega)(e\ldots e\omega)\ldots(e\ldots e\omega))\omega$$
$$= ((ace\ldots e\omega)(bde\ldots e\omega)e\ldots e)\omega$$
$$= (a + d) + (c + d).$$

Thus $(C, +)$ is entropic. Obviously, e is the identity element of $(C, +)$. Now Exercise 7.7A shows that $(C, +)$ is a commutative semigroup. To show that (7.7.3) holds, suppose that $x_1\ldots x_i e\ldots e\omega = x_1 + \cdots + x_i$. Then

$$x_1\ldots x_i x_{i+1} e\ldots e\omega$$
$$= ((x_1 e\ldots e\omega)\ldots(x_i e\ldots e\omega)(ex_{i+1}e\ldots e\omega)(e\ldots e\omega)\ldots(e\ldots e\omega))\omega$$
$$= ((x_1\ldots x_i e\ldots e\omega)(e\ldots ex_{i+1}e\ldots e\omega)(e\ldots e\omega)\ldots(e\ldots e\omega))\omega$$
$$= ((x_1 + \cdots + x_i)x_{i+1}e\ldots e)\omega$$
$$= x_1 + \cdots + x_i + x_{i+1}.$$

Thus, by induction, $x_1\ldots x_m\omega = x_1 + \cdots + x_m$. □

Lemma 7.7.4. *For any e in (C, Ω), the translations (7.7.1) are pairwise commuting endomorphisms.*

Proof. This follows by Proposition 5.6. □

Let us extend the type $\tau : \Omega \to \mathbb{Z}^+$ of (C, Ω) by taking e as the element selected by a nullary operation and by taking the translations $e_i(\omega)$ for ω in Ω and $i = 1, \ldots, \omega\tau$ as unary operations on C. In this way, one obtains an algebra (C, Ω') of type $\tau' : \Omega' \to \mathbb{Z}^+$ with $\Omega' = \Omega \bigcup \{e_i(\omega) \mid \omega \in \Omega, 1 \leq i \leq \omega\tau\} \bigcup \{e\}$.

Let $\underline{K}$ be the class of algebras of type τ' on which each $e_i(\omega)$ is an injective endomorphism.

Proposition 7.7.5. *For each $\underline{K}$-algebra (A, Ω'), there is an algebra $(\overline{A}, \Omega')$, containing a subalgebra isomorphic with (A, Ω'), such that the operations $e_i(\omega)$ are automorphisms of $(\overline{A}, \Omega')$, and such that $(\overline{A}, \Omega')$ satisfies all the quasi-identities satisfied by (A, Ω').*

Proof. First note that in the case that (A, Ω') is finite, the algebra $(\overline{A}, \Omega')$ coincides with (A, Ω'). So assume that (A, Ω') is infinite. Each operation $e_i(\omega)$ in the set $U := \{e_i(\omega) \mid \omega \in \Omega, 1 \leq i \leq \omega\tau\}$ is an injective endomorphism of (A, Ω'). Hence it is an isomorphism of (A, Ω') onto the subalgebra $Ae_i(\omega)$ of (A, Ω'). It follows that there exist an extension (AE^i_ω, Ω') of (A, Ω') and an Ω'-isomorphism $\iota^i_\omega : AE^i_\omega \to A$ extending $e_i(\omega) : A \to Ae_i(\omega)$. This isomorphism identifies each element a in $A \leq AE^i_\omega$ with the element $ae_i(\omega)$ in $Ae_i(\omega)$ and the remaining elements of AE^i_ω with the elements of $A - Ae_i(\omega)$. For a in AE^i_ω one has

$$ae_i(\omega)\iota^i_\omega = a\iota^i_\omega e_i(\omega) = a\iota^i_\omega \iota^i_\omega,$$

whence $ae_i(\omega) = a\iota^i_\omega$, and finally one has $e_i(\omega) = \iota^i_\omega$ on the set AE^i_ω. It is evident that the algebra (AE^i_ω, Ω') is in the class $\underline{K}$. Note that for any two operations $e_i(\omega)$ and $e_j(\varphi)$ in U, one has

$$\left(AE^i_\omega E^j_\varphi, \Omega'\right) \cong (AE^j_\varphi E^i_\omega, \Omega') \cong (A, \Omega').$$

By iterating the above construction, one obtains a quasi-ordered set of Ω'-algebras of the form $AE^{i_1}_{\omega_1} \ldots E^{i_n}_{\omega_n}$, each isomorphic to (A, Ω'), and such that each $AE^{i_1}_{\omega_1} \ldots E^{i_n}_{\omega_n}$ is isomorphic to

$$AE^{i_1}_{\omega_1} \ldots E^{i_{n+1}}_{\omega_{n+1}} \iota^{i_{n+1}}_{\omega_{n+1}} = AE^{i_1}_{\omega_1} \ldots E^{i_{n+1}}_{\omega_{n+1}} e_{i_{n+1}}(\omega_{n+1}),$$

a subalgebra of $AE^{i_1}_{\omega_1} \ldots E^{i_{n+1}}_{\omega_{n+1}}$.

Now consider the set $X := \{E^i_\omega \mid \omega \in \Omega, 1 \leq i \leq \omega\tau\}$, and the free commutative monoid $X^{*\kappa}$ on X, with the identity 0 denoted by $E^{i_0}_{\omega_0}$. By Example

7. SUBREDUCTS OF AFFINE SPACES

3.5.1(b), $X^{*\kappa}$ can be considered as the set of finite multisubsets of X. This set is ordered as in Section 0.7. It is easy to see that this ordering relation $\preceq$ is a (join) semilattice order, and the algebraic quasi-order of $X^{*\kappa}$. Then $(X^{*\kappa}, \preceq)$ may be considered as a small category $(X^{*\kappa})$. Let $F : (X^{*\kappa}) \to (\Omega')$ be the functor from the category $(X^{*\kappa})$ to the category (Ω') acting on objects as follows:

$$0 = E^{i_0}_{\omega_0} \mapsto (AE^{i_0}_{\omega_0}, \Omega') = (A, \Omega');$$

$$E^{i_0}_{\omega_0} E^{i_1}_{\omega_1} \ldots E^{i_n}_{\omega_n} \mapsto (AE^{i_0}_{\omega_0} E^{i_1}_{\omega_0} \ldots E^{i_n}_{\omega_n}, \Omega');$$

and on morphisms as follows:

$$\begin{array}{ccc} E^{i_0}_{\omega_0} \ldots E^{i_k}_{\omega_k} \ldots E^{i_n}_{\omega_n} & & (Ae^{i_0}_{\omega_0} \ldots E^{i_k}_{\omega_k} \ldots E^{i_n}_{\omega_n}, \Omega') \\ \uparrow & \mapsto & \uparrow \\ E^{i_0}_{\omega_0} \ldots E^{i_k}_{\omega_k} & & (AE^{i_0}_{\omega_0} \ldots E^{i_k}_{\omega_k}, \Omega') \end{array},$$

assigning to each morphism of $(X^{*\kappa})$ the corresponding embedding of algebras. The functor F defines the functorial sum $X^{*\kappa} F$ of algebras $(AE^{i_0}_{\omega_0} \ldots E^{i_n}_{\omega_n}, \Omega')$ over $X^{*\kappa}$ by the embeddings $\varphi_{I,J} : AI \to AJ$ for

$$I = E^{i_0}_{\omega_0} \ldots E^{i_k}_{\omega_k} \preceq J = E^{i_0}_{\omega_0} \ldots E^{i_k}_{\omega_k} \ldots E^{i_n}_{\omega_n},$$

defined by $x\varphi_{I,J} = xe_{i_{k+1}}(\omega_{i_{k+1}}) \ldots e_{i_n}(\omega_{i_n})$. Since the quasi-order $\preceq$ is directed, one can take the directed colimit $\varinjlim F$ as isomorphic to $(X^{*\kappa} F)^\delta$, where the congruence δ is defined as in (4.2.2). (Cf. Proposition 4.2.7 and Exercise 4.2E.) All the algebras $(AE^{i_0}_{\omega_0} \ldots E^{i_n}_{\omega_n}, \Omega')$ are isomorphic to (A, Ω'), and all are subalgebras of $\varinjlim F = (X^{*\kappa} F)^\delta$. In particular, (A, Ω') is a subalgebra of $\varinjlim F$. The directed colimit preserves all quasi-identities. Hence in $\varinjlim F$, the operations $e_i(\omega)$ are translations by e^δ and are all injective. To see that they are automorphisms of $\varinjlim F$, it is enough to show that they are surjective. So let x^δ

be in $(X^{*\kappa}F)^\delta$. We look for an element y^δ such that $y^\delta e_i(\omega) = (ye_i(\omega))^\delta = x^\delta$. First note that there is $I = E^{i_0}_{\omega_0} \ldots E^{i_n}_{\omega_n}$ in $X^{*\kappa}$ such that $x \in AI$. Let $IE^i_\omega =: J$. Then there is a sum embedding $\varphi_{I,J} : AI \to AJ$, and moreover $AI\varphi_{I,J} = AJe_i(\omega)$. Recall that $e_i(\omega) : AJ \to AJe_i(\omega)$ is an isomorphism. Take $y := x\varphi_{I,J}(e_i(\omega)^{-1})$. Then obviously $ye_i(\omega) = ye_i(\omega)\varphi_{J,J} = x\varphi_{I,J}$, whence $(ye_i(\omega))^\delta = x^\delta$. The algebra $(\overline{A}, \Omega') = \varinjlim F$ satisfies all the claims of the proposition. □

Let us go back to our cancellative mode (C, Ω). Cancellativity and Lemma 7.7.4 imply that the translations $e_i(\omega)$ are injective endomorphisms of (C, Ω'). Then Lemma 7.7.4 and Proposition 7.7.5 imply the following

Corollary 7.7.6. *The algebra (C, Ω') is isomorphic to a subalgebra of a cancellative mode $(\overline{C}, \Omega')$ on which the operations $e_i(\omega)$ are automorphisms.* □

In what follows we will identify (C, Ω') with this subalgebra of $(\overline{C}, \Omega')$. Note that each operation $e_i(\omega)$ of $(\overline{C}, \Omega')$ is also the i-th translation by e of the algebra $(\overline{C}, \Omega)$.

Lemma 7.7.7. *Let e be an element of C. For each ω in Ω, define the operation ω^* on $\overline{C}$ by*

(7.7.4) $$x_1 \ldots x_n \omega^* := (x_1 e_1(\omega)^{-1}) \ldots (x_m e_m(\omega)^{-1})\omega.$$

Let Ω^* be the set of all ω^* for ω in Ω. Then $(\overline{C}, \Omega^*)$ is entropic, and e is an identity element in $(\overline{C}, \Omega^*)$.

Proof. By Corollary 7.7.6 the translations $e_i(\omega)$, for all ω in Ω, are pairwise commuting automorphisms of $(\overline{C}, \Omega)$, and hence are all the $e_i(\omega)^{-1}$. By Exercise 7.7C, it follows that $(\overline{C}, \Omega^*)$ is entropic. Moreover

$$e \ldots exe \ldots e\omega^* = (ee_1(\omega)^{-1}) \ldots (xe_i(\omega)^{-1}) \ldots (ee_m(\omega)^{-1})\omega$$
$$= e \ldots e(xe_i(\omega)^{-1})e \ldots e\omega$$
$$= xe_i(\omega)^{-1}e_i(\omega) = x.$$

Thus e is an identity element of $(\overline{C}, \Omega^*)$. □

Proposition 7.7.8. *Let (C, Ω) be a cancellative mode, and let e be an element of C. Then (C, Ω) is a subreduct of the algebra $(\overline{C}, \Omega')$, and the following hold:*

(a) *There is a cancellative commutative monoid $(\overline{C}, +, e)$ on the set $\overline{C}$ defined by*

(7.7.5) $$x + y := (xe_1(\omega)^{-1})(ye_2(\omega)^{-1})e \ldots e\omega$$

for any ω in Ω;

(b) *The translations $e_i(\omega)$ are pairwise commuting monoid and Ω-automorphisms on $\overline{C}$;*

(c) *For each ω in Ω and $x_1, \ldots, x_m$ in $\overline{C}$, one has*

(7.7.6) $$x_1 \ldots x_m \omega = x_1 e_1(\omega) + \cdots + x_m e_m(\omega).$$

Moreover, under the operations (7.7.6), the algebra $(\overline{C}, \Omega)$ is a cancellative mode.

Proof. Define the operations ω^* as in (7.7.4). Then by Lemma 7.7.7, $(\overline{C}, \Omega^*)$ has an identity element e. Note that Lemma 7.7.3 does not require idempotency of the algebra in question. Thus it implies that there is a commutative monoid $(\overline{C}, +, e)$ such that

$$\begin{aligned} x_1 \ldots x_m \omega^* &= (x_1 e_1(\omega)^{-1}) \ldots (x_m e_m(\omega)^{-1})\omega \\ &= x_1 + \cdots + x_m \end{aligned}$$

for each (m-ary) ω in Ω. Then (7.7.6) holds. As in Lemma 7.7.3, the operation $+$ is defined by

$$x + y = xye \ldots e\omega^* = (xe_1(\omega)^{-1})(ye_2(\omega)^{-1})e \ldots e\omega,$$

whence (7.7.5) holds. The definition does not depend on the choice of ω in Ω. All the $e_i(\omega)$ are (pairwise commuting) automorphisms of $(\overline{C}, \Omega^*)$ [Exercise 7.7C]. Thus

$$(x+y)e_i(\omega) = (xye \ldots e\omega^*)e_i(\omega)$$
$$= (xe_i(\omega) \; ye_i(\omega) \; e \ldots e)\omega^*$$
$$= xe_i(\omega) + ye_i(\omega).$$

As noted before, $ee_i(\omega) = e$. Thus $e_i(\omega)$ is an automorphism of $(\overline{C}, +, e)$ for each i, and for each ω in Ω. To show that the semigroup $(\overline{C}, +)$ is cancellative, assume $x + y = x + z$ for x, y and z in $\overline{C}$. Then

$$(xe_1(\omega)^{-1})(ye_2(\omega)^{-1})e \ldots e\omega = xye \ldots e\omega^*$$
$$= x + y = x + z = xze \ldots e\omega^*$$
$$= (xe_1(\omega)^{-1})(ze_2(\omega)^{-1})e \ldots e\omega.$$

By the cancellativity of (C, Ω), it follows that $ye_2(\omega)^{-1} = ze_2(\omega)^{-1}$, whence $y = z$. □

Proposition 7.7.9. *The (cancellative) monoid $(\overline{C}, +, e)$ embeds into an abelian group $(G, +, -, 0)$. More precisely, there is a monoid monomorphism*

$$\Delta : (\overline{C}, +, e) \to (G, +, 0).$$

Proof. There is a very well-known standard construction, similar to the localization of a ring, that allows one to embed each cancellative commutative monoid into an abelian group. Take the direct power $\overline{C} \times \overline{C}$ of the set $\overline{C}$. Define the relation ρ on $\overline{C} \times \overline{C}$ by

$$((a,b),(c,d)) \in \rho :\Leftrightarrow a + d = b + c.$$

7. SUBREDUCTS OF AFFINE SPACES

It is immediate that ρ is a congruence of the semigroup $(\overline{C} \times \overline{C}, +)$, and that $G = (\overline{C} \times \overline{C}, +)^\rho$ is an abelian group. The identity of G is the diagonal $(x,x)^\rho = \{(x,x) \mid x \in \overline{C}\}$. The inverse of $(a,b)^\rho$ is $(b,a)^\rho$. The mapping

(7.7.7) $$\Delta : \overline{C} \to (\overline{C} \times \overline{C})^\rho; x \mapsto (x+x, x)^\rho$$

embeds the monoid $(\overline{C}, +, e)$ into the monoid $(G, +, 0)$ with $0 = (x,x)^\rho$. Details of the proofs are left as Exercise 7.7C. $\square$

We will identify elements x of $\overline{C}$ with $(x+x, x)^\rho$, and consider $(\overline{C}, +, e)$ as a submonoid of the monoid $(G, +, 0)$.

Lemma 7.7.10. *Let H be an abelian group, and let $f : (\overline{C}, +, e) \to (H, +, 0)$ be a monoid homomorphism. Then there is a unique group homomorphism $\overline{f} : (G, +, -, 0) \to (H, +, -, 0)$ such that $\Delta \overline{f} = f$, i.e. the diagram below is commutative:*

$$\begin{array}{ccc} \overline{C} & \xrightarrow{\Delta} & G = (\overline{C} \times \overline{C})^\rho \\ \| & & \downarrow \overline{f} \\ \overline{C} & \xrightarrow{f} & H \end{array}$$

Proof. Define the mapping

(7.7.8) $$\overline{f} : G \to H; (a,b)^\rho \mapsto af - bf.$$

The mapping is well-defined, since for $(a,b)^\rho = (a', b')^\rho$, i.e. for $a + b' = a' + b$, one has $af + b'f = (a+b')f = (a'+b)f = a'f + bf$, and hence $(a,b)^\rho \overline{f} = af - bf = a'f - b'f = (a',b')^\rho \overline{f}$. It is obviously a group homomorphism [Exercise 7.7E]. Now if $h : G \to H$ is any other group homomorphism such that $\Delta h = f$, then

$$(a,b)^\rho h = ((a,e)^\rho + (e,b)^\rho)h = (a,e)^\rho h + (e,b)^\rho h$$
$$= (a+a, a)^\rho h + (b, b+b)^\rho h$$
$$= (a+a, a)^\rho h - (b+b, b)^\gamma h$$
$$= a\Delta h - b\Delta h = af - bf = (a,b)^\rho \overline{f}.$$

Finally, note that for each a in C, one has $a\Delta \overline{f} = (a+a, a)^\rho \overline{f} = (a, e)^\rho \overline{f} = af$. Hence $\Delta \overline{f} = f$, and the diagram is commutative. □

Proof of Theorem 7.7.1. To show that the cancellative mode (C, Ω) embeds as a subreduct into an affine space, first embed (C, Ω) into the commutative monoid $(\overline{C}, +, e)$ equipped with the Ω-operations (7.6.6). Then consider the abelian group G from above as a $\mathbb{Z}$-module. We will extend the ring $\mathbb{Z}$ to a commutative ring R that will make G an R-module $(G, +, R)$, and then show that the mode (C, Ω) embeds (as a subreduct) into the affine space $(G, P, \underline{R})$.

The first aim is to define the ring R. Consider all the translations $e_i(\omega)$ (with $i = 1, \ldots, \omega\tau$) for all ω in Ω. The translations $e_i(\omega)$ are automorphisms of the monoid $(\overline{C}, +, e)$. Using Lemma 7.7.10, one can extend these automorphisms to automorphisms of the group $(G, +, -, 0)$ as shown in the following diagram:

$$\begin{array}{ccc} \overline{C} & \xrightarrow{\Delta} & G \\ e_i(\omega) \downarrow & & \downarrow \overline{e_i}(\omega) := \overline{e_i(\omega)\Delta} \\ \overline{C} & \xrightarrow{\Delta} & G \end{array}.$$

The group homomorphisms $\overline{e_i}(\omega)$ are defined by

$$(7.7.9) \quad \begin{aligned} (a, b)^\rho \overline{e_i}(\omega) &= ae_i(\omega)\Delta - be_i(\omega)\Delta \\ &= (a, e_i(\omega) + ae_i(\omega), ae_i(\omega))^\rho \\ &\quad + (be_i(\omega), be_i(\omega) + be_i(\omega))^\rho \\ &= (ae_i(\omega), e)^\rho + (e, be_i(\omega))^\rho = (ae_i(\omega), be_i(\omega))^\rho, \end{aligned}$$

and are obviously automorphisms, since the $e_i(\omega)$ are. Define the ring R to be the subring generated by all the $\overline{e_i}(\omega)$ in the ring $\text{End}(G, +)$ of group endomorphisms. Consider the R-module $(G, +, R)$, and the corresponding affine

7. SUBREDUCTS OF AFFINE SPACES   425

space $(G, P, \underline{R})$. For each ω in Ω, define an operation ω on G as follows:

(7.7.10)
$$(a_1, b_1)^\rho \ldots (a_n, b_n)^\rho \omega$$
$$:= (a_1, b_1)^\rho \overline{e_1}(\omega) + \cdots + (a_n, b_n)^\rho \overline{e_n}(\omega).$$

Now the elements $a_1, \ldots, a_m$ in $\overline{C}$ are identified under Δ with the respective elements $(a_1 + a_1, a_1)^\rho, \ldots, (a_m + a_m, a_m)^\rho$. Hence

$$(a_1\Delta) \ldots (a_m\Delta)\omega = (a_1 + a_1, a_1)^\rho \ldots (a_m + a_m, a_m)^\rho \omega$$
$$= (a_1 + a_1, a_1)^\rho \overline{e_1}(\omega) + \cdots + (a_m + a_m, a_m)^\rho \overline{e_m}(\omega)$$
$$= (a_1 e_1(\omega) + a_1 e_1(\omega), a_1 e_1(\omega))^\rho + \cdots + (a_m e_m(\omega) + a_m e_m(\omega), a_m e_m(\omega))^\rho$$
$$= a_1 e_1(\omega)\Delta + \cdots + a_m e_m(\omega)\Delta.$$

It follows that for elements of $\overline{C}$ the definition (7.7.10) of ω on G coincides with that given on $\overline{C}$ by (7.7.6). Hence $(\overline{C}, \Omega)$ embeds as a subreduct of the affine space $(G, P.\underline{R})$. □

The affine space of Theorem 7.7.1 was defined over a ring depending on the particular cancellative mode (C, Ω). However, one may also embed cancellative Ω-modes into affine spaces over the ring $R(M\tau)$ of Example 7.1.4, a ring that is independent of the particular mode being embedded.

Corollary 7.7.11. *Each cancellative mode (C, Ω) of a fixed type $\tau : \Omega \to \mathbb{Z}^+$ embeds as an Ω-subreduct into an affine space $(G, P, \underline{R(M\tau)})$ over the ring $R(M\tau)$ of Example 7.1.4, the Ω-operations on G being defined by (7.1.2).*

Proof. For x in $\overline{C}$ and ω in Ω, the equation (7.7.6) and the idempotency of $(\overline{C}, \Omega)$ yield

(7.7.11)
$$x \sum_{i=1}^{m} e_i(\omega) = x.$$

Consider an element $g = (a,b)^\rho$ of G. By (7.7.9) and (7.7.11), one has

$$g \sum_{i=1}^{m} \overline{e}_i(\omega) = (a,b)^\rho \sum_{i=1}^{m} \overline{e}_i(\omega)$$
$$= \left(a \sum_{i=1}^{m} e_i(\omega), b \sum_{i=1}^{m} e_i(\omega) \right)^\rho$$
$$= (a,b)^\rho = g,$$

so that $\sum_{i=1}^{m} e_i(\omega) = 1$ in R. It follows that the unique commutative ring homomorphism

$$\mathbb{Z}[X_{\omega i} \mid \omega \in \Omega, 1 \le i \le \omega\tau] \to R; \; X_{\omega i} \mapsto \overline{e}_i(\omega)$$

induces a unique ring homomorphism

$$R(M\tau) \to R.$$

The composite

$$R(M\tau) \to R \to \mathrm{End}(G,+)$$

then makes G into an affine space over $R(M\tau)$, and the Ω-operations on $(G, P, \underline{R(M\tau)})$ defined by (7.1.2) coincide with those on $(G, P, \underline{R})$ defined by (7.7.10). $\square$

Exercises

7.7A. Show that each entropic groupoid with an identity element is a commutative semigroup.

7.7B. Show that if (C, Ω) has an identity element e, then for each ω in Ω,

$$x_1 \ldots x_m \omega = x_{1\pi} \ldots x_{m\pi} \omega$$

for any permutation π of $\{1, \ldots, m\}$.

7.7C. Let (A, Ω) be a mode, and let $\alpha_i(\omega)$, for $i = 1, \ldots, \omega\tau$ and all ω in Ω, be pairwise commuting endomorphisms of (A, Ω). Let ω^* be operations on A defined by

$$x_1 \ldots x_m \omega^* := (x_1 \alpha_1(\omega)) \ldots (x_m \alpha_m(\omega))\omega.$$

Show that (A, Ω^*) is entropic, and that the $\alpha_i(\omega)$ are endomorphisms of (A, Ω^*).

7.7D. Complete the proof of Proposition 7.7.9.

7.7E. Show that the mapping $\overline{f} : G \to H$; $(a,b)^\rho \mapsto af - bf$ of Lemma 7.7.10 is a group homomorphism.

7.8. Embedding sums of cancellative modes

The results of the previous section will now be applied to obtain a more general result. We will show that any Lallement sum of cancellative modes over a naturally quasi-ordered mode embeds as a subreduct into a functorial sum of affine spaces. We give two proofs of this theorem. One is a direct constructive proof requiring an additional assumption, while the second is quite general, but requires some knowledge of more advanced category theory. The results extend the representations of barycentric algebras and commutative binary modes given in Theorems 7.5.11(a) and 7.5.6(a).

We consider modes of a fixed plural type $\tau : \Omega \to \mathbb{Z}^+$, and the ring $R(M\tau)$ as in the previous section.

Lemma 7.8.1. *Each homomorphism $h : (C, \Omega) \to (D, \Omega)$ between cancellative modes extends in a unique way to a homomorphism $\overline{h} : (\overline{C}, \Omega) \to (\overline{D}, \Omega)$ between cancellative extensions of (C, Ω) and (D, Ω) as given by Proposition 7.7.8.*

Proof. Consider the functors $F : (X^{*\kappa}) \to (\Omega')$ and $G : (X^{*\kappa}) \to (\Omega')$ built for algebras (C, Ω') and (D, Ω') respectively, as in the proof of Proposition 7.7.5, and the corresponding Płonka sums $X^{*\kappa}F$ with sum homomorphisms $\varphi_{I,J}$ and $X^{*\kappa}G$ with sum homomorphisms $\psi_{I,J}$. Then the algebras $(\overline{C}, \Omega)$ and $(\overline{D}, \Omega)$ can be considered as (Ω-reducts of) $\varinjlim F = X^{*\kappa}F^\delta$ and $\varinjlim G = X^{*\kappa}G^\delta$. In what follows, all the algebras are Ω-algebras and all the homomorphisms are

Ω-homomorphisms.

Now the homomorphism $h : C \to D$ can be extended to a homomorphism $h : CI \to DI$ for any I in $X^{*\kappa}$ as follows. Let $I := E_{\omega_0}^{i_0} \ldots E_{\omega_n}^{i_n}$ be in $X^{*\kappa}$, and set $e_I := e_{i_1}(\omega_1) \ldots e_{i_n}(\omega_n)$. Note that the order of the $e_{i_k}(\omega_k)$ in e_I is irrelevant, since the $e_{i_k}(\omega_k)$ are pairwise commuting endomorphisms. Recall that CIe_I and C are isomorphic via the isomorphism $\Phi_{0,I}^{-1}$. Let

$$\widetilde{h}_I := \Phi_{0,I}^{-1} h \psi_{0,I} : CIe_I \to DIe_I^{-1},$$

and let

$$h_I := e_I \widetilde{h}_I e_I^{-1} : CI \to DI.$$

To define the homomorphism $\overline{h} : \overline{C} \to \overline{D}$, we will use the universality property (2.3.9) of directed colimits. Consider the (commutative) diagram

$$\begin{array}{ccccccc}
\varinjlim F & \xleftarrow{\varphi_I} & CI & \xrightarrow{h_I} & DI & \xrightarrow{\psi_I} & \varinjlim G \\
\| & & {\scriptstyle \varphi_{I,J}} \downarrow & & \downarrow {\scriptstyle \psi_{I,J}} & & \| \\
\varinjlim F & \xleftarrow{\varphi_J} & CJ & \xrightarrow{h_J} & DJ & \xrightarrow{\psi_J} & \varinjlim G
\end{array}$$

where $I \leq J$ in $(X^{*\kappa}, \leq)$. Let $\lambda_I := h_I \psi_I$. Note that $\lambda_I = h_I \psi_I = h_I \psi_{I,J} \psi_J = \varphi_{I,J} h_J \psi_J = \varphi_{I,J} \lambda_J$. It follows that there is a uniquely defined homomorphism $\overline{h} : \varinjlim F \to \varinjlim G$ as in Definition 2.3.11. □

Proposition 7.8.2. *Let a mode (B, Ω) be a semilattice Lallement sum of cancellative modes (B_i, Ω) over an Ω-semilattice $(I, \Omega, 0)$ with smallest element 0. Then (B, Ω) is a subreduct of a Płonka sum of affine spaces over the ring $R(M\tau)$.*

Proof. By Theorem 7.4.2, the mode (B, Ω) is a subalgebra of the Płonka sum (E, Ω) of cancellative envelopes (E_i, Ω) of (B_i, Ω) over (I, Ω), by the functor $F : (I) \to (\Omega)$ defined in Theorem 4.5.17. For each i in I, the object iF is the envelope (E_i, Ω). For $j \leq k$ in $(I, \leq)$, the morphism $(j \to k)F$ is $\psi'_{j,k} : (E_j, \Omega) \to (E_k, \Omega)$.

7. SUBREDUCTS OF AFFINE SPACES

Now choose an arbitrary element e_0 in E_0, and take $e_0 \psi'_{o,i}$ for each i in I to be the constant selected in each E_i by the nullary operation e. In this way, the Ω-homomorphisms $\psi'_{j,k}$ respect the nullary operation e.

By Proposition 7.7.8, each (cancellative) mode (E_i, Ω) embeds into the cancellative mode $(\overline{E}_i, \Omega)$ that has the additional structure $(\overline{E}_i, +, e)$ of a cancellative commutative monoid. By Lemma 7.8.1, each Ω-homomorphism $\psi'_{j,k}$ extends to the Ω-homomorphism $\psi_{j,k} : \overline{E}_j \to \overline{E}_k$. By Exercise 7.8A, the homomorphisms $\psi_{j,k}$ preserve the functoriality of $\psi'_{j,k}$, so that one can consider the Płonka sum $(\overline{E}, \Omega)$ of the modes $(\overline{E}_i, \Omega)$.

We will show that the Ω-homomorphisms $\psi_{j,k}$ also respect the semigroup structure. As in the proof of Proposition 7.7.8, one has

$$x + y = xye\ldots e\omega^*$$
$$= (xe_1(\omega)^{-1})(ye_2(\omega)^{-1})e\ldots e\omega,$$

for x, y and e in E_j. Let $x = ye_i(\omega) = e\ldots eye\ldots e\omega$, so that $xe_i(\omega)^{-1} = y$. Then obviously $xe_i(\omega)^{-1}\psi_{j,k} = y\psi_{j,k}$. Moreover

$$x\psi_{j,k} = e\ldots y\ldots e\omega\psi_{j,k} = e\psi_{j,k}\ldots y\psi_{j,k}\ldots e\psi_{j,k}\omega\psi_{j,k}$$
$$= e\ldots y\psi_{j,k}\ldots e\omega = y\psi_{j,k}e_i(\omega),$$

whence $x\psi_{j,k}e_i(\omega)^{-1} = y\psi_{j,k} = xe_i(\omega)^{-1}\psi_{j,k}$. This implies the following:

$$(x+y)\psi_{j,k} = (xe_1(\omega)^{-1}ye_2(\omega)^{-1}e\ldots e\omega)\psi_{j,k}$$
$$= (xe_1(\omega)^{-1}\psi_{j,k})(ye_2(\omega)^{-1}\psi_{j,k})e\ldots e\omega$$
$$= (x\psi_{j,k}e_1(\omega)^{-1})(y\psi_{j,k}e_2(\omega)^{-1})e\ldots e\omega$$
$$= x\psi_{j,k} + y\psi_{j,k}.$$

Hence the $\psi_{j,k}$ are indeed monoid homomorphisms, and $\overline{E}$ can also be considered as the Płonka sum of the monoids $(E_i, +, e)$ by the mappings $\psi_{j,k}$ over I.

By Proposition 7.7.9, each (cancellative) monoid $(\overline{E}_i, +, e)$ embeds into the abelian group $(G_i, +, -, 0)$ by the monoid embedding $\Delta_i : E_i \to (E_i \times E_i)^{\delta}; x \mapsto (x+x, x)^{\rho}$.

Now Lemma 7.7.10 provides the following commuting diagram for each pair j, k with $j \leq k$:

(7.8.1)
$$\begin{array}{ccc} \overline{E}_k & \xrightarrow{\Delta_k} & G_k \\ \psi_{j,k} \uparrow & & \uparrow \overline{\psi}_{j,k} := \overline{\psi_{j,k}\Delta_k} \\ \overline{E}_j & \xrightarrow{\Delta_j} & G_j \end{array}$$

This shows that the monoid and Ω-homomorphisms $\psi_{j,k}$ extend to group homomorphisms $\overline{\psi}_{j,k} : G_j \to G_k$, and that the homomorphisms $\overline{\psi}_{j,k}$ satisfy the functoriality condition. Define a new functor $G : (I) \to \underline{\text{AGp}}$ from the semilattice I to the category of abelian groups. For each i in I, the object iG is the group G_i. Then $(i \to j)G := \overline{\psi}_{i,j}$. The binary operation $+$ is defined on $A := \bigcup \{G_i \mid i \in I\}$ as $x_i + y_j := x_i \overline{\psi}_{i,i+j} + y_j \overline{\psi}_{j,i+j}$ for any i, j in I, x_i in G_i and y_j in G_j. Exercise 7.8B shows that $(A, +)$ is a commutative semigroup, and that the definition of $+$ restricted to $\overline{E}$ coincides with the definition of $+$ on the subsemigroup $(\overline{E}, +)$.

In the last step of the proof we will extend the structure of the Płonka sum A of groups G_i to the Płonka sum of affine spaces G_i over I. As in the proof of Corollary 7.7.11, each group G_i can be considered as an affine space $(G_i, P, \underline{R(M\tau)})$. We will prove that the group homomorphisms $\overline{\psi}_{j,k} : G_j \to G_k$ are affine space homomorphisms. It is sufficient to show that for any x in G_j, one has $x\overline{e}_i(\omega)\overline{\psi}_{j,k} = x\overline{\psi}_{j,k}\overline{e}_i(\omega)$. This can be calculated as follows. Let

7. SUBREDUCTS OF AFFINE SPACES 431

$g = (a,b)^\rho$ be an element of G_j. Then

$$\begin{aligned}
g\overline{e}_i(\omega)\overline{\psi}_{j,k} &= (a,b)^\rho \overline{e}_i(\omega)\overline{\psi}_{j,k} \\
&= (ae_i(\omega), be_i(\omega))^\rho \overline{\psi}_{j,k} \quad \text{by (7.7.9)} \\
&= (ae_i(\omega), be_i(\omega))^\rho \overline{\psi_{j,k}\Delta_k} \\
&= ae_i(\omega)\psi_{j,k}\Delta_k - be_i(\omega)\psi_{j,k}\Delta_k \quad \text{by (7.7.8)} \\
&= (e\ldots a\ldots e\omega)\psi_{j,k}\Delta_k - (e\ldots b\ldots e\omega)\psi_{j,k}\Delta_k \quad \text{by (7.7.1)} \\
&= (a\psi_{j,k})e_i(\omega)\Delta_k - (b\psi_{j,k})e_i(\omega)\Delta_k \\
&= (2(a\psi_{j,k}e_i(\omega)), a\psi_{j,k}e_i(\omega))^\rho \\
&\quad - (2(b\psi_{j,k}e_i(\omega)), b\psi_{j,k}e_i(\omega))^\rho \quad \text{by (7.7.7)} \\
&= (a\psi_{j,k}e_i(\omega), e)^\rho + (e, b\psi_{j,k}e_i(\omega))^\rho \\
&= (a\psi_{j,k}e_i(\omega), b\psi_{j,k}e_i(\omega))^\rho \\
&= (a\psi_{j,k}, b\psi_{j,k})^\rho \overline{e}_i(\omega) \quad \text{by (7.7.9)} \\
&= (a\psi_{j,k}\Delta_k - b\psi_{j,k}\Delta_k)\overline{e}_i(\omega) \quad \text{by (7.7.7)} \\
&= (a,b)^\rho \overline{\psi}_{j,k}\overline{e}_i(\omega) = g\overline{\psi}_{j,k}\overline{e}_i(\omega).
\end{aligned}$$

It follows that the $\overline{\psi_{j,k}}$ are homomorphisms of affine $R(M\tau)$-spaces. Hence the sum A is indeed the Płonka sum $(A, P, \underline{R(M\tau)})$ of affine spaces $(G_i, P, \underline{R(M\tau)})$. By the way that the functor G was constructed, it is obvious that the mode (B, Ω) is an Ω-subreduct of the Płonka sum $(A, P, \underline{R(M\tau)})$. □

Note that the extension (B, Ω) constructed in the proof of Proposition 7.8.2 is based on the possibility of choosing the constant e in each fibre E_i in a functorial way. This is equivalent to the fact that the semilattice (I, Ω) embeds into the Płonka sum (E, Ω). In general, this is not possible for a semilattice I without 0. However, the next proposition shows that the Płonka sum over such a semilattice embeds into a Płonka sum containing I as a subalgebra.

Proposition 7.8.3. *Let $\underline{K}$ be a non-trivial prevariety of Ω-modes. Let A be* a Płonka sum of $\underline{K}$-algebras A_i over an Ω-semilattice I. Then A embeds into a Płonka sum A' of $\underline{K}$-algebras A'_i over the semilattice I in such a way that A and I are disjoint subalgebras of A'.

Proof. For each i in I, take the free product A'_i in $\underline{K}$ of the algebra A_i and the one-element algebra $\{0_i\}$, where 0_i does not lie in A'_i. By Exercise 3.3L, both A_i and $\{0_i\}$ can be considered as (disjoint) subalgebras of A'_i, and the algebra A'_i is generated by the sets A_i and $\{0_i\}$. Denote by $\varphi_{i,j}$ the Płonka homomorphism $A_i \to A_j$ for $i \leq j$, and define $0_i \varphi_{i,j} = 0_j$. Using the universality property for coproducts, extend the mapping $\varphi_{i,j}$ from the disjoint union of A_i and $\{0_i\}$ to a uniquely defined homomorphism $\overline{\varphi}_{i,j} : A'_i \to A'_j$. It is easy to see that these homomorphisms satisfy the functoriality condition, and that the Płonka sum A' they define satisfies the conditions of the proposition. In particular, $\{0_i \mid i \in I\}$ is a subalgebra of A' isomorphic with I. □

In particular, Proposition 7.8.3 holds if $\underline{K}$ is the quasivariety of cancellative Ω-modes, and if $\underline{K}$ is the variety of affine spaces over R.

Theorem 7.8.4. *Let a mode (B, Ω) be a semilattice Lallement sum of cancellative modes (B_i, Ω) over a semilattice (I, Ω). Then (B, Ω) is a subreduct of a Płonka sum of affine $R(M\tau)$-spaces.*

Proof. First embed (B, Ω) into the Płonka sum (E, Ω) as in the proof of Proposition 7.8.2. Then using Proposition 7.8.3, embed (E, Ω) into a Płonka sum (E', Ω) of cancellative modes (E_i, Ω) over I, containing copies of (B, Ω) and (I, Ω) as disjoint subalgebras. Extend the ordered set I to an order on the disjoint union I_0 of I and $\{0\}$ by setting $0 < i$ for i in I. Then define $E'_0 := \{0_0\}$ and $0_0 \varphi_{0i} := 0_i$ for i in I. Consider (E', Ω) as a Płonka sum of the (E'_i, Ω) over I_0. Then (E', Ω) satisfies the assumptions of Proposition 7.8.2, whence it embeds into the Ω-reduct of a Płonka sum of affine spaces over $R(M\tau)$. □

7. SUBREDUCTS OF AFFINE SPACES

Corollary 7.8.5. *Let a mode (B, Ω) be a semilattice Lallement sum of cancellative modes. Then (B, Ω) embeds as a subreduct into a semimodule over a ring.*

Proof. Keep notation as in the proof of Proposition 7.8.2. Note that each G_i can also be considered as a module $(G_i, +, R(M\tau))$, and that the mappings $\overline{\psi}_{j,k}$ are also (functorial) module homomorphisms. This shows that A can also be considered as a Płonka sum of the $R(M\tau)$-modules. By Exercise 4.3M, the algebra A is a semimodule over the ring $R(M\tau)$. The zero of this semimodule is the zero 0_0 of G_0, the fibre over the least element 0 of the semilattice I_0. □

Corollary 7.8.6.

(a) *Each barycentric algebra embeds as a subreduct into a semimodule.*

(b) *Each commutative binary mode embeds as a subreduct into a semimodule.* □

The most general solution of our embeddability problem can be provided using some methods of advanced category theory. For concepts not explained in this book, we refer the reader to our monograph [PMA].

Consider the quasivariety $C_l(\Omega)$ of cancellative τ-modes, and any quasivariety $\underline{\underline{Q}}$ of naturally quasi-ordered τ-modes. By Theorem 4.5.3, any $C_l(\Omega) \circ \underline{\underline{Q}}$-mode (B, Ω) is a Lallement sum of $C_l(\Omega)$-modes (B_i, Ω) over a $\underline{\underline{Q}}$-mode (I, Ω). By Theorem 7.4.2, the mode (B, Ω) embeds into a functorial sum (E, Ω) of cancellative envelopes (E_i, Ω) of (B_i, Ω) over (I, Ω) by certain homomorphisms $\varphi_{i,j}$. Denote the corresponding functor by Φ. Corollary 7.7.11 showed that a single cancellative τ-mode embeds into the Ω-reduct of an affine space over the ring $R = R(M\tau)$. The final theorem of this section shows that (B, Ω) embeds into a functorial sum of affine spaces over R.

Theorem 7.8.7. *Each $C_l(\Omega) \circ \underline{\underline{Q}}$-mode (B, Ω) embeds into a functorial sum (A, Ω) of Ω-reducts (A_i, Ω) of affine R-spaces over a $\underline{\underline{Q}}$-algebra (I, Ω).*

Proof. Consider the functor $U : \underline{R} \to \underline{M\tau}$ that maps an affine R-space $(A, P, \underline{R})$ to its Ω-reduct (A, Ω) with operations given by (7.1.2). This functor preserves underlying sets. By [PMA, Corollary IV 3.4.8], it thus has a left adjoint $F : \underline{M\tau} \to \underline{R}$. For each i in I, define the affine R-space A_i to be the image $A_i = E_i F$ of the cancellative envelope E_i under the functor F. Now by Corollary 7.7.11, there is an embedding $\iota_i : E_i \to G_i U$ of (E_i, Ω) into the Ω-reduct $G_i U = (G_i, \Omega)$ of an affine R-space $(G_i, P, \underline{R})$. Consider the component $\eta_i : E_i \to A_i U$ at E_i of the unit η of the adjunction between $U : \underline{R} \to \underline{M\tau}$ and $F : \underline{M\tau} \to \underline{R}$. Since η_i is initial in the comma category (E_i, U) [PMA, Theorem III 3.1.4], there is an $\underline{R}$-homomorphism $\theta : A_i \to G_i$ such that $\eta_i(\theta U) = z_i$. It follows that $\eta_i : E_i \to A_i U$ embeds E_i as an Ω-subreduct of the affine space $(A_i, P, \underline{R})$.

Now consider the composite functor $\Phi F : (I) \to \underline{R}$ and the corresponding functorial sum $A = \biguplus_{i \in I} A_i$. The functorial sum $E = \biguplus_{l \in I} E_i$ of Φ embeds into A via the disjoint union $\eta = \biguplus_{l \in I} \eta_i$. Then for each ω in Ω and elements e_{i_j} of E_j for $1 \leq i_j \leq m = \omega\tau$ and $i = i_1 \ldots i_m \omega$, one has

$$e_{i_1} \ldots e_{i_m} \omega \eta_i$$
$$= e_{i_1} \varphi_{i_1, i} \ldots e_{i_m} \varphi_{i_m, i} \omega \eta_i$$
$$= e_{i_1} \varphi_{i_1, i} \eta_i \ldots e_{i_m} \varphi_{i_m, i} \eta_i \omega$$
$$= e_{i_1} \eta_{i_1}(\varphi_{i_1, i} FU) \ldots e_{i_m} \eta_{i_m}(\varphi_{i_m, i} FU)$$
$$= e_{i_1} \eta_{i_1} \ldots e_{i_m} \eta_{i_m} \omega,$$

so that the embedding $\eta : E \to A$ is an Ω-homomorphism, as required. $\square$

Exercises

7.8A. Let the functors F and G be defined as in the proof of Lemma 7.8.1. Let H be a similar functor built for an algebra (E, Ω'). Finally let $h : C \to D$ and $h' : D \to E$ be Ω-homomorphisms. Show that $\overline{hh'} = \overline{h}\,\overline{h}'$.

7.8B. Show that $(A, +)$ as defined in the proof of Corollary 7.8.2 is indeed a commutative semigroup, and that the definition of $+$ restricted to $\overline{E}$ coincides with the definition of $+$ on the subsemigroup $(\overline{E}, +)$.

7.8C. Show that a Płonka sum of affine R-spaces embeds as a subreduct into a semimodule over the ring R.

7.8D. Find an example of a Płonka sum (A, Ω) of algebras (A_i, Ω) over an Ω-semilattice I such that (I, Ω) does not embed into (A, Ω).

Notes on Chapter 7.

Tensor or Kronecker products of varieties were considered in [P. Freyd 1966], p. 94. See also [MT] and [W.D. Neumann 1978]. For tensor products of rings, see [N. Jacobson 1964]. The formula (7.1.1) can also be found in [P. Freyd 1966].

The affinization of a variety of modes was first introduced in [A. Romanowska and J.D.H. Smith 1991a] under the name "linearization", and used to describe the "linearization" of the variety of differential modes. The affinizations of subvarieties of the variety $\underline{Dm}$ were described in [J. Krauze 1998]. The name was changed to "affinization" and discussed again in a more general setting in [A. Romanowska and J.D.H. Smith 1993]. Also, in this paper, the concept of a minimal cogenerator of a variety of modules was adapted to varieties of affine spaces, the concepts of finite separability and separability were introduced, and Theorem 7.3.1 proved. The proof of Theorem 7.2.3 was modelled on the proof of Theorem 261 of [MT]. The minimal cogenerator for the variety of real barycentric algebras was identified in [J. Flood 1981], while minimal cogenerators for the variety of barycentric algebras over subfields of $\mathbb{R}$ and for the variety $\underline{CQM}$ of commutative quasigroup modes were determined in [A. Romanowska 2001]. Real convex sets were characterized among barycentric algebras as finitely separable in [MT]. Also Theorem 7.2.5 appeared for the first time in [MT]. It has a precursor in Blaschke's [1956, p.111] rather vague description of AK as a "höhere konvexe Gesamtheit". Subdirectly irreducible quasi-affine modes, together with Examples 7.2.9 and 7.2.10, were described by A. Romanowska [2001].

Separable modes, and a precursor of Theorem 7.3.1, were discussed in [A. Romanowska and J.D.H. Smith 1993]. Some consequences of this theorem were discussed in [A. Romanowska 2001].

Subreducts of functorial sums of cancellative modes in general were studied in [A. Romanowska and S. Traina 1999], following an earlier characterization given for barycentric algebras in [A. Romanowska and J.D.H. Smith 1990b], and for semilattice sums in [A. Romanowska and J.D.H. Smith 1991b]. Corollary 7.4.4 was formulated for the first time in [A. Romanowska 2001].

The description of barycentric algebras in Section 7.5 is based on results of A. Romanowska and J.D.H. Smith [1990b, 1991b, 1993]. Precursors for the decomposition of real barycentric algebras over a semilattice were also considered in [J. Flood 1981], [V.V. Ignatov 1986] and [D. Pumplün and H. Röhrl 1995]. In fact, as mentioned by the last two authors, a relation equivalent to the one we have used to decompose barycentric algebras was introduced by Gleason for function algebras, and investigated in [H.S. Bear 1965], [H.S. Bear and M.L. Weiss 1967], [H. Bauer 1970], and [H. Bauer and H.S. Bear 1969], for real convex sets. They called it the *Klein-Hilbert part relation*, because "the first one introduced it in a different form in hyperbolic geometry and the second one recognized that this notion can be generalized to convex sets". (See [H.S. Bear, 1991], [R. Börges and R. Kemper 1994].) Proposition 7.5.8 goes back to [V.V. Ignatov 1986].

Varieties of real barycentric algebras were described in [W.D. Neumann 1970], and quasivarieties in [V.V. Ignatov 1985], where the result of Theorem 7.6.2 was quoted together with a short sketch of a proof. Our proof differs from his by introducing and using the characterization of relatively subdirectly irreducible algebras. The algebra T appeared independently under different

7. SUBREDUCTS OF AFFINE SPACES

names in [J. Flood 1970], [V.V. Ignatov 1984, 1986], [A. Romanowska and J.D.H. Smith 1990b], while the algebras T_n featured in [V.V. Ignatov 1984, 1986].

The Embedding Theorems 7.7.1 and 7.8.7 were proved in full generality in [A. Romanowska and J.D.H. Smith 2001]. Theorem 7.7.1 has several predecessors concerning the embeddability of (cancellative) entropic groupoids into commutative semigroups equipped with commuting endomorphisms [K. Toyoda 1941], [D.C. Murdoch 1941b], [R.H. Bruck 1944], [S.K. Stein 1957], [T. Evans 1963], [H. Ratchek 1972], [R. Strecker 1974], [T. Kepka 1979], [V. Volenec 1981], [J. Ježek and T. Kepka 1975, 1981b], (for a slightly more general version, see [Jung R. Cho 1990b]), or into entropic quasigroups [M. Sholander 1949] and [J. Ježek and T. Kepka 1983], Theorem 5.3.1. Theorem 7.8.4 and its corollaries were proved by A. Romanowska and A. Zamojska [2001].

Let us also mention some other related results not discussed in this book. The "actions of Boolean rings" of [G.M. Bergman 1991] are binary reducts of affine spaces over Boolean rings. They have an interesting connection with sheaves on Stone spaces and Boolean powers. T. Stokes [1998a,b] investigated these algebras and their applications in computer science. (See also [E.G. Manes 1993].)

Subreducts of modules in general are discussed in [R. Quackenbush 1992]. In a number of papers, A. Szendrei [1975, 1976, 1977, 1979, 1981a,b, 1982a,b] investigated aspects of reducts of affine "modules" over not necessarily commutative rings, not discussed in this book. She was interested in reducing the number and arity of basic operations, characterizing idempotent reducts of modules by certain properties of clones, and determining bases for the identities satisfied in such algebras.

Finally, let us mention a trend in universal algebra outside the scope of this book concerning the problem of embedding diagonally normal or "abelian" algebras into algebras polynomially equivalent to modules [C. Hermann 1979], [H.P. Gumm 1983], [D. Hobby and R. McKenzie 1988], [K.A. Kearnes 1995a, 1996]. Most notably, C. Hermann proved that in a congruence modular variety every "abelian" algebra is polynomially equivalent to a module.

CHAPTER VIII

BINARY MODES

Chapters 5, 6 and 7 studied modes of arbitrary type. This chapter focusses on binary (or groupoid) modes, modes with a single, binary basic operation. In a sense, such modes can be considered as a prototype for more general modes.

The first section is devoted to the study of the combinatorics of binary mode words. The main result, Theorem 8.1.10, describes a standard (or "normal") form for such words. This description is then used in the next section to find further representations of free binary modes, and to find certain classes of modes generating the variety of all binary modes. The third section is devoted to the study of simple binary modes and minimal varieties. These are the familiar varieties of semilattices, left and right zero semigroups, and varieties equivalent to varieties of affine spaces over finite fields (with the exception of $GF(2)$). The next section studies *reductive varieties*, varieties of reductive modes, generalizing the n-reductive varieties of Section 5.7. They form one of the two classes of irregular varieties of binary modes appearing in the lower part of the lattice of varieties of such algebras. The second class is formed by *affine varieties*, varieties equivalent to varieties of affine spaces. The relationship between the two classes is discussed in Section 8.5. It is shown that reductive and affine varieties are independent of each other. This provides a description of a third class of irregular varieties, consisting of joins of affine and reductive varieties. Section 8.6 is devoted to the study of certain irregular varieties that lie beyond reductive and affine varieties. Examples of modes in such varieties are obtained by taking directed colimits of infinite chains of algebras, each in one of the varieties of an infinite chain of reductive or affine varieties. The section studies the structure of right cancellative binary modes and modes equivalent to right quasigroup modes, showing how they decompose over subreducts of affine spaces. Right quasigroups (and in particular the n-cyclic binary modes) are especially interesting, since they provide one of strongest invariants of knots.

8.1. Standard form for binary mode words

In this section, we will show that for each groupoid word w there is a certain standard groupoid word s such that the identity $w = s$ is satisfied in each binary mode. In other words, each binary mode word w can be expressed in a certain standard (or "normal") form.

To describe the standard words, we need a way of encoding subwords of a given word. Groupoid words are elements of the τ-word algebra $\mathfrak{P}\Omega$ (cf. Section 1.4) for the type $\tau : \Omega = \{\cdot\} \to \{2\} \subseteq \mathbb{N}$ containing one binary operation $\cdot$. The words are defined inductively as follows:

(8.1.1) $\begin{cases} (a) & \text{each variable } x_i \text{ is a groupoid word;} \\ (b) & \text{if } u \text{ and } v \text{ are groupoid words, then so is } u \cdot v. \end{cases}$

If a string u of consecutive symbols in a groupoid word w is itself a groupoid word, then u is called a *subword* of the word w. Let $S(w)$ be the set of all subwords of w. We will represent each groupoid word w by the (directed) graph T_w of a certain (join) semilattice defined on the set $S(w)$. Such a graph is sometimes called a *parsing tree*. It is an (inverted) rooted binary tree. Its *nodes* (or *vertices*) are labelled by subwords of the word w. In particular, its *root* (the greatest element of the corresponding semilattice) is labeled by the word w itself. The minimal nodes are called *leaves*, and are labelled by the arguments of w. Each node can be reached from the root by a *geodesic*, a

8. BINARY MODES

unique path containing a finite number of *branches* (*edges*). Each node, except for the leaves, has exactly two branches, called the *left* and *right* branches. If a node is labelled by a subword $u \cdot v$ of w, then the nodes at the ends of its left and right branches are labelled by u and v respectively. Each branch is also labelled, the left branches by 0 and the right branches by 1. Each geodesic from the root to a node (or subword u) is described by the concatenated sequence of branch labels along the geodesic. Such a sequence is called the *address* of the node (or subword) u. The *level* of a subword u is the length of its address. The *depth* $d(w)$ of a word w is the maximum level of its leaves. A word w (or its parsing tree T_w) is *full* if all its arguments (or leaves) have the same level.

Example 8.1.1. Consider the word $w = (x_3 x_5 \cdot x_4) \cdot x_5 x_3$. Its parsing tree T_w can be drawn as follows:

$$
\begin{array}{ccccccccc}
x_3 & \longrightarrow & x_3 x_5 & \longrightarrow & x_3 x_5 \cdot x_4 & \longrightarrow & w & \longleftarrow & x_5 \cdot x_3 & \longleftarrow & x_3 \\
 & 0 & \uparrow & 0 & \uparrow & 0 & & 1 & \uparrow & 1 \\
 & & |1 & & |1 & & & & 0| \\
 & & x_5 & & x_4 & & & & x_5
\end{array}
$$

In the next diagram, each subword of w in T_w is replaced by its address, yielding the *address tree* of w denoted by $T_{A(w)}$.

$$
\begin{array}{ccccccccc}
000 & \longrightarrow & 00 & \longrightarrow & 0 & \longrightarrow & \Lambda & \longleftarrow & 1 & \longleftarrow & 11 \\
 & & \uparrow & & \uparrow & & & & \uparrow \\
 & & | & & | & & & & | \\
 & & 001 & & 01 & & & & 10
\end{array}
$$

The subwords $x_3 x_5 \cdot x_4$ and $x_5 \cdot x_3$ have level 1, while the subwords $x_3 x_5$ and x_5 have level 2. The depth $d(w)$ of w equals 3. $\square$

For each groupoid word w, let $a(w)$ be its address. Let $A(w)$ denote the set of addresses $a(t)$ of all subwords t of w. For each natural number m, let $A_m(w)$ denote the subset of $A(w)$ consisting of all the addresses of length m. In particular, $A_0(w) = \{\Lambda\}$ just consists of the empty address Λ. Note that each groupoid word $w = x_{i_1} \ldots x_{i_n} w$ is determined by the set

$$\{(x_{i_j}, a(x_{i_j})) \mid j = 1, \ldots, n\}$$

of its leaves together with their addresses. This observation leads to the following representation of groupoid words.

Let $\{0,1\}^*$ be the free monoid over the set $\{0,1\}$. Let $\mathbb{N}\{0,1\}^*$ denote the free (additive) commutative semigroup over the set $\{0,1\}^*$. The multiplication in the monoid $\{0,1\}^*$ extends by distributivity to $\mathbb{N}\{0,1\}^*$, making it the free semiring (with commutative addition and a multiplicative identity) over $\{0,1\}$. [Compare this construction with the construction of the free commutative semiring given in Example 3.5.1(d).] The elements of $\mathbb{N}\{0,1\}^*$ may be considered as polynomials in non-commuting indeterminates 0 and 1 with natural number coefficients. Let $X\mathbb{N}\{0,1\}^*$ denote the free $\mathbb{N}\{0,1\}^*$-semimodule over a non-empty set X. A groupoid operation $\circ$ may be defined on the set $X\mathbb{N}\{0,1\}^*$ by

$$a \circ b = a0 + b1.$$

Under this groupoid operation, the subset X of $X\mathbb{N}\{0,1\}^*$ generates a subgroupoid $\langle X \rangle$ of the groupoid $(X\mathbb{N}\{0,1\}^*, \circ)$. Exercise 8.1A shows that the groupoid $\langle X \rangle$ is isomorphic to the absolutely free groupoid $X\mathrm{Gr}$, i.e. the free groupoid over X in the variety $\underline{\mathrm{Gr}}$ of groupoids. In particular, the groupoid $\langle \mathfrak{P} \rangle$ is isomorphic to $\mathfrak{P}\mathrm{Gr}$. Note that an element of the semimodule $X\mathbb{N}\{0,1\}^*$ can be written in the form

$$\sum_{i=1}^{r} x_i e_i$$

with (not necessarily distinct) variables x_i, and with each e_i in $\{0,1\}^*$ [Exercise 8.1B]. Let $T_{\{0,1\}^*}$ be the leafless binary rooted tree. Label the two branches growing from each node with 0 and 1 respectively. Label each node by the concatenated sequence of branch labels along the unique geodesic from the node to the root. Then $\sum_{i=1}^{r} x_i e_i$ determines a multiset $\langle e_i \mid 1 \le i \le r \rangle$ of nodes of the tree. It represents a groupoid word w if and only if the multiset is the set of leaves of a rooted binary tree, a subrooted tree of $T_{\{0,1\}^*}$. For example, $x_1 001 + x_1 01$ does not represent a groupoid word.

Example 8.1.2. For $X = \{x_1, x_2, x_3, x_4, x_5\}$, the word $w = (x_3 x_5 \cdot x_4) \cdot x_5 x_3$ of Example 8.1.1 can be written as the element $x_3 000 + x_5 001 + x_4 01 + x_5 10 + x_3 11$ of the semimodule $X \mathbb{N}\{0,1\}^*$. □

Lemma 8.1.3. *Let w be a groupoid word of depth d. Then for each $m \ge d$, there is a full word v of depth m such that the identity $w = v$ holds for all idempotent groupoids.* □

The obvious proof is left as an exercise.

Example 8.1.4. Consider again the word w of Example 8.1.1. Then the variety <u>IGr</u> of idempotent groupoids satisfies the identity

$$w = (x_3 x_5 \cdot x_4) \cdot x_5 x_3 = (x_3 x_5 \cdot x_4 x_4) \cdot (x_5 x_5 \cdot x_3 x_3) = w'.$$

The parsing tree of w' has the following form

$$
\begin{array}{ccccccccc}
 & & x_5 & & & & x_4 & & \\
 & & \downarrow 1 & & & & \downarrow 0 & & \\
 & 0 & & 0 & & 1 & & 1 & \\
x_3 & \longrightarrow & x_3 x_5 & \longrightarrow & x_3 x_5 \cdot x_4 x_4 & \longleftarrow & x_4 x_4 & \longleftarrow & x_4 \\
 & & & & \downarrow 0 & & & & \\
 & & & & w' & & & & \\
 & & & & \uparrow 1 & & & & \\
 & 0 & & 0 & & 1 & & 1 & \\
x_5 & \longrightarrow & x_5 x_5 & \longrightarrow & x_5 x_5 \cdot x_3 x_3 & \longleftarrow & x_3 x_3 & \longleftarrow & x_3 \\
 & & \uparrow 1 & & & & \uparrow 0 & & \\
 & & x_5 & & & & x_3 & & \\
\end{array}
$$

The full word w' is represented as an element of the semimodule $X \mathbb{N}\{0,1\}^*$ in the form

$$x_3 000 + x_5 001 + x_4 010 + x_4 011 + x_5 100 + x_5 101 + x_3 110 + x_3 111.$$

Note that the set $A_3(w')$ consists of all 0,1-sequences of length 3, and has 2^3 elements.

For $4 > d(w) = 3$, the corresponding full word of depth 4 has the form

$$((x_3x_3 \cdot x_5x_5) \cdot (x_4x_4 \cdot x_4x_4)) \cdot ((x_5x_5 \cdot x_5x_5) \cdot (x_3x_3 \cdot x_3x_3)). \quad \square$$

The next step is to look for groupoid words that are synonymous in the variety BM of all binary modes. Since BM is an idempotent variety, each groupoid word w is synonymous in BM with a full word w' of the same depth. We thus restrict attention to full groupoid words. Note that the address tree $T_{A(w)}$ of a full groupoid word w is a subrooted tree of the leafless rooted binary tree $T_{\{0,1\}^*}$ with nodes labelled by the elements of the free monoid $\{0,1\}^*$.

First we collect some basic properties of groupoid words and their addresses. For each word w and address $\alpha \in A(w)$, let w_α denote the subword of w at the address α. If β is another address in $A(w)$, then $\varepsilon_{\alpha\beta}(w)$ denotes the word obtained from w by interchanging the subwords w_α and w_β.

Lemma 8.1.5. *Let u, v, w and z be groupoid words. Let α and β be elements of the free monoid $\{0,1\}^*$, and let m be a positive integer. Then the following holds.*

(a) $0\alpha \in A_{m+1}(uv)$ if and only if $\alpha \in A_m(u)$.

(b) $1\beta \in A_{m+1}(uv)$ if and only if $\beta \in A_m(v)$.

(c) *If $\alpha, \beta \in A_m(u)$, then*

$$\varepsilon_{\alpha,\beta}(u)v = \varepsilon_{0\alpha,0\beta}(uv)$$

and

$$v\varepsilon_{\alpha,\beta}(u) = \varepsilon_{1\alpha,1\beta}(vu).$$

(d) *If $i, j, k, \ell \in \{0,1\}$ and $ij\alpha, k\ell\beta \in A_m(uv \cdot wz)$, then the identity*

(8.1.2) $$\varepsilon_{ij\alpha,k\ell\beta}(uv \cdot wz) = \varepsilon_{ji\alpha,\ell k\beta}(uw \cdot vz)$$

holds in all entropic groupoids.

Proof. (a) If w is a subword of u, then the address of w in uv is simply the address of w in u preceded by 0. Conversely, the address $a(w)$ in $A(u)$ is obtained from the address $a(w)$ in $A(uv)$ by removing the first digit.

(b) is analogous to (a).

(c) Let s and t be subwords of u at the same level. Then s and t are subwords of uv. If w is the word obtained from u by interchanging s and t, then wv is the word obtained from uv by interchanging s and t. Hence (a) and (b) imply (c).

(d) First note that if s is a subword of any of u, v, w or z, and similarly t is a subword of u, v, w or z, then s and t are subwords of $uv \cdot wz$ and $uw \cdot vz$. Moreover, in entropic groupoids $uv \cdot wz = uw \cdot vz$. Next note that if $ij\alpha \in A(uv \cdot wz)$, then by (a) and (b), $\alpha \in A(t)$, where $t \in \{u, v, w, z\}$. A similar remark holds for $kl\beta$. If s and t are both subwords of the same word among u, v, w, z, or if s has an address of the form $ii\alpha$ and t has an address of the form $(1-i)(1-i)\beta$, then obviously (8.1.2) holds in each entropic groupoid. To prove that (8.1.2) holds in the remaining cases, it suffices to note that if the address of v in $uv \cdot wz$ is 01α then the address of v in $uw \cdot vz$ is 10α. Similarly, if the address of w in $uv \cdot wz$ is 10β, then the address of w in $uw \cdot vz$ is 01β. Let u' be obtained from u by replacing u_α with v_β. Let v' be obtained from v by replacing v_β with u_α. Then $\varepsilon_{0\alpha,1\beta}(uv) = u'v'$, and by (c), one has

$$\varepsilon_{00\alpha,01\beta}(uv \cdot wz) = \varepsilon_{0\alpha,1\beta}(uv) \cdot wz = u'v' \cdot wz = u'w \cdot v'z = \varepsilon_{00\alpha,10\beta}(uw \cdot vz). \quad \square$$

Proposition 8.1.6. *Let w be a full groupoid word of depth d. Assume that $\alpha, \beta \in A_d(w)$ and both have the same number of 1's. Then the identity*

$$w = \varepsilon_{\alpha,\beta}(w)$$

holds in the variety <u>EGr</u> *of entropic groupoids.*

Proof. The proof is by induction on the depth d of full groupoid words. Each full word of depth 2 has the form $xy \cdot zt$. The only addresses in $A_2(w)$ with the same number of 1's are $a(y) = 01$ and $a(z) = 10$. Then $\varepsilon_{01,10}(w) = \varepsilon_{10,01}(w) = xz \cdot yt$, and by the entropic law

$$w = xy \cdot zt = xz \cdot yt = \varepsilon_{01,10}(w) = \varepsilon_{10,01}(w).$$

Now assume that the proposition holds for all full words of depth not greater than d, and consider a word w of depth $d+1$. Note that for $m \leq d+1$, the m-th level of the address tree $T_{A(w)}$ of w consists of all elements of $A_m(w)$ (the addresses of length m), and has 2^m elements. Suppose that the addresses in $A_m(w)$ are ordered in sequence in the same way as the corresponding subwords of w from left to right. Finally, let 1_m denote the sequence of numbers of 1's in the consecutive addresses of the sequence $A_m(w)$. Note that the word w can be written as $w = uv \cdot tz$, or more explicitly as

(8.1.3)
$$(x_1 \ldots x_{2^{d-1}} u \cdot x_{2^{d-1}+1} \ldots x_{2^d} v) \cdot (x_{2^d+1} \ldots x_{2^d+2^{d-1}} t \cdot x_{2^d+2^{d-1}+1} \ldots x_{2^{d+1}} z).$$

Obviously, each address $\alpha \in A_m(w)$ grows two branches with nodes 0α and 1α. Hence 0α and α have the same number of 1's, whereas 1α has one more 1 than α. It follows that for $i = 1, \ldots, 2^m$ (and in particular for $m = d$) one has

(8.1.4a) $$\begin{cases} 1_m(i) = 1_{m+1}(i), \\ 1_{m+1}(i) + 1 = 1_{m+1}(i + 2^m), \end{cases}$$

and for $i = 2^{m-1} + 1, \ldots, 2^m$, one has

(8.1.4b) $$1_{m+1}(i) = 1_{m+1}(i + 2^{m-1}).$$

The depth $d(uv)$ of uv and the depth $d(tz)$ of tz are both equal to d. By the induction hypothesis, for any addresses $\alpha, \beta \in A_d(uv)$, and similarly $\gamma, \delta \in A_d(tz)$, with the same number of 1's, the identities

$$uv = \varepsilon_{\alpha,\beta}(uv) \text{ and } tz = \varepsilon_{\gamma,\delta}(tz)$$

8. BINARY MODES

hold in the variety EGr. Hence by Lemma 8.1.5, the following identities also hold in EGr:

(8.1.5) $$uv \cdot tz = \varepsilon_{0\alpha,0\beta}(uv \cdot tz) = \varepsilon_{1\gamma,1\delta}(uv \cdot tz).$$

Evidently 0α and 0β have the same number of 1's precisely when α and β have. Similar, 1γ and 1δ have the same number of 1's precisely when γ and δ have. As the addresses of the variables of uv in w are all written in the form 0α, where α can be $0\alpha'$ or $1\alpha''$, the addresses of variables of u can be written as $00\alpha'$, and those of v as $01\alpha''$. A similar remark holds for 0β with the variables of u written as $00\beta'$ and of v as $01\beta''$. For the word tz, the addresses of the corresponding variables in t and z are written as $10\gamma'$ and $11\gamma''$, and as $10\delta'$ and $11\delta''$, respectively. Lemma 8.1.5(d) also shows that the identities

$$\varepsilon_{01\alpha'',10\delta}(uv \cdot tz) = \varepsilon_{10\alpha'',01\delta}(ut \cdot vz)$$

hold in EGr.

Note that (8.1.5) takes care of comparisons of addresses with the same number of 1's for variables belonging to uv and for variables belonging to tz. To compare the remaining addresses of variables in w, first note that by (8.1.4), the consecutive addresses of variables in t have the same number of 1's as the consecutive addresses of variables in v. Choose any variable x_{2^d+k}, for $k \leq 2^{d-1}$, with an address 10φ. Use the entropic law to obtain the word $ut \cdot vz$ from $uv \cdot tz$. Then the address $a(x_{2^d+k})$ in $ut \cdot vz$ is 01φ. By (8.1.5),

$$ut \cdot vz = \varepsilon_{00\psi,01\varphi}(ut \cdot vz)$$

holds in EGr for any address 00ψ of a variable in u having the same number of 1's as 01φ. Using the entropic law and (8.1.5) again one obtains the identity

(8.1.6) $$uv \cdot tz = \varepsilon_{00\psi,10\varphi}(uv \cdot tz)$$

true in EGr. Similar reasoning also shows that identities of the form

(8.1.7) $$uv \cdot tz = \varepsilon_{10\psi, 11\varphi}(uv \cdot tz)$$

hold for appropriate φ and ψ. Now (8.1.3)–(8.1.7) together show that the proposition holds. $\square$

For a permutation σ of $A_d(w)$, let $\sigma(w)$ be the word obtained from w by replacing w_α with $w_{\sigma(\alpha)}$ for all α in $A_d(w)$.

Corollary 8.1.7. *Let w be a full groupoid word of depth d. Let σ be a permutation of $A_d(w)$ such that α and $\sigma(\alpha)$ have the same number of 1's for $\alpha \in A_d(w)$. Then the identity*

$$w = \sigma(w)$$

holds in the variety EGr. $\square$

For a full groupoid word w of depth d, a variable x and a natural number k with $0 \leq k \leq d$, let $N_k(w, x)$ denote the number of x's in w with addresses in $A_d(w)$ having precisely k ones.

Corollary 8.1.8. *Let w and t be full groupoid words of depth d with (not necessarily distinct) variables $x_1, \ldots, x_{2^d}$. Let k be a natural number with $0 \leq k \leq d$. If $N_k(w, x_i) = N_k(t, x_i)$ for each $i = 1, \ldots, 2^d$, then the identity*

$$w = t$$

holds in the variety BM. $\square$

Let W_d be the set of full groupoid words w of depth d such that, if two variables of w have addresses with the same number of ones, then the variable on the left has index not greater than the index of the variable on the right. The elements of W_d are called *standard binary mode words* (of depth d).

Corollary 8.1.9. *Let w and t be words in W_d. Consider a natural number k with $0 \le k \le d$. If $N_k(w, x_i) = N_k(t, x_i)$ for all $i = 1, \ldots, 2^d$, then w and t coincide.* □

Now we are ready to describe the standard form for binary mode words.

Theorem 8.1.10. *Let w be a groupoid word of depth d. Then there exists a standard binary mode word s in W_d such that the identity*

$$w = s$$

holds in the variety $\underline{BM}$ of all binary modes.

Proof. By Lemma 8.1.3, there is a full word w' of depth d such that the identity

(8.1.8) $$w = w'$$

holds in $\underline{BM}$. Order the addresses of $A_d(w')$ as in the proof of Proposition 8.1.6. Define two addresses α and β of $A_d(w')$ to be related by θ iff they have the same number of ones. The relation θ is an equivalence relation on $A_\alpha(w')$. For each θ-class C, let σ_c be the permutation of $A_d(w')$ such that $\sigma_c(\alpha) = \alpha$ for $\alpha \in A_d(w') - C$, and otherwise, if $\alpha, \beta \in C$ with $\alpha < \beta$, then the variable at the address $\sigma_c(\alpha)$ has index not greater than that of the variable at the address $\sigma_c(\beta)$. Let σ be the product of all these permutations σ_c. Then $\sigma(w')$ is in W_d, and for all $\alpha \in A_d(w')$ the addresses α and $\sigma(\alpha)$ have the same number of ones. Define s to be $\sigma(w')$. Then by Corollary 8.1.7 and (8.1.8),

$$w = w' = \sigma(w') = s$$

is an identity holding in $\underline{BM}$. □

Example 8.1.11. Consider the word $w = (x_3 x_5 \cdot x_4) \cdot x_5 x_3$ of Examples 8.1.1 and 8.1.4. The third level of the address tree $T_{A(w')}$ forms the sequence

$(000, 001, 010, 011, 100, 101, 110, 111)$. The relation σ of Theorem 8.1.6 partitions $A_3(w')$ into four classes $\{000\}, \{001, 010, 100\}, \{011, 101, 110\}, \{111\}$. The standard word $s = \sigma(w')$ has the form

$$s = (x_3 x_4 \cdot x_5 x_3)(x_5 x_4 \cdot x_5 x_3),$$

and obviously $w = s$ is an identity true in $\underline{\mathrm{BM}}$. □

Let $s_d := s$ be the standard word obtained from a word w of depth d as in Theorem 8.1.10. Using the idempotent law, one can easily find other standard words s_m, of any depth $m > d$, so that the identities $w = s_m = s_d$ hold in $\underline{\mathrm{BM}}$. We call all such words s_m *standard* (binary mode) *forms* of the word w.

Corollary 8.1.12. *Let w and t be groupoid words. Then the identity $w = t$ is satisfied in the variety $\underline{\mathrm{BM}}$ if and only if there is a natural number d such that the words w and t have the same standard form of depth d.* □

The next corollary follows directly from the results of Section 3.5.

Corollary 8.1.13. *In the free binary mode XBM over a non-empty set X, each further element may be expressed as a standard binary mode word of minimal depth.* □

Exercises.

8.1A. Show that the subgroupoid $\langle X \rangle$ of $(X\mathbb{N}\{0,1\}^*, \circ)$ generated by a non-empty set X is isomorphic to the absolutely free groupoid $X\Omega$ for $\Omega = \{\cdot\}$ consisting of a unique binary multiplication.

8.1B. Show that the semimodule $X\mathbb{N}\{0,1\}^*$ is indeed the free semimodule over X, and that each of its elements may be written as $\sum_{i=1}^{r} x_i e_i$ for $x_i \in X$ and $e_i \in \{0,1\}^*$.

8.1C. Consider the full word (8.1.3). Show that the addresses of $x_{2^d+1}, \ldots, x_{2^{d+1}}$ are the numbers $2^d, \ldots, 2^{d+1} - 1$ written in the binary system, whereas the addresses of $x_1, \ldots, x_{2^d}$ are the numbers $0, \ldots, 2^d - 1$ written in the binary system with additional zeroes at the beginning.

8.1D. Find the standard binary mode form of the word

$$w = x_5 x_2 \cdot [(x_2 x_3 \cdot x_1) \cdot x_2] \quad .$$

8.2. Generators of the variety $\underline{BM}$

The description of standard binary mode words obtained in the previous section will now be used to find further representations of free binary modes, and to find certain classes of modes generating the variety $\underline{BM}$.

Let $(G, \cdot)$ be a binary mode. Assume that $(G, \cdot)$ is a subreduct of an affine space, i.e. there is an affine R-space $(A, P, \underline{R})$ and r in R such that $(G, \cdot) = (G, \underline{r}) \leq (A, \underline{r})$. By Example 7.1.4, the ring $R(BM)$ of the affinization of $\underline{BM}$ is (isomorphic to) the (commutative) ring $\mathbb{Z}[X_1, X_2]/\langle 1 - X_1 - X_2 \rangle \cong \mathbb{Z}[M]$. It follows that we may always take $R = \mathbb{Z}[M]$ as the ring R of $(A, P, \underline{R})$, and the indeterminate M as the element r. Note however that in the case where $(G, \cdot)$ is a member of a proper subvariety $\underline{V}$ of $\underline{BM}$, the ring $R(V)$ of the affinization $R(V)$ of $\underline{V}$ is a quotient $\mathbb{Z}[M]/I$ of $\mathbb{Z}[M]$. Then r is just a coset $M + I$, later usually identified with M.

Now each full groupoid word w_d of depth d determines the corresponding derived operation on $(G, \cdot)$ that can be described as in the proof of Theorem 6.4.1. Since

(8.2.1) $\quad \begin{cases} x_1 x_2 w_1 = x_1 x_2, \\ x_1 \ldots x_{2^{d+1}} w_{d+1} = x_1 \ldots x_{2^d} W_d \cdot x_{2^d+1} \ldots x_{2^{d+1}} w_d \end{cases}$

and

(8.2.2) $\quad x_1 x_2 w_1 = x_1 x_2 \underline{r} = x_1(1 - r) + x_2 r,$

it follows that

(8.2.3) $$x_1 \ldots x_{2^d} w_d = \sum_{k=0}^{d}(x_{j_1} + \cdots + x_{j_s})r^k(1-r)^{d-k},$$

where $s = \binom{d}{k}$.

As described in Section 8.1, the groupoid words can be represented as elements $\sum_{k=1}^{r} x_i e_i$ of the free semimodule $X\mathbb{N}\{0,1\}^*$ over the free semiring $\mathbb{N}\{0,1\}^*$. Note that the universality property for free semirings furnishes a unique semiring homomorphism

$$f : \mathbb{N}\{0,1\}^* \to \mathbb{Z}[M]$$

extending the mapping $0 \mapsto 1 - M$ and $1 \mapsto M$. It follows that the free $\mathbb{Z}[M]$-module $X\mathbb{Z}[M]$ over X may equivalently be considered as a semimodule over the semiring $\mathbb{N}\{0,1\}^*$. Now the universality property for free semimodules over the semiring $\mathbb{N}\{0,1\}^*$ furnishes a unique semimodule homomorphism

$$h' : X\mathbb{N}\{0,1\}^* \to X\mathbb{Z}[M]$$

extending the identity mapping on X. Define a groupoid operation $\circ$ on $X\mathbb{N}\{0,1\}^*$ as in Section 8.1, i.e.

$$a \circ b := a0 + b1,$$

and on $X\mathbb{Z}[M]$ by

$$a \circ b := a(1 - M) + bM.$$

Then obviously h' is also a homomorphism between the corresponding groupoids $(X\mathbb{N}\{0,1\}^*, \circ)$ and $(X\mathbb{Z}[M], \circ)$. Let $\langle X \rangle_{\mathbb{N}\{0,1\}^*}$ and $\langle X \rangle_{\mathbb{Z}[M]}$ be the subgroupoids generated by X in $X\mathbb{N}\{0,1\}^*$ and $X\mathbb{Z}[M]$, respectively. Then the restriction

$$h := h'|_{\langle X \rangle_{\mathbb{N}\{0,1\}^*}} : \langle X \rangle_{\mathbb{N}\{0,1\}^*} \to \langle X \rangle_{\mathbb{Z}[M]}$$

8. BINARY MODES

is a surjective groupoid homomorphism. Note that if two words $w = \sum_{i=1}^{r} x_i e_i$ and $t = \sum_{j=1}^{s} x_j e_j$ in $\langle X \rangle_{\mathbb{N}\{0,1\}}$ have the same standard form, then $wh = th$. It follows that the congruence $\ker h$ identifies words having the same standard form. Let w represent a standard binary mode word w_d of depth d. Write w as

$$\sum_{i=1}^{2^d} x_i e_i.$$

If a variable x_i has an address e_i with k instances of 1, then in wh the variable $x_i = x_i h$ has the coefficient $(1-M)^{d-k} M^k$. It follows that the operation (8.2.3) on $\langle X \rangle_{\mathbb{Z}[M]}$ can be rewritten as

$$x_1 \ldots x_{2^d} w_d$$

(8.2.4)
$$= x_1 \left(\sum_{k=0}^{d} N_k(w_d, x_1)(1-M)^{d-k} M^k \right)$$
$$+ \cdots +$$
$$x_{2^d} \left(\sum_{k=0}^{d} N_k(w_d, x_{2^d})(1-M)^{d-k} M^k \right),$$

where $N_k(w_d, x_i) = 1$ if the variable x_i appears with an address with k instances of 1, and otherwise $N_k(w_d, x_i) = 0$. If not all the variables of w_d are distinct, and $w_d = w$ with $w = x_1 \ldots x_j w$ for $j < 2^d$, then (8.2.4) implies that

(8.2.5) $$w_d = x_1 \ldots x_j w = \sum_{i=1}^{j} x_i \left(\sum_{k=0}^{d} N_k(w, x_i)(1-M)^{d-k} M^k \right).$$

This proves the following.

Lemma 8.2.1. *Each full groupoid word w of depth d determines a derived operation on the groupoid $\langle X \rangle_{\mathbb{Z}[M]}$ that may be written as (8.2.5).* □

Call the expression on the right hand side of (8.2.5) the *polynomial form of the word w*, or merely a *polynomial word*. Let w_m, for $m > d$, be another standard form of the word w. Then idempotency of the operation $\underline{M}$ implies that w_m also has a polynomial form like (8.2.5), but with m replacing d. If w is a standard word of minimal depth d, then its polynomial form can also be written in an obvious standard form of minimal depth. Moreover, there is an obvious one-to-one correspondence between the set of standard binary mode words of minimal depths and the set of standard polymomial words of minimal depth. This implies the following.

Proposition 8.2.2. *The subgroupoid $\langle X \rangle_{\mathbb{Z}[M]}$ of the groupoid $(X\mathbb{Z}[M], \circ)$ is (isomorphic to) the free binary mode XBM over X.*

Proof. This follows directly by Corollary 8.1.13. □

Note also that by Proposition 5.3.1, the groupoid $\langle X \rangle_{\mathbb{Z}[M]}$ is a subreduct of the free affine $\mathbb{Z}[M]$-space $(X\mathbb{Z}[M], P, \underline{\mathbb{Z}[M]})$.

Corollary 8.2.3. *Let w and t be full groupoid words of depth d. Then the identity $w = t$ is satisfied in the groupoid $\langle \mathfrak{P} \rangle_{\mathbb{Z}[M]}$ if and only if*

$$(8.2.6) \qquad \sum_{k=0}^{d}[N_k(w, x_i) - N_k(t, x_i)](1-M)^{d-k}M^k = 0$$

for all x_i in $\mathfrak{P}$. □

Note that the condition (8.2.6) carries over to all binary modes $(G, \cdot) = (G, \underline{M})$ that are subreducts of affine R-spaces with $R = \mathbb{Z}[M]/I$.

Corollary 8.2.4. *The free binary mode XBM over X is cancellative.*

Proof. This follows by Exercise 5.3J and the freeness of the ring $\mathbb{Z}[M]$. □

Corollary 8.2.5. *Each binary mode is a homomorphic image of a cancellative binary mode. In particular, the variety $\underline{BM}$ is generated by the quasivariety of cancellative binary modes.* □

It is evident (cf. Section 3.5) that the variety $\underline{BM}$ is generated by the free binary mode $\mathfrak{P}BM$. We will look now for other generators of the variety $\underline{BM}$.

Consider the rings $\mathbb{Z}_p$ for prime numbers $p \neq 2$, and the binary modes $(\mathbb{Z}_p, \cdot) = (\mathbb{Z}_p, \underline{r})$, where r is any element of $\mathbb{Z}_p$ different from 0 and 1. Let $\mathfrak{F}_{pr}$ be the set of all groupoids $(\mathbb{Z}_p, \cdot)$ for prime numbers $p \neq 2$, with arbitrary $r \neq 0, 1$ in $\mathbb{Z}_p$. It is clear that the set $\mathrm{Id}(\mathfrak{F}_{pr})$ of identities true in $\mathfrak{F}_{pr}$ contains the set $\mathrm{Id}(\underline{BM})$ of identities true in the variety $\underline{BM}$. We will show that the converse inclusion holds as well.

Lemma 8.2.6. *Let $p > m > 2$ be a prime number, and let $a_0, a_1, \ldots, a_m$ be integers not all equal to $0 \pmod p$. Then the polynomial*

$$P(x) = \sum_{j=0}^{m} a_j (1-x)^{m-j} x^j$$

is a non-trivial polynomial of degree m with coefficients in $\mathbb{Z}_p$, and hence has at most m roots in $\mathbb{Z}_p$.

Proof. Let k be the smallest integer such that $a_k \neq 0$. Then

$$P(x) = x^k Q(x),$$

where

$$Q(x) = \sum_{i=0}^{m-k} a_{k+i} (1-x)^{m-k-i} x^i.$$

As $Q(0) = a_k \neq 0$, the polynomial $Q(x)$ is non-trivial. Thus the polynomial $P(x)$ is non-trivial, too. $\square$

Lemma 8.2.7. *Let w and t be full groupoid words of depth d. Then the identity $w = t$ holds in $\mathfrak{F}_{pr}$ precisely if*

(8.2.7) $$N_k(w, x_i) = N_k(t, x_i)$$

for all natural k and x_i in $\mathfrak{P}$.

Proof. First note that all $\mathbb{Z}_p$, for $p > 2$, are quotients of $\mathbb{Z}[M]$. Then by Corollary 8.2.3, the identity $w = t$ is satisfied in $\mathfrak{F}_{pr}$ if and only if the condition

(8.2.6) holds. Hence the equality (8.2.7) trivially implies that $w = t$ holds in $\mathfrak{F}_{pr}$.

Now suppose that the identity $w = t$ is true in $\mathfrak{F}_{pr}$, but $N_k(w,x) \neq N_k(t,x)$ for some k and $x = x_i$ in $\mathfrak{P}$. Take $p > 2^d + 2$. Then $N_k(w,x) \neq N_k(t,x)$ (mod p) since $N_k(w,x) < p$ and $N_k(t,x) < p$. Hence the polynomial

$$P(x) = \sum_{k=0}^{d}(N_k(w,x) - N_k(t,x))(1-2)^{d-j}z^j$$

is non-trivial, and by Lemma 8.2.6 has at most d roots in $\mathbb{Z}_p$. Since $p > 2^d + 2 \geq d + 2$, there is r in $\mathbb{Z}_p$ such that $r \neq 0, r \neq 1$ and $P(r) \neq 0$. This however means that the identity $w = t$ does not hold in $\mathfrak{F}_{pr}$, providing a contradiction. Thus (8.2.7) holds for all k and x_i in $\mathfrak{P}$. □

Theorem 8.2.8. *The variety* $V(\mathfrak{F}_{pr})$ *of binary modes generated by the set* $\mathfrak{F}_{pr}$ *and the variety* $\underline{BM}$ *of all binary modes coincide.*

Proof. We will show that the two varieties in question satisfy precisely the same sets of identities. We already know that $\mathrm{Id}(\underline{BM}) \subseteq \mathrm{Id}(\mathfrak{F}_{pr})$. So we only need to show that the converse inclusion holds.

Let $w = t$ be an identity true in $\mathfrak{F}_{pr}$. By Lemma 8.1.3, there are full words w' and t' of the same depth d such that the identities $w' = w = t = t'$ hold in $\mathfrak{F}_{pr}$. By Lemma 8.2.7, the equality $N_k(w', x_i) = N_k(t', x_i)$ holds for all x_i and all natural k. By Corollary 8.1.8, the identity $w' = t'$ holds in $\underline{BM}$. Since $w = w'$ and $t = t'$ hold in $\underline{BM}$, too, it follows that $t = w$ is an identity true in $\underline{BM}$. Consequently $\mathrm{Id}(\mathfrak{F}_{pr}) \subseteq \mathrm{Id}(\underline{BM})$, and the theorem holds. □

Corollary 8.2.9. *The variety* $\underline{BM}$ *of binary modes is generated by the set* $\mathfrak{F}_{pr}$. □

Exercises.

8.2A. Show that the free binary mode on $n+1$ generators is isomorphic to the subgroupoid of the groupoid $(\mathbb{Z}[M]^n, \cdot) = (\mathbb{Z}[M]^n, \underline{M})$ generated by the (canonical) basis of the corresponding $\mathbb{Z}[M]$-module $(\mathbb{Z}[M]^n, +, \mathbb{Z}[M])$. [Hint: Note that $\mathbb{Z}[M]^n$ is a free $\mathbb{Z}[M]$-module on n generators, and at the same time a free affine $\mathbb{Z}[M]$-space on $n+1$ generators.]

8.2B. Deduce from 8.2A that the subgroupoid of $(\mathbb{Z}[M], \underline{M})$ generated by 0 and 1 is the free binary mode on two generators.

8.2C. [Faitlowicz, Mycielski 1974] Prove that the variety $\underline{BM}$ is generated by the binary mode $(\mathbb{R}, \cdot) = (\mathbb{R}, \underline{r})$ if and only if r is transcendental.
[Hint: Show that if r is an algebraic number, then for all n large enough there are integers $a_0, \ldots, a_n$ not all equal to 0 such that

$$\sum_{k=0}^{n} a_k r^k (1-r)^{n-k} = 0 \text{ and } |a_k| \leq \binom{n}{k}$$

for $k = 0, \ldots, n$. Then use the method of the proof of Theorem 8.2.8.]

8.3. Simple binary modes and minimal varieties

In this section, we will describe simple binary modes and determine the minimal varieties of binary modes.

We already know that the 2-element semilattice, 2-element left zero band and 2-element right zero band are simple. Each of them generates a minimal variety of binary modes. These are respectively the variety $\underline{Sl}$ of semilattices, the variety $\underline{Lz}$ of left zero bands and the variety $\underline{Rz}$ of right zero bands. (Cf. Exercise 4.3C and Section 3.5.) The next series of simple binary modes is given by the set $\mathfrak{F}_{pr}$ of groupoids $(\mathbb{Z}_p, \underline{r})$ considered in Section 8.2. Since they are equivalent to the corresponding simple affine spaces $(\mathbb{Z}_p, \mathbb{Z}_p)$, they are simple, too. A similar argument shows more generally that the binary mode reducts $(GF(p^n), \underline{r})$ of affine spaces $(GP(p^n), \underline{GF(p^n)})$ for $p^n \geq 3$ are simple for generators r of the ring $GF(p^n)$ [Exercise 8.3B]. Let $\mathfrak{F}$ be the set of all such groupoids. We will show that the binary modes described above are in fact all the simple binary modes.

Let $(G,\cdot)$ be a non-trivial binary mode. Define the following subsets of G:

(8.3.1)
$$\begin{cases} D_l := \{x \in G \mid xa = xb \text{ for all } a,b \in G\}, \\ D_r := \{x \in G \mid ax = bx \text{ for all } a,b \in G\}, \\ C_l := \{x \in G \mid xa = xb \to a = b \text{ for all } a,b \in G\}, \\ C_r := \{x \in G \mid ax = bx \to a = b \text{ for all } a,b \in G\}. \end{cases}$$

Note that always

(8.3.2)
$$C_l \cap D_l = \varnothing.$$

Indeed, if $a \in C_l \cap D_l$, then the definition of D_l implies that $ac = ad$ for all $c,d \in G$. Then the definition of C_l implies that $c = d$, contradicting the non-triviality of G. A similar argument shows that

(8.3.3)
$$C_r \cap D_r = \varnothing.$$

If the intersection $D_l \cap D_r$ is nonempty, say $x \in D_l \cap D_r$, then

(8.3.4)
$$D_l \cap D_r = \{x\},$$

and $\{x\}$ is a sink in $(G,\cdot)$, whence x is a zero of $(G,\cdot)$. Indeed, for each a in G one has

$$xa = xx = x = xx = ax.$$

Moreover, if y is another element of $D_l \cap D_r$, then

$$y = yx = xy = x,$$

implying that $x = y$. Note also that any of the sets D_l, D_r, C_l, C_r may be empty.

Lemma 8.3.1. *Let $(G,\cdot)$ be a (non-trivial) binary mode. Assume that $G = C_l \cup D_l = C_r \cup D_r$. Then one of the following cases occurs:*

(a) $(G,\cdot)$ *is a left zero band;*

(b) $(G,\cdot)$ *is a right zero band;*

(c) $(G,\cdot)$ *is cancellative;*

(d) $(G,\cdot)$ *is a cancellative mode extended by zero.*

8. BINARY MODES

Proof. We have to consider several cases depending on which of the sets (8.3.1) are empty.

First suppose that $C_l = D_r = \emptyset$. Then $G = D_l = C_r$, whence for all a and b in G, one has $ab = aa = a$. This shows that $(G, \cdot)$ is a left zero band.

In a similar fashion one may show that if $C_r = D_l = \emptyset$, then $(G, \cdot)$ is a right zero band.

Now assume that $C_l = C_r = \emptyset$. Then $G = D_l = D_r$. Since $D_l \cap D_r = D_l = D_r$ is a sink consisting of a unique element, it follows that G is trivial, a contradiction.

If $D_l = D_r = \emptyset$, then $G = C_l = C_r$. This shows that $(G, \cdot)$ is cancellative.

The two remaining cases concern the situation where both D_l and C_l, or both D_r and C_r, are non-empty. As both these cases have similar proofs, and both imply that $(G, \cdot)$ is a cancellative mode extended by zero, it is sufficient to consider only one of them.

Assume that $D_l \neq \emptyset \neq C_l$. Suppose first that $G = C_r$. Let $a \in C_l$ and $b \in D_l$. If $ab \in D_l$ then
$$ab = ab \cdot ab = ab \cdot b.$$
Since $G = C_r$, it follows that $a = ab \in D_l$, a contradiction to (8.3.2). Hence $ab \in C_l$. Since $b \in D_l$, one has
$$ab \cdot a = a \cdot ba = a \cdot bb = ab = ab \cdot ab.$$
Since $ab \in C_l$, it follows that $a = aa = ab$. Since also $a \in C_l$, one obtains $a = b$, again a contradiction. Consequently G cannot coincide with C_r, and both C_r and D_r are non-empty. Moreover (8.3.3) shows that in fact $G = C_r \cup D_r$, the disjoint union of C_r and D_r. In particular, there is a zero x in G, and by (8.3.4), one has $D_l \cap D_r = \{x\}$. If v is another element of D_l, and $u \in C_r$, then
$$vu = vx = xx = x = xu$$

implies that $v = x$ and $D_l = \{x\}$. Since also $C_l \neq \varnothing$, a similar argument shows that $D_r = \{x\} = D_l$. Hence $x = b$.

Now if $c, d \in G - \{b\}$ and $cd \notin G - \{b\}$, then $cd = b = cb$. Since $c \in C_l$, this implies $b = d$, a contradiction. It follows that $cd \in G - \{b\}$, i.e. $G - \{b\} \leq G$ and $G - \{b\} = C_l = C_r$ is cancellative. □

Corollary 8.3.2. *Each simple binary mode* $(G, \cdot)$ *with at least three elements is cancellative.*

Proof. For each $x \in G$ define a binary relation γ_x on G by

$$(a, b) \in \gamma_x \text{ iff } xa = xb.$$

Obviously γ_x is an equivalence relation. Since $xa = xb$ and $xa' = xb'$ imply that

$$x \cdot aa' = xa \cdot xa' = xb \cdot xb' = x \cdot bb',$$

it follows that γ_x is a congruence relation. As G is simple, each relation γ_x is the universal congruence G^2 or the diagonal $\widehat{G}$. If $\gamma_x = G^2$, then for all $a, b \in G$ one has $xa = xb$, i.e. $x \in D_l$. If $\gamma_x = \widehat{G}$ and $xa = xb$ then $a = b$, i.e. $x \in C_l$. In other words, $G = C_l \cup D_l$, and the assumptions of Lemma 8.3.1 are satisfied. If either (a) or (b) of Lemma 8.3.1 holds, then G has two elements, since each equivalence relation on G is a congruence in these cases. If (d) holds and G has at least three elements, then G is not simple. Indeed, in this case G has a zero, say z, and a non-trivial congruence θ defined by

$$(a, b) \in \theta \text{ iff } a = b = z \text{ or } a, b \in G - \{z\}.$$

It follows that the groupoid G is cancellative. □

As the two-element semilattice, two-element left zero band, and two-element right zero band are all simple two-element binary modes [Exercise 8.3A], we

8. BINARY MODES

are left with the following question: What are the simple cancellative binary modes?

To answer this question, let us first return to Example 7.1.4. It was shown there that the affinization of the variety $\underline{M\tau}$ of modes of type $\tau : \Omega \to \mathbb{N} - \{0,1\}$ is the variety $\underline{R(M\tau)}$ of affine spaces over the ring $R(M\tau) = R/\langle 1 - \sum_{i=1}^{\omega\tau} X_{\omega\tau} \mid \omega \in \Omega \rangle$, where

(8.3.5) $$R = \mathbb{Z}[X_{\omega i} \mid \omega \in \Omega,\ 1 \leq i \leq \omega\tau].$$

For n-ary ω in Ω, the corresponding operation of an affine $R(M\tau)$-space was defined by
$$x_1 \ldots x_n \omega = \sum_{i=1}^{n} x_i X_{\omega i}.$$
Now the affine $R(M\tau)$-spaces can also be considered as affine R-spaces [cf. Section 5.3]. By Theorem 7.7.1, each cancellative mode (C, Ω) of type τ is an Ω-subreduct of an affine $R(M\tau)$-space, or equivalently of an affine R-space. By Theorem 7.2.3, (C, Ω) is an Ω-subreduct of a power of the minimal cogenerator $F = F_R$. By Corollary 7.2.6, if such an algebra is subdirectly irreducible (in particular simple), then it is an Ω-subreduct of F.

A description of the simple $\underline{M\tau}$-algebras equivalent to affine spaces may be deduced from the known description of simple R-modules. Each such algebra (A, Ω) is polynomially equivalent to an R-module $(A, +, R)$, and for each e in A, the algebra (A, Ω, e) is equivalent to the module $(A, +, R)$. It follows that the affine space $(A, P, \underline{R})$ and the module $(A, +, R)$ have the same congruences. In particular, $(A, P, \underline{R})$ is subdirectly irreducible or simple if and only if $(A, +, R)$ has the same property. Now an R-module is simple if and only if it is isomorphic to the quotient R/I for some maximal ideal I of the ring R. (See e.g. [Hungerford 1989, Chapter IX, Theorem 1.3].) Recall that the ring R/I is a field if and only if the ideal I is maximal. One immediately obtains the following.

Proposition 8.3.3. *Each simple $\underline{M\tau}$-algebra that is equivalent to an affine space is isomorphic to a simple affine space R/I, where I is a maximal ideal of the ring R.* □

If the set Ω is finite, then the Hilbert Basis Theorem implies that the ring R defined by (8.3.5) is Noetherian. Since a homomorphic image of a Noetherian ring is again Noetherian, it follows that the ring $R(M\tau)$ is always a commutative Noetherian ring. It is known that in the case $R = \mathbb{Z}[X_1, \ldots, X_n]$, the quotient ring R/I by a maximal ideal I is a finite field. (See e.g. Białynicki-Birula 1987, Chapter VII, §4.) It follows that the simple R-modules are precisely the finite fields. This implies the following.

Proposition 8.3.4. *For a finite type τ, each simple $\underline{M\tau}$-algebra equivalent to an affine space is equivalent to a simple affine space $(GF(p^n), \underline{GF(p^n)})$.* □

Corollary 8.3.5. *Each simple binary mode equivalent to an affine space is isomorphic to some groupoid $(GF(p^n), \cdot) = (GF(p^n), \underline{r})$, where $p^n \geq 3$ and $r \neq 1$ is a generator of the ring $GF(p^n)$.*

Proof. This follows by Proposition 8.3.4 and Theorem 6.4.1. Note that for a generator r of the ring $GF(p^n)$, one has $(GF(p^n), \underline{r}) \simeq (GF(p^n), \underline{GF(p^n)})$. □

To describe simple cancellative binary modes that are not equivalent to affine spaces, we will need more information about the minimal cogenerator $F = F_R$ of the variety $\underline{R}$ of affine spaces over the ring R. We briefly recall the necessary results from module theory, referring the reader for details to the monographs [Anderson, Fuller 1992] and [Golan, Head 1991, Chapters 7 and 8].

Assume now that R is a commutative ring with identity. Consider the variety (or category) $\underline{\mathrm{Mod}}_R$ of R-modules. An R-module C is *injective* if for each submodule A of an R-module B, each module homomorphism $h: A \to C$ extends to a module homomorphism $\bar{h}: B \to C$, i.e. the following diagram

commutes.
$$\begin{CD} C @= C \\ @AhAA @AA\bar{h}A \\ A @>>> B \end{CD}$$

It is known that an R-module C is injective if and only if for each ideal I of R, each module homomorphism $h : I \to C$ extends to a module homomorphism $\bar{h} : R \to C$. If R is Noetherian, the word "ideal" here can be replaced by the word "prime ideal". For any ring R, each R-module can be embedded into an injective R-module. For a given R-module A, there is a certain special injective module containing it that is, in a sense, minimal. A submodule A of an R-module B is *large* in B if $A \cap A' \neq \{0\}$ for any nonzero submodule A' of B. If A is large in B, then B is also called an *essential extension* of A. An injective R-module C is called an *injective hull* of an R-module A if A is a large submodule of C. It is known that each module has an injective hull, and that the injective hull is unique up to isomorphism. Moreover, if an R-module A is contained in an injective hull $H(A)$ and C is any injective R-module containing A, then the inclusion map $\iota : A \to C$ can be extended to a module monomorphism $\bar{\iota} : H(A) \to C$,

$$\begin{CD} C @= C \\ @A\iota AA @AA\bar{\iota}A \\ A @>>> H(A). \end{CD}$$

The concept of an injective hull allows one to describe the minimal cogenerator of the variety $\underline{\mathrm{Mod}}_R$.

Let $\mathcal{S}$ be a set of representative simple R-modules, one from each isomorphism class. For a simple R-module T, let $H(T)$ be its injective hull. Then it is known that the minimal cogenerator F is isomorphic to the direct sum of all $H(T)$, i.e.

(8.3.6) $$F \cong \bigsqcup_{T \in \mathcal{S}} H(T).$$

Note also that if R is defined by (8.3.5) for a finite type τ, then the cogenerator F is injective. (Cf. [Golan, Head 1991, Chapter 8, §4, Observation 3].)

Theorem 8.3.6. *Each simple cancellative $\underline{M\tau}$-mode (A, Ω) embeds as an Ω-subreduct into a simple affine space $(R/I, P, \underline{R})$, where I is a maximal ideal of the ring R defined by (8.3.5).*

Proof. By Corollary 7.2.3, the mode (A, Ω) embeds as an Ω-subreduct into the minimal cogenerator $(F, P, \underline{R})$. By (8.3.6) the module F is the direct sum $\bigsqcup_{T \in \mathcal{S}} H(T)$ of injective hulls $H(T)$ of simple modules. The decomposition of F into this direct sum determines a congruence of the module (and at the same time of the affine space) F that restricts to a congruence of (A, Ω). Since (A, Ω) is simple, it follows that A embeds into one of the $H(T)$. Now the submodule T of the module $H(T)$ determines a congruence of the module $H(T)$ that restricts to a congruence of (A, Ω). It follows that A embeds into a coset of T. Since each coset of T is a simple affine space over R, the theorem follows by Proposition 8.3.3. □

Corollary 8.3.7. *Each simple cancellative $\underline{M\tau}$-mode is an Ω-subreduct of an affine space $(F, \underline{F})$, where F is a field.*

Now let us return to binary modes. In this case, the ring R of (5.8.3) is just the ring $\mathbb{Z}[X]$, and the corresponding minimal cogenerator F of (8.3.6) is a direct sum of injective hulls of finite fields $GF(p^n)$.

Corollary 8.3.8. *Each simple cancellative binary mode $(A, \cdot)$ is isomorphic to a groupoid $(GF(p^n), \cdot) = (GF(p^n), \underline{r})$, where $p^n \geq 3$ and $r \neq 1$ is a generator of the ring $GF(p^n)$.*

Moreover, if s is a generator of the ring $GF(q^m)$, then the binary modes $(GF(p^n), \underline{r})$ and $(GF(q^m), \underline{s})$ are isomorphic if and only if $p^n = q^m$ and $s = r^{p^k}$ for some integer k such that $0 \leq k < n$.

8. BINARY MODES

Proof. By Theorem 8.3.6 and Corollary 8.3.5, the mode $(A, \cdot)$ is a subreduct $(A, \underline{r})$ of an affine space $(GF(p^n), \underline{GF(p^n)})$. Since $(A, \cdot)$ is cancellative, $p^n \neq 2$. The element r of the field $GF(p^n)$ generates a subfield of $GF(p^n)$. Now, it is well-known that every subfield of $GF(p^n)$ has order p^m, where m divides n, and there is exactly one subfield of $GF(p^n)$ with p^m elements. (See [Lidl, Niederreiter 1983, Theorem 2.6].) By Theorem 6.4.1, the groupoid $(GF(p^n), \underline{r})$ and the affine space $(GF(p^m), \underline{GF(p^m)})$ are equivalent. By Exercise 8.3B, the mode $(GF(p^m), \underline{r})$ is plain. Hence $(A, \underline{r})$ must be isomorphic to $(GF(p^m), \underline{r})$.

Now it is clear that $p^n = q^m$ for isomorphic groupoids $(GF(p^n), \underline{r})$ and $(GF(q^m), \underline{s})$. Suppose that r and s are generators of the ring $GF(p^n)$. Since both these groupoids are equivalent to the affine space $(GF(p^n), \underline{GF(p^n)})$, the isomorphisms between these groupoids correspond precisely to the automorphisms of the field $GF(p^n)$. More exactly, $(GF(p^n), \underline{r}) \cong (GF(p^n), \underline{s})$ iff there is an automorphism α of the field $GF(p^n)$ such that $r\alpha = s$. Now, it is known that the distinct automorphisms of the field $GF(p^n)$ are exactly the k-th powers of the Frobenius automorphism defined by $a \mapsto a^p$ for $0 \leq k < n$. (See [Lidl, Niederreiter 1983, Chapter 2, §2].) □

The results of this section may be summarized as follows.

Theorem 8.3.9. *The following groupoids are (up to isomorphism) the only simple binary modes:*

(a) *the two-element semilattice;*

(b) *the two-element left zero band;*

(c) *the two-element right zero band;*

(d) *the groupoids $(GF(p^n), \cdot) = (GF(p^n), \underline{r})$, where $p^n \geq 3$ and r is a generator of the ring $GF(p^n)$.*

The simple modes $(GF(p^n), \underline{r})$ and $(GF(p^n), \underline{s})$ are isomorphic if and only if $s = r^{p^k}$ for $0 \leq k < n$. □

Since any simple mode $(GF(p^n), \underline{r})$ is plain, the variety it generates is a minimal variety of binary modes. Moreover, this variety is equivalent to the variety $\underline{GF(p^n)}$ of affine $GF(p^n)$-spaces. This gives the following description of minimal varieties of binary modes.

Theorem 8.3.10. *The following varieties are the only minimal varieties of binary modes:*

(a) *the variety* $\underline{Sl}$ *of semilattices;*

(b) *the variety* $\underline{Lz}$ *of left zero bands;*

(c) *the variety* $\underline{Rz}$ *of right zero bands;*

(d) *the varieties of binary modes equivalent to varieties* $\underline{GF(p^n)}$ *of affine spaces for* $p^n \geq 3$.

Exercises

8.3A. Show that the two-element semilattice, the two-element left zero band, and the two-element right zero band are the only two-element simple binary modes.

8.3B. Show that groupoid reducts $(GF(p^n), \underline{r})$ of the affine space $(GF(p^n), \underline{GF(p^n)}$ where $p^n \geq 3$ and r is a generator of the ring $GF(p^n)$,

(a) are simple,

(b) have no proper subalgebras.

8.3C. Find all non-isomorphic simple binary modes $(GF(p^2), \underline{r})$ for generators r of the ring $GF(p^2)$.

8.3D. Show that the two-element Ω-semilattice, the two-element left zero Ω-band, and the two-element right zero Ω-band are the only two-element simple Ω-modes.

8.4. Reductive modes

The aim of this section is to describe an important subclass of the variety of binary modes consisting of subvarieties containing the varieties of left or right zero bands, but not containing varieties equivalent to varieties of affine spaces. Amongst them are the familiar varieties $\underline{\underline{Dm}}$ of differential modes considered in Section 5.6, and the more general varieties $\underline{\underline{R}}_n$ of n-step left reductive (or n-differential) binary modes of Section 5.7. Each variety $\underline{\underline{R}}_n$ clearly contains the variety $\underline{\underline{Lz}}$ as a subvariety. In particular, all the identities satisfied in $\underline{\underline{R}}_n$ are left regular. (Cf. Section 1.5.) On the other hand, Exercise 6.4O shows that the variety $\underline{\underline{Lz}}$ is not equivalent to any variety of affine spaces. Now if $\underline{\underline{R}}_n$ contained a subvariety $\underline{\underline{V}}$ equivalent to a variety $\underline{\underline{S}}$ of affine S-spaces, then each $\underline{\underline{R}}_n$-groupoid contained in $\underline{\underline{V}}$ would have a Mal'cev operation P satisfying the identity $xxyP = y$ that is not left regular. It follows that the variety $\underline{\underline{R}}_n$ cannot contain a subvariety equivalent to a variety of affine spaces.

The variety $\underline{\underline{R}}_n^{OP}$ of binary modes opposite to the variety $\underline{\underline{R}}_n$ consists of n-step right reductive or briefly right n-reductive modes defined by the identity

(R_n^{OP}) $(\ldots(yx_n \cdot x_{n-1})\ldots x_2)x_1 = (\ldots(x_n x_{n-1})\ldots x_2)x_1.$

This identity is equivalent to the following:

(nD^{OP}) $yx^n := (\ldots(yx \cdot x)\ldots x)x = x.$

A binary mode is *right reductive* if it is right n-reductive for some n. All the facts formulated for left reductive binary modes may easily be reformulated in the opposite way for right reductive binary modes. In particular, the relation ρ_k on an $\underline{\underline{R}}_n^{OP}$-groupoid $(A, \cdot)$ defined by

$$(a,b) \in \rho_k \Leftrightarrow \forall x \in A, \ ax^k = bx^k$$

is a congruence of $(A, \cdot)$ such that each ρ_k-class is in $\underline{\underline{R}}_k^{OP}$ and the quotient $(A^{\rho_k}, \cdot)$ is in $\underline{\underline{R}}_{n-k}^{OP}$. The variety $\underline{\underline{R}}_n^{OP}$ coincides with the Mal'cev product $\underline{\underline{R}}_k^{OP} \circ \underline{\underline{R}}_{n-k}^{OP} = (\underline{\underline{Rz}})^n$. Each variety $\underline{\underline{R}}_n^{OP}$ contains the variety $\underline{\underline{Rz}}$ of right zero bands, and cannot contain a subvariety equivalent to a variety of affine spaces.

In this section we fully describe the Mal'cev product $\underline{R}_m \circ \underline{R}_n^{OP}$ (relative to the variety BM of binary modes). We show that it coincides with the independent join of $\underline{R}_m$ and $\underline{R}_n^{OP}$, and also provide some further characterizations. We start with some technical lemmata concerning identities satisfied by binary modes.

Lemma 8.4.1. *The following identities hold in each binary mode for any positive integers m and n:*

(a) $x^m(zx^n) = (x^m z)x^n$;

(b) $z(xy)^n = zx^{n-1} \cdot xy^n$;

(c) $(xy)^m(xy)^n = x^m z x^{n-1} \cdot y^m x y^n$.

Proof. (a) For $m = n = 1$, the identity in (a) is just the partial associativity (or di-associativity) of binary mode operations. If $x \cdot yx^n = (xy)x^n$ holds, then distributivity and di-associativity imply that $x \cdot yx^{n+1} = x(yx^n)x = (xy)x^{n+1}$ holds as well. Hence, by distributivity again, one gets $x^2(yx^n) = x \cdot (xy)x^n = (x^2 y)x^n, \ldots, x^m(yx^n) = (x^m y)x^n$.

(b) By the entropic law for $\cdot$, one obtains $z(xy)^n = (zx \cdot xy^2)(xy)^{n-2} = \cdots = zx^{n-1} \cdot xy^n$.

(c) By (a), (b) and entropicity the following holds:

$$(xy)^m z(xy)^n = (xy)^m [z(xy)^n] = (xy)^m [zx^{n-1} \cdot xy^n]$$
$$= (xy)^{m-1}[xzx^{n-1} \cdot yxy^n] = \cdots = x^m z x^{n-1} \cdot y^m x y^n. \quad \square$$

In what follows, we define $xy^0 := x$ and $x^0 y := y$, and we write $x^m z x^n$ for $x^m(zx^n) = (x^m z)x^n$.

Proposition 8.4.2. *The following identities are equivalent in binary modes for any positive integers m and n:*

(a) $x^m z x^n = x$;

(b) $xy^k = (x^m z x^{n-k})y^k$, where $k = 0, 1, \ldots, n-1$;

(c) $y^k x = y^k(x^{m-k} z x^n)$, where $k = 0, 1, \ldots, m-1$;

(d) $(x^m z)y^n = xy^n$;

(e) $x^m(zy^n) = x^m y$.

Proof. (a)$\Rightarrow$(b) Assume that the identity of (a) holds. Note that this is the identity of (b) for $k = 0$. Now assume that the identity of (b) holds for some k. Then substituting xy for x we get

$$xy^{k+1} = (xy)y^k = ((xy)^m z(xy)^{n-k})y^k$$
$$= (x^m z x^{n-k-1} \cdot y^m x y^{n-k})y^k$$
$$= (x^m z x^{n-k-1})y^k \cdot y^m x y^n$$
$$= (x^m z x^{n-k-1})y^{k+1}.$$

The third identity holds by Lemma 8.4.1(c), the fourth by distributivity, and the fifth by (a).

(a)$\Rightarrow$(c) This can be proved by an argument dual to the previous one.

(b)$\Rightarrow$(d) The proof goes like the proof of (a)$\Rightarrow$(b).

Assume that the identity of (b) holds for $k = n - 1$, and substitute xy for x in it. Then we get

$$xy^n = (xy)y^{n-1} = ((xy)^m(z \cdot xy))y^{n-1}$$
$$= (x^m z \cdot y^m xy)y^{n-1}$$
$$= x^m z y^{n-1} \cdot y^m x y^n$$
$$= x^m z y^n.$$

(c)$\Rightarrow$(e) The proof is dual to the proof of (b)$\Rightarrow$(d).

(d)$\Rightarrow$(a) and (e)$\Rightarrow$(a) are immediate. $\square$

Let $\underline{R}_{m,n}$ denote the variety of binary modes defined by any of the equivalent identities of Proposition 8.4.2. Algebras belonging to any of the varieties $\underline{R}_{m,n}$ will be called *reductive modes*.

Recall that two varieties $\underline{V}_1$ and $\underline{V}_2$ of binary modes are independent if there is a decomposition word $x_1 \circ x_2$ such that the identity $x_1 \circ x_2 = x_i$ holds in $\underline{V}_i$ for $i = 1, 2$. Whenever the varieties $\underline{V}_1$ and $\underline{V}_2$ are independent, each algebra $(A, \cdot)$ in their join $\underline{V} = \underline{V}_1 \vee \underline{V}_2$ is isomorphic to a product $(A_1, \cdot) \times (A_2, \cdot)$, with $(A_i, \cdot)$ in $\underline{V}_i$ for $i = 1, 2$, and the algebras $(A_i, \cdot)$ are determined up to isomorphism. We denote the independent join $\underline{V}$ of $\underline{V}_i$ by $\underline{V}_1 \otimes \underline{V}_2$ (cf. Section 3.5).

Theorem 8.4.3. *The following conditions are equivalent for any variety $\underline{V}$ of binary modes and any positive integers m, n:*

(a) $\underline{V} = \underline{R}_m \vee \underline{R}_n^{OP}$;
(b) $\underline{V} = \underline{R}_m \circ \underline{R}_n^{OP}$;
(c) $\underline{V} = \underline{R}_{m,n}$;
(d) $\underline{V} = \underline{R}_m \otimes \underline{R}_n^{OP}$.

Proof. (b)$\Rightarrow$(c) Let $(A, \cdot)$ be in $\underline{R}_m \circ \underline{R}_n^{OP}$. This means that there is a congruence, say α, on $(A, \cdot)$ such that the quotient $(A^\alpha, \cdot)$ is in $\underline{R}_n^{OP}$ and for each a in A, the α-class a^α is in $\underline{R}_m$. Hence for any a, b in A, $(ab^n)^\alpha = a^\alpha (b^\alpha)^n = b^\alpha$, and consequently $(b, ab^n) \in \alpha$. Since a^α is in $\underline{R}_m$, it follows that $b^m(ab^n) = b$.

(c)$\Rightarrow$(b) Assume that $(A, \cdot)$ is in $\underline{R}_{m,n}$. Note that the relation λ_m of Theorem 5.7.4 is a congruence on any binary mode. (Cf. Exercise 8.4A.) Then for each a in A, the λ_m-class a^{λ_m} is in $\underline{R}_m$. Since for any x, a, b in A, $x^m(ab^n) = x^m b$, it follows that $(b, ab^n) \in \lambda_m$, and hence $a^{\lambda_m}(b^{\lambda_m})^n = (ab^n)^{\lambda_m} = b^{\lambda_m}$. This implies that $(A^{\lambda_m}, \cdot)$ is in $\underline{R}_n^{OP}$. Consequently, $(A, \cdot)$ is in $\underline{R}_m \circ \underline{R}_n^{OP}$.

(a)$\Rightarrow$(d) It is enough to show that the varieties $\underline{R}_m$ and $\underline{R}_n^{OP}$ are independent. Let $(A, \cdot)$ be in $\underline{R}_m$. Consider the congruence λ_{m-1} on $(A, \cdot)$. By Theorem 5.7.4, the quotient $(A^{\lambda_{m-1}}, \cdot)$ is in $\underline{R}_1$. In particular, for any a and b in A, $(a, ab) \in \lambda_{m-1}$, whence for any positive integer j one also has $(a, ab^j) \in \lambda_{m-1}$. It follows that $(a, ab^n) \in \lambda_{m-1}$, i.e. $aA^n \subseteq a^{\lambda_{m-1}}$. Recall

8. BINARY MODES

that the congruences λ_i form a chain $\widehat{A} = \lambda_0 \leq \lambda_1 \leq \cdots \leq \lambda_m = A \times A$, and since $(a^{\lambda_k}, \cdot)$ is in $\underline{\underline{R}}_k$, the quotient $(a^{\lambda_k}, \cdot)^{\lambda_{k-1}}$ is in $\underline{\underline{R}}_1$. As before, we get $a(aA^n)^n \subseteq a^{\lambda_{m-2}}, a[a(aA^n)^n]^n \subseteq a^{\lambda_{m-3}}$ etc., and finally in the m-th step

$$a(\ldots(a(aA^n)^n)^n \ldots)^n \subseteq a^{\lambda_{m-m}} = a^{\lambda_0} = \{a\}.$$

It follows that $(A, \cdot)$ satisfies the identity

$$x * y := x(\ldots x(x(xy^n)^n)^n \ldots)^n = x.$$

On the other hand, if $(A, \cdot)$ is in $\underline{\underline{R}}_n^{OP}$, then it obviously satisfies $x * y = x(\ldots x(x(xy^n)^n)^n \ldots)^n = x(\ldots x(xy^n)^n \ldots)^n = \cdots = xy^n = y$. Consequently $\underline{\underline{R}}_m$ and $\underline{\underline{R}}_n^{OP}$ are independent.

(a)⇔(c) Let $(A, \cdot)$ be in $\underline{\underline{R}}_{m,n}$. Consider again the congruences λ_m and ρ_n. Then by Theorem 5.7.4 and its dual, for each a in $A, (a^{\lambda_m}, \cdot)$ is in $\underline{\underline{R}}_m$ and $(a^{\rho_n}, \cdot)$ is in $\underline{\underline{R}}_n^{OP}$. Hence $(a^{\lambda_m \wedge \rho_n}, \cdot)$ is in $\underline{\underline{R}}_m \wedge \underline{\underline{R}}_n^{OP}$. However, since $\underline{\underline{R}}_m$ and $\underline{\underline{R}}_n^{OP}$ are independent, it follows that $\underline{\underline{R}}_m \wedge \underline{\underline{R}}_n^{OP}$ is the trivial variety, whence $a^{\lambda_m \wedge \rho_n} = \{a\}$ and $\lambda_m \wedge \rho_n = 0$, the equality relation.

Note that if both λ_m and ρ_n are different from 0, then $(A, \cdot)$ is a subdirect product of $(A^{\lambda_m}, \cdot)$ and $(A^{\rho_n}, \cdot)$. If they are both 0, then for each x in $A, x^m a = x^m b$ implies that $a = b$, and similarly $ax^n = bx^n$ implies that $a = b$. It follows that $a \neq b$ implies $x^m a \neq x^m b$, which in turn implies $x^m a x^n \neq x^m b x^n$. But this gives a contradiction, since $(A, \cdot)$ is in $\underline{\underline{R}}_{m,n}$.

The previous paragraph shows that if $(A, \cdot)$ is subdirectly irreducible, then exactly one of λ_m and ρ_m is 0, and the other is different from 0. Since $(A^{\lambda_m}, \cdot)$ is in $\underline{\underline{R}}_n^{OP}$, it follows that if $\lambda_m = 0$, then $(A, \cdot) \cong (A^{\lambda_m}, \cdot)$ is in $\underline{\underline{R}}_n^{OP}$ and if $\rho_n = 0$, then $(A, \cdot) \cong (A^{\rho_n}, \cdot)$ is in $\underline{\underline{R}}_m$. Since each subdirectly irreducible $\underline{\underline{R}}_{m,n}$-algebra is either in $\underline{\underline{R}}_m$ or $\underline{\underline{R}}_n^{OP}$, it follows that $\underline{\underline{R}}_{m,n} \subseteq \underline{\underline{R}}_m \vee \underline{\underline{R}}_n^{OP}$. The opposite inclusion is obvious, since both varieties $\underline{\underline{R}}_m$ and $\underline{\underline{R}}_n^{OP}$ satisfy the identity defining $\underline{\underline{R}}_{m,n}$. □

Lemma 8.4.4. *The variety* $\underline{\underline{R}}_{m,n}$ *satisfies the identity*

$$y^m x y^s = y^m (x^k z x^{n+l-s}) y^s$$

for any $s < n + l$.

Proof. Compare the proof of the first implication of Proposition 8.4.2. Assume that the identity holds for s, and substitute xy for x. Then we get the following:

$$\begin{aligned} y^m x y^{s+1} &= y^m [(xy)^k z (xy)^{n+l+s}] y^s \\ &= y^m [x^k z x^{n+l-s-1} \cdot y^k z y^{n+l-s}] y^s \\ &= y^m [x^k z x^{n+l-s-1}] y^s \cdot y^{m+k} z y^{n+l} \\ &= y^m [x^k z x^{n+l-s-1}] y^{s+1}. \end{aligned}$$

The second identity holds by Lemma 8.4.1(c), the third by distributivity, and the fourth by the defining identity of $\underline{\underline{R}}_{m,n}$. □

Extend the definition of $\underline{\underline{R}}_{m,n}$ by setting $\underline{\underline{R}}_{m,0} := \underline{\underline{R}}_m$ and $\underline{\underline{R}}_{0,n} := \underline{\underline{R}}_n^{OP}$. Recall that $x^0 y := y$ and $xy^0 := x$.

Lemma 8.4.5. *For any positive integers* m, n, k *and* l,

$$\underline{\underline{R}}_{m,n} \circ \underline{\underline{R}}_{k,l} = \underline{\underline{R}}_{m+k,n+l}.$$

Proof. Let $(A, \cdot)$ be in $\underline{\underline{R}}_{m,n} \circ \underline{\underline{R}}_{k,l}$. Then there is a congruence α of $(A, \cdot)$ such that $(A^\alpha, \cdot)$ is in $\underline{\underline{R}}_{k,l}$ and for each a in $A, (a^\alpha, \cdot)$ is in $\underline{\underline{R}}_{m,n}$. It follows that for a, b in $A, (a^\alpha)^k b^\alpha (a^\alpha)^l = (a^k b a^l)^\alpha = a^\alpha$, and hence $(a, a^k b a^l) \in \alpha$. Consequently $a^m (a^k b a^l) a^n = a^{m+k} b a^{l+n} = a$, i.e. $(A, \cdot)$ is in $\underline{\underline{R}}_{m+k,n+l}$.

Assume now that $(A, \cdot)$ is in $\underline{\underline{R}}_{m+k,n+l}$. Define the relation $\theta_{m,n}$ on A by

$$(a, b) \in \theta_{m,n} :\Leftrightarrow \forall x \in A,\ x^m a x^n = x^m b x^n.$$

It is easy to see that $\theta_{m,n}$ is a congruence on $(A, \cdot)$, and that for each a in A, the $\theta_{m,n}$-class $a^{\theta_{m,n}}$ is in $\underline{\underline{R}}_{m,n}$. By Lemma 8.4.4, for any x, a, b in $A, x^m (a^k b a^l) x^n = x^m a x^n$. It follows that $(a, a^k b a^l) \in \theta_{m,n}$, whence $(A^{\theta_{m,n}}, \cdot)$ is in $\underline{\underline{R}}_{k,l}$. Consequently $(A, \cdot)$ is in $\underline{\underline{R}}_{m,n} \circ \underline{\underline{R}}_{k,l}$. □

8. BINARY MODES

Theorem 8.4.6. *An iterated Mal'cev product of m copies of $\underline{R}_1$ and n copies of $\underline{R}_1^{OP}$, with the product performed in any order, is the variety $\underline{R}_{m,n}$.*

Proof. Let $\underline{K}$ be any iterated Mal'cev product of m copies of $\underline{R}_1$ and n copies of $\underline{R}_1^{OP}$. Then $\underline{K}$ can be written as $\underline{K} = \underline{K}_1 \circ \underline{K}_2$. Now if $\underline{K}_1$ and $\underline{K}_2$ have the required form, and $\underline{K}_1 = \underline{R}_{s,t}$ and $\underline{K}_2 = \underline{R}_{k,l}$ for some natural numbers s, t, k and l, then obviously $m = s + k$ and $n = t + l$. By Theorem 5.7.4, Corollary 5.7.5, Theorem 8.4.3 and Lemma 8.4.5, $\underline{K}$ has the required form. $\square$

In the last part of this section we will give a detailed description of the lattice $\mathcal{L}(\underline{Dm})$ of subvarieties of the variety $\underline{Dm} = \underline{R}_2$ of differential (or left 2-reductive) modes. Then Theorems 8.4.3 and 8.4.6 provide some information about the variety $\underline{R}_2 \vee \underline{R}_2^{OP}$.

We start with a discussion of free differential modes and some properties of identities satisfied in $\underline{Dm}$.

Theorem 8.4.7. *In the free differential mode $X\mathrm{Dm}$ on a non-empty set X, each element may be expressed in the standard form*

$$(8.4.1) \qquad x_1 x_2^{k_2} \ldots x_s^{k_s},$$

where $x_j \in X$ for each $j = 1, \ldots, s$, and $x_i \neq x_j$ for $i \neq j$.

Proof. Let $x_1 \ldots x_s w$ be a groupoid word with variables in the set X. The proof goes by induction on s. For $s = 2$, the word $x_1 x_2 w$ is $x_1 x_2$ or $x_2 x_1$, and it is already in the standard form. Now suppose the theorem holds for $2 \leq s < n$. By induction, a word $x_1 \ldots x_n w$ not in standard form may be expressed as

$$x_1 \ldots x_n w = y_1 \ldots y_p w_1 \cdot z_1 \ldots z_q w_2$$
$$= y_1 y_2^{i_2} \ldots y_p^{i_p} \cdot z_1 z_2^{j_2} \ldots z_q^{j_q},$$

a product of standard forms, with all y_i and z_j in X and $p < n$ and $q < n$. By the reduction law (Rd) of Section 5.6,

$$x_1 \ldots x_n w = y_1 y_2^{i_2} \ldots y_p^{i_p} z_1.$$

If $z_1 = y_m$ for some $m = 1, \ldots, p$, the left normal law (Ln) of Section 5.6 implies that

$$x_1 \ldots x_n w = y_1 y_1^{i_2} \ldots y_m^{i_m+1} \ldots y_p^{i_p}.$$

This shows that $x_1 \ldots x_n w$ may be written in the standard form. □

By Exercise 8.4B, the order of the variables $x_2, \ldots, x_s$ in the operation (8.4.1) in any differential mode is irrelevant.

For $i = 0, 1, \ldots$ and $p = 1, 2, \ldots$, let $\underline{\underline{D}}_{i,i+p}$ be the subvariety of $\underline{\underline{Dm}}$ defined by the identity

$(d_{i,i+p})$ $\qquad xy^i = xy^{i+p}.$

Corollary 8.4.8. *In the free $\underline{\underline{D}}_{i,i+p}$-mode $XD_{i,i+p}$, each element may be expressed in the standard form (8.4.1) with $k_j < i + p$ for all $j = 2, \ldots, s$.* □

Recall that each identity true in the variety $\underline{\underline{Dm}}$ is necessarily left regular.

Lemma 8.4.9. *In the variety $\underline{\underline{Dm}}$:*
(a) The identity

(8.4.2) $\qquad x\, x_1^{j_1} \ldots x_m^{j_m} y_{m+1}^{j_{m+1}} \ldots y_r^{j_r} = x\, x_1^{j_1} \ldots x_m^{j_m} z_{m+1}^{k_{m+1}} \ldots z_s^{k_s},$

where any of the sets $\{x_1, \ldots, x_m\}$ and $\{z_{m+1}, \ldots z_s\}$ may be empty, is equivalent to the identity

(8.4.3) $\qquad xy^k = x,$

where $k = GCD(j_{m+1}, \ldots, j_r, k_{m+1}, \ldots, k_s)$.
(b) The identity

$$x\, x_1^{j_1} \ldots x_r^{j_r} = x\, x_1^{k_1} \ldots x_r^{k_r}$$

is equivalent to the set of identities

$$x \, x_p^{j_p} = x \, x_p^{k_p},$$

where $p = 1, \ldots, r$.

Proof. Substituting x for all variables different from y_p in (8.4.2) one gets

(8.4.4) $$x \, y_p^{j_p} = x$$

for all $p = m+1, \ldots, r$. In similar fashion, (8.4.2) implies

(8.4.5) $$x \, z_q^{k_q} = x$$

for all $q = m+1, \ldots, r$. Exercise 8.4C shows that (8.4.2) is equivalent to the set of identities of (8.4.4) and (8.4.5), and this set is equivalent to the unique identity of (8.4.3).

The proof of (b) is similar to that of (a), and is left as Exercise 8.4D. □

Lemma 8.4.10. *If the identity*

(8.4.6) $$xy^{i+r} = xy^i$$

holds in the variety $\underline{Dm}$, *then so does the identity*

(8.4.7) $$xy^{j+kr} = x \, y^j$$

for each $j \geq i$.

Proof. For $i = j$, the proof goes by induction on k. The identity (8.4.6) implies that $xy^{i+2r} = xy^{i+r}y^r = xy^i y^r = xy^{i+r} = xy^i$. Now suppose (8.4.7) holds. Then by induction $xy^{i+kr+r} = x \, y^{i+kr} y^r = xy^i y^r = xy^{i+r} = xy^i$.

Now let $j > i$. Multiplying the identity (8.4.6) by y^{j-i} on the right one obtains the identity

$$xy^{j+r} = xy^j.$$

The first part of the proof then shows that if this identity holds in $\underline{Dm}$, then so does (8.4.7). □

Lemma 8.4.11. *The variety* $\underline{Dm}$ *satisfies the identities*

(8.4.8) $$xy^{i+r} = xy^i \quad \text{and} \quad xy^{j+s} = xy^j$$

for $i \leq j$ if and only if it satisfies the identity

(8.4.9) $$xy^{i+GCD(r,s)} = xy^i.$$

Proof. ($\Leftarrow$) This follows by Lemma 8.4.10.
($\Rightarrow$) Let $r < x$. By the Euclidean Algorithm:

$$s = rq_1 + p_1$$
$$r = p_1 q_2 + p_2$$
$$p_1 = p_2 q_3 + p_3$$
$$\vdots$$
$$p_{k-2} = p_{k-1} q_k + p_k$$
$$p_{k-1} = p_k q_{k+1}$$

with $p_k = GCD(r,s)$. By Lemma 8.4.10, the identities (8.4.8) for $i = j$ imply the satisfaction of the identities

$$xy^{i+p_1} = xy^i$$
$$xy^{i+p_2} = xy^i$$
$$\vdots$$
$$xy^{i+p_k} = xy^i,$$

so that (8.4.9) holds if $i = j$.

Now assume that $i < j$. If $r|s$, then $GCD(r,s) = r$ and (8.4.9) holds. Suppose $r \nmid s$. By Lemma 8.4.10, the first identity of (8.4.8) implies that

(8.4.10) $$xy^{j+r} = xy^j$$

holds in $\underline{Dm}$. By the first part of the proof, it follows that the identities (8.4.10) and the second identity of (8.4.8) together are equivalent to

(8.4.11) $$xy^{j+GCD(r,s)} = xy^j.$$

Let $j = i + ar + b$ with $a \geq 0$ and $0 \leq b < r$. Multiplying (8.4.11) by y^{r-b} on the right and using (8.4.8) and Lemma 8.4.10, one gets (8.4.9). This completes the proof. □

Proposition 8.4.12. *In the variety $\underline{Dm}$, the set of identities*

$$x\, y^{i_k + r_k} = x\, y^{i_k},$$

where $k = 1, \ldots, q$ and $0 \leq i_1 \leq i_2, \ldots, i_q$, is equivalent to the unique identity

$$x\, y^{i_1 + GCD(r_1, \ldots, r_q)} = x\, y^{i_1}.$$

Proof. This follows immediately by Lemma 8.4.11. □

Theorem 8.4.7 and Lemma 8.4.9 show that each proper subvariety $\underline{V}$ of the variety $\underline{Dm}$ is contained in some variety $\underline{D}_{i,j}$. We will show that each non-trivial $\underline{V}$ coincides with some $\underline{D}_{i,j}$.

Proposition 8.4.13. *If $\underline{V}$ is a subvariety of $\underline{D}_{j,j+r}$, then either $\underline{V}$ is trivial or else $\underline{V} = \underline{D}_{i,i+p}$ for some $i \leq j$. Moreover, $\underline{D}_{i,i+p} \subseteq \underline{D}_{j,j+r}$ if and only if $i \leq j$ and $p|r$.*

Proof. Assume that $\underline{V}$ is a non-trivial proper subvariety of $\underline{D}_{j,j+r}$. By Corollary 8.4.8 and Lemma 8.4.9, the set of those identities satisfied in $\underline{V}$ that are not consequences of the axioms of $\underline{Dm}$ is equivalent to a set I of identities

$$x\, y^{j_k} = x\, y^{r_k},$$

where $j_k + r_k < j + r$. Evidently this set is finite, say with ℓ elements, and one of the j_k, say j_1, is less than all the others. Hence, by Proposition 8.4.12, the set I is equivalent to the unique identity

$$xy^i = xy^{i+p},$$

where $i = j_1$ and $p = GCD(r_1, \ldots, r_l)$.

If $i \leq j$ and $p|r$, then by Lemma 8.4.10, $\underline{D}_{i,i+p} \subseteq \underline{D}_{j,j+r}$. If $\underline{D}_{i,i+p} \subseteq \underline{D}_{j,j+r}$, then $j = i + k$, and by Proposition 8.4.12, $\underline{D}_{i,i+p} \cap \underline{D}_{i+k,i+k+r} = \underline{D}_{i,i+GCD(p,r)} = \underline{D}_{i,i+p}$. Hence $GCD(p,r) = p$, and consequently $p|r$. □

Let $\mathbb{N}_d$ be the lattice of positive natural numbers with division. (Cf. Exercise 1.1F.) Let $\mathbb{N}_c$ be the chain of natural numbers equipped with the usual ordering. Let L be the product $\mathbb{N}_d \times \mathbb{N}_c$ of the lattices $\mathbb{N}_d$ and $\mathbb{N}_c$. The lattice $L_{0,1}$ is obtained from L by adding two new elements, an element 1 that is greater than all the elements of L, and an element 0 smaller than all the elements of L.

Theorem 8.4.14. *The lattice* $\mathcal{L}(\underline{\text{Dm}})$ *of subvarieties of the variety* $\underline{\text{Dm}}$ *is isomorphic to the lattice* $L_{0,1}$. *The isomorphism is given by the mapping*

$$h : \mathcal{L}(\underline{\text{Dm}}) \to L_{0,1},$$

where $\underline{\text{Tr}} \mapsto 0$, $\underline{D}_{i,i+p} \mapsto (i,p)$, *and* $\underline{\text{Dm}} \mapsto 1$.

Proof. It is evident that the mapping h surjects. By Proposition 8.4.13, it is also injective.

By Exercise 1.2R, it suffices to show that h and h^{-1} are monotone mappings. But this follows immediately by the second part of Proposition 8.4.13. □

Note also that any non-trivial subvarieties, one of $\underline{\text{Dm}}$ and one of $\underline{\text{Dm}}^{OP}$, are independent. Consider subvarieties $\underline{D}_{i,m}$ of $\underline{R}_2 = \underline{\text{Dm}}$ and $\underline{D}_{j,n}^{OP}$ of $\underline{R}_2^{OP} = \underline{\text{Dm}}^{OP}$. The decomposition operation for $\underline{R}_{2,2}$ can be used to show that $\underline{D}_{i,m}$

and $\underline{D}^{OP}_{j,n}$ are independent, and hence $\underline{D}_{i,m} \vee \underline{D}^{OP}_{j,n} = \underline{D}_{i,m} \times \underline{D}^{OP}_{j,n}$. Exercise 3.5I can be used to find a basis for the identities true in such a variety.

Similar remarks hold for the subvarieties $\underline{D}^m_{i,k}$ of $\underline{R}_m$ defined by $(d_{i,k})$, and for the subvarieties $(\underline{D}^n_{j,l})^{OP}$ of $\underline{R}^{OP}_n$ defined by the identity opposite to $(d_{j,l})$. However, we do not know if such subvarieties comprise all the subvarieties of $\underline{R}_m$ and $\underline{R}^{OP}_n$ respectively.

Exercises

8.4A. For a type $\tau : \Omega \to \mathbb{N}$ of modes, let $x \cdot y$ be a fixed Ω-word. An Ω-mode (A, Ω) is said to be (left or right) n-reductive if the groupoid $(A, \cdot)$ is respectively (left or right) n-reductive.

  (a) Show that the relation λ_k defined by (5.7.2), and similarly the relation ρ_k of this section, are congruences on any mode (A, Ω). Further, show that each λ_k-class (a^{λ_k}, Ω) is left k-reductive, and that each ρ_k-class (a^{ρ_k}, Ω) is right k-reductive.

  (b) Show that if $k < n$ and a mode (A, Ω) is left n-reductive, then the quotient (A^{λ_k}, Ω) is left $(n - k)$-reductive. Formulate the corresponding property of right reductive modes.

  (c) Show that Theorem 5.7.4 and Corollary 5.7.5 hold for left reductive Ω-modes. Formulate corresponding theorems for right reductive modes.

  (d) Show that Theorems 8.4.3 and 8.4.6 hold for reductive Ω-modes.

8.4B. Show that the identity

$$x_1 x_2^{k_2} \ldots x_s^{k_s} = x_1 x_{2\varphi}^{k_{2\varphi}} \ldots x_{s\varphi}^{k_{s\varphi}}$$

is satisfied in $\underline{Dm}$ for any permutation φ of the set $\{2, \ldots, s\}$.

8.4C. Show that the identity (8.4.2) is equivalent to the set of identities in (8.4.4) and (8.4.5), and that this set is equivalent to the unique identity (8.4.3).

8.4D. Prove Lemma 8.4.9(b).

8.4E. [Pilitowska, Romanowska, Roszkowska 1996] Find the identities defining the variety $\underline{D}_{0,k}^{m} \times (\underline{D}_{0,l}^{n})^{OP}$.

8.4F. [Pilitowska, Romanowska, Roszkowska 1996] Consider the class $CA(\underline{D}_{2,2})$ of complex algebras of submodes of $\underline{D}_{2,2}$-modes. (Cf. Section 5.1.)

  (a) Show that $CA(\underline{D}_{2,2})$-modes are differential, but do not necessarily satisfy the identity $(d_{0,2})$.

  (b) Describe the free $\underline{D}_{2,2}$-groupoids $F_{n+1} = \{x, y_1, \ldots, y_n\} D_{2,2}$ on $n+1$ generators, and their complex algebras $(F_{n+1}S, \cdot)$.

  (c) Show that in the differential mode $(F_{n+1}S, \cdot)$, none of the identities $xy^n = xy^j$, for $j = 0, 1, \ldots, n-1$, is satisfied.

  (d) Deduce that the variety $\underline{Dm}$ of differential modes is generated by the class $CA(\underline{D}_{2,2})$.

8.5. Reductive and affine binary mode varieties

This section gives a general approach to the study of the lattice $\mathcal{L}(\underline{BM})$ of varieties of binary modes. By the results of Section 3.5, the lattice $\mathcal{L}(\underline{BM})$ is a complete lattice. By Corollary 3.5.9, it is atomic, i.e. each variety in $\mathcal{L}(\underline{BM})$ contains an atom, a minimal subvariety of $\underline{BM}$. The minimal subvarieties of $\underline{BM}$, described in Theorem 8.3.9, consist of $\underline{Sl}, \underline{Lz}, \underline{Rz}$, and varieties equivalent to the varieties $\underline{GF(q)}$, for $q \neq 2$, of affine spaces over finite fields. Note that, by Exercise 4.3C, the varieties $\underline{Lz}, \underline{Rz}$ and $\underline{Sl}$ generate all eight varieties of normal bands (semigroup modes).

The lattice $\mathcal{L}(\underline{V})$ of subvarieties of a given variety $\underline{V}$ may be very complicated. So it is often useful to partition it into fragments that are easier to investigate, or of special interest. A first natural partition of the variety $\mathcal{L}(\underline{BM})$ is into two classes: the class $\mathcal{L}_r(\underline{BM})$ of regular and the class $\mathcal{L}_{ir}(\underline{BM})$ of irregular subvarieties of $\underline{BM}$. The trivial variety $\underline{Tr}$ of binary modes is evidently the least element of $\mathcal{L}_{ir}(\underline{BM})$, and by Corollary 8.2.9, $\mathcal{L}_{ir}(\underline{BM})$ does not have a largest element. The class $\mathcal{L}_r(\underline{BM})$ is bounded above by the variety $\underline{BM}$, and by Exercise 4.2D(a) it has the variety $\underline{Sl}$ of semilattices as its least element.

Proposition 8.5.1. *The classes $\mathcal{L}_{ir}(\underline{BM})$ and $\mathcal{L}_r(\underline{BM})$ are both sublattices of the lattice $\mathcal{L}(\underline{BM})$.*

Proof. By the results of Section 3.5, it is clear that $\mathcal{L}_{ir}(\underline{BM})$ is closed under arbitrary intersections. To show that it is closed under binary joins, let $\underline{V}$ and $\underline{W}$ be two varieties in $\mathcal{L}_{ir}(\underline{BM})$. By Lemma 4.3.1, the variety $\underline{V}$ is defined by regular identities and one irregular identity, say $x \cdot y = x$. Similarly $\underline{W}$ is defined by regular identities and one irregular identity, say $x \circ y = x$. Each of these irregular identities implies that the identity

$$x \cdot (x \circ y) = x \circ (x \cdot y) = x$$

is also satisfied both in $\underline{V}$ and in $\underline{W}$, and hence also in their join $\underline{V} \vee \underline{W}$. This shows that the join belongs to $\mathcal{L}_{ir}(\underline{BM})$.

The class $\mathcal{L}_r(\underline{BM})$ is a sublattice of $\mathcal{L}(\underline{BM})$, since all consequences of regular identities are regular. □

Note that the variety $\underline{Sl}$ of semilattices is the unique minimal subvariety of $\underline{BM}$ contained in $\mathcal{L}_r(\underline{BM})$. Note also that $\mathcal{L}_r(\underline{BM})$ contains all regularizations $\underline{\widetilde{V}}$ of irregular varieties $\underline{V}$ in $\mathcal{L}_{ir}(\underline{BM})$. Recall that by Theorem 4.3.4, the lattice $\mathcal{L}(\underline{\widetilde{V}})$ of subvarieties of the regularization $\underline{\widetilde{V}}$ of $\underline{V}$ in $\mathcal{L}_{ir}(\underline{BM})$ is isomorphic to the direct product $\mathcal{L}(\underline{V}) \times \underline{2}$ of the lattice $\mathcal{L}(\underline{V})$ of subvarieties of $\underline{V}$ and the two-element lattice $\underline{2}$. By Płonka's Theorem 4.3.2, each $\underline{\widetilde{V}}$-mode is a Płonka sum of $\underline{V}$-modes. The more important algebraic properties of the regularization $\underline{\widetilde{V}}$ may easily be deduced from the corresponding properties for $\underline{V}$, using the results of Section 4.3. In particular, Theorem 4.3.2 shows how to find a basis for the identities holding in $\underline{\widetilde{V}}$, given a basis for the identities holding in $\underline{V}$. Note however that there are regular subvarieties of $\underline{\mathrm{BM}}$ which are not regularizations. An obvious example is given by the variety $\underline{\mathrm{BM}}$. Further examples will be provided in the next section.

All these facts show that the problem of classifying varieties of binary modes can be reduced to the study of irregular varieties and regular varieties that are not regularizations. While we still do not know much about the latter class, we do have some knowledge concerning the lattice $\mathcal{L}_{ir}(\underline{\mathrm{BM}})$, especially its "lower" part.

Amongst the irregular subvarieties of $\underline{\mathrm{BM}}$ are all the varieties equivalent to varieties of affine spaces, in particular the varieties $\underline{\mathrm{V}}_{n,m}$, introduced in Example 6.4.2, that are defined by the identities

(8.5.1) $$x^n y = y = yx^m.$$

Recall that, by Corollary 6.4.2.1, the varieties $\underline{\mathrm{V}}_{n,m}$ contain all the binary modes equivalent to affine spaces over finite commutative rings, in particular over all finite fields different from GF(2). It follows that the varieties $\underline{\mathrm{V}}_{n,m}$ contain all the minimal subvarieties of $\underline{\mathrm{BM}}$ with the exception of $\underline{\mathrm{Lz}}$, $\underline{\mathrm{Rz}}$, and $\underline{\mathrm{Sl}}$. Furthermore, the join of all the $\underline{\mathrm{V}}_{n,m}$ is the variety $\underline{\mathrm{BM}}$ of all binary modes.

Note that the identity (8.5.1) implies the following:

(8.5.2) $$x^n y x^m = y.$$

Recall that, by Lemma 8.4.1, $x^n y x^m = (x^n y) x^m = x^n (y x^m)$. Also the identity (8.5.2) defines a variety of binary modes equivalent to a variety of affine spaces, obviously having $\underline{\mathrm{V}}_{n,m}$ as a subvariety.

Proposition 8.5.2. *The variety* $\underline{\mathrm{V}}^*_{n,m}$ *of binary modes defined by the identity (8.5.2) is equivalent to the variety* $\underline{\mathrm{R}}^*_{n,m}$ *of affine spaces, where*

$$R^*_{n,m} := \mathbb{Z}[X]/\langle X^n(1-X)^m - 1\rangle.$$

Proof. First note that the operation

$$xyzP = (y^n xy^{m-1})(y^{n-1} zy^m)$$

is a Mal'cev operation. Indeed,

$$xxzP = x^n zx^m = z,$$

$$xzzP = z^n xz^m = x.$$

For any $\underline{V}^*_{n,m}$-groupoid $(G, \cdot) = (G, \underline{r})$ with r a generator of $R^*_{n,m}$,

$$x^n y = x(1 - r^n) + yr^n,$$

$$yx^m = y(1-r)^m + x[1 - (1-r)^m].$$

(Cf. Sections 5.7 and 5.5.) Hence

$$\begin{aligned} x^n y x^m &= [x(1-r^n) + yr^n](1-r)^m + x[1-(1-r)^m] \\ &= x[(1-r^n-1)(1-r)^m + 1] + yr^n(1-r)^m \\ &= x(-r^n(1-r)^m + 1) + yr^n(1-r)^m = y. \end{aligned}$$

Equating coefficients of x shows that $x(r^n(1-r)^m - 1) = 0$, so that $R^*_{n,m}$ is the quotient $\mathbb{Z}[X]/\langle X^n(1-X)^m - 1\rangle$. □

Since one of the Mal'cev identities is left regular, and the other is right regular, none of the varieties $\underline{V}^*_{n,m}$ can contain a reductive variety as a subvariety. One immediately obtains the following.

Theorem 8.5.3. *Let m, n, k, l be natural numbers different from 0 and 1. Then each non-trivial subvariety $\underline{A}$ of any variety $\underline{V}^*_{n,m}$ of affine spaces is independent of each non-trivial subvariety $\underline{R}$ of any reductive variety $\underline{R}_{k,l}$. In particular,*

$$\underline{A} \cap \underline{R} = \underline{\mathrm{Tr}},$$

$$\underline{A} \vee \underline{R} = \underline{A} \,\underline{\diamondsuit}\, \underline{R} = \underline{A} \times \underline{R}.$$

Proof. Note that the variety $\underline{V}^*_{n,m}$ is defined by the identity

$$x^n y x^m = y,$$

while the variety $\underline{R}_{k,l}$ is defined by the identity

$$x^k y x^l = x.$$

It follows that $\underline{V}^*_{n,m}$ also satisfies the identity

$$x^{LCM(n,k)} y x^{LCM(m,l)} = y,$$

and that $\underline{R}_{k,l}$ satisfies the identity

$$x^{LCM(n,k)} y x^{LCM(m,l)} = x.$$

Hence the varieties $\underline{V}^*_{n,m}$ and $\underline{R}_{k,l}$ are indeed independent. □

Example 8.5.4. Note that the identity $y = yx^m$ of (8.5.1) is the m-cyclic (or m-symmetric) identity. Recall that the symmetric identity (S) of Section 5.5.1 coincides with the 2-cyclic (or 2-symmetric) identity. The variety $\underline{V}_{0,m}$ will be called an m-cyclic variety. Fix $m \geq 2$. It is easy to see that the m-cyclic variety contains infinitely many subvarieties $\underline{V}_{n,m}$ equivalent to varieties of affine spaces. In fact, $\underline{V}_{n,m} = \underline{R}_{n,m}$ for the ring $R_{n,m} \cong \mathbb{Z}[X]/\langle x^n - 1, (1 - X)^m - 1\rangle$. On the other hand, the variety $\underline{V}_{0,m}$ also contains infinitely many subvarieties $\underline{R}_n \cap \underline{V}_{0,m}$ of left n-reductive binary modes. The variety $\underline{V}_{0,m}$ itself is neither reductive, nor equivalent to a variety of affine spaces. Any two varieties $\underline{R}_n \cap \underline{V}_{0,m}$ and $\underline{V}_{n,m}$ are independent, so that each groupoid in the join of $\underline{R}_n \cap \underline{V}_{0,m}$ and $\underline{V}_{n,m}$ is a direct product of an $\underline{R}_n \cap \underline{V}_{0,m}$-mode and an $\underline{V}_{n,m}$-mode. More general m-cyclic modes will be discussed in Section 8.6. □

Theorem 8.5.3 has the following generalization.

8. BINARY MODES

Theorem 8.5.5. *Each variety of binary modes equivalent to a variety of affine spaces is independent of each reductive variety.*

Proof. Let $\underline{R}$ be any reductive and $\underline{A}$ any affine variety of binary modes. Let S be the ring of the variety $\underline{A}$, so that $\underline{A} = \underline{S}$. Then for each $\underline{A}$-algebra $(G, \cdot)$, the multiplication can be written as $\cdot = \underline{r}$ for some r in S. Moreover, the Mal'cev operation P, as a derived operation on $(G, \cdot)$, can be written as $xyzP = xyzw$ for some groupoid word w. In particular, $xxyP = xxyw$. Without loss of generality, one may assume that x is the leftmost variable of $xxyw$. (Otherwise take $yxxP$ instead of $xxyP$.) Let $x \circ y := xxyPxxP$. Then in each $\underline{A}$-algebra, one has $x \circ y = y$ as a true identity, and consequently also

$$(8.5.3) \qquad x \circ (x \circ \ldots (x \circ y) \ldots) = y$$

for any number k of repetitions of x.

Now we will show that any left m-reductive mode $(A, \cdot)$ satisfies the identity

$$(8.5.4) \qquad x^m \circ y := x \circ (x \circ \ldots (x \circ y) \ldots) = x,$$

where x is repeated m times. In fact, we will show more, namely that the identity (8.5.4) holds for any binary word $x \circ y$. Let a, b be in A, and consider the congruences λ_i of Section 5.7. Recall that in a left m-reductive mode $\widehat{A} = \lambda_0 \leq \lambda_1 \leq \cdots \leq \lambda_m = A \times A$. Since $(A^{\lambda_{m-1}}, \cdot)$ is a left zero band, it follows that $a^{\lambda_{m-1}} \circ b^{\lambda_{m-1}} = a^{\lambda_{m-1}}$. Hence $(a \circ b, a) \in \lambda_{m-1}$. Since $(a^{\lambda_k}, \cdot)$ is left k-reductive, the quotient $(a^{\lambda_k}, \cdot)^{\lambda_{k-1}}$ is a left zero band. As before, one obtains $(a^k \circ b, a) \in \lambda_{m-k}$, etc., and finally in the m-th step

$$(a^m \circ b, a) \in \lambda_0 = \widehat{A},$$

i.e. $a^m \circ b = a$. It follows that $x^m \circ y$ is a decomposition operation for the varieties $\underline{A}$ and $\underline{R}_m$.

Similarly, one shows that $\underline{\underline{A}}$ and $\underline{\underline{R}}_n^{OP}$ are independent. Let $y * x := xxyPxxP$. Then evidently each $\underline{\underline{A}}$-algebra satisfies the identity $y * x = y$, and hence also
$$y * x^n := (\ldots((y * x) * x)\ldots) * x = y,$$
where x is repeated n times. The proof that each right n-reductive mode $(A, \cdot)$ satisfies the identity
$$y * x^n = x$$
is dual to the corresponding proof for $\circ$.

Now consider the word $(x^m \circ y) * x^n$. It is clear that the variety $\underline{\underline{A}}$ satisfies the identity
$$(x^m \circ y) * x^n = y.$$
Both reductive varieties $\underline{\underline{R}}_m$ and $\underline{\underline{R}}_n^{OP}$, and hence also the variety $\underline{\underline{R}}_{m,n} = \underline{\underline{R}}_m \times \underline{\underline{R}}_n^{OP}$, satisfy the identity
$$(x^m \circ y) * x^n = x.$$
Consequently, $\underline{\underline{A}}$ and $\underline{\underline{R}}_{m,n}$ are independent. □

As we already know, the variety $\underline{\underline{BM}}$ is generated by the class of binary modes equivalent to affine spaces. In fact, $\underline{\underline{BM}}$ is generated by the set $\mathfrak{F}_{pr}$ considered in Section 8.2. Reductive binary modes provide another class of generators.

Let q be a positive integer not equal to 1. Consider the following sequence
$$\mathbb{Z}_q \le \mathbb{Z}_{q^2} \le \cdots \le \mathbb{Z}_{q^n} \le \ldots$$
of $\mathbb{Z}$-modules and its directed colimit $\mathbb{Z}_{q^\infty} = \varinjlim \mathbb{Z}_{q^n}$. This is the colimit of the functor $F : (\mathbb{Z}^+, \le) \to \underline{\underline{\mathrm{Mod}}}_\mathbb{Z}$ that acts on morphisms as follows:

$$\begin{array}{ccc} n+1 & & \mathbb{Z}/q^{n+1}\mathbb{Z} \cong \mathbb{Z}_{q^{n+1}} \\ \uparrow & \mapsto & \uparrow \\ n & & \mathbb{Z}/q^n\mathbb{Z} \cong \mathbb{Z}_{q^n} \end{array}$$

with the $\mathbb{Z}$-module embeddings

$$\mathbb{Z}/q^n\mathbb{Z} \to \mathbb{Z}/q^{n+1}\mathbb{Z}; \quad x + q^n\mathbb{Z} \mapsto qx + q^{n+1}\mathbb{Z}.$$

For each positive integer n, consider the reducts

$$(\mathbb{Z}_{q^n}, \cdot) = (\mathbb{Z}_{q^n}, \underline{q}) \text{ and } (\mathbb{Z}_{q^\infty}, \cdot) = (\mathbb{Z}_{q^\infty}, \underline{q}).$$

Since

$$x^n y = xy\underline{q}^n$$

(cf. Section 6.4), it follows that the groupoid $(\mathbb{Z}_{q^n}, \cdot)$ satisfies the identity $x^n y = x$, and hence is left n-reductive. Note that $(\mathbb{Z}_{q^n}, \cdot)$ does not satisfy the identity $x^{n-1} y = x$, and that for each positive n, the groupoid $(\mathbb{Z}_{q^n}, \cdot)$ is a subgroupoid of $(\mathbb{Z}_{q^\infty}, \cdot)$. It follows that for no positive n is the groupoid $(\mathbb{Z}_{q^\infty}, \cdot)$ left n-reductive. Let $\underline{\underline{V}}_q$ be the variety of binary modes generated by $(\mathbb{Z}_{q^\infty}, \cdot)$.

Lemma 8.5.6. *The variety $\underline{\underline{V}}_q$ of binary modes generated by $(\mathbb{Z}_{q^\infty}, \cdot)$ contains varieties equivalent to varieties $\underline{\mathrm{GF}(p)}$ of affine spaces for all prime $p > q > 1$.*

Proof. First note that all the groupoids $(\mathbb{Z}_{q^n}, \cdot)$ are in $\underline{\underline{V}}_q$, and hence also their product $\prod(\mathbb{Z}_{q^n} \mid n \in \mathbb{Z}^+)$. The universality property for products implies that the diagram below is commutative.

$$\begin{array}{ccc} \prod(\mathbb{Z}_{q^n} \mid n \in \mathbb{Z}^+) & \xrightarrow{\pi_n} & \mathbb{Z}/q^n\mathbb{Z} \\ {\scriptstyle f}\uparrow & & \| \\ \mathbb{Z} & \xrightarrow{f_n} & \mathbb{Z}/q^n\mathbb{Z} \end{array}$$

In the diagram, the mappings f_n are defined by $z \mapsto z + q_n\mathbb{Z}$, and the π_n are projections. The unique homomorphism f that makes the diagram commutative is an embedding, so that we can consider $(\mathbb{Z}, \cdot) = (\mathbb{Z}, \underline{q})$ to be a subalgebra of $\prod(\mathbb{Z}_{q^n} \mid n \in \mathbb{Z}^+)$, and hence a member of the variety $\underline{\underline{V}}_q$.

It follows that for each $n \geq 2$, the quotient $(\mathbb{Z}/n\mathbb{Z}, \underline{q}) \cong (\mathbb{Z}_n, \underline{q}) = (\mathbb{Z}_n, \cdot)$ is also in the variety $\underline{V}_q$. In particular, if n is a prime number p and $0, 1 \neq q < p$, then $(\mathbb{Z}_p, \cdot) = (\mathbb{Z}_p, \underline{q})$ are in $\underline{V}_q$. Now each $(\mathbb{Z}_p, \cdot)$ is equivalent to an affine space $(\mathbb{Z}_p, \underline{\mathbb{Z}}_p)$. It follows that $\underline{V}_q$ contains the varieties equivalent to $\underline{\mathrm{GF(p)}}$ for $p > q$. □

Theorem 8.5.7. *The variety* $\underline{\mathrm{BM}}$ *of binary modes is generated by the left reductive subvarieties. (Similarly, it is generated by the right reductive subvarieties.)*

Proof. Let $\underline{V}$ be a subvariety of $\underline{\mathrm{BM}}$ containing the varieties $\underline{R}_m$. Then $\underline{V}$ contains all the varieties $\underline{V}_q$ for any positive number $q \geq 2$, and hence each variety generated by the groupoid $(\mathbb{Z}_p, \underline{q})$, where p is a prime number and $0, 1 \neq q < p$. By Theorem 8.2.8, it follows that the variety $\underline{V}$ coincides with the variety $\underline{\mathrm{BM}}$. □

Exercises

8.5A. Use Exercise 8.4A to generalize Theorem 8.5.5 for modes of any type.

8.5B. Show that the variety $\underline{V}_2$ is contained in the variety $\underline{S}$ of symmetric binary modes.

8.5C. Show that each dually n-cyclic variety $\underline{V}_{n,0}$ contains infinitely many subvarieties equivalent to varieties of affine spaces, and infinitely many subvarieties of right m-reductive binary modes. Provide examples of $\underline{V}_{n,0}$-modes that are not direct products of an affine space and a reductive groupoid.

8.6. Beyond reductive and affine binary modes

We have seen in the previous section that the joins of certain chains of varieties of reductive modes may be neither reductive nor affine. The aim of this section is to provide more information about binary modes that lie in varieties beyond the joins of reductive and affine varieties.

8. BINARY MODES

We start with a discussion of certain congruences on binary modes. It will provide a tool to describe a structure of certain non-reductive and non-affine binary modes.

In particular, we will investigate *n-cyclic* (or *n-symmetric*) modes. More generally, we will discuss right quasigroup modes and groupoids equivalent to them.

Let $(A, \cdot)$ be a binary mode. Consider two familiar chains of equivalences on A, the chain

$$\lambda_0 \leq \lambda_1 \leq \cdots \leq \lambda_k \leq \cdots$$

of Theorem 5.7.4, defined by

$$(a,b) \in \lambda_k :\Leftrightarrow \forall x \in A,\ x^k a = x^k b,$$

and the chain

$$\rho_0 \leq \rho_1 \leq \cdots \leq \rho_k \leq \cdots$$

of Section 8.4, defined by

$$(a,b) \in \rho_k :\Leftrightarrow \forall x \in A,\ ax^k = bx^k.$$

Note that all the λ_k and ρ_k are congruences on $(A, \cdot)$. (Cf. Exercise 8.4A.) Moreover, for each a in A, the λ_k-class a^{λ_k} is left k-reductive and the ρ_k-class a^{ρ_k} is right k-reductive. (Cf. Exercise 8.4A.) Define

$$\lambda := \sum_{k \in \mathbb{N}} \lambda_k \quad \text{and} \quad \rho := \sum_{k \in \mathbb{N}} \rho_k.$$

Note that, by the results of Section 3.2, one has

$$\sum_{k \in \mathbb{N}} \lambda_k = \bigcup_{k \in \mathbb{N}} \lambda_k \quad \text{and} \quad \sum_{k \in \mathbb{N}} \rho_k = \bigcup_{k \in \mathbb{N}} \rho_k.$$

In particular, $(a,b) \in \lambda$ iff there is $j \in \mathbb{N}$ such that $(a,b) \in \lambda_j$. A similar remark holds for ρ. Note also that

$$(a,b) \in \lambda_1 \text{ iff } R_a = R_b.$$

Similarly,
$$(a,b) \in \rho_1 \text{ iff } L_a = L_b.$$

Here, L_a and R_a are the left and right multiplications by an element a, as defined in Example 4.1.3. Define two further sequences of equivalences on A as follows:

(8.6.1) $$\varphi_0 = \psi_0 = \lambda_0 = \rho_0 = \widehat{A},$$

and if $k \neq 0$, then

$(a,b) \in \varphi_k :\Leftrightarrow \forall (x_1, \ldots, x_k) \in A^k, \; x_1(\ldots(x_{k-1} \cdot x_k a)\ldots) = x_1(\ldots(x_{k-1} \cdot x_k b)).$
$(a,b) \in \psi_k :\Leftrightarrow \forall (x_1, \ldots, x_k) \in A^k, \; (\ldots(ax_1 \cdot x_2)\ldots)x_k = (\ldots(bx_1 \cdot x_2)\ldots)x_k.$

Note that $\varphi_1 = \lambda_1$ and $\psi_1 = \rho_1$. Moreover,

(8.6.2) $$\varphi_0 \leq \varphi_1 \leq \cdots \leq \varphi_k \leq \cdots,$$

(8.6.3) $$\psi_0 \leq \psi_1 \leq \cdots \leq \psi_k \leq \cdots.$$

Define
$$\varphi = \bigcup_{k \in \mathbb{N}} \varphi_k \text{ and } \psi = \bigcup_{k \in \mathbb{N}} \psi_k.$$

Lemma 8.6.1. *Let k be a natural number. Let $(A, \cdot)$ be a binary mode. Then the following hold:*

(a) *All the relations $\varphi_k, \psi_k, \varphi$ and ψ are congruences on $(A, \cdot)$;*

(b) *For each a in A, the φ_k-class a^{φ_k} is left k-reductive and the ψ_k-class a^{ψ_k} is right k-reductive;*

(c) *For each a in A and $i < k$, the quotient $(a^{\varphi_k})^{\varphi_i}$ is left $k - i$-reductive and the quotient $(a^{\psi_k})^{\psi_i}$ is right $k - i$-reductive.*

(d) *The relations φ_k, λ_1 and ψ_k, ρ_1 are related as follows:*

(8.6.4) $$A^{\varphi_k} \cong (\ldots((A^{\lambda_1})^{\lambda_1})\ldots)^{\lambda_1} \text{ and } A^{\psi_k} \cong (\ldots((A^{\rho_1})^{\rho_1})\ldots)^{\rho_1},$$

where each of λ_1 and ρ_1 appears k times on the right hand side.

Proof. (a) The easy proof is left as Exercise 8.6A.

(b) It is evident that for each positive k, one has $\varphi_k \leq \lambda_k$ and $\psi_k \leq \rho_k$. So the claim of (b) follows by the fact that the λ_k-classes are left k-reductive and the ρ_k-classes are right k-reductive.

(c) Let $(a,b) \in \varphi_k$. Then for each $(x_1,\ldots,x_k)$ in A^k, one has

$$x_1(\ldots(x_{k-1} \cdot x_k a)\ldots) = x_1(\ldots(x_{k-1} \cdot x_k b)\ldots).$$

In particular, for $x_{i+1},\ldots,x_k$ equal to a, one has

$$x_1(\ldots(x_{i-1} \cdot x_i a)\ldots) = x_1(\ldots(x_{i-1} \cdot x_i(a^{k-i}b))\ldots).$$

It follows that $(a, a^{k-i}b) \in \varphi_k$, whence

$$a^{\varphi_k} = (a^{\varphi_k})^{k-i} b^{\varphi_k}.$$

This shows that the quotient $(a^{\varphi_k})^{\varphi_i}$ of a^{φ_k}, by the restriction of φ_i to this subgroupoid, is left $k - i$-reductive. The proof for ψ_k is obtained dually.

(d) Clearly $\lambda_1 = \varphi_1$ and $\rho_1 = \psi_1$. Now suppose that (8.6.4) holds for k. We show by induction that $A^{\varphi_{k+1}}$ has the required form. (The proof for $A^{\psi_{k+1}}$ is similar.) First note that $(a^{\varphi_k}, b^{\varphi_k}) \in \lambda_1$ iff for each x^{φ_k} one has

$$(xa)^{\varphi_k} = x^{\varphi_k} a^{\varphi_k} = x^{\varphi_k} b^{\varphi_k} = (xb)^{\varphi_k},$$

i.e. for each $(x_1,\ldots,x_k)$ in A^k one has

$$x_1(\ldots(x_k \cdot xa)\ldots) = x_1(\ldots(x_k \cdot xb)\ldots).$$

But this holds precisely when $(a,b) \in \varphi_{k+1}$. Consequently, by induction,

$$A^{\varphi_{k+1}} \cong (A^{\varphi_k})^{\lambda_1} = (\ldots((A^{\lambda_1})^{\lambda_1})\ldots)^{\lambda_1},$$

with $k+1$ appearances of λ_1 on the right hand side. $\square$

A binary mode is *left cancellative* if it satisfies the quasi-identity

$$xy = xz \to y = z,$$

and is *right cancellative* if it satisfies the quasi-identity

$$yx = zx \to y = z.$$

Obviously, cancellative binary modes are both left and right cancellative.

Proposition 8.6.2. *Let $(A, \cdot)$ be a binary mode such that there is a positive k with $\varphi_{k+1} = \varphi_k$ so that*

$$\varphi_0 < \varphi_1 < \cdots < \varphi_k = \varphi_{k+1}.$$

Then either $(A, \cdot)$ is left k-reductive, or, for each $a \in A$, the φ_k-class a^{φ_k} is left k-reductive and the quotient A^{φ_k} is left cancellative.

Proof. We have to consider two cases: $\varphi_k = A \times A$ and $\varphi_k < A \times A$.

If $\varphi_k = A \times A$, then for all a and b in A, one has $(a,b) \in \varphi_k$. This means that for each $(x_1, \ldots, x_k)$ in A^k,

$$x_1(\ldots(x_k a)\ldots) = x_1(\ldots(x_k b)\ldots).$$

However, by Lemma 5.7.2, this is equivalent to the fact that the identity

$$x^k y = x$$

holds for the groupoid $(A, \cdot)$, whence $(A, \cdot)$ is left k-reductive.

Now assume that $\varphi_k < A \times A$. By Lemma 8.6.1

$$A^{\varphi_{k+1}} \cong (A^{\varphi_k})^{\lambda_1} = A^{\varphi_k}.$$

In particular, $(a^{\varphi_k}, b^{\varphi_k}) \in \lambda_1$ iff for each x^{φ_k}, one has $x^{\varphi_k} a^{\varphi_k} = x^{\varphi_k} b^{\varphi_k}$. But since λ_1 is the equality relation on A^{φ_k}, it follows that $a^{\varphi_k} = b^{\varphi_k}$, whence A^{φ_k} is left cancellative. By Lemma 8.6.1, each φ_k-class a^{φ_k} is left k-reductive. □

Note that the condition of Proposition 8.6.2 holds in any binary mode satisfying the quasi-identity

$$x_1(\ldots(x_k \cdot xy)\ldots) = x_1(\ldots(x_k \cdot xz)\ldots) \to x_1(\ldots(x_ky)\ldots) = x_1(\ldots(x_kz)\ldots).$$

The dual proposition holds for the congruences ψ_i.

The definition of the congruences φ_i (and dually ψ_i) may easily be extended as follows. Let $(A, \cdot)$ be a binary mode, and let α be an ordinal number. Define the relation φ_α on the set A as follows. For $\alpha = 0$, it is the familiar $\varphi_0 = \lambda_0$. If $\alpha \geq 0$, then

$$(a, b) \in \varphi_{\alpha+1} :\Leftrightarrow \forall x \in A, \ (xa, xb) \in \varphi_\alpha,$$

and for a limit ordinal α,

$$\varphi_\alpha := \bigcup_{\beta < \alpha} \varphi_\beta.$$

Note that $\varphi_\alpha \leq \varphi_\beta$ whenever $\alpha \leq \beta$, and that there is an ordinal γ such that $\varphi_\gamma = \varphi_{\gamma+1}$. Denote this relation φ_γ by $\bar\varphi$. The relations ψ_α and $\bar\psi$ are defined dually. Exercise 8.6E shows that the relations $\varphi_\alpha, \bar\varphi, \psi_\alpha$ and $\bar\psi$ are all congruence relations on $(A, \cdot)$. The proof of the following corollary is left as Exercise 8.6F.

Corollary 8.6.3. *For any binary mode $(A, \cdot)$, the quotients A^{φ_α} and $A^{\bar\varphi}$ are left cancellative.* □

Let $\underline{LCl}, \underline{RCl}$ and $\underline{Cl}$ denote the quasivarieties of left cancellative, right cancellative, and cancellative binary modes respectively. Proposition 8.6.2 shows that each binary mode $(A, \cdot)$ with a finite chain of different φ_i is either a member of some variety $\underline{R}_k$, or is a member of the Mal'cev product $\underline{R}_k \circ \underline{LCl}$. If a non-reductive $\underline{R}_k \circ \underline{LCl}$-mode $(A, \cdot)$ has a right cancellative quotient A^{φ_k}, then $(A, \cdot)$ is a member of $\underline{R}_k \circ \underline{Cl}$. By Theorem 7.7.1, this means that the quotient A^{φ_k} embeds as a subreduct into an affine space. We will show that this happens in particular if the mode $(A, \cdot)$ is right cancellative.

Lemma 8.6.4. *If $(A, \cdot)$ is a right cancellative binary mode, then each quotient A^{φ_α}, along with $A^{\bar\varphi}$, is also right cancellative.*

Proof. First, we will show that A^{φ_1} is right cancellative. Note that right cancellativity of A^{φ_1} means that for x, y, a in A, the equality $x^{\varphi_1} a^{\varphi_1} = (xa)^{\varphi_1} = (ya)^{\varphi_1} = y^{\varphi_1} a^{\varphi_1}$ implies the equality $x^{\varphi_1} = y^{\varphi_1}$. This is equivalent to the satisfaction of the implication

$$z \cdot xa = z \cdot ya \to zx = zy$$

in $(A, \cdot)$. But the implication obviously holds in $(A, \cdot)$, since $(A, \cdot)$ is right cancellative and $z \cdot xa = zx \cdot za = zy \cdot za = z \cdot ya$.

Now assume that A^{φ_α} is right cancellative for some ordinal α. Then by Lemma 8.6.1 (d), one has $A^{\varphi_{\alpha+1}} \cong (A^{\varphi_\alpha})^{\varphi_1}$. Hence, by the first part of the proof, $A^{\varphi_{\alpha+1}}$ is also right cancellative.

If α is a limit ordinal and for all $\beta < \alpha$, the quotient A^{φ_β} is right cancellative, then $(a, b) \in \varphi_\alpha$ iff there is $\beta < \alpha$ such that $(a, b) \in \varphi_\beta$ with A^{φ_β} right cancellative. In particular, $\beta + 1 < \alpha$, and $A^{\varphi_{\beta+1}}$ is right cancellative with $(a, b) \in \varphi_{\beta+1}$. This means that for any $z \in A$, one has

$$(za, zb) \in \varphi_\beta.$$

Now assume that for x, y, a in A,

$$x^{\varphi_\alpha} a^{\varphi_\alpha} = y^{\varphi_\alpha} a^{\varphi_\alpha},$$

i.e. $(xa, ya) \in \varphi_\alpha$. This means that for some $\beta < \alpha$, one has $(xa, ya) \in \varphi_{\beta+1}$, whence for any z in A,

$$(z \cdot xa, z \cdot ya) \in \varphi_\beta.$$

In other words,

$$z^{\varphi_\beta} x^{\varphi_\beta} \cdot z^{\varphi_\beta} a^{\varphi_\beta} = (z \cdot xa)^{\varphi_\beta} = (z \cdot ya)^{\varphi_\beta} = z^{\varphi_\beta} y^{\varphi_\beta} \cdot z^{\varphi_\beta} a^{\varphi_\beta}.$$

By the right cancellativity of A^{φ_β}, this implies that

$$z^{\varphi_\beta} x^{\varphi_\beta} = z^{\varphi_\beta} y^{\varphi_\beta},$$

i.e. that $(x,y) \in \varphi_{\beta+1}$, And since $\varphi_{\beta+1} < \varphi_\alpha$, this implies that

$$x^{\varphi_\alpha} = y^{\varphi_\alpha},$$

yielding the right cancellativity of A^{φ_α}. The lemma follows by the Principle of Transfinite Induction 0.8.1. □

Corollary 8.6.5. *For each right cancellative binary mode $(A, \cdot)$, the quotients A^{φ_α} and $A^{\bar{\varphi}}$ are cancellative.* □

The dual statement holds for left cancellative binary modes.

In particular, it follows that each of A^{φ_α} and $A^{\bar{\varphi}}$ embeds as a subreduct into an affine space. By Theorem 3.4.1, Corollary 3.7.14, and the results of Section 7.7, it is clear that each right cancellative binary mode A always possesses a congruence, say γ, such that the quotient A^γ is the replica of A in an appropriate quasivariety of subreducts of affine spaces.

Recall that a left (right) division binary mode $(A, \cdot)$ is a mode such that the mapping L_a (the mapping R_a) is surjective, i.e. such that for all a and b in A there is x in A with $ax = b$ (with $xa = b$ respectively). A division mode is both a left and a right division mode. A right cancellative and right division binary mode is a right quasigroup. (Similarly a left cancellative and left division binary mode is a left quasigroup.) Exercise 8.6D shows that the three "right notions" ("left notions") coincide for finite binary modes. In particular, a finite cancellative binary mode is a quasigroup, and hence (is equivalent to) an affine space.

Corollary 8.6.6. *Each finite right quasigroup mode $(A, \cdot)$ is either left k-reductive for some natural number k, or else, for each a in A, the φ_k-class a^{φ_k} is left k-reductive and the quotient A^{φ_k} is a quasigroup.* □

Note that the quasigroup A^{φ_k} is naturally quasi-ordered. Hence the (finite) right quasigroup $(A, \cdot)$ can be reconstructed from its k-reductive fibres and the quotient A^{φ_k} by means of the Lallement sum construction. However, at the time of writing, we do not yet know if all the cancellative subreducts of infinite affine spaces are naturally quasi-ordered.

Let us return to finite right quasigroup modes. If both the divisions of the quasigroup A^{φ_k} of Corollary 8.6.6 can be defined in terms of multiplication, we can give a more detailed description of the finite right quasigroup mode $(A, \cdot)$. First, note the following.

Lemma 8.6.7. *Each finite groupoid $(G, \cdot)$ satisfies the identities*

(8.6.5) $$xy^{m+j} = xy^m \quad \text{and} \quad y^{n+l}x = y^n x$$

for some natural numbers j, l, m, n.

Proof. First note that for each y in G, the right multiplication

$$R_y : G \to G; \; g \mapsto gy$$

generates a finite cyclic monoid. Hence there is an index i and a period p such that $R_y^{i+p} = R_y^i$. This implies that for each x in G,

$$xy^{i+p} = xy^i,$$

and consequently for x, y in G one has

$$xy^{m+j} = xy^m,$$

where m is maximal amongst all the indices and j is the least common multiple of all the periods.

Analogously, there are n and l such that the second identity holds. □

Proposition 8.6.8. *Each finite quasigroup satisfies the identities*

(8.6.6) $$xy^j = x \text{ and } y^l x = x$$

for some natural numbers j and l.

Proof. By Lemma 8.6.7, the quasigroup $(Q, \cdot)$ satisfies the identities (8.6.5) for some m, n, j, l. The identities (8.6.6) are obtained by dividing the first of (8.6.5) m times from the right and the second of (8.6.5) n times from the left. □

Similarly, one proves the following.

Proposition 8.6.9. *Each finite right quasigroup satisfies the identity $xy^j = x$ for some natural number j, i.e. it is a j-cyclic mode.* □

On the other hand, each j-cyclic mode $(A, \cdot)$ is a right quasigroup. Indeed, for any a, b in A, the equation $xa = b$ has the (unique) solution $x = ba^{j-1}$. Moreover, the right division is defined as $b/a = ba^{j-1}$.

Corollary 8.6.10. *Each finite j-cyclic mode is a member of some quasivariety $\underline{R}_k \circ \underline{V}_{l,j}$, and a Lallement sum of k-reductive modes over a $\underline{V}_{l,j}$-space.* □

Note however that there are j-cyclic modes that lie in varieties beyond the reductive and affine varieties. For example, the symmetric mode $(\mathbb{Z}, \cdot) = (\mathbb{Z}, \underline{2})$ is a reduct of the affine space $(\mathbb{Z}, P, \underline{\mathbb{Z}})$, but is neither reductive nor equivalent to an affine space. (Cf. Section 6.4.) More generally, the variety $\underline{S}$ of symmetric binary modes lie beyond the reductive and affine varieties.

Right quasigroup modes are also known as "abelian quandles". General "quandles" are defined as idempotent distributive right quasigroups. They form one of the strongest invariants of knots. (See [Joyce 1982a,b] and [Winker 1984].) An *abelian quandle* or *quandle mode* can then be defined as a mode $(A, \cdot, /)$ with two binary operations satisfying the identities

(8.6.7) $$(x/y)y = x = (xy)/y.$$

(Cf. Exercise 1.6C.) Note that for an (abelian) quandle $(Q, \cdot, /)$, the right multiplications

$$(8.6.8) \qquad R_q := R(q) : Q \to Q;\ x \mapsto xq,$$

defined in Example 1.7.3 are automorphisms [Exercise 8.6G]. The set $R(Q) = \{R_q \mid q \in Q\}$ of all right multiplications generates a subgroup of the group $\operatorname{Aut} Q$ of all automorphisms of the quandle Q. This subgroup is called the *right multiplication group* [Smith 1992] or the *inner automorphism group* [Joyce 1982a] of Q, and is denoted by $RMlt\, Q$ or $\operatorname{Inn} Q$, respectively. It has a subgroup $\operatorname{Trans} Q$, called the *transvection group*, generated by the elements of the form $R_q R_s^{-1}$.

Proposition 8.6.11. *A quandle Q is abelian (i.e. it is a mode quandle) if and only if the group $\operatorname{Trans} Q$ is abelian.*

Proof. First note that the entropic law $wx \cdot yz = wy \cdot xz$ for Q can be equivalently written as

$$(8.6.9) \qquad R_x R_z^{-1} R_y = R_y R_z^{-1} R_x.$$

Indeed, by (8.6.7),

$$(wx/z)y \cdot z = (wx/z)z \cdot yz = wx \cdot yz$$
$$= wy \cdot xz = (wy/z)z \cdot xz = (wy/z)x \cdot z,$$

whence again by (8.6.7)

$$(wx/z)y = (wy/z)x.$$

Since $R_z^{-1} : Q \to Q$ is defined by $a \mapsto a/z$, this means that (8.6.9) holds.

On the other hand, $\operatorname{Trans} Q$ is abelian iff the following holds:

$$R_x R_z^{-1} R_y R_t^{-1} = R_y R_t^{-1} R_x R_z^{-1}.$$

For $t = z$, this implies (8.6.9), and shows that if $\operatorname{Trans} Q$ is abelian then Q is abelian. Conversely, if Q is abelian, then

$$R_x R_z^{-1} R_y R_t^{-1} = R_y R_z^{-1} R_x R_t^{-1}$$
$$= R_y R_t^{-1} R_x R_z^{-1},$$

which implies that $\operatorname{Trans} Q$ is abelian. $\square$

Let Q be an abelian quandle, and let e be a fixed element of Q. Consider the mapping

(8.6.10) $\qquad \Phi : Q \to \operatorname{Trans} Q; \; q \mapsto R_q R_e^{-1}.$

It is easy to see that

$$\ker \Phi = \lambda_1 = \varphi_1.$$

Consider the inner automorphism

$$\iota_e : \operatorname{Inn} Q \to \operatorname{Inn} Q; \; x \mapsto R_e^{-1} x R_e$$

of the group $\operatorname{Inn} Q$. By Exercise 8.6M, one has

$$\iota_e(\operatorname{Trans} Q) = \operatorname{Trans} Q.$$

Then by Exercise 8.6J, the set $\operatorname{Trans} Q$ becomes a quandle $(\operatorname{Trans} Q, \circ, /)$, where

$$x \circ y = \iota_e(xy^{-1})y = R_e^{-1}(xy^{-1})R_e y,$$
$$x/y = \iota_e^{-1}(xy^{-1})y = R_e^{-1}(yx^{-1})R_e y.$$

Exercise 8.6N shows that i_e is a group and a quandle automorphism on $\operatorname{Trans} Q$.

Lemma 8.6.12. *The mapping Φ of (8.6.10) is a quandle homomorphism.*

Proof. Let q and s be in Q. Then

$$\begin{aligned}
q\Phi \circ s\Phi &= R_q R_e^{-1} \circ R_s R_e^{-1} \\
&= R_e^{-1}(R_q R_e^{-1} R_e R_s^{-1}) R_e R_s R_e^{-1} \\
&= R_e^{-1} R_q R_s^{-1} R_e R_s R_e^{-1} \\
&= R_e^{-1} R_e R_s^{-1} R_q R_s R_e^{-1} \qquad \text{(by 8.6.9)} \\
&= R_s^{-1} R_q R_s R_e^{-1} \\
&= R_{qs} R_e^{-1} = (qs)\Phi,
\end{aligned}$$

since for each $x \in Q$, one has

$$\begin{aligned}
xR_s^{-1} R_q R_s &= (x/s)q \cdot s = (x/s)s \cdot qs \\
&= x \cdot qs = xR_{qs}. \quad \square
\end{aligned}$$

Recall that for an abelian quandle Q, the group $\operatorname{Trans} Q$ is abelian. So let us use an additive notation, denoting its basic binary operation by $+$ and the neutral element by 0. The automorphisms ι_e and ι_e^{-1} generate a (commutative) ring R turning the abelian group $\operatorname{Trans} Q$ into an affine space over this ring. The quandle operations $\circ$ and $/$ on $\operatorname{Trans} Q$ can be now written as

$$x \circ y = x\iota_e - y\iota_e + y = y(1 - \iota_e) + x\iota_e = yx_{\underline{L_e}},$$
$$x/y = x\iota_e^{-1} - y\iota_e^{-1} + y = y(1 - \iota_e^{-1}) + x\iota_e^{-1} = yx_{\underline{L_e}}^{-1}.$$

This proves the following.

Theorem 8.6.13. *Let Q be an abelian quandle. Let Φ be the quandle homomorphism defined by (8.6.10). Then the $\ker \Phi$-classes are left zero bands with both operations $\cdot$ and $/$ equal, and the image $\Phi(Q)$ is a subreduct of an affine space defined on the set $\operatorname{Trans} Q$.* $\square$

Corollary 8.6.14. *Each abelian quandle is a Lallement sum of left zero bands over a subreduct of an affine space.* □

Exercises

8.6A. Prove Lemma 8.6.1(a).

8.6B. Formulate and prove the dual form of Proposition 8.6.2 and Corollary 8.6.3.

8.6C. Show that each left n-reductive and left cancellative binary mode is necessarily trivial. Formulate the dual form of this statement.

8.6D. Let $(A, \cdot)$ be a finite binary mode. Show that the following conditions are equivalent:

(a) $(A, \cdot)$ is right (left) cancellative;

(b) $(A, \cdot)$ is right (left) division;

(c) $(A, \cdot)$ is a right (left) quasigroup. (Cf. Exercises 1.6C and 1.6D.)

8.6E. Show that the relations $\varphi_\alpha, \bar\varphi, \psi_\alpha$ and $\bar\psi$ are congruence relations.

8.6F. Use the Principle of Transfinite Induction 0.8.1 to prove Corollary 8.6.3.

8.6G. Show that for each quandle $(Q, \cdot, /)$, the right multiplications R_q of (8.6.6) are quandle automorphisms.

8.6H. Show that for a quandle $(Q, \cdot, /)$, idempotence and entropicity of the multiplication imply that $(Q, \cdot, /)$ is a mode. [Cf. Exercise 5.4A.]

8.6I. Let $(Q, \cdot, /)$ and $(Q', \cdot, /)$ be right quasigroups. Show that each mapping $f : Q \to Q'$ preserving the multiplication is a right quasigroup homomorphism.

8.6J. [Joyce 1982a] Let G be a group. Define two binary operations $\circ$ and $/$ on the set G as follows:

$$x \circ y := y^{-1}xy \text{ and } x/y := yxy^{-1}.$$

(a) Show that $(G, \circ, /)$ is a quandle.

(b) Show that for a quandle $(Q, \cdot, /)$, the mapping
$$R : Q \to \operatorname{Inn} Q;\ q \mapsto R_q$$
is a homomorphism from the quandle $(Q, \cdot, /)$ to the quandle $(\operatorname{Inn} Q, \circ, /)$.

(c) Show that $\operatorname{Trans} Q$ is a normal subgroup of the group $\operatorname{Inn} Q$, and that the quotient $\operatorname{Inn} Q / \operatorname{Trans} Q$ is a cyclic group.

8.6K. [Joyce 1982a] Let h be an automorphism of a group G. Define binary operations $\circ$ and $/$ on G as follows:
$$x \circ y := h(xy^{-1})y \text{ and } x/y := h^{-1}(xy^{-1})y.$$
Show that $(G, \circ, /)$ is a quandle. In particular, if h is an inner automorphism
$$\iota_a : G \to G;\ g \mapsto a^{-1}ga,$$
then

(8.6.11) $$x \circ y = a^{-1}(xy^{-1})ay \text{ and } x/y = a^{-1}(yx^{-1})ay.$$

8.6L. Show that each element of the group $\operatorname{Trans} Q$ has the form $R_{x_1}^{\varepsilon_1} \ldots R_{x_n}^{\varepsilon_n}$, where $\varepsilon_i = 1$ or $\varepsilon_1 = -1$ and $\varepsilon_1 + \cdots + \varepsilon_n = 0$.
[Hint: Show that $R_q^{-1} R_s = R_s R_{qs}^{-1}$.]

8.6M. Show that for each element e of a quandle Q, with corresponding inner automorphism ι_e of the group $\operatorname{Trans} Q$,
$$\iota_e(\operatorname{Trans} Q) = \operatorname{Trans} Q.$$

8.6N. Define a quandle structure on the group $\operatorname{Trans} Q$ by (8.6.11). Show that for any e in Q, the automorphism ι_e of $\operatorname{Trans} Q$ is also a quandle automorphism.

Notes on Chapter 8.

Our approach to the analysis of binary mode words in Sections 8.1 and 8.2 is based on (though slightly different from) the work of [S. Fajtlowicz and J. Mycielski 1974] (with some improvements given by [Jung R. Cho 1990b]) and of [J.D.H. Smith 1991a]. Some results were extended to modes with one n-ary basic operation in [Jung R. Cho 1988, 1990a]. For other characterizations of free binary modes, see also [Jung R. Cho 1986] and [J. Ježek and T. Kepka 1983b]. For general combinatorial properties of words, see [H. Minc 1957] and [M. Lothaire 1983]. For free groupoids in the variety generated by cancellative entropic groupoids, see [J. Ježek and T. Kepka 1981a]. A basis for the identities satisfied by such groupoids was found in [G. Pollák and A. Szendrei 1981].

Simple binary modes equivalent to affine spaces appeared first in [J.P. Soublin 1971], although in a different setting. Then J. Ježek and T. Kepka [1974] provided a description of all the simple distributive groupoids, and an equational characterization of all the minimal varieties of distributive groupoids. As a consequence one obtains (with slightly simpler proofs) the characterization of all simple binary modes and the description of all minimal varieties of binary modes in Section 8.3. The characterization of simple binary modes equivalent to affine spaces in this section was made possible by subsequent developments of module theory. The characterization of simple binary modes was extended to general modes in [K. Kearnes 1996] (albeit without reference to any of the earlier work). See also Chapter 10.

The characterization of reductive varieties comes from [A. Pilitowska and A. Romanowska 1998]. The lattice of subvarieties of the variety of differential modes was first described in [A. Romanowska and B. Roszkowska 1987], and the lattice of varieties of symmetric binary modes in [B. Roszkowska 1987a]. Again, see also Chapter 10.

The congruences described in Section 8.6 have appeared in several papers, e.g. [J. Ježek, T. Kepka and P. Němec 1981], [J. Ježek and T. Kepka 1983b] and [A. Pilitowska and A. Romanowska 1998]. Quandles (and in particular abelian quandles) were formulated by D. Joyce [1982a,b]. In particular, Proposition 8.6.11 goes back to [D. Joyce 1982b]. For general properties of quandles, see also [S. Winker 1984] and [P. Lizak 1991]. Right quasigroups were investigated in [T.S.R. Fuad and J.D.H. Smith 1996] and [J.D.H. Smith 1991b]. So far as we know Theorem 8.6.13 was not published in this form before. However, we have used some ideas from [D. Joyce 1982b] and [K. Kearnes 1995a] in parts of our proof.

CHAPTER IX

HIERARCHICAL STATISTICAL MECHANICS

This chapter presents one sample application of the theory of modes, to the study of statistical mechanics. The modes involved are real barycentric algebras. Sections 5.8 and 7.5 have discussed the relationship of barycentric algebras to convex sets. On the other hand, the role of convexity in classical statistical mechanics has been eloquently delineated by Wightman in his introduction to Israel's thesis [Israel 1979]. The motivation for the ideas of the current chapter came from Brooks' and Wiley's bold attempt to use information-theoretical statistical mechanics as a basis for a unified theory of biology [Brooks and Wiley 1986]. Traditional statistical mechanics has already found various applications in biology (e.g. [Demetrius 1997], [Kerner 1972]), and there has been some feedback to mathematics (e.g. [Arnold, Gundlach, Demetrius 1994]). But Brooks and Wiley proposed a dramatic step forward. Realizing that the essence of biological systems is their complex, hierarchical organization, spanning such diverse levels as the atomic, macromolecular, cellular, organismic, demographic and ecological, they sought a genuinely hierarchical information-theoretical statistical mechanics. (This is not to be confused with compartmental-model based approaches such as [Auger 1989], where the hierarchy is only quantitative, not qualitative.) Now if classical statistical mechanics is based on convex sets, then hierarchical statistical mechanics is based on whole systems of convex sets, indexed by an ordered set specifying the hierarchy. Barycentric algebras possess such a structure (Theorem 7.5.10), and thus form an ideal foundation for hierarchical statistical mechanics. The purpose of the current chapter is to establish this theory: hierarchical statistical mechanics founded on barycentric algebras.

9.1. Plan of the chapter

Classical statistical mechanics uses classical probability distributions. Section 9.2 introduces a new notation for derived operations in barycentric algebras that serves to facilitate the transition from these classical distributions. Section 9.3 examines preduals of barycentric algebras. For a barycentric algebra A, the predual A^* is the algebra of homomorphisms β from A to the extended reals $\mathbb{R}^\infty$. Elements of A^* play a similar role to that of linear functionals in analysis and linear algebra. Section 9.4 defines concepts of concavity for extended-real valued functions, and logarithmic convexity for non-negative real valued functions, on a barycentric algebra A. These concepts reduce to their traditional counterparts from analysis if A is a convex set (and the codomain is suitably restricted). Section 9.5 presents a strengthened version of the Structure Theorem 7.5.10 for barycentric algebras, as required for the proof of the major Theorem 9.8.1. Section 9.6 begins the hierarchical statistical mechanics. There is a finite set X, and a function $f : X \to A$ whose codomain is a barycentric algebra generated by the image of f. The set X is the set of observable macrostates of a system, while the values of f represent the values of observations or measurements made on X. In the very classical case of equilibrium thermodynamics (Example 9.9.3), the function f measures energy, and the barycentric algebra A is just the convex hull of the set of observable energy values. For each "potential" β in the predual A^*, a corresponding canonical distribution q^β is defined on X (Definition 9.6.1). In equilibrium thermodynamics, a finite β corresponds to a fixed temperature, and q^β describes the equilibrium distribution of macrostates at that temperature [Grandy 1987,

9. HIERARCHICAL STATISTICAL MECHANICS

Ch. 3]. More generally, if A is a convex set and β is finite, then q^β is a well-known entropy maximizer [Garrett 1991] [Grandy 1987, Ch. 2]. Using the added power offered by barycentric algebras, in this case the extended reals $\mathbb{R}^\infty$, Example 9.6.2 shows that every distribution on the finite set X is canonical, taking f to be the log-odds function and β as a subalgebra embedding. In the general situation, the inherent hierarchy is represented by the semilattice replica A^ρ of A (9.5.2), the largest semilattice quotient of A. (The classic treatments really deal with the open convex set A°, and then the semilattice replica or hierarchical structure is trivial.) In the context of Example 9.6.2, realizing an arbitrary distribution on X as canonical, the inherent hierarchy separates those points in the support of the distribution from those points outside it.

The normalizing factor (9.6.8) in the canonical distribution with potential β is the partition function $Z(\beta)$. Section 9.7 studies the properties of the non-negative real valued partition function on A^*, showing that it is surjective and logarithmically convex. In the classical situation, namely A convex and β finite, logarithmic convexity of the partition function is usually derived by identifying the Hessian of $-\log Z(\beta)$ as a negative semidefinite covariance matrix [Garrett 1991]. Of course, this method is no longer available in the general case (or even the case of [Grandy 1987, Problem 2.10] for a properly supported distribution).

Each element α of the barycentric algebra A generated by the image of f may be expressed in various ways as the expected value of a probability distribution on the generators. The entropy $H(\alpha)$ of the element α is defined (9.6.14) to be the supremum of the information-theoretic entropies of these various probability distributions yielding α. Theorem 9.8.2 then shows that the entropy is a non-negative real-valued concave function on A. Moreover, the entropy function and the negated logarithm of the partition function are mutual Legendre transforms. (Compare [Grandy 1987, §2C] for the classical case.)

The last part of the chapter offers some examples illustrating the scope of the theory. Section 9.9 presents three elementary examples. Example 9.9.1 shows how partial sums of the Riemann zeta function appear as partition functions, and examines the corresponding entropy functions. Example 9.9.2 shows how the character theory of (unpointed) semilattices [Hofmann, Mislove and Stralka 1974] [Romanowska and Smith 1997, §2] fits into the current theory. This example raises the general problem of finding a good duality theory for barycentric algebras. Example 9.9.3 makes the connection with equilibrium thermodynamics.

Sections 9.10 and 9.11 apply the current theory to an important model in mathematical biology, Eigen's phenomenological rate equations under constant total organization [Eigen 1971]. (Although classical techniques may be used on the equations, they fail to respect or reveal the underlying biological structure, and may become numerically unstable in the critical cases where this structure is non-trivial.) Example 9.11.2 presents the simplest case of the emergence of a natural hierarchy. Consider two biological species in competition for limited food resources. One of the two species appears in two distinct stages. Within the current theory, the biological system is modelled by the barycentric algebra T described in Exercise 5.8D. The hierarchy inherent in the barycentric algebra [namely its semilattice replica (9.11.11)] has two essential parts: the crossbar of the T subordinate to the vertical of the T. In the model, the crossbar of the T represents the demographic level, concerning itself with interactions between the two stages of the stage-structured species. The vertical of the T represents the ecological level, concerning itself with the competition between the two species for the limited food resources. It is curious to contemplate that the barycentric algebra T first arose in a purely universal-algebraic context, namely the classification of quasivarieties of barycentric algebras (Section 7.6). This is yet another case illustrating Wigner's doctrine of "the unreasonable

9. HIERARCHICAL STATISTICAL MECHANICS

effectiveness of mathematics in the natural sciences" [Wigner 1960].

9.2. Barycentric operations in sum notation

By Corollary 5.8.3 and Theorem 5.8.4, finitely generated free barycentric algebras are just simplices, the free generators appearing as the vertices of the simplex. By Sections 1.4 and 3.3, each element of a free algebra corresponds to a derived operation. These facts are used to give a convenient notation for derived operations in barycentric algebras, emphasizing the connection with classical finite probability distributions. Thus an n-ary derived operation on barycentric algebras may be written in the form

$$(9.2.1) \qquad \sum_{x \in X} p_x x$$

for a set X of cardinality n, with $p_x \geq 0$ and $\sum_{x \in X} p_x = 1$. For instance, each side of the skew associative law (SA) may be written in the form

$$(9.2.2) \qquad p'q'x + pq'y + pqz.$$

In this way, derived operations in barycentric algebras correspond to finite probability distributions. If a coefficient p_x in (9.2.1) is zero, then the corresponding element x of X is understood not to appear as a (necessarily fictitious) argument in the derived operation. For example, with $X = \{x_1, x_2, x_3\}$ and $(p_1, p_2, p_3) = (0, \frac{1}{2}, \frac{1}{2})$, the image of $(\infty, 0, 2)$ in $\mathbb{R}^\infty$ under the operation $\sum_{i=1}^{3} p_i x_i$ would be 1.

9.3. The predual of a barycentric algebra

Let A and B be barycentric algebras. By Corollary 5.2, the set $\underline{B}(A, B)$ of barycentric homomorphisms from A to B itself forms a barycentric algebra. For elements f and g of $\underline{B}(A, B)$, and for p in $I°$, the function $fg\underline{p} : A \to B$ is defined by

$$(9.3.1) \qquad a(fg\underline{p}) = a^f a^g \underline{p}$$

for each a in A. Now by Example 7.1.7, the ring $R(\underline{B})$ is just $\mathbb{R}$ (cf. Example 7.2.9). If a barycentric algebra A is an affine space, then the barycentric algebra $\underline{B}(A,\mathbb{R})$ is the algebra of affine functionals from A to $\mathbb{R}$ (i.e. linear functionals with respect to the vector space structure fixed by choice of an origin, followed by a translation). Recall that a barycentric algebra A is said to be *finitely separable* iff

(9.3.2) $$\forall\, a \neq b \in A,\ \exists \gamma \in \underline{B}(A,\mathbb{R}).\quad a\gamma \neq b\gamma.$$

Thus affine spaces are finitely separable. More generally, a barycentric algebra is a convex set iff it is finitely separable (Corollary 7.2.4). Given a convex set C, one may consider affine functionals $\gamma : C \to \mathbb{R}$, e.g. factorizing γ as the embedding of C in its affine hull H followed by an affine functional on H.

The only barycentric homomorphisms from a semilattice to $\mathbb{R}$ are constant. For a general barycentric algebra A, define the *predual* to be the barycentric algebra

(9.3.3) $$A^* = \underline{B}(A, \mathbb{R}^\infty).$$

Then each barycentric algebra is *separable* (in the variety $\underline{B}$), in the sense that

(9.3.4) $$\forall\, a \neq b \in A,\ \exists\, \beta \in A^*.\quad a\beta \neq b\beta$$

(Theorem 7.5.11).

Example 9.3.1. Consider the interval I as a barycentric algebra, namely as a subalgebra of its affine hull $\mathbb{R}$. For real numbers s (the *scaling factor*) and t (the *translation*), there are elements

(9.3.5) $$\beta_{s,t} : I \to \mathbb{R};\ a \mapsto as + t$$

of I^*, indeed of $\underline{B}(I,\mathbb{R})$. Other elements of I^* are given by the disjoint union of $0 \mapsto t$ with $]0,1] \to \{\infty\}$, or by the disjoint union of $1 \mapsto t$ with $[0,1[\to \{\infty\}$. Finally, there is the constant ∞ in I^*. $\square$

Example 9.3.2. Consider the two-element join semilattice S with Hasse diagram $0 \to \infty$. The only non-constant elements of S^* send 0 to a finite real and ∞ to ∞. □

Example 9.3.3. Let T be the subalgebra $(\{0\} \times I) \cup (]0,1] \times \{\infty\})$ of the product algebra $\mathbb{R} \times \mathbb{R}^\infty$. The first uniand is called the "crossbar", and the second is called the "vertical". Note that

(9.3.6) $\quad\quad \forall\, x, y \in \{0\} \times I,\ \forall\, z \in\,]0,1] \times \{\infty\},\ \forall\, p \in I^\circ,\ xz\underline{p} = yz\underline{p}.$

Consider an element β of T^* that is finite on all elements z of the interior $I^\circ \times \{\infty\}$ of the vertical. Then β is necessarily finite on the crossbar. Indeed, for an element (p^2, ∞) of $I^\circ \times \{\infty\}$, one has $x(p, \infty)\underline{p} = (p^2, \infty)$ for any x in the crossbar, and $x\beta = \infty$ would then imply $(p^2, \infty)\beta = \infty$, a contradiction. But now β has to be constant on the crossbar, since $x\beta \neq y\beta$ for elements x, y of the crossbar would imply the contradiction $(p^2, \infty)\beta = x(p, \infty)\underline{p}\beta = x^\beta(p, \infty)^\beta \underline{p} \neq y^\beta(p, \infty)^\beta \underline{p} = (p^2, \infty)\beta$. □

Exercises

9.3A Show that the barycentric algebra T of Example 9.3.3 is isomorphic to the algebra T_2 of (7.6.1).

9.3B Use the Structure Theorem 7.5.10 to decompose the predual I^* of the closed unit interval I.

9.3C Show that $\mathbb{R}^\infty$ is a subalgebra of the predual of each non-empty barycentric algebra.

9.4. Concavity and logarithmic convexity

Let A be a barycentric algebra. Let $f : A \to \mathbb{R}^\infty$ be a function. Note that f is a barycentric homomorphism iff its *graph*

(9.4.1) $\quad\quad\quad\quad \{(a, r) \in A \times \mathbb{R}^\infty \mid af = r\}$

is a subalgebra of $A \times \mathbb{R}^\infty$. Now the extended reals are totally ordered by the usual order relation $\leq$ on $\mathbb{R}$ together with $\forall\, r \in \mathbb{R}^\infty$, $r \leq \infty$. Define the *hypograph*

$$(9.4.2) \qquad \{(a,r) \in A \times \mathbb{R}^\infty \mid af \geq r\}$$

of f. Then the function $f : A \to \mathbb{R}^\infty$ is said to be *concave* iff its hypograph is a subalgebra of $A \times \mathbb{R}^\infty$. Note that this definition subsumes the usual definition for finite-valued functions with convex domain.

Proposition 9.4.1. *Let $f : A \to \mathbb{R}^\infty$ be a function. Suppose that there is a subset S of A^* such that*

$$(9.4.3) \qquad \forall\, a \in A,\ af = \inf\{a\beta \mid \beta \in S\}.$$

Then f is concave.

Proof. For a in A, r in $\mathbb{R}^\infty$, (9.4.3) implies

$$(9.4.4) \qquad af \geq r \Leftrightarrow \forall\, \beta \in S,\ a\beta \geq r.$$

Suppose (a, r) and (b, s) are elements of the hypograph of f. Consider $\underline{p}$ in I°. Then for a homomorphism β in S, one has $ab\underline{p}\beta = a^\beta b^\beta \underline{p} \geq rb^\beta \underline{p} \geq rs\underline{p}$, the inequalities following by (9.4.4) and the monotonicity of the binary operation $\underline{p}$ of $\mathbb{R}^\infty$ in each of its arguments. Using (9.4.4) again, one concludes that $(ab\underline{p}, rs\underline{p}) = (a,r)(b,s)\underline{p}$ is also an element of the hypograph of f. $\square$

For a finite, non-negative real number r, set

$$(9.4.5) \qquad -\log r = \text{ if } r = 0 \text{ then } \infty \text{ else } -\log_e r.$$

This defines a function

$$(9.4.6) \qquad -\log : [0, \infty[\, \to \mathbb{R}^\infty.$$

A finite, non-negative valued function

(9.4.7) $$f : A \to [0, \infty[$$

is then said to be *logarithmically convex* if the composite

(9.4.8) $$-\log f : A \to \mathbb{R}^\infty; \ a \mapsto -\log a^f$$

is concave. Note that this subsumes the usual definition for the case where A is convex and f takes only positive finite values.

9.5. The Structure Theorem for barycentric algebras

The main results of Section 9.8 below require an extended Structure Theorem for barycentric algebras, going beyond the presentation of Section 7.5. This Extended Structure Theorem 9.5.2 is formulated as an equivalence between the category $\underline{B}$ of real barycentric algebras and a category $\underline{T}$ of "barycentric structures," representing the hierarchically ordered systems of convex sets that underlie hierarchical statistical mechanics. Readers who are just interested in the applications, and who are willing to take the proof of Theorem 9.8.1 on trust, may wish to skip this section.

We begin by summarizing the construction techniques of Sections 4.5 and 7.5 as they apply to an individual barycentric algebra.

Definition 9.5.1. Let $(H, +)$ be a (join) semilattice. Suppose given an envelope E_h of a barycentric algebra C_h for each element h of H. For each $h \leq k$ in H (i.e. each morphism in the small category H), suppose given a barycentric homomorphism $\varphi_{h,k} : C_h \to E_k$ such that:

(a) $\varphi_{h,h}$ is the inclusion of C_h in E_h;
(b) $\forall\, \underline{p} \in I^\circ,\ (C_h \varphi_{h,h+h'})(C_{h'} \varphi_{h',h+h'})\underline{p} \subseteq C_{h+h'}$;
(c) $\forall\, h + h' \leq k,\ \forall\, a \in C_h,\ \forall b \in C_{h'},\ \forall \underline{p} \in I^\circ$,
$$(a\varphi_{h,h+h'})(b\varphi_{h',h+h'})\underline{p}\varphi_{h+h',k} = (a\varphi_{h,k})(b\varphi_{h',k})\underline{p};$$
(d) $\forall k \in H,\ E_k = \{a\varphi_{h,k} \mid h \leq k,\ a \in C_h\}$.

Then the disjoint union $B = \bigcup_{h \in H} C_h$, equipped with the operations

(9.5.1) $\qquad \underline{p} : C_h \times C_{h'} \to C_{h+h'}; (a,b) \mapsto (a\varphi_{h,h+h'})(b\varphi_{h',h+h'})\underline{p}$

for each p in $I°$, is called the *semilattice Lallement sum* of the algebras $(C_h, I°)$ over the semilattice H by the mappings $\varphi_{h,k}$, or more briefly a *semilattice Lallement sum*. □

In the general context of Definition 9.5.1, the extra condition
(9.5.2) $\qquad \forall\, h, h' \leq k \leq l \in H,\ \forall\, a \in C_h,\ \forall\, b \in C_{h'},\ a\varphi_{h,k} = b\varphi_{h',k} \Rightarrow a\varphi_{h,l} = b\varphi_{h',l}$

is also required below.

The structure of barycentric algebras is then described with the help of a category $\underline{T}$ *of barycentric structures*. Its objects are *barycentric structures*

(9.5.3) $\qquad \Phi_H = (\varphi_{h,k} : C_h \to E_k \mid h \leq k \in H).$

They consist of a (join) semilattice H, an algebraically open convex set C_h for each h in H, a convex set envelope E_h of C_h for each h in H, and a barycentric homomorphism $\varphi_{h,k} : C_h \to E_k$ for each $h \leq k$ in H, such that (9.5.2) and the conditions (a)–(d) of Definition 9.5.1 are satisfied. The morphisms $f : \Phi_H \to \Phi_{H'}$ of $\underline{T}$ are semilattice homomorphisms $f : H \to H'$, with *component* barycentric homomorphisms $f_h : E_h \to E'_{hf}$ for each h in H, restricting to barycentric homomorphisms $f : C_h \to C'_{hf}$, such that the diagrams

(9.5.4)
$$\begin{array}{ccc} C_h & \xrightarrow{f_h} & C'_{hf} \\ {\scriptstyle \varphi_{h,l}} \downarrow & & \downarrow {\scriptstyle \varphi'_{hf,lf}} \\ E_l & \xrightarrow{f_l} & E'_{lf} \end{array}$$

commute for each $h \leq l$ in H.

A *Lallement sum functor* $L : \underline{T} \to \underline{B}$ is defined by sending a barycentric structure $\Phi_H = (\varphi_{h,k} : C_h \to E_k \mid h \leq k \in H)$ to the semilattice Lallement

9. HIERARCHICAL STATISTICAL MECHANICS

sum $\Phi_H L$ of the C_h over H by the $\varphi_{h,k}$. Given a $\underline{T}$-morphism $f : \Phi_H \to \Phi_{H'}$, a mapping $fL : \Phi_H L \to \Phi_{H'} L$ is defined as the disjoint union of the components $f_h : C_h \to C'_{hf}$ over all h in H. Conversely, a *decomposition functor* $D : \underline{B} \to \underline{T}$ is defined. Let A be a barycentric algebra. The underlying semilattice H of the barycentric structure A^D is the semilattice replica A^ρ of A. For an element $k = a^\rho$ of H, the convex set C_k is the equivalence class a^ρ, a subalgebra of A. Define the *pre-envelope* $P_k = [a]$, the principal wall generated by a. The envelope E_k is defined as the quotient of P_k by the congruence

$$(9.5.5) \qquad \mu_k = \bigcap \{\ker(P_k \to C_k; x \mapsto xb\underline{p}) \mid b \in C_k\}$$

for any p in I°. For $h \leq k$ in H, define

$$(9.5.6) \qquad \varphi_{h,k} : C_h \to E_k; \; x \mapsto x^\mu.$$

This establishes the barycentric structure A^D. For a barycentric homomorphism $\theta : A \to A'$, the $\underline{T}$-morphism $\theta^D = (f : \Phi_H \to \Phi_{H'})$ is defined as follows. The semilattice homomorphism $f : H \to H'$ is the image of the $\underline{B}$-morphism θ under the semilattice replica functor $\underline{B} \to \underline{Sl}$; $(A, I^\circ) \mapsto (A^\rho, +)$. The component barycentric homomorphisms $f_h : E_h \to E_{hf}$ are given by the commutative diagrams

$$(9.5.7) \qquad \begin{array}{ccc} P_h & \xrightarrow{\theta} & P'_{hf} \\ \text{nat}\mu_h \downarrow & & \downarrow \text{nat}\mu_{hf} \\ E_h & \xrightarrow{f_h} & E'_{hf} \end{array}$$

in $\underline{B}$, where the vertical arrows are the natural projections given by the congruences (9.5.5). The extended structure theorem is then formulated as follows [Romanowska and Smith 1990b, Th. 4.6]. The proof is relegated to the exercises.

Theorem 9.5.2 (**Extended Structure Theorem for Barycentric Algebras**). *The decomposition functor $D : \underline{B} \to \underline{T}$ and Lallement sum functor $L : \underline{T} \to \underline{B}$ yield an equivalence of categories.* □

Exercises

9.5A Write out the barycentric structure T^D determined by the barycentric algebra T of Example 9.3.3.

9.5B Let A be a barycentric algebra. Let C_h and C_k, for elements h and k of the semilattice replica H of A, be convex sets in the barycentric structure A^D. For elements a of C_h and b of C_k, show that the product $ab\underline{p}$ in the barycentric algebra A^{DL} coincides with the original product $ab\underline{p}$ in A. Conclude that A is naturally isomorphic to the barycentric algebra A^{DL}.

9.5C For the barycentric structure (9.5.3), show that H is the semilattice replica of the barycentric algebra $\Phi_H L$. [Hint: Apply Proposition 7.5.2.]

9.5D Let k be an element of the semilattice H of the barycentric structure (9.5.3).

   (a) Show that
   $$\bigcup \{C_h \mid h \leq_+ k\}$$
   is the pre-envelope P_k in the barycentric algebra $\Phi_H L$.

   (b) Show that the mapping
   $$\bigcup_{h \leq_+ k} (\varphi_{h,k} : C_h \to E_k) : P_k \to E_k$$
   is a surjective barycentric algebra homomorphism.

   (c) Let λ be the kernel of the homomorphism of (b). Show that P_k^λ is isomorphic to E_k, and that λ preserves C_k.

   (d) Show that λ coincides with the congruence (9.5.5).

(e) Show that the homomorphisms nat $\mu : C_h \to P_k^\mu$ built in $\Phi_H LD$ agree with the homomorphisms nat $\lambda = \phi_{h,k} : C_h \to E_k$ of the barycentric structure Φ_H.

9.5E Use Exercises 9.5B and 9.5D to complete the proof of Theorem 9.5.2.

9.6. Canonical distributions and entropy

Let X be a finite set. The elements x of X are called *macrostates*. Let A be a (finitely generated) barycentric algebra. Let

(9.6.1) $$f : X \to A$$

be a function such that

(9.6.2) $$A = \langle Xf \rangle,$$

i.e. A is generated by the image of f. Fix a barycentric homomorphism

(9.6.3) $$\beta : A \to \mathbb{R}^\infty.$$

In the present context, the element (9.6.3) of A^* is known as the *potential*. Given (9.6.1) and (9.6.3), define the *partition function* (or *Zustandssumme* or "sum over (macro-)states")

(9.6.4) $$Z(\beta) = \sum_{x \in X} e^{-xf\beta},$$

with $e^{-xf\beta} = 0$ for $xf\beta = \infty$.

Definition 9.6.1. The *canonical distribution* on X with potential β (and function f) is the probability distribution

(9.6.5) $$q^\beta : X \to [0, 1]; x \mapsto q_x^\beta$$

given by

(9.6.6) $$-\log q_x^\beta = \log Z(\beta) + xf\beta.$$

If β is constant (i.e. $A\beta$ is a singleton subset of $\mathbb{R}^\infty$), then q^β is uniform on X. This includes the case $A\beta = \{\infty\}$. If $A\beta$ properly contains $\{\infty\}$, and $xf\beta = \infty$, then $q_x^\beta = 0$. □

Example 9.6.2 (**every distribution is canonical** [Athreya and Smith, 2000]). Let

(9.6.7) $$p: X \to [0,1]; x \mapsto p_x$$

be a probability distribution on X. Define the *log odds function*

(9.6.8) $$f: X \to A; \ x \mapsto -\log p_x,$$

where A is the subalgebra of the barycentric algebra $\mathbb{R}^\infty$ generated by $\{-\log p_x \mid x \in X\}$. Then the arbitrary distribution p is the canonical distribution with potential $\beta: A \hookrightarrow \mathbb{R}^\infty; \ a \mapsto a$ and log odds function (9.6.8). □

Given a distribution (9.6.7) on X, and a function (9.6.1), the *expected value* of f subject to p is the element

(9.6.9) $$\alpha = \sum_{x \in X} p_x(xf)$$

of A. If f in (9.6.9) is the log odds function (9.6.8), then (9.6.9) becomes the *entropy* of the distribution p. The entropy is a non-negative (finite) real number.

Fix a function (9.6.1) and potential (9.6.3). Let α be the expected value of f subject to the canonical distribution q^β. Corollary 9.6.7 below shows that, of all distributions (9.6.7) on X satisfying (9.6.9), the canonical distribution q^β is the unique distribution maximizing the entropy. For the proof of this fact, the following auxiliary result is needed.

9. HIERARCHICAL STATISTICAL MECHANICS

Proposition 9.6.3. *Let p and q be arbitrary distributions on X. Then for each x in X, one has*

$$(9.6.10) \qquad p_x - q_x \leq p_x \log p_x - p_x \log q_x,$$

equality holding if and only if $p_x = q_x$.

Proof. If $p_x = 0$, then (9.6.10) reduces to $-q_x \leq 0$, certainly true, with equality holding iff $q_x = 0 = p_x$. Otherwise, if $q_x = 0$, then (9.6.10) reduces to $p_x \leq \infty$. The equality certainly holds strictly, while $p_x = q_x$ is incompatible with this case ($p_x \neq 0$, $q_x = 0$). Otherwise, both p_x and q_x are non-zero, and (9.6.10) may be rewritten in the form

$$(9.6.11) \qquad 0 \leq \frac{q_x}{p_x} - 1 - \log \frac{q_x}{p_x}.$$

Consider the auxiliary function

$$(9.6.12) \qquad \varphi:]0, \infty[\to \mathbb{R}; \; y \mapsto y - 1 - \log y.$$

Then $\varphi'(y) = 1 - y^{-1}$ is negative on $]0, 1[$ and positive on $]1, \infty[$. Thus φ has a unique global minimum of 0 at $y = 1$. Setting $y = q_x/p_x$ verifies (9.6.11), equality holding iff $1 = q_x/p_x$. $\square$

Theorem 9.6.4. *On the finite set X of macrostates, suppose given a function (9.6.1) such that (9.6.2) holds. Then there is a function*

$$(9.6.13) \qquad H: A \to [0, \log |X|]$$

defined by

$$(9.6.14) \qquad H(\alpha) = \sup\{-\sum_{x \in X} p_x \log p_x \mid \alpha = \sum_{x \in X} p_x(xf)\}$$

for $\alpha \in A$.

Proof. Fix α in A. Since A is the barycentric algebra generated by Xf, there is certainly one distribution p on X such that (9.6.9) holds. Thus the right hand side of (9.6.14) is taking the supremum of a non-empty set. The elements of this set, as entropies of distributions on X, are non-negative (finite) real numbers. Let p be a distribution satisfying (9.6.9), and let $q : X \to \{|X|^{-1}\}$ be the uniform distribution on X. Summing (9.6.10) over all elements of X, one obtains

$$(9.6.15) \qquad 0 \leq \sum_{x \in X} p_x \log p_x + \log |X|.$$

Since the elements of the set on the right hand side of (9.6.14) are thus bounded above by $\log |X|$, the supremum exists and lies in the interval $[0, \log |X|]$. □

Definition 9.6.5. The function (9.6.13) is called the *entropy function* determined by $f : X \to \langle Xf \rangle$. □

Theorem 9.6.6. *On the finite set X of macrostates, fix a function (9.6.1) such that (9.6.2) holds. Then*

$$(9.6.16) \qquad \forall\, \alpha \in A,\ \forall\, \beta \in A^*,\ H(\alpha) \leq \log Z(\beta) + \alpha\beta.$$

Proof. In (9.6.10), let p satisfy (9.6.9), and let q be the canonical q^β of (9.6.5). Sum (9.6.10) over all x in X. Using (9.6.6) for the first equality, one obtains

$$0 \leq \sum_{x \in X} p_x \log p_x - \sum_{x \in X} p_x \log q_x^\beta$$
$$= \sum_{x \in X} p_x \log p_x + \log Z(\beta) + \sum_{x \in X} p_x (xf\beta)$$
$$= \sum_{x \in X} p_x \log p_x + \log Z(\beta) + \left[\sum_{x \in X} p_x (xf)\right] \beta$$
$$= \sum_{x \in X} p_x \log p_x + \log Z(\beta) + \alpha\beta.$$

Here the second equality holds since $\beta : A \to \mathbb{R}^\infty$ is a barycentric homomorphism, while the last equality holds by (9.6.9). Thus

$$(9.6.17) \qquad -\sum_{x \in X} p_x \log p_x \leq \log Z(\beta) + \alpha\beta.$$

9. HIERARCHICAL STATISTICAL MECHANICS

On taking the supremum over all distributions p subject to (9.6.9), one obtains (9.6.16). □

Corollary 9.6.7. *Fix (9.6.1) with (9.6.2), and β in A^*. Suppose*

$$(9.6.18) \qquad \alpha = \sum_{x \in X} q_x^\beta(xf).$$

Then the supremum in (9.6.14) is attained uniquely by q^β, and one then has

$$(9.6.19) \qquad H(\alpha) = \log Z(\beta) + \alpha\beta.$$

Proof. By Proposition 9.6.3, equality obtains in (9.6.17) if and only if $p_x = q_x^\beta$. □

Remark 9.6.8. Taking the potential $\beta : A \to \mathbb{R}^\infty$ to be the constant function ∞, Definition 6.1 interpreted q^∞ to be the uniform distribution on X. Defining

$$(9.6.20) \qquad \alpha_f := \sum_{x \in X} |X|^{-1}(xf)$$

as the *center of gravity of A under f*, consistency of (9.6.16) and Corollary 9.6.7 yield

$$(9.6.21) \qquad H(\alpha_f) = \log |X|$$

and $H(\alpha_f) = \log Z(\infty) + \alpha_f \infty$, so by convention

$$(9.6.22) \qquad \log Z(\infty) + \alpha\infty = \log |X|$$

for all α in A. □

9.7. Properties of the partition function

In the context of (9.6.1) with (9.6.2), the partition function is defined by (9.6.4) as

$$(9.7.1) \qquad Z : A^* \to [0, \infty[; \ \beta \mapsto \sum_{x \in X} e^{-xf\beta}.$$

In this section, it will be shown that the partition function is surjective and logarithmically convex. Indeed, it will be shown that

$$(9.7.2) \qquad -\log Z : A^* \to \mathbb{R}^\infty$$

is a surjective, concave function.

Lemma 9.7.1. *For* $f : X \to A = \langle Xf \rangle$ *with* $|X| < \infty$,

$$(9.7.3) \qquad \forall \alpha \in A, \ \forall \beta \in A^*, \ \alpha\beta \geq \min\{xf\beta \mid x \in X\}.$$

Proof. For $\alpha \in A = \langle Xf \rangle$, there is a distribution p on X such that (9.6.9) holds. For a barycentric homomorphism β in A^*, one then has $\alpha\beta = \left[\sum_{x \in X} p_x(xf)\right]\beta = \sum_{x \in X} p_x(xf\beta) \geq \min\{xf\beta \mid x \in X\}$. $\square$

Theorem 9.7.2. *On the finite set X of macrostates, fix a function (9.6.1) such that (9.6.2) holds. Then the negated logarithm of the partition function (9.7.1) yields a function (9.7.2) well-defined by*

$$(9.7.4) \qquad -\log Z(\beta) = \inf\{\alpha\beta - H(\alpha) \mid \alpha \in A\}$$

for β in A^*.

Proof. For α in A, (9.6.13) yields $-H(\alpha) \geq -\log|X|$. Adding this inequality to (9.7.3), one obtains

$$(9.7.5) \qquad \alpha\beta - H(\alpha) \geq \min\{xf\beta \mid x \in X\} - \log|X|.$$

9. HIERARCHICAL STATISTICAL MECHANICS

For fixed β in A^*, the (non-empty) set on the right hand side of (9.7.4) is thus bounded below. By (9.6.16), one has

$$(9.7.6) \qquad -\log Z(\beta) \leq \alpha\beta - H(\alpha)$$

for each α in A. On the other hand, for α as in (9.6.18), equation (9.6.19) yields equality in (9.7.6). This verifies equation (9.7.4). □

Corollary 9.7.3. *The partition function (9.7.1) is logarithmically convex, i.e. (9.7.2) is concave.*

Proof. For each α in A, the function

$$(9.7.7) \qquad A^* \to \mathbb{R}^\infty; \; \beta \mapsto \alpha\beta - H(\alpha)$$

is a barycentric homomorphism. By (9.7.4), $-\log Z : A^* \to \mathbb{R}^\infty$ is the infimum of these homomorphisms. Proposition 9.4.1 then shows that $-\log Z$ is concave. □

Corollary 9.7.4. *For β in A^*, one has*

$$(9.7.8) \qquad \min\{xf\beta \mid x \in X\} \leq \log|X| - \log Z(\beta) \leq \frac{1}{|X|} \sum_{x \in X} xf\beta.$$

Proof. The left hand inequality follows by (9.7.4) and (9.7.5). The right hand inequality follows by (9.6.16) with α taken to be the center of gravity (9.6.20) of A under f. □

Corollary 9.7.5. *The partition function (9.7.1) is surjective.*

Proof. It will be shown that (9.7.2) is surjective. First, note that $-\log Z(\infty) = \infty$. Now let r be a finite real number. Let $\beta : A^* \to \mathbb{R}^\infty$ be the constant function with value $\log|X| + r$. Then $-\log Z(\beta) = r$, by (9.7.8) or by direct computation. □

9.8. Canonical separability and the Legendre transform

In the context of (9.6.1) with (9.6.2), the entropy function $H : A \to [0, \log |X|]$ was defined by (9.6.14). In this section, it will be shown that the entropy function is concave, and indeed is the Legendre transform

$$(9.8.1) \qquad H(\alpha) = \inf\{\alpha\beta + \log Z(\beta) \mid \beta \in A^*\}$$

of the concave function $-\log Z : A^* \to \mathbb{R}^\infty$. The key step, invocation of Corollary 9.6.7, is achieved by a new form of separability. (Recall that classical separability of A, namely (9.3.4), entailed an adequate supply of elements of A^*.) The algebra A (generated by Xf for a function $f : X \to A$ with finite domain) is said to be *canonically separable* (with respect to f) if

$$(9.8.2) \qquad \forall\, \alpha \in A,\ \exists\, \beta \in A^*.\ \ \alpha = \sum_{x \in X} q_x^\beta(xf).$$

Theorem 9.8.1 below shows that each algebra A is canonically separable. (The proof of the theorem involves the structure theory of barycentric algebras.)

Theorem 9.8.1. *On the finite set X of macrostates, fix a function $f : X \to A$ such that $A = \langle Xf \rangle$. Then A is canonically separable.*

Proof. Fix an element α of A, lying in an algebraically open convex set C_k that is the class α^ρ of the semilattice replica congruence ρ on A. Let E_k be the envelope of C_k in the structural decomposition of A. Since A is finitely generated, E_k is a convex polytope. For each element x of $f^{-1}[\alpha]$, the inverse image of the principal wall of A generated by α, let

$$(9.8.3) \qquad \varphi_x : xf^\rho \to E_k$$

be the barycentric homomorphism to E_k from the semilattice replica class of xf appearing as part of the barycentric structure A^D.

Consider the problem of maximizing the entropy

(9.8.4) $$-\sum_{x \in f^{-1}[\alpha]} p_x \log p_x$$

of a probability distribution on $f^{-1}[\alpha]$, subject to the normalization

(9.8.5) $$1 = \sum_{x \in f^{-1}[\alpha]} p_x$$

and the condition

(9.8.6) $$\alpha = \sum_{x \in f^{-1}[\alpha]} p_x(xf\varphi_x).$$

The Lagrangean of the problem is

(9.8.7) $$L(p_x, \lambda, g) = -\sum_{x \in f^{-1}[\alpha]} p_x \log p_x + \lambda \left(1 - \sum_{x \in f^{-1}[\alpha]} p_x\right) + \left[\alpha - \sum_{x \in f^{-1}[\alpha]} p_x(xf\varphi_x)\right] \gamma$$

with an affine functional $\gamma : E_k \to \mathbb{R}$. Now the entropy (9.8.4) is strictly concave in the p_x. The constraints (9.8.5) and (9.8.6) are linear. Of course, there are also non-negativity constraints $\forall\, x \in f^{-1}[\alpha]$, $p_x \geq 0$. Since α lies in E_k, the feasible set for the maximization problem is bounded and non-empty. It follows [Lancaster 1968, §5.7] that there is an affine functional

(9.8.8) $$\gamma : E_k \to \mathbb{R}$$

yielding a solution to the maximization problem. Since α lies in the interior C_k of the feasible set E_k, the solution is given by a critical point of the Lagrangean (9.8.7). Thus for the maximization, one has

(9.8.9) $$0 = \frac{\partial L}{\partial p_x} = -\log p_x - (1 + \lambda) - xf\varphi_x\gamma.$$

for each element x of $f^{-1}[\alpha]$, whence

(9.8.10) $$p_x = \widetilde{Z}(\gamma)^{-1} \exp(-xf\varphi_x\gamma)$$

with

(9.8.11) $$\widetilde{Z}(\gamma) = \sum_{x \in f^{-1}[\alpha]} \exp(-xf\varphi_x\gamma).$$

A barycentric homomorphism $\beta : A \to \mathbb{R}^\infty$ will now be constructed. Recall that the semilattice replica of $\mathbb{R}^\infty$ is the two-element semilattice with Hasse diagram $\mathbb{R} \to \{\infty\}$. The $\underline{B}$-morphism β will be specified, using the Extended Structure Theorem 9.5.2, as the image under the Lallement sum functor $L : \underline{T} \to \underline{B}$ of a $\underline{T}$-morphism b. The semilattice homomorphism part of this $\underline{T}$-morphism is given by

(9.8.12) $$b : h \mapsto \text{if } h \leq k \text{ then } \mathbb{R} \text{ else } \{\infty\}.$$

For an element h of A^ρ, the component barycentric homomorphism b_h of the $\underline{T}$-morphism b is given by

(9.8.13) $$b_h = \text{if } h \leq k \text{ then } \varphi_{h,k}\gamma \text{ else } \infty,$$

where the barycentric homomorphism $\varphi_{h,k} : C_h \to E_k$ (for $h \leq k$) is part of the barycentric structure A^D, while γ is the barycentric homomorphism (9.8.8). The commutativity conditions (9.5.4) are readily checked. Thus $b : AD \to \mathbb{R}^\infty D$ is a $\underline{T}$-morphism, yielding the barycentric homomorphism $\beta = bL$.

Now consider the canonical distribution q^β on X with potential β. The partition function is $Z(\beta) = \sum_{x \in X} e^{-xf\beta} = \sum_{x \in f^{-1}[\alpha]} \exp(-xf\varphi_x\gamma) = \widetilde{Z}(\gamma)$ as in (9.8.11). For $x \notin f^{-1}[\alpha]$, one has $q_x^\beta = \widetilde{Z}(\gamma)^{-1}\exp(-\infty) = 0$. For $x \in f^{-1}[\alpha]$, one has $-\log q_x^\beta = \log Z(\beta) + xf\beta = \log \widetilde{Z}(\gamma) + xf\varphi_x\gamma$. Then
$$\sum_{x \in X} q_x^\beta(xf) = \sum_{x \in f^{-1}[\alpha]} q_x^\beta(xf) = \sum_{x \in f^{-1}[\alpha]} \widetilde{Z}(\gamma)^{-1}\exp(-xf\varphi_x\gamma)(xf\varphi_x) = \alpha.$$

Here the penultimate equality holds by the structure of A, while the last equality holds since (9.8.10) satisfies the constraint (9.8.6). The canonical separability of A follows. □

The canonical separability of A with respect to $f : X \to A$ enables one to recognize the Legendre transform relationship between H and $-\log Z$. (For the classical Legendre transform in convexity theory, see [Fenchel 1949].)

Theorem 9.8.2. *On the finite set X of macrostates, fix a function $f : X \to A$ such that $A = \langle Xf \rangle$. Consider the entropy function $H : A \to [0, \log |X|]$ of (9.6.14) and the function $-\log Z : A^* \to \mathbb{R}^\infty$ of (9.7.2).*

(a) $\forall\, \alpha \in A,\ \forall\, \beta \in A^*,\ \alpha\beta \geq H(\alpha) - \log Z(\beta)$.

(b) *The function $-\log Z : A^* \to \mathbb{R}^\infty$ is concave, with*

(9.8.14) $\qquad \forall\, \beta \in A^*,\ -\log Z(\beta) = \inf\{\alpha\beta - H(\alpha) \mid \alpha \in A\}.$

(c) *The function $H : A \to [0, \log |X|]$ is concave, with*

(9.8.15) $\qquad \forall\, \alpha \in A,\ H(\alpha) = \inf\{\alpha\beta + \log Z(\beta) \mid \beta \in A^*\}.$

Proof. (a) is just a restatement of (9.6.16) from Theorem 9.6.6.
(b) is just a restatement of (9.7.4) from Theorem 9.7.2, together with Corollary 9.7.3. Towards (c), note that $H(\alpha)$ is a lower bound for the set on the right hand side of (9.8.15), by (9.6.16). Now consider a fixed element α of A. By Theorem 9.8.1, there is a homomorphism β in A^* such that (9.6.18) holds. By (9.6.19) of Corollary 9.6.7, equality in (9.8.15) follows. Finally, for each element β of A^*, the function

(9.8.16) $\qquad A \to \mathbb{R};\ \alpha \mapsto \alpha\beta + \log Z(\beta)$

is a barycentric homomorphism. By (9.8.15), $H : A \to \mathbb{R}$ is the infimum of these homomorphisms. Then by Proposition 9.4.1, it is concave. □

9.9. Elementary examples

This section gives three simple examples illustrating the scope of the concepts that have been introduced. A deeper example, from mathematical biology, is presented in the two subsequent sections.

Example 9.9.1 (**the Riemann zeta function**). Take (9.6.1) to be the logarithm function

$$\text{(9.9.1)} \qquad \log : \{1, 2, \ldots, N\} \to [0, \log N]$$

for a positive integer N, the codomain being a subalgebra of the affine space $\mathbb{R}$. For a real number s, consider the barycentric homomorphism

$$\text{(9.9.2)} \qquad s : [0, \log n] \to \mathbb{R}^\infty; \ r \mapsto rs.$$

Then the partition function (9.6.4) gives the partial sum

$$\text{(9.9.3)} \qquad Z(s) = \sum_{n=1}^{N} n^{-s}$$

of the Riemann zeta function $\zeta(s)$ for $s > 1$ [Hardy and Wright 1968, (17.2.1)]. By Theorem 9.8.2, the concave function $-\log Z(s)$ Legendre-transforms to a surjective, concave entropy function

$$\text{(9.9.4)} \qquad H : [0, \log N] \to [0, \log N].$$

Note that $H(0) = H(\log N) = 0$. The entropy function attains its maximum value of $\log N$ at the center of gravity

$$\text{(9.9.5)} \qquad N^{-1} \log \Gamma(N+1)$$

of $[0, \log N]$ under the log function (9.9.1). Stirling's formula gives the approximate value

$$\text{(9.9.6)} \qquad \log N - 1 + \frac{\log 2\pi N}{2N}$$

of (9.9.5) for large N. Note the asymmetry: the maximum is well to the high side of the interval $[0, \log N]$ for large N. □

9. HIERARCHICAL STATISTICAL MECHANICS

Example 9.9.2 (**characters of semilattices**). Take (9.6.1) to be the identity function

(9.9.7) $$1_A : A \to A$$

on a finite (join) semilattice A. Consider the 2-element semilattice S with Hasse diagram $0 \to \infty$ as a subalgebra of $\mathbb{R}^\infty$. A *character* of A is a semilattice homomorphism

(9.9.8) $$\chi : A \to S.$$

(Compare the concept of a character $\chi : A \to S^1$ of an abelian group A, a homomorphism into the circle group $S^1 = \{z \in \mathbb{C} \mid z\bar{z} = 1\}$. A unified treatment of the character theory of semilattices and abelian groups is provided by *regularized Pontryagin duality* [Romanowska and Smith 1997, §8].) For an element α of A, the principal wall $[\alpha]$ is just the down-set

(9.9.9) $$[\alpha] = \{x \in A \mid x \le \alpha\}$$

of α. Define the character $\widehat{\alpha} : A \to S$ by

(9.9.10) $$\widehat{\alpha} : x \mapsto \text{ if } x \le \alpha \text{ then } 0 \text{ else } \infty.$$

Noting that $\alpha = \sum_{x \in A} q_x^{\widehat{\alpha}} x$, one obtains

(9.9.11) $$H(\alpha) = \log Z(\widehat{\alpha}) = \log |[\alpha]|.$$

For example, the concavity of $-\log Z : A^* \to \mathbb{R}^\infty$ and $H : A \to [0, \log |A|]$ correspond respectively to the left and right inequalities in

(9.9.12) $$|[a] \cap [b]| \le |[a]|^{1-p} |[b]|^p \le |[a+b]|$$

for p in I° and a, b in A. □

Example 9.9.3 (**equilibrium thermodynamics**). Let X be the finite set of observable macrostates of a physical system. (The finiteness may be justified by limitations on the range and precision of the available measuring devices.) Take (9.6.1) to be the energy function

$$(9.9.13) \qquad E : X \to A,$$

where A is the convex hull in the affine space $\mathbb{R}$ of the set of observable energy values. Thus the energy of the system in macrostate x is $E(x)$ joules. For a fixed temperature T degrees Kelvin, consider the barycentric homomorphism

$$(9.9.14) \qquad \beta : A \to \mathbb{R}^\infty; \ a \mapsto a/kT,$$

where k is the numerical value 1.38×10^{-23} of Boltzmann's constant in joules per degree Kelvin. If the physical system is in equilibrium at the temperature $T°K$, then the probability that it is in macrostate x is q_x^β. (One may make the preceding statement tautologous by using it as a definition of "equilibrium".) Consider the expected value

$$(9.9.15) \qquad U = \sum_{x \in X} q_x^\beta E(x)$$

of the observed energy (in joules) at temperature $T°K$. Define the *physical entropy* (in units of joules per degree Kelvin) to be

$$(9.9.16) \qquad S = kH(U).$$

Define the *Helmholtz free energy* (in joules) to be

$$(9.9.17) \qquad F = -kT \log Z(\beta).$$

[Grandy 1987, (3–8)]. Then (9.6.19), in the form $H(U) = \log Z(\beta) + U/kT$, reduces to the classical relationship

$$(9.9.18) \qquad F = U - TS$$

[Israel 1979, p. xxiii] [Rumer and Ryvkin 1980, (19.5)]. □

9.10. Competition between species

Eigen's phenomenological rate equations [Eigen 1971] under constant total organization are useful tools for considering growth and selection of competing species. They are capable of interpretation at various levels, from the submolecular to the ecological. Although Eigen originally offered explicit solutions only in the absence of mutability, exact solutions in the presence of mutability were soon given by [Thompson and McBride 1974], [Jones, Enns and Rangnekar 1976], and others. Some twenty years later, solution methods using canonical distributions were studied [Smith 1996]. Consider first the case without mutability. Let $X = \{1, \ldots, r\}$ be a set of species. Suppose that the i-th species has a known net growth rate of E_i. In other words, if a population of n_i units of the i-th species is allowed to develop without constraint, its net rate of change is given by

$$(9.10.1) \qquad \dot{n}_i = E_i n_i.$$

Now suppose that the r species are brought together in a joint population maintained at a constant count N (e.g. by control of a common food supply). Thus the birth of one individual has to be compensated by the death of another, not necessarily of the same species. If n_i now represents the number of individuals of species i present in the joint population, its net rate of change is given by

$$(9.10.2) \qquad \dot{n}_i = (E_i - U) n_i.$$

Here U is the death rate required to keep the total population constant. This rate is common to all the species, but varies with time. Indeed, summing (9.10.2) over X, one obtains

$$(9.10.3) \qquad U = \sum_{i \in X} \frac{n_i}{N} E_i.$$

Note the similarity of (9.10.3) with (9.9.15), although the units for (9.10.3) are e.g. "per annum" rather than joules.

One may consider the evolving population as a biological system that one wishes to study. To observe the system, catch an individual and determine its species. The observed species represents a macrostate of the system. In the affine space $\mathbb{R}$, let A be the convex hull of the set of growth rates. Consider the function

(9.10.4) $$f : X \to A; \; i \mapsto E_i.$$

For a time t (measured e.g. in years), consider the barycentric homomorphism

(9.10.5) $$\beta : A \to \mathbb{R}^*; \; a \mapsto -at.$$

Assuming the probability of macrostate i to be the relative frequency n_i/N of species i in the population, the canonical distribution q^β on X yields

(9.10.6) $$n_i = \frac{N \exp(E_i t)}{\sum_{x \in X} \exp(E_x t)},$$

and in particular

(9.10.7) $$n_i(0) = N/r$$

for each i. It is important to note that the "initial condition" (9.10.7) was not imposed separately, but emerged automatically. In the analysis of viable existing biological systems, it is usually unrealistic to speak of an "initial state" of the system. On the other hand, one may solve the initial value problem (9.10.2), (9.10.3), (9.10.7) classically using the Ansatz

(9.10.8) $$n_i = N l_i \exp\left[-\int_0^t U(\tau)\delta\tau\right]$$

9. HIERARCHICAL STATISTICAL MECHANICS

to reduce to the linear system

(9.10.9) $$\dot{l}_i = E_i l_i.$$

The solution is

(9.10.10) $$n_i = \frac{N \exp(E_i t)}{r \exp \int_0^t U(\tau)d\tau}.$$

Comparing (9.10.10) with (9.10.6), one sees that the canonical distribution has already solved the initial value problem, with

(9.10.11) $$Z(\beta) = r \exp \int_0^t U(\tau)d\tau.$$

Remark 9.10.1. One may derive (9.10.11) directly, without resorting to the classical solution (9.10.10). Indeed, (9.6.19) becomes $H(\alpha) = \log Z(\beta) - \alpha \tau$. Taking the partial derivative with respect to τ yields $U(\tau) = \alpha = \partial \log Z(\beta)/\partial \tau$. Recalling $Z(0) = r$, the Fundamental Theorem of Calculus gives

(9.10.12) $$\log Z(\beta) - \log Z(0) = \int_0^t U(\tau)d\tau,$$

whence (9.10.11) follows by exponentiation. Note that (9.10.12) represents the total number of individuals lost to population control over the time period $[0,t]$. □

9.11. Competition with mutability

Writing

(9.11.1) $$\underline{n} = \begin{bmatrix} n_1 \\ \vdots \\ n_r \end{bmatrix}$$

for the population vector describing a population composed of species from the set $X = \{1, \ldots, r\}$, Eigen's phenomenological rate equations under constant total organization

$$(9.11.2) \qquad N = \sum_{x \in X} n_x$$

take the form

$$(9.11.3) \qquad \dot{\underline{n}} = (A - U)\underline{n},$$

with $r \times r$ matrix A and scalar matrix U, for the case of mutability. (The case (9.10.2) without mutability corresponds to diagonal A.) Assuming that the matrix A has r distinct real eigenvalues $\lambda_1, \ldots, \lambda_r$, the canonical technique of the preceding section yields a partial solution to (9.11.3), (9.11.2). The partiality of the solution is actually very significant, since it reflects the hierarchical structure of the underlying biology. This will be apparent in the Example 9.11.2 below.

Let Q be an $r \times r$ matrix whose columns $\underline{Q}_j$, for $1 \leq j \leq r$, are eigenvectors of A, say

$$(9.11.4) \qquad A\underline{Q}_j = \lambda_j \underline{Q}_j.$$

Normalize the eigenvectors as follows. If $\sum_{i=1}^{r} Q_{ij} \neq 0$, choose $\underline{Q}_j$ with $\sum_{i=1}^{r} Q_{ij} = 1$. If $\sum_{i=1}^{r} Q_{ij} = 0$, choose $\underline{Q}_j$ with $\sum_{i=1}^{r} |Q_{ij}| = 1$ and with the first non-zero term of the sequence $Q_{1j}, \ldots, Q_{rj}$ being positive. (This normalization of eigenvectors $\underline{Q}_j$ with $\sum_{i=1}^{r} Q_{ij} = 0$ is purely conventional.) Since the columns of Q form a basis of $\mathbb{R}^r$, there are unique scalar functions $m_1(t), \ldots, m(t)$ of time t such that

$$(9.11.5) \qquad \underline{n} = \sum_{j=1}^{r} m_j(t) \underline{Q}_j.$$

9. HIERARCHICAL STATISTICAL MECHANICS

This expression is to be contrasted with the expression

(9.11.6) $$\underline{n} = \sum_{j=1}^{r} n_j(t) \underline{I}_j$$

of the vector $\underline{n}$ as a linear combination of the columns $\underline{I}_j$ of the identity matrix. The column $\underline{I}_j$ represents one individual of the species j. By analogy, the column $\underline{Q}_j$ of Q will be considered to represent one individual of *virtual species* j. A given population vector $\underline{n}$ may then be described either by the presence of n_j individuals of each species j via (11.6), or by the presence of m_j individuals of each virtual species j via (11.5). It is this latter description that facilitates application of the canonical technique. The virtual species j is said to be *substantial* if and only if $\sum_{i=1}^{r} Q_{ij} = 1$ according to the normalization. Now substitution from (9.11.5) into (9.11.3) gives

$$\sum_{j=1}^{r} \dot{m}_j(t) \underline{Q}_j = \underline{\dot{n}} = (A - U)\underline{n}$$

$$= \sum_{j=1}^{r} m_j(t)(A - U)\underline{Q}_j$$

$$= \sum_{j=1}^{r} m_j(t)(\lambda_j - U)\underline{Q}_j.$$

Equating coefficients of $\underline{Q}_j$ yields

(9.11.7) $$\dot{m}_j = (\lambda_j - U)m_j$$

for each virtual species j. On the other hand, substitution from (9.11.5) into (9.11.2) gives

$$N = \sum_{i=1}^{r} n_i = \sum_{i=1}^{r} \sum_{j=1}^{r} m_j(t) Q_{ij} = \sum_{j=1}^{r} m_j(t) \sum_{i=1}^{r} Q_{ij}$$

or

(9.11.8) $$N = \sum_{y \in Y} m_y$$

with $Y = \{1 \leq j \leq r \mid \underline{Q}_j \text{ substantial}\}$.

Summarizing, one has the following

Theorem 9.11.1. [Smith 1996, p.79]. *Eigen's phenomenological rate equation (9.11.3) for the evolution of species in the presence of mutation, and subject to constant overall organization (9.11.2), yields the phenomenological rate equation (9.11.7) for the evolution of substantial virtual species in the absence of mutation, subject to the same constant overall organization (9.11.8).* □

Example 9.11.2. Consider a set $X = \{1, 2, 3\}$ of three species, with complementary recognition between the first two [Eigen 1971, §IV.2] [Thompson and McBride]. In other words, take

$$A = \begin{bmatrix} -d & a & 0 \\ a & -d & 0 \\ 0 & 0 & e \end{bmatrix}$$

in (9.11.3), with $a > d > 0$ and $e > 0$. The virtual species are

(9.11.9) $\quad \underline{Q}_1 = \begin{bmatrix} \frac{1}{2} \\ \frac{1}{2} \\ 0 \end{bmatrix}, \ \underline{Q}_2 = \begin{bmatrix} 0 \\ 0 \\ 1 \end{bmatrix}, \ \underline{Q}_3 = \begin{bmatrix} \frac{1}{2} \\ -\frac{1}{2} \\ 0 \end{bmatrix};$

the first two are substantial. Theorem 9.11.1 yields the reduced system

(9.11.10)
$$\dot{m}_1 = [(a-d) - U]m_1$$
$$\dot{m}_2 = [e - U]m_2$$

subject to $m_1 + m_2 = N$, a constant. The "presence of an individual from virtual species 1" really represents the presence of an individual from one of the two complementary species 1 or 2, without regard to which of these it actually belongs. The canonical technique applied to (9.11.10) describes the competition between the complementary pair 1,2 and the other species 3. (What happens depends on the comparison between $(a-d)$ and e.) It does not concern itself with the "fine structure within the complementary pair", i.e. with the relative proportions between species 1 and 2.

Using the theory developed in this chapter, one may now refine the canonical technique. Consider the barycentric algebra T of Example 9.3.3, and the map

$f : X \to T$ with $1f = (0,0)$, $2f = (0,1)$, and $3f = (1,\infty)$. Note that Xf generates T. The canonical technique applied to (9.11.10) corresponds to barycentric homomorphisms $\beta : T \to \mathbb{R}^\infty$ that are finite on the vertical. The corresponding q^β assign equal probability to the complementary species 1 and 2. To study the fine structure canonically, one would have to use a barycentric homomorphism that was infinite on the vertical. The hierarchical structure of the barycentric algebra T is reflected in its non-trivial semilattice replica:

(9.11.11)
$$\{1f\} \to (\{0\} \times I^\circ) \leftarrow \{2f\}$$
$$\downarrow$$
$$(I^\circ \times \{\infty\})$$
$$\uparrow$$
$$\{3f\}$$

This structure corresponds exactly to the hierarchy of interactions in the model biological system: species 1 and 2 do not interact directly (by competition) with species 3, but only in common. As far as the competition for the limited resources is concerned, there is no distinction between species 1 and 2. If the "species" 1 and 2 in Eigen's terminology are actually two stages of the same biological species, competing with "species" 3 as another biological species, then one may regard interactions described by the crossbar $\{0\} \times I^\circ$ in (9.11.11) as belonging to the demographic level, while the interactions described by the vertical $I^\circ \times \{\infty\}$ in (9.11.11) belong to the ecological level. □

Exercises

9.11A Consider the system (9.11.3) with
$$A = \begin{bmatrix} -d_1 & a_1 & 0 & 0 \\ a_1 & -d_1 & 0 & 0 \\ 0 & 0 & -d_2 & a_2 \\ 0 & 0 & a_2 & -d_2 \end{bmatrix},$$

where $(a_1 - d_1) > (a_2 - d_2) > 0$. Determine the virtual species, classifying each as substantial or insubstantial.

9.11B Describe the barycentric algebra corresponding to the system of Exercise 9.11A. What is its semilattice replica?

CHAPTER X

RECENT DEVELOPMENTS AND OPEN PROBLEMS

This final chapter will briefly survey further aspects of the theory of modes for which space considerations prevent a more detailed coverage. In particular, we will present some of the topics of current research.

10.1. The structure of binary and general modes

The results of Chapters 5, 6 and 8 show that the basic binary modes are equivalent to sets (as left and right zero semigroups), to semilattices, or to (reducts of) affine spaces. Many other binary modes are obtained as different types of sum (possibly iterated) of such basic algebras over other basic algebras. This structure is conjectured to be characteristic of all binary modes, although at present no proof is known. A typical general tool for the structural description of modes is the Lallement sum. If possible, however, descriptions using strict sums are always much more informative. (This is apparent, for example, in the cases of differential groupoids and commutative modes.) Another example of such a sum was given in [Roszkowska 1987b, 1999a], used to characterize all symmetric binary modes in a very elegant way. This sum was called an AG-sum (or "abelian group" sum). The indexing groupoid $(I, \cdot)$ of such a sum $(A, \cdot)$ is a left zero band, and the summands $(A_i, \cdot)$, for $i \in I$, are $\underline{2}$-reducts $(A_i, \underline{2})$ of affine $\mathbb{Z}$-spaces. The sum extensions $(E_i, \cdot)$ coincide with the sum fibres $(A_i, \cdot)$. For all $(i, j) \in I \times I$, there are homomorphisms

$\varphi_{i,j} : A_i \to A_j$ such that

(10.1.1) $$a_i\varphi_{i,i} = a_i 2,$$
(10.1.2) $$(-a_i + b_j\varphi_{j,i})\varphi_{i,n} = -a_i\varphi_{i,n} + b_j\varphi_{j,n} 2.$$

The multiplication in the sum $A = \bigcup_{l \in I} A_i$ is defined by

$$a_i b_j := -a_i + b_j\varphi_{j,i}.$$

The condition (10.1.2) may be rewritten as

$$(a_i b_j)\varphi_{i,n} = a_i\varphi_{i,n} \cdot b_j\varphi_{j,n}.$$

Theorem 10.1.1 [Roszkowska]. *Each symmetric binary mode is an AG-sum of $\underline{2}$-reducts of affine $\mathbb{Z}$-spaces over a left zero band.* □

We do not know if a similar characterization is possible for n-cyclic modes when n is greater than 2.

As for general modes, we can also distinguish modes equivalent to sets, to semilattices, and to (reducts) of affine spaces. More complex modes may then be composed from these basic types by means of constructions analogous to those used in the binary case. The phenomenon appears throughout this book, for example in the description of general reductive modes given in Exercise 8.4A (compare also [Pilitowska and Romanowska 1998]).

However, the general case is much more complicated than the binary case. There exist modes having a semilattice operation together with other operations which do not belong to one of the basic types described above. Simple examples are provided by idempotent distributive semirings in which a semilattice addition distributes over a rectangular multiplication [Romanowska 1982b,c], [Pastijn and Romanowska 1982]. Other examples may be obtained by taking the set J of ideals of a commutative ring R, defining the semilattice addition

10. RECENT DEVELOPMENTS AND OPEN PROBLEMS

$I + J$ of two ideals to be their intersection, and then defining a unary operation f_r by $If_r = \{s \in R \mid rs \in I\}$ for each $r \in R$. The general case was investigated by Kearnes [1995c] under the name "semilattice modes." We will return to these in Section 10.2 when discussing modes that are subreducts of semimodules.

The conjecture that all binary modes may be constructed by means of sums from the various basic binary modes finds support in [Kearnes 1995c, 1996]. Here the author claims to have an unpublished result showing that each locally finite variety $\underline{V}$ of modes decomposes as

$$(10.1.3) \qquad \underline{V} = (\underline{V}_1 \vee \underline{V}_2) \circ \underline{V}_5 ,$$

where ∘ denotes the Mal'cev product relative to $\underline{V}$. The numbering of the subvarieties $\underline{V}_1, \underline{V}_2$ and $\underline{V}_5$ is a convention of the tame congruence theory expounded in [Hobby and McKenzie 1988]. For more information, we refer the reader to [Kearnes 1996]. Here we merely indicate what these subvarieties are. In the binary case, the variety $\underline{V}_1$ is a variety of reductive modes. (Note however that, in general, reductive varieties are not necessarily locally finite.) The variety $\underline{V}_5$ is the variety $\underline{Sl}$ of semilattices. In the general case, $\underline{V}_1$ is a so-called "locally strongly solvable" variety. For each finite algebra A of the type of $\underline{V}_1$, and for a minimal congruence α on A, the mode A is in $\underline{V}_1$ if and only if A^α is in $\underline{V}_1$ and each α-class is equivalent to a set. The variety $\underline{V}_2$ is equivalent to a variety of affine spaces, and the variety $\underline{V}_5$ is equivalent to a variety of semilattice modes. In [Kearnes 1995c], the author remarks that his decomposition is obtained by showing that each finite mode A has a congruence ρ such that A^ρ is a semilattice mode and each ρ-class is a direct product of a $\underline{V}_1$-mode and a $\underline{V}_2$-mode. Note that the proof of Theorem 8.5.5 may easily be extended to show that the varieties $\underline{V}_1$ and $\underline{V}_2$ are independent. The structure of the algebras in the Mal'cev product of (10.1.3) depends on the structure of the algebras in $\underline{V}_5$. For $\underline{V}_5$ equivalent to $\underline{Sl}$, it will be given by corresponding Płonka or semilattice Lallement sums. If $\underline{V}_5$ consists of naturally quasi-ordered algebras, then $\underline{V}$-algebras will be Lallement sums of $\underline{V}_1 \bigcirc\!\!\!\!\vee \underline{V}_2$-algebras. However, it is an open question whether semilattice modes are necessarily naturally quasi-ordered.

10.2. Semi-affine spaces

A natural generalization of affine spaces is given by idempotent reducts of semimodules over commutative semirings. Such algebras are called *semi-affine spaces*. Their axioms may differ according to variations in the definitions of semirings and semimodules. The additive reduct of a semiring and the corresponding semimodules is usually a commutative semigroup, but may also be a semigroup mode. In the commutative case, one may or may not demand the existence of a zero and an identity.

For algebras equivalent to semimodules, see [Csákány 1963] and [Ježek and Kepka 1998]. Little is known about the structure of semi-affine spaces. On the other hand, we do know that modes in many classes (not necessarily in Mal'cev varieties) may be characterized as subreducts of semimodules over commutative semirings. Barycentric algebras and commutative binary modes have this property, for example, as well as general semilattice Lallement sums of cancellative modes (cf. Corollaries 7.8.5 and 7.8.6). In the case of binary modes, the following was proved in [Ježek and Kepka 1981b and 1983b, Theorem 3.3.1, Corollary 3.3.10].

Theorem 10.2.1 [Ježek and Kepka]. *A groupoid $(G, \cdot)$ is a binary mode if and only if there exists a commutative semigroup $(S, +)$ equipped with two commuting endomorphisms f and g satisfying $xf + xg = x$ such that $(G, \cdot)$ is a subreduct of the algebra $(S, +, f, g)$ with $x \cdot y := xf + yg$.* □

Unfortunately, the complicated proof of Theorem 10.2.1 is not completely

10. RECENT DEVELOPMENTS AND OPEN PROBLEMS

transparent, and it would be good to find a clearer, more direct proof. Note that the endomorphisms f and g of Theorem 10.2.1 generate a commutative subsemigroup $(A, \cdot)$ of the semigroup $\text{End}(S, +)$ of endomorphisms of $(S, +)$. With pointwise addition $x(h + k) = xh + xk$, this semigroup becomes a commutative semiring $(A, +, \cdot)$ acting on the semigroup $(S, +)$. This shows that the groupoid $(G, \cdot)$ is in fact a subreduct of the A-semimodule $(S, +, A)$.

It is an open question if each general mode is a subreduct of some semimodule. The answer is "yes" in the case of semilattice modes, thoroughly investigated in [Kearnes 1995c,d]. One of the main results can be summarized as follows.

Theorem 10.2.2 [Kearnes]. *Each semilattice mode is a subreduct of a semimodule over a commutative semiring with identity, satisfying the identities $0 \cdot x = 0$ and $1 + x = 1$.* □

Note that such semirings satisfy all the usual defining axioms for rings which do not refer to negation. In fact, to each variety $\underline{V}$ of semilattice modes, one can assign a unique commutative semiring $S(\underline{V})$ acting as the semiring of Theorem 10.2.2 on all $\underline{V}$-algebras. This semiring determines many properties of the variety $\underline{V}$. The underlying set of the semiring $S(\underline{V})$ may be taken as the subset $\{t \mid y \leq t \leq x + y\}$ of the free $\underline{V}$-algebra on the two generators x and y. The semiring addition is the semilattice addition, while the multiplication $\circ$ is defined by $s \circ t = xytys$. The zero element is y, and the identity element is $x + y$. The rôle of $S(\underline{V})$ for the variety $\underline{V}$ is reminiscent of the rôle of the ring $\{0, 1\}V$ for a Mal'cev variety $\underline{V}$. In particular, the following hold:

Theorem 10.2.3 [Kearnes]. *Each variety $\underline{V}$ of semilattice modes is generated by the free $\underline{V}$-algebra on two generators. In particular, $\underline{V}$ is defined by the defining identities for modes together with additonal binary identities.* □

Theorem 10.2.4 [Kearnes]. *Each variety $\underline{V}$ of semilattice modes has a cogenerating algebra C such that $\underline{V} = \mathrm{SP}(C)$.* □

A subset of $S(\underline{V})$ is said to be an *ideal* of $S(\underline{V})$ if it is a lower set of the semilattice $S(\underline{V})$ that is closed under addition. The underlying semilattice of the algebra C is then given by the operation of intersection on the set of ideals of $S(\underline{V})$.

Theorem 10.2.5 [Kearnes]. *The lattice $\mathcal{L}(\underline{V})$ of subvarieties of a semilattice mode variety $\underline{V}$ is dually isomorphic to the congruence lattice $\mathrm{Cg}\, S(\underline{V})$ of the semiring $S(\underline{V})$.* □

For more information about semilattice modes, in particular concerning the semirings of varieties of semilattice modes, we refer the reader to the aforementioned papers.

There are further possible generalizations of the notion of a semi-affine space. For any type $\tau : \Omega \to \mathbb{N}$, an Ω-*ring* is an algebra $(R, \cdot, 1, \Omega)$ such that $(R, \cdot, 1)$ is a monoid and for each (n-ary) ω in Ω the identities

$$x(y_1 \ldots y_n \omega) = (xy_1) \ldots (xy_n)\omega,$$

$$(y_1 \ldots y_n \omega)x = (y, x) \ldots (y_n x)\omega$$

are satisfied. A (right unital) R-*act* is an algebra (A, Ω, R) such that for each r in R, there is an operation $r : A \to A;\ a \mapsto ar$ such that for each ω in Ω the following are satisfied:

$$(ar)s = a(rs);$$

$$a(r_1 \ldots r_n \omega) = (ar_1) \ldots (ar_n)\omega;$$

$$(a_1 \ldots a_n \omega)r = (a_1 r) \ldots (a_n r)\omega;$$

$$a1 = a.$$

One usually also assumes that the Ω-reducts of the Ω-ring and R-act are entropic. (See [Fleischer 1975], [Sokratova 2000].) For commutative Ω-rings

R, i.e. Ω-rings with commutative monoid reduct, the appropriate generalizations of semi-affine spaces would comprise idempotent reducts of R-acts. Typical examples of commutative Ω-rings are provided by the dual monoids $(\{0,1\}V, \cdot, 1)$ of Theorem 5.9.5, equipped with the Ω-mode operations of the free $\underline{V}$-mode $\{0,1\}V$. Typical examples of general Ω-rings are provided by the sets $\text{End}(A,\Omega)$ of endomorphisms of an Ω-mode (A,Ω), with Ω-operations defined as in Proposition 5.1, and using composition of endomorphisms as multiplication. Note finally that an R-act (A,Ω,R) may also be defined as an entropic Ω-algebra (A,Ω) together with an Ω-ring homomorphism

$$h : (R, \cdot, 1, \Omega) \to (\text{End}(A,\Omega), \cdot, 1, \Omega);$$
$$r \mapsto (h_r : (A,\Omega) \to (A,\Omega); a \mapsto ar).$$

10.3. Simple and subdirectly irreducible modes

Simple binary modes are characterized by Theorem 8.3.9. As for general simple modes, note first that Theorem 8.3.6 and its proof yield the following:

Proposition 10.3.1. *Each simple, quasi-affine $\underline{M_T}$-mode (A,Ω) that is not equivalent to a set is an Ω-subreduct of an affine space $(F,\underline{F})$, where F is a field. In particular, $(F,\underline{F})$ itself is simple.* □

Theorem 10.3.2 [Kearnes 1996]. *Each simple mode is equivalent to one of the following:*

(a) *the two-element semilattice;*

(b) *the two-element set;*

(c) *a simple subreduct of an affine space $(F,\underline{F})$ over a field F.* □

It would be interesting to find a direct proof showing that the algebras of Theorem 10.3.2 are all the simple modes. Note however that the proof used in the binary case in Section 8.3 does not extend to the general case. Another interesting fact discovered in [Kearnes 1996] is that the simple modes of Proposition 10.3.1 (not equivalent to affine spaces) are binary reducts of affine spaces over strongly archimedean ordered fields (cf. [Fuchs 1963]).

The structure of subdirectly irreducible algebras that are Płonka or Lallement sums was discussed in Chapter 4, and obviously applies to general modes. Chapter 7 provided a characterization of subdirectly irreducible quasi-affine modes and subdirectly irreducible sums of quasi-affine modes. In particular, a direct characterization of subdirectly irreducible barycentric algebras and commutative binary modes was given there.

There are two other classes of binary modes for which such a direct characterization of subdirectly irreducible members is known. In [Romanowska and Roszkowska 1989], the subdirectly irreducible differential n-cyclic binary modes were described as follows, using the concept of an Lz-Lz-sum as defined in Exercise 5.6F. Let $n = p_1^{r_1} \ldots p_m^{r_m}$, where $p_1, \ldots, p_m$ are prime numbers. If $(G, \cdot)$ is a subdirectly irreducible groupoid in this class, then $(G, \cdot)$ is $p_i^{r_i}$-cyclic for some $i = 1, \ldots, m$. Moreover $(G, \cdot)$ is the Lz-Lz-sum of left zero bands $G_0 := \{0, 1, \ldots, p_i^{r_i} - 1\}$ and $G_i := \{g_i\}$ for $i = 1, \ldots, k$ with $1 \leq k \leq p_i^{r_i}$. For each $x \in G_0$ one has $g_i x = g_i$ and $x g_i = (x + i) \bmod p^r$. These groupoids are the only subdirectly irreducible members of the class in question. Note that the monolith μ of the congruence lattice of $(G, \cdot)$ has each of the G_i as a one-element μ-class, and the blocks of the monolith of the (subdirectly irreducible) group $\mathbb{Z}_{p^r}$ as the other μ-classes. The groupoid $(G, \cdot)$ is an envelope of $(G_0, \cdot)$.

The second class is the class of n-reductive symmetric binary modes. The subdirectly irreducible members of this class were characterized by Roszkowska [1987b, 1999b], and can be described as follows. Each such groupoid $(G, \cdot)$ that is not a left zero band is an AG-sum of two or three subgroupoids $G_0 \biguplus G_1$ or $G_0 \biguplus G_1 \biguplus G_2$, where $G_0 = \{0, 1, \ldots, 2^k - 1\}$ and $G_1 = \{0, 1, \ldots, 2^{k-1} - 1\} = G_2$ with $k \in \{1, \ldots, n - 1\}$. Denote an element i of G_j by i_j. Then the multiplication in the sums is defined as follows:

$$m_i \cdot n_j := \begin{cases} ((-m+2n) \bmod 2^{r_i})_i & \text{if } i = j \text{ or } |j-i| = 2, \\ ((-m+2n+1) \bmod 2^{r_i})_i & \text{if } i = 1,\ j = 2 \text{ or } i = 3,\ j = 2, \\ ((-m+2n-1) \bmod 2^{r_i})_i & \text{if } i = 2,\ j = 1 \text{ or } i = 2,\ j = 3, \end{cases}$$

and $r_i = k$ for $i = 0$, while $r_i = k-1$ for $i = 1, 2$. Thus the groupoid $(G, \cdot)$ is a sum of $\underline{2}$-reducts of integral affine spaces $\mathbb{Z}_{2^k}$ and $\mathbb{Z}_{2^{k-1}}$. At the same time, it is an envelope of the $\underline{2}$-reduct $(G_0, \cdot)$ of $\mathbb{Z}_{2^k}$. Note that the characterization above together with the knowledge of the lattice $\mathcal{L}(\underline{S})$ of symmetric binary mode varieties (see Section 10.4) provides a description of all subdirectly irreducible symmetric binary modes in non-trivial subvarieties of $\underline{S}$ [Roszkowska 1999b]. At the present, we do not know a similar characterization of the subdirectly irreducible (left or right) reductive or n-cyclic binary modes.

Subdirectly irreducible semilattice modes were characterized in [Kearnes 1995c]. Let $S(\underline{V})$ be the semiring of a semilattice mode variety. Each binary word s in $S(\underline{V})$ determines a derived operation $s: A \times A \to A$ in each $\underline{V}$-algebra A. If a $\underline{V}$-algebra A is subdirectly irreducible, then the semilattice $(A, +)$ has a least element 0. Let $\bar{s}: A \to A; a \mapsto s(a, 0)$. Then $(A, +, 0, \{\bar{s}\}_{s \in S(\underline{V})})$ is a semimodule over the semiring $S(\underline{V})$, and the $\underline{V}$-algebra A is polynomially equivalent to this semimodule. It is also equivalent to the reduct $(A, \{b_s\}_{s \in S(\underline{V})})$, where $b_s: A \times A \to A;\ (a, b) \mapsto a\bar{s} + b$. General subdirectly irreducible modes were considered in [Kearnes 1999].

10.4. More about varieties and quasivarieties of binary modes

As already mentioned in Chapter 8, we do not know if the irregular varieties of binary modes encountered in this book comprise all the irregular varieties of binary modes. We do not even know whether the affine varieties of Chapter 8 comprise all the varieties of binary modes that are equivalent to varieties of affine spaces. The lattice $\mathcal{L}(\underline{R}_2)$ of subvarieties of the variety $\underline{R}_2 = \underline{\text{Dm}}$, and hence of $\underline{R}_2 \vee \underline{R}_2^{OP}$, has been determined (cf. Section 8.4), but we do not know if a similar description holds for the varieties $\underline{R}_n$ with $n > 2$. As for the m-cyclic varieties $\underline{V}_{0,m}$, a detailed description is known only for the lattice $\mathcal{L}(\underline{V}_{0,2})$ of varieties of symmetric binary modes [Roszkowska 1987]. The varieties presented in Example 8.5.4 under the case $m = 2$ are in fact all the subvarieties of the variety $\underline{V}_{0,2} = \underline{S}$. The subvarieties equivalent to affine spaces are the varieties $\underline{S}_{2k+1}$ of Section 6.4.5. The reductive varieties form the infinite chain

$$\underline{S}_1 = \underline{\mathrm{Tr}} < \underline{S}_2 = \underline{R}_1 \cap \underline{S} < \underline{S}_{2^2} = \underline{R}_2 \cap \underline{S} < \cdots < \underline{S}_{2^n} = \underline{R}_n \cap \underline{S} < \ldots.$$

The remaining (proper) subvarieties are of the form

$$\underline{S}_{2^n(2k+1)} = \underline{S}_{2^n} \lor \underline{S}_{2k+1}.$$

Each of these subvarieties is defined by one binary identity. The variety $\underline{S}_n$ is defined by the identity $xyw_n = x$, where the words w_n are as at the beginning of Example 6.4.5. For $n = 2^k$, such an identity is equivalent to the n-reductive identity $x^n y = x$. For $n = 2k+1$, one obtains the identity (s_{2k+1}) of Example 6.4.5. The lattice of proper subvarieties of $\underline{S}$ is isomorphic to the lattice of natural numbers with division. We do not know if the varieties mentioned in Example 8.5.4 are all the subvarieties of the variety $\underline{V}_{0,m}$ for $m > 2$, but we think that this is possible, and that each $\underline{V}_{0,m}$ is both the join of its subvarieties of affine spaces, and the join of its reductive subvarieties.

A second question concerns the regular joins of infinite families of irregular varieties. Two examples of such joins are known. These are the variety $\underline{\mathrm{BM}}$ of all binary modes, and the variety $\underline{C}$ of commutative binary modes. The lattice $\mathcal{L}(\underline{C})$ was described in [Ježek and Kepka 1975a]. (See also [MT, Section 4.5].) The irregular subvarieties consist of the varieties $\underline{C}_{2k+1}$ of Section 6.6, which are equivalent to the varieties $\underline{Z}_{2k+1}$ of affine spaces. The regular subvarieties are precisely the regularizations $\underline{\widetilde{C}}_{2k+1}$ of the $\underline{C}_{2k+1}$. Each subvariety is defined by one binary identity. The variety $\underline{C}_{2k+1}$ is defined by the

10. RECENT DEVELOPMENTS AND OPEN PROBLEMS

identity $xyw_{2k+1} = x$ of (6.6.1), and its regularization $\widetilde{\underline{C}}_{2k+1}$ by the identity $xyw_m = xyw'_m$, where w'_m represents the quotient $m2^{-l(m)-1}$. (Cf. Section 6.6 and [MT, Section 4.5].) The lattice of proper subvarieties of $\underline{C}$ is isomorphic to the product of the lattice of odd natural numbers with division and the two-element lattice. The variety $\underline{C}$ is the join of its subvarieties of affine spaces, and is not the regularization of any proper subvariety. We do not know if there are any other regular varieties of this type, but think that this is possible. Note that the variety $\underline{C}$ is in fact generated by the union of the (infinitely many) minimal subvarieties of affine spaces. The same is true for the variety $\underline{S}$, although $\underline{S}$ is not regular. We do not know if two different infinite families of minimal varieties of affine spaces generate different varieties, but think that it is likely. If this were true, it would show that the lattice $\mathcal{L}(\underline{BM})$ of binary mode varieties is uncountable.

Another question is whether each binary mode non-semigroup variety is defined by identities involving only two variables.

It is apparent in many cases throughout the book that one may provide better, more transparent descriptions of the structure of algebras in quasivarieties than in varieties. However, it is usually more difficult to describe subquasivarieties of a given variety than subvarieties. We were able to describe the lattice of subquasivarieties of the regularizations of some irregular varieties in Section 4.4, and the lattice of subquasivarieties of the variety of barycentric algebras over a subfield of $\mathbb{R}$. We do not know much about quasivarieties of binary modes. Hogben and Bergman [1985] have shown that each variety $\underline{CQ}_{2k+1}$ of commutative quasigroup modes is *deductive*, i.e. it has no subquasivarieties that are not varieties. (See also [Bergman 1988].) The same holds for the equivalent varieties $\underline{Z}_{2k+1}$, $\underline{C}_{2k+1}$, and $\underline{S}_{2k+1}$. This leaves us with the following question: Do the varieties $\underline{CQM}, \underline{C}$ and $\underline{S}$ have subquasivarieties that are not subvarieties? The answer is yes. For example, for each subvariety $\underline{C}_{2k+1}$

of $\underline{\underline{C}}$ (defined by the one additional identity $xyw_{2k+1} = x$), one may consider the subquasivariety $\underline{\underline{Q}}_{2k+1}$ defined by the quasi-identity

$$(xyw_{2k+1} = x) \to (x = y).$$

One may show that the $\underline{\underline{Q}}_{2k+1}$ are not varieties, and that all are contained in the quasivariety of cancellative commutative binary modes. (Note the difference from the case of barycentric algebras.) However, it may happen that some of the $\underline{\underline{Q}}_{2k+1}$ coincide. There are other examples of subquasivarieties of $\underline{\underline{C}}$.

10.5. Modes of submodes

As shown in Section 5.1, for any Ω-mode A, the algebra AS of non-empty submodes and the algebra AP of non-empty finitely generated submodes of A are again Ω-modes satisfying each linear identity holding in A. This "self-replicating property," by which various sets of submodes of a given mode again form modes under complex products, is one of the most important consequences of the combination of idempotence and entropicity in modes. In particular, for a real vector space E, the algebra $(ES, \underline{I}^\circ)$ is the mode of all convex subsets of E, while the algebra $(EP, \underline{I}^\circ)$ is the mode of all polytopes. If a variety $\underline{V}$ of modes is defined by linear identities, then the modes of subalgebras of $\underline{V}$-algebras are again in $\underline{V}$. For example, modes of submodes of commutative binary modes are again commutative binary modes. By Lemma 5.7.1 and Proposition 5.7.3, the mode of submodes of any n-reductive binary mode is again n-reductive. By Corollary 5.8.5, the mode of submodes of any barycentric algebra is again a barycentric algebra. On the other hand, subalgebras of affine spaces do not form affine spaces. As shown in [Pilitowska 1996, 1998], it is usually very difficult to determine which (non-linear) identities satisfied in a mode A are also satisfied in the algebras AS and AP of submodes. This is in contrast with the complex algebra of non-empty subsets of the set

A under complex operations. As shown in [Grätzer and Lakser 1988] (see also [Brink 1993]), the identities satisfied by the complex algebras of the algebras in a given variety $\underline{V}$ are precisely the consequences of the linear identities holding in $\underline{V}$. A similar characterization for the algebras AS of submodes of algebras in a given variety $\underline{V}$ of modes is not known. For some partial results, in particular concerning affine spaces, see [Pilitowska 1998]. There it is conjectured that for an idempotent variety $\underline{V}$, the class of (non-empty) subalgebras of $\underline{V}$-algebras satisfies precisely the consequences of idempotence and the linear identities holding in $\underline{V}$.

A second problem concerning modes of submodes is to describe their general structure. The most interesting known results concern affine spaces and their reducts. First consider a real vector space E. The corresponding projective space is the join semilattice $(L(E), +)$ of non-empty vector subspaces of E. The affine space $(E, \mathbb{R})$ has a convex set reduct $(E, \underline{I}°)$. The following theorem of Romanowska and Smith [1985b] provides an invariant, purely algebraic passage from real affine to projective geometry.

Theorem 10.5.1 [Romanowska, Smith]. *The projective space* $(L(E), \underline{I}°)$ *is the semilattice replica of the $\underline{I}°$-reduct $(ES, \underline{I}°)$ of the algebra $(ES, \mathbb{R})$ of affine subspaces of the affine space $(E, \mathbb{R})$.* □

This result was generalized to affine spaces over any field R, with a certain subset Ω_R of $\{P\} \cup \underline{R}$ replacing $\underline{I}°$. Considerations of convexity suggested choices for the set Ω_R of operations, and it was shown that the algebra (ES, Ω_R) is a Płonka sum of quotient spaces of the affine space E over the Ω_R-semilattice $(L(E), \Omega_R)$ corresponding to the projective space $(L(E), +)$. A similar problem for an arbitrary commutative ring R with identity was considered in [Pilitowska, Romanowska and Smith 1995]. The rôle of Ω_R was taken there by the subset $\underline{J}_R° \cup \{P\}$ of $\underline{R} \cup \{P\}$, where $\underline{J}_R°$ comprises the set of units r of R such that $1 - r$ is also invertible. This made it possible to provide an adequate description of the structure of $(ES, P, \underline{R})$ for certain affine spaces.

Theorem 10.5.2 [Pilitowska, Romanowska, Smith]. *For an affine space $(E, P, \underline{R})$ in $\underline{R}$, each algebra (ES, Ω), where $\Omega \subseteq \underline{J}_R^\circ \cup \{P\}$, is a Płonka sum of Ω-reducts of affine R-spaces E/U over the projective space $(L(E), +)$ by the functor $F : (L(E)) \to (\Omega)$ with object part $UF = \{x + U \mid x \in F\}$ and morphism part $(U \to V)F : UF \to VF;\ x + U \mapsto x + V$.* □

In the case of a variety $\underline{V}$ of Ω-modes equivalent to a variety $\underline{R}$ of affine R-spaces, this theorem implies that for any $\underline{V}$-algebra A, the algebra (AS, Ω) is a Płonka sum of $V(A)$-algebras, equivalent to affine $R(A)$-spaces, over the semilattice $(AS, +)$. Examples include quasigroup modes, pairs of mutually orthogonal quasigroup modes, and Mal'cev varieties of binary modes.

In [Pilitowska, Romanowska and Roszkowska 1996], these results were extended to varieties that are independent joins of some other mode varieties, amongst which at least one is a variety of affine spaces. For other generalizations to certain other reducts of affine spaces see [Pilitowska 1999]. Even so, the same method did not work for describing the full structure of the algebra AS of affine subspaces of an affine R-space A. To overcome this difficulty, Pilitowska [1996] introduced a different (though equivalent) description of affine R-spaces, as certain ternary reducts $(E, \bar{R})$ of corresponding R-modules $(E, +, R)$, and then considered the algebras $((E, \bar{R})S, \bar{R})$ of (non-empty) subalgebras of $(E, \bar{R})$ instead of $(E, P, \underline{R})$. Note that a submodule of the module $(E, +, R)$ also forms a ternary $\bar{R}$-algebra $(ESM, \bar{R})$, a subalgebra of $(ES, \bar{R})$. The full class of algebras $(ESM, \bar{R})$ generates a variety of semilattice modes. The semiring of this variety is isomorphic to the semiring of finitely generated ideals of the ring R. Pilitowska generalized the concept of a functorial sum by introducing certain sets of $\omega\tau$ homomorphisms for each operation ω in Ω, and showed that for a ternary affine space $(E, \bar{R})$, each algebra $((E, \bar{R})S, \bar{R})$ is such a sum of ternary affine R-spaces $(E/V, \bar{R})$ over the semilattice mode $(ESM, \bar{R})$. For related results, see also [Pilitowska 2001].

10.6. Modals

The sets $AS \cup \{\varnothing\}$ and $AP \cup \{\varnothing\}$ of subalgebras of a mode (A, Ω) also carry the usual lattice structure, with intersection inherited from the power set of A. From these two semilattice structures, the join semilattices $(AS, +)$ and $(AP, +)$ on the sets of non-empty submodes, with $S_1 + S_2 = \langle S_1 \cup S_2 \rangle$, play a special rôle. (Note also that $(AP, +)$ is a subsemilattice of $(AS, +)$.) The Ω-operations *distribute* over the semilattice operation $+$, i.e. for each (n-ary) ω in Ω and $a_1, \ldots, a_j, \ldots a_n, a'_j$ in A, for $1 \leq j \leq n$,

$$(10.6.1) \qquad a_1 \ldots (a_j + a'_j) \ldots a_n \omega = a_1 \ldots a_j \ldots a_n \omega + a_1 \ldots a'_j \ldots a_n \omega.$$

In this way, one obtains algebras $(AS, +, \Omega)$ and $(AP, +, \Omega)$ that provide basic examples of modals. In general, a *modal* is an algebra $(A, +, \Omega)$ such that (A, Ω) is a mode, $(A, +)$ is a (join) semilattice, and the operations Ω distribute over $+$. Given a mode variety $\underline{V}$, a modal $(A, +, \Omega)$ is called a $\underline{V}$-*modal* if the mode reduct (A, Ω) of the modal lies in $\underline{V}$. The name "modal" was intended both to refer to the relationship with modes, and to suggest the analogy with "modules," which are also algebras $(E, +, R)$ with a set R of operations distributing over $+$. Examples of modals beyond $(AP, +, \Omega)$ and $(AS, +, \Omega)$ include semilattices (where $\Omega = \varnothing$), distributive lattices, dissemilattices (or meet-distributive bisemilattices) – algebras $(A, +, \cdot)$ with two semilattice structures $(A, +)$ and $(A, \cdot)$ in which the operation $\cdot$ distributes over the operation $+$, semilattice ordered binary modes, and the algebra $(\mathbb{R}, \max, \underline{I}^\circ)$. Modals were investigated in [MT], [Romanowska and Smith 1981, 1989a,b], and [Smith 1986, 1999]. The fundamental elementary properties of a modal $(A, +, \Omega)$ of type τ may be summarized as follows.

Monotonicity Lemma 10.6.1. *Each basic operation*

$$\omega : (A^{\omega \tau}, \leq_+) \to (A, \leq_+)$$

is monotone. □

Convexity Lemma 10.6.2. *For each positive integer r, the join*

$$\Sigma_r : (A^r, \Omega) \to (A, +, \Omega); \quad (x_1, \ldots, x_r) \mapsto x_1 + \cdots + x_r$$

is convex, i.e. $a_1 \ldots a_{\omega\tau} \omega \Sigma_r \leq a_1 \Sigma_r \ldots a_{\omega\tau} \Sigma_r \omega$ for each operation ω in Ω and $\omega\tau$-tuple $(a_1, \ldots, a_{\omega\tau})$ of elements of $A^{\omega\tau}$. □

Sum-Superiority Lemma 10.6.3. *For each basic operation ω in Ω, one has*

$$\omega \leq \Sigma_{\omega\tau}.\quad □$$

The construction AP of finitely generated non-empty subalgebras (or *polytopes*) of (A, Ω), and the construction AS of non-empty subalgebras of (A, Ω), yield two covariant functors: the *polytope functor P* and the *submode functor S*. For a variety $\underline{D}$ of $\underline{V}$-modals, P is defined as follows:

(10.6.2) $\quad P : \underline{V} \to \underline{D}; \; (f : A \to B) \mapsto (fP : AP \to BP; \; X \mapsto Xf).$

There is also a forgetful functor

(10.6.3) $\quad U : \underline{D} \to \underline{V}; \; (f : (D, +, \Omega) \to (E, +, \Omega)) \mapsto (fU : (D, \Omega) \to (E, \Omega))$

forgetting the semilattice structure of modals.

Theorem 10.6.4. *The polytope functor P is left adjoint to the forgetful functor U. For a $\underline{V}$-mode A, the unit*

$$\eta_A : A \to APU; \quad a \mapsto \{a\}$$

embeds A as the algebra of singletons. For a $\underline{V}$-modal $(D, +, \Omega)$, the counit

$$\varepsilon_D : DUP \to D; \quad \langle d_1, \ldots, d_n \rangle \mapsto d_1 + \cdots + d_n$$

10. RECENT DEVELOPMENTS AND OPEN PROBLEMS

sums the generators of a polytope. $\square$

The latter theorem was formulated in its present form in [Smith 1999] using some results of [MT, Chapter 3]. It was then extended to the submode functor S and to the contravariant functor T assigning to each algebra A the *totality*, the mode of all subalgebras of A. The definition of the functor S requires introduction of the concepts of a complete (join) semilattice (having joins or suprema of all non-empty subsets), of complete distributivity of Ω over $+$ (see [MT] and [Smith 1999] for the definition), and of a complete modal $(A, +, \Omega)$ (with a complete semilattice $(D, +)$ and Ω completely distributive over $+$). Then the variety $\underline{D}$ is replaced by the category $\underline{E}$ of complete $\underline{V}$-modals and homomorphisms, while the counit ε_D assigns to each nonempty subalgebra X the supremum of its one-element subalgebras. For the extension of Theorem 10.6.4 to the functor T, see [Smith 1999, Theorem 7.6]. Theorem 10.6.4 and its extension to the functor S are closely connected to a description of certain types of free modals and representation theorems for modals [MT, Section 3.5]. A typical representation theorem is given by the following.

Modal Representation Theorem 10.6.5. *Let $\underline{D}$ be a variety of $\underline{V}$-modals. Then each $\underline{D}$-modal $(D, +, \Omega)$ is a quotient of the modal DUP of finitely generated nonempty submodes of its mode reduct $DU = (D, \Omega)$.* $\square$

The modal DUP is "free" over the mode (D, Ω). For a similar representation, but using the functor S instead of P, see [MT, Theorem 3.5.6].

In [MT, Section 6.4], and [Romanowska and Smith 1981, 1989a,b], the structure of modals is examined by means of various decompositions, together with corresponding construction methods for recovering the algebras from their decompositions. The decomposition is always over an algebra equivalent to a dissemilattice, whereas the construction methods generalize the concept of a semilattice Lallement sum investigated and applied in this book. Dissemilattices themselves have quite an extensive literature. Their algebraic theory (varieties, free algebras, subdirectly irreducible members, representations, etc.) was investigated in [Płonka 1967a], [Kalman 1971], [Balbes 1970], [Knoebel 1976], [Nieminen 1977], [McKenzie and Romanowska 1979], [Romanowska 1980a,b,c, 1982a,d, 1983, 1984, 1994, 1995], [Romanowska and Smith 1981, 1985a, 1989a,b], [Gierz and Romanowska 1991]. For some applications, see e.g. [Libkin 1993], [Pilitowska 1991], [Puhlmann 1993], [Romanowska 1997], [Romanowska and Smith 1996], [Romanowska and Trakul 1989]. For related topics see [Hałkowska 1976], [Dudek and Romanowska 1983], [Bandelt 1981], [Bandelt and Hedliková 1983], [Knoebel and Romanowska 1991], [Romanowska and Smith 1985a] and [Zając 1988, 1990, 1991, 1992, 1994].

A sample application of the theory of modals in analysis is offered by the support functions of arbitrary convex sets. The concept of the support function of a non-empty compact convex set was introduced in [Minkowski 1911]. Since then it has played a vital rôle in many of the applications of convexity, from optimization theory to the geometry of numbers. Support functions of non-empty compact convex subsets of a finite-dimensional Euclidean space $\mathbb{R}^d$ are characterized as positively homogeneous convex real-valued functions on $\mathbb{R}^d$, and the compact convex subsets are determined uniquely by their support functions [Bonnesen and Fenchel 1948]. One of the research programmes in applied modal theory has been to extend the support function concept from compact to arbitrary convex subsets of finite-dimensional Euclidean spaces. The extension to bounded (i.e. not necessarily closed) convex sets was carried out in [Romanowska and Smith 1989c], and to arbitrary convex sets in [Choi 1998] and [Choi and Smith 2001]. For a short summary, see also [Smith 1999].

The basic idea is to find a suitable codomain D_d, replacing the codomain $\mathbb{R}$ of Minkowski's support functions, so that convex subsets of $\mathbb{R}^d$ are determined uniquely by their D_d-valued support functions defined on $\mathbb{R}^d$. Conditions are found, analogous to Minkowski's positive homogeneity and convexity, characterizing the support functions amongst all the D_d-valued functions on $\mathbb{R}^d$. The set of support functions should be closed under an appropriate algebraic structure, and should reflect comparable algebraic structure on the set of convex subsets of $\mathbb{R}^d$. The correct algebraic structure on D_d for use in the context of support functions comprises convex combinations forming a barycentric algebra and distributing over a join semilattice structure so that the two structures combine to form a modal. The support functions then form a submodal of the induced modal structure on the full set of functions, and the modal structure on the support functions reflects exactly the modal structure on the convex sets given by convex combinations and convex hulls of unions.

10.7. Dualities and equivalences

Proposition 5.1 and Corollary 5.2 show that for any prevariety $\underline{K}$ of modes, and for any two $\underline{K}$-algebras A and B, the morphism set $\underline{K}(A, B)$ is again a $\underline{K}$-mode. More precisely, $\underline{K}(A, B)$ is a subalgebra of the product algebra B^A in which the Ω-operations on B^A are defined as in the proof of Proposition 5.1. In particular, if the prevariety $\underline{K}$ is generated by a unique $\underline{K}$-algebra T, then for each $\underline{K}$-algebra A, the morphism set $\underline{K}(A, T)$ is again a $\underline{K}$-algebra. This fact offers a way of representing modes as modes of functions, and may lead to a duality theory for $\underline{K}$-modes.

Recall from Section 2.2 that two categories $\mathcal{C}$ and $\mathcal{D}$ are said to be equivalent if there are two covariant functors $F : \mathcal{C} \to \mathcal{D}$ and $G : \mathcal{D} \to \mathcal{C}$ such that $FG \cong 1_{\mathcal{C}}$ and $GF \cong 1_{\mathcal{D}}$. If the two functors are contravariant, one speaks of a *dual equivalence* or *duality* between the categories $\mathcal{C}$ and $\mathcal{D}$. Frequently, the category $\mathcal{C}$ is a category $\mathcal{A}$ of algebras with corresponding homomorphisms, while $\mathcal{D}$ is a category $\mathcal{X}$ of "representation spaces," geometric spaces with structure-preserving maps.

Dualities between categories, if they exist, may be obtained in many different ways. One of them leads to what some authors name the theory of "natural dualities" (e.g. [Clark and Davey 1998]). In this case the category $\mathcal{A}$ is a category of algebras of a given type τ, while the category $\mathcal{X}$ is a category of relational topological spaces satisfying certain special conditions. There is a finite set T which is described as being *schizophrenic*. In one of its "personalities" it carries algebraic structure so that it appears as the object $\underset{\sim}{T}$ of $\mathcal{A}$. In its other "personality" it carries relational and topological structures so that it appears as the object $\underset{\sim}{T}$ of $\mathcal{X}$. The category $\mathcal{A}$ is the class $\mathrm{SP}(\underset{\sim}{T})$, while the category $\mathcal{X}$ is the class $\mathrm{SCP}^+(\underset{\sim}{T})$ of isomorphic copies of closed substructures of non-empty powers of $\underset{\sim}{T}$. The functors F and G are the hom functors $F = \mathcal{A}(\_, T)$ and $G = \mathcal{S}(\_, \underset{\sim}{T})$ (cf. Example 2.2.7), defined by

$$
\begin{array}{ccccc}
A & & \mathcal{A}(\_, \underset{\sim}{T}) & & f\alpha \\
f\downarrow & \vec{F} & \uparrow fF & & \uparrow \\
B & & \mathcal{A}(\_, \underset{\sim}{T}) & & \alpha \\
\\
X & & \mathcal{X}(\_, \underset{\sim}{T}) & & g\alpha \\
g\downarrow & \vec{G} & \uparrow gG & & \uparrow \\
Y & & \mathcal{X}(\_, \underset{\sim}{T}) & & \alpha
\end{array}
$$

There is a naturally defined dual adjunction between $\mathcal{A}$ and $\mathcal{X}$. For each A in $\mathcal{A}$ and X in $\mathcal{X}$, there are *evaluation* maps

$$e_A : A \to AFG = \mathcal{X}(\mathcal{A}(A, \underset{\sim}{T}), \underset{\sim}{T}); \ a \mapsto (ae : x \mapsto ax),$$
$$\varepsilon_X : X \to XGF = \mathcal{A}(\mathcal{X}(X, \underset{\sim}{T}), \underset{\sim}{T}); \ x \mapsto (x\varepsilon : \alpha \mapsto x\alpha),$$

the components of natural transformations $e : 1_\mathcal{A} \overset{\cdot}{\to} FG$ and $\varepsilon : GF \overset{\cdot}{\to} 1_\mathcal{X}$. It may be shown that for each a in A and x in X, the evaluation maps e_A and ε_X are embeddings. If all the e_A are isomorphisms, then in the language of "natural dualities" one says that there is a *dual representation* between $\mathcal{A}$ and $\mathcal{X}$ and that $\underset{\sim}{T}$ *dualizes* $\underline{T}$ (or that $\underline{T}$ *is dualizable*). If further the ε_X are all isomorphisms, then one says that there is a *natural full duality* between $\mathcal{A}$ and $\mathcal{X}$. We refer the reader to [Clark and Davey 1998] for further information and numerous examples of dual representations and natural dualities. Here we are interested in natural and general dualities for modes.

Such dualities exist for all (non-trivial) semigroup mode varieties. The varieties <u>Lz</u> and <u>Rz</u> are both isomorphic to the category <u>Set</u>, and there is a well known *Lindenbaum-Tarski duality* between the category <u>Set</u> of sets and the category <u>St B</u> of Stone topological Boolean algebras, isomorphic to the category of complete atomic Boolean algebras. (See [Banaschewski 1971], [Davey and Werner 1983], [Johnstone 1982, VI. 4.6(b)], [Romanowska and Smith 1996].) (Recall that a *Stone* (or *Boolean*) topological space is one having a compact Hausdorff zero-dimensional topology.) The schizophrenic two-element object 2 appears in <u>Lz</u> as the band $\underline{2} = (\{0,1\}, *)$ and in <u>St B</u> as the Boolean algebra $\underset{\sim}{2} = (\{0,1\}, +, \cdot, ', 0, 1, \mathcal{T})$ equipped with the discrete topology $\mathcal{T}$. Since the variety <u>Re</u> of rectangular bands can be described as the product of the categories <u>Lz</u> and <u>Rz</u> (cf. [Herrlich and Strecker 1973]), one may obtain a duality between <u>Re</u> and the product of <u>St B</u> with itself using the Lindenbaum-Tarski duality for <u>Set</u> on each factor. (A "full natural duality" for <u>Re</u> is described in [Clark and Davey 1998, 4.4.9]. There the ring $\mathbb{Z}_6$ with the discrete topology dualizes the product of a 2-element left zero and 3-element right zero band that is used as a generator for <u>Re</u>. Of course, such a duality is "natural" only in the current purely technical sense.)

A ("full natural") duality for the variety of semilattices with neutral element was first found in [Hofmann, Mislove and Stralka 1974]. A small modification gives a duality between the variety <u>Sl</u> of semilattices and the class of Stone topological semilattices with their least and greatest elements selected by nullary operations. The two-element object is again schizophrenic, appearing once as the two-element semilattice $(\{0,1\},\cdot)$ and again as a bounded two-element semilattice $(\{0,1\},\cdot,0,1,\mathcal{T})$ with the discrete topology [Davey and Werner 1983].

Each of the three remaining semigroup mode varieties $\underline{Ln}, \underline{Rn}$ and $\underline{NB}$ consists of Płonka sums of semigroups in corresponding irregular varieties. The papers [Romanowska and Smith 1996, 1997] presented a method for "lifting" a known duality for a (strongly) irregular variety $\underline{V}$ to its regularization $\underline{\widetilde{V}}$. In particular, if there is a duality between $\underline{V}$ and $\mathcal{X}$ realized by means of a schizophrenic object T, then the disjoint union T^∞ of T with a singleton $\{\infty\}$ furnishes a schizophrenic object for a duality between the regularization $\underline{\widetilde{V}}$ and a certain class $\widetilde{\mathcal{X}}$ of representation spaces "regularizing" $\mathcal{X}$. This yields dualities for $\underline{Ln}, \underline{Rn}$ and $\underline{NB}$. (For an explicit description of such a duality for $\underline{Ln}$, see [Romanowska and Smith 1996]. For a translation into the context of "natural dualities," see [Davey and Knox 1999].)

As for the other classes of modes, a ("full natural") duality is known for each variety of affine spaces over a finite field $GF(q)$ with $q \neq 2^n$ [Pszczoła 2001]. The class $\mathcal{X}$ of representation spaces consists of Stone topological affine $GF(q)$-spaces with constant operations, one for each element of the field $GF(q)$. The schizophrenic object is the field $GF(q)$ appearing as the affine space $(GF(q), \underline{GF(q)})$ in $\underline{GF(q)}$ and as the affine space with constants $(GF(q), \underline{GF(q)}, \{\widetilde{k} \mid k \in GF(q)\}, \mathcal{T})$, where $\mathcal{T}$ is the discrete topology on $GF(q)$. This duality also yields dualities for varieties of binary and quasigroup modes equivalent to varieties $\underline{GF(q)}$. A different dual representation for quasigroup modes appears in [Suvorov 1969]. See also [Cecil 1975] and [Das 1976, 1977] for related considerations. We do not know if a similar type of duality exists for varieties of affine spaces over other commutative rings.

Not much is known about duality for modals, with one exception due to

10. RECENT DEVELOPMENTS AND OPEN PROBLEMS

[Gierz and Romanowska 1991]. This paper was a precursor of [Romanowska and Smith 1996, 1997] and [Davey and Knox 1999]. It gave an explicit ("full natural") duality for the variety of distributive bisemilattices, the regularization using a 3-element algebra as a schizophrenic object. As a distributive bisemilattice, this is the Płonka sum $\underline{2}^{\infty}$ of a two-element and a one-element lattice. The class of representation spaces is described as the class of ordered topological left normal bands with three constant operations.

We conclude with brief mention of varieties categorically equivalent to varieties of modes. While a variety equivalent to a variety of modes is again a variety of modes, this is no longer true for varieties categorically equivalent to varieties of modes. On the other hand, if two idempotent varieties are categorically equivalent, then they are equivalent [Ježek 1982]. Similarly, if they are entropic and categorically equivalent, then they are also equivalent [Wraith 1970]. The paper [Bergman and Berman 1999] gives a characterization (up to equivalence) of those varieties that are categorically equivalent to a fixed finitely generated variety $\underline{V}$ of modes. This was done by expanding the set of basic operations and adding new identities to an equational basis for $\underline{V}$. In particular, varieties categorically equivalent to varieties of semilattices, left normal bands, rectangular bands and affine spaces were described in detail. The techniques used were based on a universal-algebraic characterization of categorical equivalence worked out in [McKenzie 1996]. For another characterization of varieties categorically equivalent to the variety $\underline{Sl}$ of semilattices, see [Davey and Werner 1983].

BIBLIOGRAPHY

Aczél, J.

[1948] On mean values, *Bull. Amer. Math. Soc.* **54**, 392–400.

Adámek, T.

[1990] How many variables does a quasivariety need?, *Algebra Universalis* **27**, 44–48.

Aizenstat, A. J. and Boguta, B. K.

[1979] On the lattice of varieties of semigroups, *Semigroup Varieties and Endomorphism Semigroups*, Leningr. Gos. Ped. Inst., pp. 3–46, in Russian.

Albert, A. A.

[1943] Quasigroups I, *Trans. Amer. Math. Soc.* **54**, 507–519.

Anderson, F. W. and Fuller, K. R.

[1992] *Rings and Categories of Modules*, Springer Verlag, New York, New York.

Arnold, L., Gundlach, V. M. and Demetrius, L.

[1994] Evolutionary formalism for products of positive random matrices, *Ann. Appl. Prob.* **4**, 859–901.

Arworn, S. and Denecke, K.

[1999a] Left-edge solid varieties of differential groupoids, *Demonstratio Math.* **32**, 1–11.
[1999b] Hyperidentities and hypersubstitutions in the variety of symmetric, idempotent, entropic groupoids, *Demonstratio Math.* **32**, 677–686.

Athreya, K. B. and Smith, J. D. H.

[2000] Canonical distributions and phase transitions, *Discuss. Math. Prob. and Stat.* **20**, 167–176.

Auger, P.

[1989] *Dynamics and Thermodynamics in Hierarchically Organized Systems*, Pergamon, Oxford.

Balbes, R.

[1970] A representation theory for distributive quasilattices, *Fund. Math.* **68**, 207–214.

Banaschewski, B.

[1971] Projective covers in categories of topological spaces and topological algebras, in *General Topology and its Relations to Modern Analysis and Algebra*, III, Academia, Prague, pp. 63–91.

Banaschewski, B. and Herrlich, H.

[1976] Subcategories defined by implications, *Houston J. Math.* **2**, 149–171.

Bandelt, H. J.

[1981] Płonka sums of complete lattices, *Simon Stevin* **55**, 169–171.

Bandelt, H. J. and Hedlíková, J.

[1983] Median algebras, *Discrete Math.* **45**, 1–30.

Barnhardt, J.

[1978] Distributive groupoids and biassociativity, *Aequationes Math.* **18**, 304–321.

Bauer, H.

[1970] An open mapping theorem for convex sets with only one part, *Aequationes Math.* **4**, 332–337.

Bauer, H. and Bear, H. S.

[1969] The part metric for convex sets, *Pacific J. Math.* **30**, 15–33.

Bear, H. S.

[1965] A geometric characterisation of Gleason parts, *Proc. Amer. Math. Soc.* **16**, 407–412.
[1991] Part metric and hyperbolic metric, *Math. Monthly* **30**, 109–123.

Bear, H. S. and Weiss, M. L.

[1967] An intrinsic metric for parts, *Proc. Amer. Math. Soc.* **18**, 812–817.

Belousov, V. D.

[1958] Transitive distributive quasigroups, *Ukrain. Mat. Zh.* **10**, 13–22, in Russian.
[1965] Systems of quasigroups with generalized identities, *Uspekhi Mat. Nauk* **20 (121)**, 75–146, in Russian. English translation: Russian Math. Surveys 20, 75–143.
[1967] *Foundations of the Theory of Quasigroups and Loops*, Nauka, Moscow, in Russian.
[1972] *N-ary Quasigroups*, Štiinca, Kishinev, in Russian.

Bénéteau, L.

[1971] Classification des espaces barycentrés et des espaces planairement affines, *C. R. Acad. Sc. Paris Sér. A* **281**, 9–11.

Bergman, C.

[1988] Structural completeness in algebra and logic, *Colloq. Math. Soc. János Bolyai* **54**, 59–73.

Bergman, C. and Berman, J.

[1999] Categorical equivalence of modes, *Discuss. Math. Algebra & Stochastic Methods* **19**, 41–62.

Bergman, C. and McKenzie, R.

[1990] Minimal varieties and quasivarieties, *J. Austral. Math. Soc. Ser. A* **48**, 133–147.

Bergman, C. and Romanowska, A.

[1996] Subquasivarieties of regularized varieties, *Algebra Universalis* **36**, 536–563.

Bergman, G. M.

[1991] Actions of Boolean rings on sets, *Algebra Universalis* **28**, 153–187.
[1998] *An Invitation to General Algebra and Universal Constructions*, H. Helson, Berkeley, California.

Białynicki-Birula, A.

[1987] *An Outline of Algebra*, PWN, Warsaw, in Polish.

Birkhoff, G.

[1935] On the structure of abstract algebras, *Proc. Camb. Phil. Soc.* **31**, 433–454.
[1944] Subdirect unions in universal algebra, *Bull. Amer. Math. Soc* **50**, 764–768.
[1967] *Lattice Theory*, 3rd ed., Amer. Math. Soc., Providence, Rhode Island.

Blaschke, W.

[1956] *Kreis und Kugel*, de Gruyter, Berlin.

Bogdanović, S. and Ćirić, M.

[1998] Quasi-orders and semilattice decompositions of semigroups (a survey), in *Semigroups (Kunming 1995)*, Springer Verlag, Singapore, pp. 27–56.

Bonnesen, T. and Fenchel, W.

[1948] *Theorie der konvexen Körper*, Chelsea, New York, New York.

Bos, W. and Wolff, G.

[1978a,b] Affine Räume I,II, *Mitt. Math. Sem. Giessen* **129**, 1–115, **130**, 1–83.

Börger, R. and Kemper, R.

[1994] Cogenerators for convex spaces, *Appl. Categ. Structures* **2**, 1–11.

Brink, C.

[1993] Power structures, *Algebra Universalis* **30**, 177–216.

Brønsted, A.

[1983] *An Introduction to Convex Polytopes*, Springer Verlag, New York, New York.

Brooks, D. R. and Wiley, E. O.

[1986] *Evolution as Entropy – Towards a Unified Theory of Biology*, University of Chicago Press, Chicago, Illinois.

Bruck, R. H.

[1944] Some results in the theory of quasigroups, *Trans. Amer. Math. Soc.* **55**, 19–52.
[1971] *A Survey of Binary Systems*, 3rd ed. Springer Verlag, Berlin.

Burmistrovič, I. E.

[1965] Commutative bands of cancellative semigroups, *Sibirsk. Mat. Zh.* **6**, 284–299, in Russian.

Burris, S. and Sankappanavar, H. P.

[1981] *A Course in Universal Algebra*, Springer Verlag, New York, New York.

Cecil, D. R.

[1975] Duality and reflexivity for medial systems, *Tex. J. Sci.* **26**, 61–65.

Chein, O.

[1990] Examples and methods of constructions, in O. Chein, H. O. Pflugfelder and J. D. H. Smith (eds), *Quasigroups and Loops: Theory and Applications*, Heldermann Verlag, Berlin, pp. 27–94.

Chein, O., Pflugfelder, H. O. and Smith, J. D. H. (eds.)

[1990] *Quasigroups and Loops: Theory and Applications*, Heldermann Verlag, Berlin.

Chick, H. L. and Gardner, J.

[1987] The preservation of some ring properties by semilattice sums, *Comm. Algebra* **15**, 1017–1038.

Cho, Jung R.

[1986] *Varieties of Medial Algebras*, Ph.D. Thesis, Emory University, Atlanta.
[1988] Idempotent medial n-groupoids defined on fields, *Algebra Universalis* **25**, 235–246.
[1990a] A note on n-groupoids linearly defined on fields, *Algebra Universalis* **27**, 587–594.
[1990b] Representations of certain medial algebras, *J. Korean Math. Soc.* **27**, 69–76.
[1996] Abstract differentiation on certain groupoids, *J. Korean Math. Soc.* **11**, 925–932.

Cho, Jung R. and Dudek, J.

[2000] Medial idempotent groupoids III, *J. Austral. Math. Soc.* **68**, 312–320.

Choi, Dug-Hwan

[1998] *Support Functions of Convex Subsets of a Finite-Dimensional Real Space*, Ph.D. Thesis, Iowa State University, Ames, Iowa.

Choi, Dug-Hwan and Smith, J. D. H.

[2001] Support functions of general convex sets, *Algebra Universalis*, to appear.

Chrapek, Z.

[1987] *Algebras of Dyadic Numbers*, Master's Thesis, Warsaw University of Technology, in Polish.

Chrislock, J. L.

[1969] On medial semigroups, *J. Algebra* **12**, 1–9.

Ćirić, M. and Bogdanović, S.

[1996] Semilattice decomposition of semigroups, *Semigroup Forum* **52**, 119–132.

Ćirić, M. , Petković, T. and Bogdanović, S.

[2001] Sums and limits of generalized direct families of algebras, *Southeast Asian Bull. Math.* **25**, 47–60.

Clark, D. M. and Davey, B. A.

[1998] *Natural Dualities for the Working Algebraist*, Cambridge University Press, Cambridge.

Clifford, A. H.

[1941] Semigroups admitting relative inverses, *Ann. Math.* **42**, 1037–1049.
[1954] Bands of semigroups, *Proc. Amer. Math. Soc.* **5**, 499–504.

Cohn, P. M.

[1981] *Universal Algebra*, 2nd ed., Reidel, Dordrecht.

Csákány, B.

[1963] Primitive classes of algebras which are equivalent to classes of semimodules and modules, *Acta Sci. Math.* **24**, 157–164, in Russian.
[1964] Abelian properties of primitive classes of universal algebras, *Acta Sci. Math.* **25**, 202–208, in Russian.
[1975a] Varieties of affine modules, *Acta Sci. Math.* **37**, 3–10.
[1975b] On affine spaces over prime fields, *Acta Sci. Math.* **37**, 33–36.
[1975c] Varieties of modules and affine modules, *Acta Math. Acad. Sci. Hungar.* **26**, 263–266.

Csákány, B. and Megyesi, L.

[1975] Varieties of idempotent medial quasigroups, *Colloq. Math.* **37**, 17–23.
[1979] Varieties of idempotent medial n-quasigroups, *Colloq. Math.* **42**, 45–52.

Curtis, R. T.

[1979] A classification of Howard Eves' 'equihoops', preprint, Bowdoin College, Brunswick.

Czédli, G. and Smith, J. D. H.

[1979] On the uniqueness of Mal'cev polynomials, *Colloq. Math. Soc. János Bolyai* **28**, 127–145.

Das, P.

[1976] Note on the character quasigroup of an abelian topological quasigroup, *Indian J. Pure Appl. Math.* **7**, 421–424.

Das, P., *ctd.*

[1977] Isotopy of abelian quasigroups, *Proc. Amer. Math. Soc.* **63**, 317–323.

Davey, B. A. and Davis, G.

[1985] Tensor products and entropic varieties, *Algebra Universalis*, **21**, 65–88.

Davey, B. A. and Knox, B. J.

[1999] Regularising natural dualities, *Acta Math. Univ. Comenian.* **68**, 295–318.

Davey, B. A. and Priestley, H. A.

[1990] *Introduction to Lattices and Order*, Cambridge University Press, Cambridge.

Davey, B. A. and Werner, H.

[1983] Dualities and equivalences for varieties of algebras, *Colloq. Math. Soc. János Bolyai* **33**, 101–275.

Demetrius, L.

[1997] Directionality principles in thermodynamics and evolution, *Proc. Nat. Acad. Sci. USA* **94**, 3491–3498.

Denes, J. and Keedwell, A. D.

[1972] On P-quasigroups and decompositions of complete undirected graphs, *J. Combin. Theory Ser. B* **13**, 270–275.

Donnell, W.

[1974] A note on entropic groupoids, *Portug. Math.* **33**, 77–78.

Doraczyńska, E.

[1974] A complete theory of the midpoint operation, *Bull. Acad. Pol. Sci., Math. Phys.* **22**, 1195–1200.

Dudek, J.

[1974] A characterization of some idempotent abelian groupoids, *Colloq. Math.* **30**, 219–222.
[1981] Medial groupoids and Mersenne numbers, *Fund. Math.* **114**, 109–112.
[1986a] A polynomial characterization of affine spaces over GF(3), *Colloq. Math.* **50**, 167–171.
[1986b] Polynomial characterization of some idempotent algebras, *Acta Sci. Math.* **50**, 39–49.
[1989] Polynomials in idempotent commutative groupoids, *Dissertationes Math.* **286**.
[1990] The unique minimal clone with three essentially binary operations, *Algebra Universalis* **27**, 261–269.
[1991] Medial idempotent groupoids I, *Czechoslovak Math J.* **41**, 249–259.
[1992] On Csákány problem concerning affine spaces, *Acta Sci. Math.* **56**, 3–13.
[1994a] Medial idempotent groupoids II, *Contributions to General Algebra* **9**, 133–150.
[1994b] On varieties of groupoid modes, *Demonstratio Math.* **27**, 815–828.

Dudek, J. and Graczyńska, E.

[1981] The lattice of varieties of algebras, *Bull. Acad. Polon. Sci., Sér. Sci. Math.* **29**, 337–340.

Dudek, J. and Romanowska, A.

[1983] Bisemilattices with four essentially binary polynomials, *Colloq. Math. Soc. János Bolyai* **33**, 337–360.

Dudek, J. and Tomasik, J.

[1996] Affine spaces over GF(4), *Algebra Universalis* **36**, 279–285.

Duplák, J.

[1974] Certain permutations of a medial quasi-group (Russian), *Mat. Čas.* **24**, 315–324.

Eigen, M.

[1971] Self-organization of matter and the evolution of biological macromolecules, *Naturwissenschaften* **58**, 465–523.

Endres, N.

[1991] Group related symmetric groupoids, *Demonstratio Math.* **24**, 63–74.
[1993] Die Algebraische Struktur der Homotopiemenge von Produkt-abbildungen auf Sphären, *Geometriae Dedicata* **48**, 267–294.
[1994] Various applications of symmetric groupoids, *Demonstratio Math.* **27**, 673–685.
[1995] Homotopiemengen von Multiplikationen auf Sphären und ihre algebraische Struktur, *Geometriae Dedicata* **54**, 57–85.
[1996a,b] Idempotent and distributive group related groupoids I, II, *Demonstratio Math.* **29**, 270–290, 291–308.

Etherington, I. M. H.

[1949] Non-associative arithmetics, *Proc. Roy. Soc. Edinburgh Sect. A* **62**, 442–453.
[1958a] Entropic functions for linear algebras, *Proc. Roy. Soc. Edinburgh Sect. A* **65**, 84–108.
[1958b] Groupoids with additive endomorphisms, *Amer. Math. Monthly* **65**, 596–601.
[1964/65] Quasigroups and cubic curves, *Proc. Edinburgh Math. Soc.* **14**, 273–291.

Evans, T.

[1961] A condition for a group to be commutative, *Amer. Math. Monthly* **68**, 898–899.
[1962a] Endomorphisms of abstract algebras, *Proc. Roy. Soc. Edinburgh Sect. A* **66**, 53–64.
[1962b] Properties of algebras almost equivalent to identities, *J. London Math. Soc.* **37**, 53–59.

Evans, T., *ctd.*

[1963] Abstract mean values, *Duke Math. J.* **30**, 331–349.
[1971] The lattice of semigroup varieties, *Semigroup Forum* **2**, 1–43.
[1976] The construction of orthogonal k-skeins and latin k-cubes, *Aequationes Math.* **14**, 485–491.
[1979] Universal algebra and Euler's officer problem, *Amer. Math. Monthly* **86**, 466–473.
[1982a] Finite representations of two-variable identities or why are finite fields important in combinatorics?, *Ann. Discrete Math.* **15**, 135–141.
[1982b] Universal-algebraic aspects of combinatorics, *Colloq. Math. Soc. János Bolyai* **29**, 241–266.

Fajtlowicz, S. and Mycielski, J.

[1974] On convex linear forms, *Algebra Universalis* **4**, 244–249.

Fenchel, W.

[1949] On conjugate convex functions, *Canad. J. Math.* **1**, 23–27.

Fisher, E. R.

[1977] Abstract 742-08-4: Vopěnka's Principle, category theory and universal algebra, *Notices Amer. Math. Soc.* **24**, A-44.

Fleischer, V.

[1975] Ω-rings over which all acts are n-free, *Acta et Comment. Univ. Tartuensis* **390**, 56–83.

Flood, J.

[1981] Semiconvex geometry, *J. Austral. Math. Soc.* **30**, 496–510.

Frayne, T., Morel, A. C. and Scott, D. S.

[1962/63] Reduced direct products, *Fund. Math.* **51**, 195–228.

Freese, R. and McKenzie, R.

[1987] *Commutator Theory for Congruence Modular Varieties*, Cambridge University Press, Cambridge.

Freyd, P.

[1966] Algebra valued functors in general and tensor products in particular, *Colloq. Math.* **14**, 89–106.

Frink, O.

[1955] Symmetric and self-distributive systems, *Amer. Math. Monthly* **62**, 697–707.

Fuad, T. S. R. and Smith, J. D. H.

[1996] Quasigroups, right quasigroups and category coverings, *Algebra Universalis* **35**, 233–248.

Fuchs, L.

[1963] *Partially Ordered Algebraic Systems*, Pergamon Press, New York, New York.

Fulton, C. and Stein, S.

[1957] The passage from geometry to algebra, *Math. Ann.* **134**, 140–142.

Ganter, B. and Werner, H.

[1975] Equational classes of Steiner systems, *Algebra Universalis* **5**, 125–140.

Gardner, B. J.

[1974] Radicals of supplementary semilattice sums of associative rings, *Pacific J. Math.* **58**, 387–392.

[1989] Płonka-type sums over free products, *Math. Nachr.* **141**, 161–175.

Garrett, A. J. M.

[1991] Macroirreversibility and microreversibility reconciled, in *Maximum Entropy in Action*, Clarendon Press, Oxford, pp. 139–170.

Garrison, G. N.

[1940] Quasigroups, *Ann. Math.* **41**, 474–487.

Gatial, J.

[1969] Some geometrical examples of an IMC-quasigroup, *Mat. Čas.* **19**, 292–298.
[1971] Einige Eigenschaften der Symmetrien der IMC-Quasigruppe, *Sb. Elektrotech. Fak. SVŠT Bratislava* pp. 183–185.

Gierz, G. and Romanowska, A.

[1991] Duality for distributive bisemilattices, *J. Austral. Math. Soc.* **51**, 247–275.

Golan, J. S.

[1992] *The Theory of Semirings with Applications in Mathematics and Theoretical Computer Science*, Longman Scientific & Technical, Harlow.

Golan, J. S. and Head, T.

[1991] *Modules and the Structure of Rings. A Primer*, Dekker, New York, New York.

Gorbunov, V. A.

[1994] The structure of the lattices of quasivarieties, *Algebra Universalis* **32**, 493–530.
[1998] *Algebraic Theory of Quasivarieties*, Consultants Bureau, New York, New York.

Graczyńska, E.

[1981] On regular and symmetric identities, *Polish. Acad. Sci. Inst. Philos. Sociol. Bull. Sect. Logic* **10**, 104–107.
[1983a] On regular and symmetric identities II, *Polish. Acad. Sci. Inst. Philos. Sociol. Bull. Sect. Logic.* **11**, 100–102.
[1983b] On regular identities, *Algebra Universalis* **17**, 369–375.
[1987] The word problem for regular identities, *Beiträge Algebra Geom.* **24**, 41–49.
[1989a] On normal and regular identities and hyperidentities, in K. Hałkowska and B. Stawski (eds.), *Universal and Applied Algebra*, World Scientific Publishing, New Jersey, pp. 107–135.
[1989b] On a problem of bases for the regular extensions of varieties of algebras, *Studia Math. Hungar.* **24**, 37–42.
[1990] On normal and regular identities, *Algebra Universalis* **27**, 387–397.

Graczyńska, E., Kelly, D. and Winkler, P.

[1986] On the regular part of varieties of algebras, *Algebra Universalis* **23**, 77–84.

Graczyńska, E. and Pastijn, F.

[1982] Proofs of regular identities, *Houston J. Math.* **8**, 61–67.

Graczyńska, E. and Wroński, A.

[1978] On normal Agassiz systems of algebras, *Colloq. Math.* **40**, 1–8.

Grandy, W. T., Jr.

[1987] *Foundations of Statistical Mechanics, Vol. I*, Reidel, Dordrecht.

Grätzer, G.

[1979] *Universal Algebra*, 2nd edn., Springer Verlag, New York, New York.

Grätzer, G. and Lakser, H.

[1988] Identities for globals (complex algebras) of algebras, *Colloq. Math.* **56**, 19–29.

Grätzer, G., Lakser, H. and Płonka, J.

[1969] Joins and direct products of equational classes, *Canad. Math. Bull.* **12**, 741–744.

Grätzer, G. and Padmanabhan, R.

[1971] On idempotent, commutative and nonassociative groupoids, *Proc. Amer. Math. Soc.* **28**, 75–80.

Grätzer, G. and Płonka, J.

[1970] A characterization of semilattices, *Colloq. Math.* **22**, 21–24.

Grätzer, G. and Sichler, J.

[1974] Agassiz sums of algebras, *Colloq. Math.* **30**, 57–59.

Grätzer, G. and Whitney, S.

[1984] Infinitary varieties of structures closed under the formation of complex structures, *Colloq. Math.* **48**, 485–488.

Grillet, P. A.

[1969] The tensor product of semigroups, *Trans. Amer. Math. Soc.* **138**, 267–280.

Grünbaum, B.

[1967] *Convex Polytopes*, Interscience, London.

Gudder, S. P.

[1973] Convex structures and operational quantum mechanics, *Comm. Math. Phys.* **29**, 249–264.
[1977] Convexity and mixtures, *SIAM Rev.* **19**, 221–240.
[1978] Erratum: Convexity and mixtures, *SIAM Rev.* **20**, 837.

Gumm, H. P.

[1976] Congruence equalities and Mal'cev conditions in regular equational classes, *Acta. Sci. Math.*, 265–272.

[1979] Algebras in permutable varieties: Geometrical properties of affine algebras, *Algebra Universalis* **9**, 8–34.

[1980] An easy way to the commutator in modular varieties, *Arch. Math. (Basel)*, **34**, 220–228.

[1983] *Geometrical Methods in Congruence Modular varieties*, Memoirs of the Amer. Math. Soc. **45**, no. 286.

Hagemann, J. and Herrmann, C.

[1979] A concrete ideal multiplication for algebraic systems and its relation to congruence distributivity, *Arch. Math. (Basel)*, **32**, 234–245.

Hałkowska, K.

[1976] On a certain equationally defined class of algebras, *Zeszyty Nauk. Wyż. Szk. Ped. im. Powst. Śląskich w Opolu, Mat.* **19**, 127–136, in Polish.

Hardy, G. H. and Wright, E. M.

[1968] *An Introduction to the Theory of Numbers*, Clarendon Press, Oxford.

Head, T. J.

[1968] The variety of commutative monoids, *Nieuw Archief voor Wiskunde* **16**, 203–206.

Hermann, C.

[1979] Affine algebras in congruence modular varieties, *Acta Sci. Math.* **41**, 119–125.

Herrlich, H. and Strecker, G. E.

[1973] *Category Theory*, Allyn and Bacon, Boston, Massachusetts.

Hobby, D. and McKenzie, R.

[1988] *The Structure of Finite Algebras*, American Mathematical Society, Providence, Rhode Island.

Hofmann, K. H., Mislove, M. and Stralka, A.

[1974] *The Pontryagin Duality of Compact 0-Dimensional Semilattices and its Applications*, Springer Lecture Notes in Mathematics **396**.

Hogben, L. and Bergman, C.

[1985] Deductive varieties of modules and universal algebras, *Trans. Amer. Math. Soc.* **285**, 303–320.

Howie, J. M.

[1976] *An Introduction to Semigroup Theory*, Academic Press, London.

Howroyd, T.

[1973] Cancellative medial groupoids and arithmetic means, *Bull. Austral. Math. Soc.* **8**, 17–21.

Hungerford, T. W.

[1989] *Algebra*, Springer Verlag, New York, New York.

Ignatov, V. V.

[1984] *The structure of convexors*, Ph.D. Thesis, Moscow University, in Russian.
[1985] Quasivarieties of convexors, *Izv. Vyssh. Uchebn. Zaved. Mat.* **29**, 12–14, in Russian.
[1986] The structure of convexors, *Vestnik Moskov. Univ. Ser. I. Mat. Mekh.* **3**, 29–33, in Russian.

Israel, R. B.

[1979] *Convexity in the Theory of Lattice Gases*, Princeton University Press, Princeton, New Jersey.

Jacobson, N.

[1964] *Structure of Rings*, American Mathematical Society, Providence, Rhode Island.

Jamison-Waldner, R. E.

[1974] Functional representations of algebraic intervals, *Pacific J. Math.* **53**, 399–423.

Ježek, J.

[1982] A note on isomorphic varieties, *Comment. Math. Univ. Carolin.* **23**, 579–588.

Ježek, J. and Kepka, T.

[1974] Atoms in the lattice of varieties of distributive groupoids, *Colloq. Math. Soc. János Bolyai* **14**, 185–194.
[1975a] The lattice of varieties of commutative abelian distributive groupoids, *Algebra Universalis* **5**, 225–237.
[1975b] Semigroup representations of commutative idempotent abelian groupoids, *Comment. Math. Univ. Carolin.* **16**, 487–500.
[1976] Free commutative idempotent abelian groupoids and quasigroups, *Acta Univ. Carolin. Math. Phys.* **17/2**, 13–19.
[1977] Varieties of abelian quasigroups, *Czechoslovak Math. J.* **27**, 473–503.
[1981a] Free entropic groupoids, *Comment. Math. Univ. Carolin.* **22**, 223–233.
[1981b] Semigroup representations od medial groupoids, *Comment. Math. Univ. Carolin.* **22**, 513–524.
[1983a] Ideal-free CIM-groupoids and open convex sets, *Springer Lecture Notes in Mathematics* **1004**, 166–176.
[1983b] *Medial Groupoids*, Academia, Praha.
[1983c] Equational theories of medial groupoids, *Algebra Universalis* **17**, 174–190.

Ježek, J., Kepka, T. and Němec, P.

[1981] *Distributive Groupoids*, Academia, Praha.

Ježek, J. and Quackenbush, R.

[1990] Directoids: algebraic models of up-directed sets, *Algebra Universalis* **27**, 49–69.

John, B.

[1976] On classes of algebras defined by regular equations, *Colloq. Math.* **36**, 17–21.

Johnstone, P. T.

[1982] *Stone Spaces*, Cambridge University Press, Cambridge.

Jones, B. L., Enns, R. H. and Rangnekar, S. S.

[1976] On the theory of selection of coupled macromolecular systems, *Bull. Math. Biol.* **38**, 15–28.

Jónsson, B. and Nelson, E.

[1974] Relatively free products in regular varieties, *Algebra Universalis* **4**, 14–19.

Joyce, D.

[1982a] A classifying invariant of knots, the knot quandle, *J. Pure Appl. Algebra* **23**, 37–65.
[1982b] Simple quandles, *J. Algebra* **79**, 307–318.

Kalman, J.

[1971] Subdirect decomposition of distributive quasilattices, *Fund. Math.* **71**, 161–163.

Kearnes, K. A.

[1995a] A quasi-affine representation, *Internat. J. Algebra Comput.* **5**, 673–702.
[1995b] Minimal clones with abelian representations, *Acta. Sci Math.* **61**, 59–76.

Kearnes, K. A., *ctd.*

[1995c] Semilattice modes I: the associated semiring, *Algebra Universalis* **34**, 220–272.
[1995d] Semilattice modes II: the amalgamation property, *Algebra Universalis* **34**, 273–303.
[1996] Idempotent simple algebras, in *Logic and Algebra, Proceedings of the Magari Memorial Conference, Siena, 1994*, Dekker, New York, New York, pp. 529–572.
[1999] Subdirectly irreducible modes, *Discuss. Math. Algebra & Stochastic Methods* **19**, 129–145.

Kearnes, K. A. and Szendrei, A.

[1999] The classification of commutative minimal clones, *Discuss. Math. Algebra & Stochastic Methods* **19**, 147–178.

Kepka, T.

[1975] Quasigroups which satisfy certain generalized forms of the abelian identity, *Čas. Pěst. Mat.* **100**, 46–60.
[1976] Structure of triabelian quasigroups, *Comment. Math. Univ. Carolin.* **17**, 229–240.
[1978a] A note on WA-quasigroups, *Acta Univ. Carolin. Math. Phys.* **19/2**, 61–62.
[1978b] Structure of weakly abelian quasigroups, *Czechoslovak Math. J.* **28**, 181–188.
[1979] Medial division groupoids, *Acta Univ. Carolin. Math. Phys.* **20/1**, 41–60.

Kerner, E. H.

[1972] *Gibbs Ensemble: Biological Ensembles*, Gordon and Breach, London.

Kimura, N.

[1958] The structure of idempotent semigroups I, *Pacific J. Math.* **8**, 257–275.

Kikkawa, M.

[1973] On some quasigroups of algebraic models of symmetric spaces, *Mem. Fac. Lit. Sci., Shimane Univ. Nat. Sci.* **6**, 9–13.
[1974] On some quasigroups of algebraic models of symmetric spaces, *Mem. Fac. Lit. Sci., Shimane Univ. Nat. Sci.* **7**, 29–35.
[1975] On some quasigroups of algebraic models of symmetric spaces, *Mem. Fac. Lit. Sci., Shimane Univ. Nat. Sci.* **9**, 7–12.

Klukovits, L.

[1973] On commutative universal algebras, *Acta Sci. Math.* **34**, 171–174.
[1983] On the independence of the mediality conditions, *Contributions to General Algebra* **2**, 209–214.

Knoebel, A.

[1976] A comment on Balbes' representation theorem for distributive quasilattices, *Fund. Math.* **90**, 187–188.

Knoebel, A. and Romanowska, A.

[1991] Distributive multisemilattices, *Dissertationes Math.* **309**.

Kogalovskiĭ, S. R.

[1965] On the theorem of Birkhoff, *Uspekhi Mat. Nauk.* **20**, 206–207, in Russian.

Kotzig, A.

[1970] Groupoids and partitions of complete graphs, in *Combinatorial Structures and Their Applications*, Gordon and Breach, New York, New York, pp. 215–221.

Krauze, J.

[1998] *Differential groupoids*, Master's Thesis, Warsaw University of Technology, in Polish.

Kuras, J.

[1984] Agassiz bands of algebras, *Bull. Acad. Polon. Sci. Math.* **32**, 643–645.

[1985] *Application of Agassiz systems to representation of sums of equationally defined clases of algebras*, Ph.D. Thesis, M. Kopernik University, Toruń, in Polish.

[1987] Even equations and Agassiz sums, *Colloq. Math.* **53**, 9–16.

Kurpiel, A.

[1987] A monadic approach to the modal theory, *Demonstratio Math.* **20**, 247–258.

Lakser, H., Padmanabhan, R. and Platt, C. R.

[1972] Subdirect decomposition of Płonka sums, *Duke Math. J.* **39**, 485–488.

Lallement, G.

[1967] Demi-groupes réguliers, *Ann. Mat. Pura Appl.* **77**, 47–129.

Lancaster, K.

[1968] *Mathematical Economics*, Macmillan, New York, New York.

Lévai, L. and Pálfy, P.

[1996] On binary minimal clones, *Acta Cybernet.* **12**, 279–294.

Liber, S. A.

[1974] On free algebras of normal closures of varieties, *Ordered Sets and Lattices, Izdat. Saratov. Univ.*, 51–53.

Libkin, L.

[1993] Algebraic characterization of edible powerdomains, preprint.

Lidl, R. and Niederreiter, H.

[1983] *Finite Fields*, Addison-Wesley Publishing Company, Reading, Massachusetts.

Lindner, C. C. and Mendelsohn, N. S.

[1973] Construction of perpendicular Steiner quasigroups, *Aequationes Math.* **9**, 150–156.
[1976] Construction of n-cyclic quasigroups and applications, *Aequationes Math.* **14**, 111–121.

Linton, F. E. J.

[1965] Autonomous categories and duality of functors, *J. Algebra* **2**, 315–349.
[1966] Autonomous equational categories, *J. Mathematics and Mechanics* **15**, 637–642.

Lizak, P.

[1991] *Quandles, Algebraic Structures and Applications*, Master's Thesis, Warsaw Technical University, in Polish.

Loos, D.

Symmetric Spaces, Benjamin, New York, New York.

Lothaire, M.

[1983] *Combinatorics on Words*, Addison–Wesley Publishing Company, Reading, Massachusetts.

Mac Lane, S.

[1971] *Categories for the Working Mathematician*, 1st ed., Springer Verlag, New York, New York.

Mac Lane, S. and Birkhoff, G.

[1967] *Algebra*, Macmillan, New York, New York.

Magari, R.

[1969] Una demonstrazione del fatto che ogni varietá ammette algebre simplici, *Ann. Univ. Ferrara Sez. VII (N.S.)* **14**, 1–4.

Mal'cev, A. I.

[1954] On the general theory of algebraic systems, *Mat. Sb.* **35**(77), 3–20, in Russian. English translation in: A.M.S. Transl. Ser. 2, Vol. 27, 125–142, Providence, Rhode Island, 1963.
[1966] Several remarks on quasivarieties of algebraic systems, *Algebra i Logika Sem.* **3, 5**, 3–9, in Russian.
[1967] Multiplication of classes of algebraic systems, *Sibirsk. Mat. Zh.* **8**, 346–365, in Russian. English translation in: *The Metamathematics of Algebraic Systems. Collected Papers: 1936–1967.*
[1971] *The Metamathematics of Algebraic Systems. Collected Papers: 1936–1967*, North-Holland Publishing Co., Amsterdam.
[1973] *Algebraic Systems*, Springer Verlag, New York, New York.
[1975] *Foundations of Linear Algebra*, Nauka, Moscow, in Russian.

Manes, E. G.

[1993] Adas and the equational theory of if-then-else, *Algebra Universalis* **30**, 373–394.

McKenzie, R.

[1996] An algebraic version of categorical equivalence for varieties and more general algebraic categories, in *Logic and Algebra, Proceedings of the Magari Memorial Conference, Siena, 1994*, Dekker, New York, New York, pp. 211–243.

McKenzie, R. N., McNulty, G. F. and Taylor, W. A.

[1987] *Algebras, Lattices, Varieties*, Vol. 1, Wadsworth & Brooks/Cole, Monterey, California.

McKenzie, R. and Romanowska, A.

[1979] Varieties of ·-distributive bisemilattices, *Contributions to General Algebra* **1**, 213–218.

McLean, D.

[1954] Idempotent semigroups, *Amer. Math. Monthly* **61**, 110–113.

Mel'nik, I. I.

[1969] Varieties of Ω-algebras, *Studies in Algebra 1, Izdat. Saratov. Univ.*, 32–40, in Russian.

[1970a] Regular closure of varieties of semigroups that are equipped with commuting idempotents, *Studies in Algebra 2, Izdat. Saratov. Univ.*, 58–66, in Russian.

[1970b] Varieties and lattices of varieties of semigroups, *Studies in Algebra 2, Izdat. Saratov. Univ.*, 47–57, in Russian.

[1971] Normal closures of perfect varieties of universal algebras, *Ordered Sets and Lattices 1, Izdat. Saratov. Univ.*, 56–65, in Russian.

Mendelsohn, N. S.

[1970] Orthogonal Steiner systems, *Aequationes Math.*, **5**, 268–272.

Merriell, D.

[1970] An application of quasigroups to geometry, *Amer. Math. Monthly* **77**, 44–46.

Minc, H.

[1957] Polynomials and bifurcating root-trees, *Proc. Roy. Soc. Edinburgh Sec. A* **64**, 319–341.

Minkowski, H.

[1911] *Gesammelte Abhandlungen*, 2. Band, Teubner, Leipzig.

Mitchell, B.

[1965] *Theory of Categories*, Academic Press, New York, New York.

Mitschke, A.

[1973] On a representation of groupoids as sums of directed systems, *Colloq. Math.* **28**, 11–18.
[1977] Summen von gerichteten Systemen von Algebren, *Mitt. Math. Sem. Giessen* **124**, 1–56.

Mitschke, A. and Werner, H.

[1973] On groupoids representable by vector spaces over finite fields, *Arch. Math. (Basel)* **24**, 14–20.

Mituhisa, T.

[1943] Abstractions of symmetric functions, *Tôhoku Math. J.* **49**, 145–207, in Japanese.

Morgado, J.

[1966] Entropic groupoids and abelian groups, *Gaz. Math.* **27**, 8–10.
[1967] A theorem on entropic groupoids, *Portugal. Math.* **26**, 449–452.

Morgenstern, D. and Neumann, J.

[1953] *Theory of Games and Economics Behavior*, 3rd ed., Princeton Univ. Press., Princeton, New Jersey.

Movsisyan, Yu. M.

[1998] Hyperidentities in algebras and varieties, *Uspekhi Mat. Nauk* **53 (319)**, 61–114.

Murdoch, D. C.

[1939] Quasigroups which satisfy certain generalized associative laws, *Amer. J. Math.* **61**, 509–522.
[1941a] Note on normality in quasigroups, *Bull. Amer. Math. Soc.* **47**, 134–138.
[1941b] Structure of abelian quasigroups, *Trans. Amer. Math. Soc.* **49**, 392–409.

Neumann, W. D.

[1970] On the quasivariety of convex subsets of affine spaces, *Arch. Math. (Basel)* **21**, 11–16.

[1978] Mal'cev conditions, spectra and Kronecker product, *J. Austral. Math. Soc.* **25**, 103–117.

Nieminen, J.

[1977] Ideals in distributive quasilattices, *Studia Univ. Babeş – Bolyai Math.* **22**, 6–11.

Ostermann, F. and Schmidt, J.

[1966] Der baryzentrische Kalkül als axiomatische Grundlage der affinen Geometrie, *J. Reine Angew. Math.* **224**, 44–57.

Padmanabhan, R.

[1969] A note on inverse binary operation in abelian groups, *Fund. Math.* **65**, 61–63.

[1971] Regular identities in lattices, *Trans. Amer. Math. Soc.* **158**, 179–188.

Padmanabhan, R. and Płonka, J.

[1980] Idempotent reducts of abelian groups, *Algebra Universalis* **11**, 7–11.

Pastijn, F. and Reynaerts, H.

[1977] Semilattices of modules, Springer Lecture Notes in Mathematics, **586**, pp. 145–155.

Pastijn, F. and Romanowska, A.

[1982] Idempotent distributive semirings, *Acta Sci. Math.* **44**, 239–253.

Pelling, M. J. and Rogers, D. G.

[1978] Stein quasigroups I: Combinatorial Aspects, *Bull. Austral. Math. Soc.* **18**, 221–236.

Pelling, M. J. and Rogers, D. G., *ctd.*

[1979] Stein quasigroups II: Algebraic Aspects, *Bull. Austral. Math. Soc.* **20**, 321–344.

Petrich, M.

[1964] The maximal semilattice decomposition of a semigroup, *Math. Z.* **85**, 68–82.
[1973] *Introduction to Semigroups*, Charles E. Merrill (Bell & Howell), Columbus, Ohio.
[1974] The structure of completely regular semigroups, *Trans. Amer. Math. Soc.* **189**, 211–236.
[1977] *Lectures in Semigroups*, Akademie-Verlag, Berlin.

Pierce, R. S.

[1978] Symmetric groupoids, *Osaka J. Math.* **15**, 51–76.
[1979] Symmetric groupoids II, *Osaka J. Math.* **16**, 317–348.

Pigozzi, D.

[1979] Minimal locally finite varieties that are not finitely axiomatizable, *Algebra Universalis* **9**, 374–390.

Pilitowska, A.

[1991] Free P-bilattices, *Demonstratio Math.* **24**, 121–127.
[1996] *Modes of Submodes*, Ph.D. Thesis, Warsaw University of Technology.
[1998] Identities for classes of algebras closed under the complex structures, *Discuss. Math. Algebra & Stochastic Methods* **18**, 85–109.
[1999] Enrichment of affine spaces and algebras of subalgebras, *Discuss. Math. Algebra & Stochastic Methods* **19**, 207–225.
[2001] The lattice of subvarieties of the variety of some ternary modes, *Contributions to General Algebra* **13**, 265–274.

Pilitowska, A. and Romanowska, A.

[1998] Reductive modes, *Periodica Math. Hungarica* **36**, 67–78.

Pilitowska, A., Romanowska, A. and Roszkowska, B.

[1996] Products of mode varieties and algebras of subalgebras, *Math. Slovaca* **46**, 497–514.

Pilitowska, A., Romanowska, A. and Smith, J. D. H.

[1995] Affine spaces and algebras of subalgebras, *Algebra Universalis* **34**, 527–540.

Płonka, J.

[1966a] Diagonal algebras, *Fund. Math.* **58**, 309–321.
[1966b] Remarks on diagonal and generalized diagonal algebras, *Colloq. Math.* **15**, 19–23.
[1967a] On distributive quasi-lattices, *Fund. Math.* **60**, 191–200.
[1967b] On a method of construction of abstract algebras, *Fund. Math.* **61**, 183–189.
[1967c] A representation theorem for idempotent medial algebras, *Fund. Math.* **61**, 191–198.
[1967d] Sums of direct systems of abstract algebras, *Bull. Acad. Polon. Sci., Sér. Sci Math. Astronom. Phys.* **15**, 133–135.
[1967e] On some properties and applications of a notion of the sum of a direct system of abstract algebras, *Bull. Acad. Polon. Sci., Sér. Sci Math. Astronom. Phys.* **15**, 681–682.
[1968a] On distributive n-lattices and n-quasilattices, *Fund. Math.* **62**, 293–300.
[1968b] Some remarks on sums of direct systems of algebras, *Fund. Math.* **62**, 301–308.
[1969a] On equational classes of abstract algebras defined by regular equations, *Fund. Math.* **64**, 241–247.
[1969b] On sums of direct systems of Boolean algebras, *Colloq. Math.* **20**, 209–214.
[1970] On the arity of idempotent reducts of groups, *Colloq. Math.* **21**, 35–37.
[1971a] On algebras with n distinct n-ary operations, *Algebra Universalis* **1**, 73–79.
[1971b] On free algebras and algebraic decomposition of algebras from some equational classes defined by regular equations, *Algebra Universalis* **1**, 261–264.
[1973a] On the sum of a direct system of relational systems, *Bull. Acad. Polon. Sci., Sér. Sci Math. Astronom. Phys.* **21**, 595–597.

Płonka, J., ctd.

[1973b] R-prime idempotent reduct of abelian groups, *Arch. Math. (Basel)* **24**, 129–132.

[1974a] On connection between the decomposition of an algebra into sums of direct systems of subalgebras, *Fund. Math.* **84**, 237–244.

[1974b] On groups in which idempotent reducts form a chain, *Colloq. Math.* **29**, 87–91.

[1984a] On the sum of a direct system of abstract algebras with nullary polynomials, *Algebra Universalis* **19**, 197–207.

[1984b] On the sum of a $\langle t_\omega, t_1 \rangle$ semilattice ordered system of algebras, *Bull. Polish Acad. Sci. Math* **32**, 255–259.

[1985a] On k-cyclic groupoids, *Math. Japon.* **30**, 371–382.

[1985b] On the sum of a i-semilattice ordered system of algebras, *Studia Sci. Math. Hungar.* **20**, 301–307.

[1989a] Minimal generics of regular varieties, in K. Hałkowska and B. Stawski (eds), *Universal and Applied Algebra*, World Scientific Publishing, Teaneck, New Jersey, pp. 227–234.

[1989b] On some decomposition of the sum of a direct system of algebras into a subdirect products, *Demonstratio Math.* **22**, 235–239.

[1990] Some decompositions of the sum of a semilattice ordered system of algebras, in R. Mlitz (ed.), *General Algebra*, Elsevier Science Publishers B.V., North-Holland, Amsterdam, pp. 213–219.

Płonka, J. and Romanowska, A.

[1992] Semilattice sums, in A. Romanowska and J. D. H. Smith (eds), *Universal Algebra and Quasigroup Theory*, Heldermann Verlag, Berlin, pp. 123–158.

Pollák, G. and Szendrei, A.

[1981] Independent basis for the identities of entropic groupoids, *Comment. Math. Univ. Carolin.* **22**, 71–85.

Pöschel, R. and Reichel, M.

[1993] Projection algebras and rectangular algebras, in K. Denecke and H. J. Vogel (eds), *General Algebra and Applications*, Heldermann Verlag, Berlin, pp. 180–194.

Pszczoła, K.

[2001] Duality for affine spaces over finite fields, *Contributions to General Algebra* **13**, 285–293.

Pucharev, N. K.

[1968] Geometric problems concerning some medial quasigroups, *Sibirsk. Mat. Zh.* **9**, 891–897, in Russian.

Puhlmann, H.

[1993] The snack powerdomain for database semantics, preprint TU Darmstadt.

Pumplün, D.

[1984] Regularly ordered Banach spaces and positively convex spaces, *Res. Math.* **7**, 85–112.
[1985] The Hahn–Banach theorem for totally convex spaces, *Demonstratio Math.* **18**, 567–588.

Pumplün, D. and Röhrl, H.

[1985a] Banach spaces and totally convex spaces I, *Comm. Algebra* **12**, 953–1019.
[1985b] Banach spaces and totally convex spaces II, *Comm. Algebra* **13**, 1047–1113.
[1990] Congruence relations in totally convex spaces, *Comm. Algebra* **18**, 1469–1496.
[1991] Convexity theories II. The Hahn–Banach theorem for real convexity theories, in H. Herrlich and H. Porst (eds), *Category Theory at Work*, Heldermann Verlag, Berlin, pp. 387–395.
[1995] Convexity theories IV. Klein–Hilbert parts in convex modules, *Appl. Categ. Structures* **3**, 173–200.

Putcha, M. S.

[1973] Semilattice decomposition of semigroups, *Semigroup Forum* **6**, 12–34.
[1978] On the maximal semilattice decomposition of the power semigroup of a semigroup, *Semigroup Forum* **15**, 263–267.

Putcha, M. S., *ctd.*

[1981b] Rings which are semilattices of archimedean semigroups, *Semigroup Forum* **23**, 1–5.

Quackenbush, R.

[1988] Completeness theorem for universal and implicational logics of algebras via congruences, *Proc. Amer. Math. Soc.* **103**, 1015–1021.
[1992] Quasi–affine algebras, *Algebra Universalis* **20**, 318–327.

Raftery, J. G.

[2001] A characterization of varieties, *Algebra Universalis* **45**, 449–450.

Ratschek, H.

[1972] Eine Bemerkung über entropische Gruppoide, *Math. Nachr.* **52**, 141–146.

Robinson, D.

[1962] Concerning two questions of S. K. Stein, *Notices of the Amer. Math. Soc.* **9**, 149.

Romanowska, A.

[1978] On free algebras in some equational classes defined by regular equations, *Demonstratio Math.* **11**, 1131–1137.
[1980a] Free bisemilattices with one distributive law, *Demonstratio Math.* **13**, 565–572.
[1980b] On bisemilattices with one distributive law, *Algebra Universalis* **10**, 36–47.
[1980c] Subdirectly irreducible ··-distributive bisemilattices I, *Demonstratio Math.* **13**, 757–785.
[1982a] Building bisemilattices from lattices and semilattices, *Contributions to General Algebra* **2**, 343–358.
[1982b] Free idempotent distributive semirings with a semilattice reduct, *Math. Japon.* **27**, 467–481.
[1982c] Idempotent distributive semirings with a semilattice reduct, *Math. Japon.* **27**, 483–493.

Romanowska, A., ctd.

[1982d] On distributivity of bisemilattices with one distributive law, *Colloq. Math. Soc. János Bolyai* **29**, 653–661.

[1983] Algebras of functions from partially ordered sets into distributive lattices, *Springer Lecture Notes in Mathematics* **1004**, 245–256.

[1984] On some constructions of bisemilattices, *Demonstratio Math.* **17**, 1011–1022.

[1985] Constructing and reconstructing of algebras, *Demonstratio Math.* **18**, 209–230.

[1986] On regular and regularized varieties, *Algebra Universalis* **23**, 215–241.

[1988] On some representations of groupoid modes satisfying the reduction law, *Demonstratio Math.* **21**, 943–960.

[1992] Mal'cev modes, affine spaces and barycentric algebras, in A. Romanowska, J. D. H. Smith (eds.), *Universal Algebra and Quasigroup Theory*, Heldermann Verlag, Berlin, pp. 173–194.

[1994] Dissemilattices with multiplicative reducts chains, *Demonstratio Math.* **27**, 843–857.

[1995] Subdirectly irreducible meet–distributive bisemilattices, II, *Discuss. Math. Algebra & Stochastic Methods* **15**, 1–17.

[1997] From dissemilattices to snack algebras, *Fund. Inform.* **31**, 65–77.

[1999] Sums of algebras, subdirect products and subdirectly irreducible algebras, *Contributions to General Algebra* **11**, 181–190.

[2000] Barycentric algebras, in K. Denecke and H. J. Vogel (eds), *General Algebra and Applications*, Shaker Verlag, Aachen, pp. 167–181.

[2001] Notes on sums of algebras, subdirect products and subdirectly irreducible algebras, *Algebra Universalis* **46**, 321–341.

Romanowska, A. and Roszkowska, B.

[1987] On some groupoid modes, *Demonstratio Math.* **20**, 277–290.
[1989] Representations of n-cyclic groupoids, *Algebra Universalis* **26**, 7–15.
[1997] Remarks on k-sums of reducts of affine spaces, preprint.

Romanowska, A. and Smith, J. D. H.

[1981] Bisemilattices of subsemilattices, *J. Algebra* **70**, 78–88.

[1985a] Distributive lattices, generalisation, and related non-associative structures, *Houston J. Math.* **11**, 367–383.

[1985b] From affine spaces to projective geometry via convexity, *Springer Lecture Notes in Mathematics* **1149**, 255–269.

Romanowska, A. and Smith, J. D. H., *ctd.*

[1985c] *Modal Theory – an Algebraic Approach to Order, Geometry and Convexity*, Heldermann Verlag, Berlin.
[1989a] Subalgebra systems of idempotent entropic algebras, *J. Algebra* **120**, 247–262.
[1989b] On the structure of subalgebra systems of idempotent entropic algebras, *J. Algebra* **120**, 263–283.
[1989c] Support functions and ordinal products, *Geom. Dedicata* **30**, 281–296.
[1990a] Communicating processes, and entropic algebras, *Fund. Inform.* **13**, 263–274.
[1990b] On the structure of barycentric algebras, *Houston J. Math.* **16**, 431–448.
[1991a] Differential groupoids, *Contributions to General Algebra* **7**, 283–290.
[1991b] On the structure of semilattice sums, *Czechoslovak Math J.* **41**, 24–43.
[1993] Separable modes, *Algebra Universalis* **30**, 61–71.
[1996] Semilattice based dualities, *Studia Logica* **56**, 225–261.
[1997] Duality for semilattice representations, *J. Pure Appl. Algebra* **115**, 289–308.
[2001] Embedding sums of cancellative modes into functorial sums of affine spaces, in J. M. Abe and S. Tanaka (eds), *Unsolved Problems on Mathematics for the 21st Century, a Tribute to Kiyoshi Iseki's 80th Birthday*, IOS Press, Amsterdam.

Romanowska, A. and Traina, S.

[1999] Algebraic quasi-orders and sums of algebras, *Discuss. Math. Algebra & Stochastic Methods* **19**, 239–263.

Romanowska, A. and Trakul, A.

[1989] On the structure of some bilattices, *in* K. Hałkowska and B. Stawski (eds.), *Universal and Applied Algebra*, World Scientific Publishing, New Jersey, pp. 235–253.

Romanowska, A. and Zamojska–Dzienio, A.

[2001] Embedding semilattice sums of cancellative modes into semimodules, *Contributions to General Algebra* **13**, 295–303.

Roszkowska, B.

[1987a] The lattice of varieties of symmetric idempotent entropic groupoids, *Demonstratio Math.* **20**, 259–275.

[1987b] *Symmetric idempotent entropic groupoids*, Ph.D. Thesis, Warsaw Technical University, in Polish.

[1989] On some varieties of symmetric idempotent entropic groupoids, in K. Hałkowska and B. Stawski (eds), *Universal and Applied Algebra*, World Scientific Publishing, New Jersey, pp. 254–274.

[1999a] A representation of symmetric idempotent entropic groupoids, *Demonstratio Math.* **32**, 247–262.

[1999b] Subdirectly irreducible symmetric idempotent entropic groupoids, *Demonstratio Math.* **32**, 469–484.

Röhrl, H.

[1991a] Convexity theories I. Γ-convex spaces, *Constantin Carathéodory: An International Tribute*, World Scientific Publishing, Teaneck, New Jersey, pp. 1175–1209.

[1991b] Convexity theories $\mho$ – back to the future, in H. Herrlich and H. E. Porst (eds), *Category Theory at Work*, Heldermann Verlag, Berlin, pp. 321–324.

[1993] Convexity theories III. Classification of certain real convexity theories, *Geom. Dedicata* **45**, 323–340.

[1994] Convexity theories 0. Foundations, *Appl. Categ. Structures* **2**, 13–43.

Rumer, Yu. B. and Ryvkin, M. Sh.

[1980] *Thermodynamics, Statistical Physics, and Kinetics*, tr. S. Emyonov, Mir, Moscow.

Sade, A.

[1957] Quasigroupes obéissant á certaines lois, *Rev. Fac. Sci. Univ. Istanbul* **22**, 151–184.

[1959] Entropie demossienne de multigroupöides et des quasigroupes, *Ann. Soc. Sci. Bruxelles Ser. I* **73**, 302–309.

Salii, V. N.

[1969a] Equationally normal varieties of semigroups, *Izv. Vyssh. Uchebn. Zaved. Mat.* **84**, 61–68, in Russian.
[1969b] Equationally normal varieties of universal algebras, *Works of Young Scientists: Mathematics and Mechanics* **2**, 124–130, in Russian.
[1971] A theorem on homomorphisms of strong semilattices of semigroups, in V. V. Vagner (ed.), *Theory of Semigroups and Applications* **2**, Izd. Saratov. Univ., Saratov, pp. 69–74, in Russian.

Schein, B. H.

[1963] On the theory of generalized groups, *Dokl. Akad. Nauk SSSR* **153**, 296–299.
[1974] Bands of monoids, *Acta Sci. Math.* **36**, 145–154.
[1996] Bands of semigroups: variations on a Clifford theme, in *Semigroup Theory and its Applications*, Cambridge University Press, Cambridge, pp. 53–80.

Shafaat, A.

[1969] On implicationally defined classes of algebras, *J. London Math. Soc.* **44**, 137–140.
[1974] On varieties closed under the construction of power algebras, *Bull. Austral. Math. Soc.* **11**, 213–218.

Sholander, M.

[1949] On the existence of the inverse operation in alternation groupoids, *Bull. Amer. Math. Soc.* **55**, 746–757.

Skornyakov, L. A.

[1981] Convexors, *Studia Sci. Math. Hungar.* **16**, 25–34.
[1982a] The algebra of stochastic distributions, *Izv. Vyssh. Uchebn. Zaved. Mat.* **25**, 59–67, in Russian.
[1982b] Stochastic acts and conmodules, *Semigroup Forum* **25**, 269–282.
[1985] Stochastic algebras, *Izv. Vyssh. Uchebn. Zaved. Mat.* **29**, 3–11.

Smith, J. D. H.

[1976] *Mal'cev Varieties*, Springer Lecture Notes in Mathematics **554**.

[1979] Finite equationally complete entropic quasigroups, *Contribution to General Algebra* **1**, pp. 345–356.

[1986] Modal theory, partial orders and digital geometry, *Springer Lecture Notes in Computer Science* **239**, 308–323.

[1990a] Centrality, in O. Chein, H. O. Pflugfelder and J. D. H. Smith (eds), *Quasigroups and Loops: Theory and Applications*, Heldermann Verlag, Berlin, pp. 95–114.

[1990b] Entropy, character theory and centrality of finite quasigroups, *Math. Proc. Cambridge Philos. Soc.* **108**, 435–443.

[1991a] Finite codes and groupoid words, *Europ. J. Combinatorics* **12**, 331–339.

[1991b] Skein polynomials and entropic right quasigroups, *Demonstratio Math.* **24**, 241–246.

[1992] Quasigroups and quandles, *Discrete Math.* **109**, 277–282.

[1996] Competition and the canonical ensemble, *Math. Biosci.* **133**, 69–83.

[1997] Homotopy and semisymmetry of quasigroups, *Algebra Universalis* **38**, 175–184.

[1998] Barycentric algebras, canonical distributions and Legendre transforms, *ISU Mathematics Reports* **M 98-01**, 1–36.

[1999] Modes and modals, *Discuss. Math. Algebra & Stochastic Methods* **19**, 9–40.

Smith, J. D. H. and Romanowska, A.

[1999] *Post-Modern Algebra*, Wiley & Sons, New York, New York.

Sobczyńska, J.

[1988] *Barycentric Algebras*, Master's Thesis, Warsaw University of Technology, in Polish.

Sokratova, O.

[2000] Ω*-rings, their flat and projective acts with some applications*, Ph.D. Thesis, University of Tartu.

Soublin, J. P.

[1966] Mediations, *C. R. Acad. Sci. Paris Ser. A-B*, **263**, A 49–50, A 115–117.
[1971] Etude algébrique de la notion de moyenne, *J. Math. Pures et Appl.* **50**, 53–264.

Stein, S. K.

[1956] Foundations of quasigroups, *Proc. Nat. Acad. Sci. USA* **42**, 545–546.
[1957] On the foundations of quasigroups, *Trans. Amer. Math. Soc.* **85**, 228–256.

Stokes, T.

[1998a] Sets with B-action and linear algebra, *Algebra Universalis* **39**, 31–43.
[1998b] Radical classes of algebras with B-action, *Algebra Universalis* **40**, 73–85.

Stone, M. H.

[1949] Postulates for the barycentric calculus, *Annali di Matematica* **29**, 25–30.

Strecker, R.

[1974] Über entropische Gruppoide, *Math. Nachr.* **64**, 361–371.

Sushkievich, A. K.

[1937] *The Theory of Generalized Groups*, Gos. Nauč.-Tech. Izd. Ukrainny, Charkov–Kiev, in Russian.

Suvorov, N. M.

[1969] A duality theorem for distributive transitive topological quasigroups, *Matem. Issled.* **4**, 162–166, in Russian.

Świrszcz, T.

[1974] Monadic functors and convexity, *Bull. Acad. Polon. Sci., Sér. Sci Math. Astronom. Phys.* **22**, 39–42.

[1975] *Monadic functors and categories of convex sets*, Mathematical Institute, Polish Academy of Sciences, preprint no. 70.

Szendrei, A.

[1975] On affine modules, *Colloq. Math. Soc. János Bolyai* **17**, 457–464.
[1976] Idempotent reducts of abelian groups, *Acta Sci. Math.* **38**, 171–182.
[1977] On the arity of affine modules, *Colloq. Math.* **38**, 1–4.
[1979] On modules in which idempotent reducts form a chain, *Colloq. Math.* **40**, 191–196.

[1981a] Identities satisfied by convex linear forms, *Algebra Universalis* **12**, 103–122.

[1981b] Identities in idempotent affine algebras, *Algebra Universalis* **12**, 172–199.

[1982a,b] On the idempotent reducts of modules I, II *Colloq. Math. Soc. János Bolyai* **29**, 753–768, 769–780.

[1986] *Clones in Universal Algebra*, les Presses de l'Université de Montréal, Montreal, Quebec.

Szmielew, W.

[1983] *From Affine to Euclidean Geometry, an Axiomatic Approach*, PWN, Warsaw.

Tamura, T.

[1964] Another proof concerning the greatest semilattice decomposition of a semigroup, *Proc. Japan Acad. Sci.* **40**, 777–780.

[1968] Notes on medial archimedian semigroups without idempotents, *Proc. Japan Acad. Sci.* **44**, 776–778.

[1972] On Putcha's theorem concerning semilattice of Archimedean semigroups, *Semigroup Forum* **4**, 83–86.

[1975] Quasi-orders, generalized Archimedeanness and semilattice decompositions, *Math. Nachr.* **68**, 201–220.

Tamura, T. and Kimura, N.

[1954] On decomposition of a commutative semigroup, *Kodai Math. Sem. Rep.* **4**, 109–112.
[1955] Existence of the greatest decomposition of a semigroup, *Kodai Math. Sem. Rep.* **7**, 83–84.

Taylor, M. A.

[1978] On the generalized equations of associativity and bisymmetry, *Aequationes Math.* **17**, 154–163.

Taylor, W.

[1981] Hyperidentities and hypervarieties, *Aequationes Math.* **23**, 30–49.
[1982] Some applications of the term condition, *Algebra Universalis* **14**, 11–25.

Thompson, C. J. and McBride, J. L.

[1974] On Eigen's theory of the self-organization of matter and the evolution of biological macromolecules, *Math. Biosci.* **21**, 127–142.

Toyoda, K.

[1940a] On axioms of mean transformations and automorphic transformations of abelian groups, *Tôhoku Math. J.* **46**, 239–251.
[1940b] On affine geometry of abelian groups, *Proc. Imp. Acad. Tokyo* **16**, 161–164.
[1941] On axioms of linear functions, *Proc. Imp. Acad. Tokyo* **17**, 221–227.

Urbanik, K.

[1959/60] A representation theorem for Marczewski's algebras, *Fund. Math.* **48**, 147–167.

van Lint, J. H. and Wilson, R. M.

[1992] *A Course in Combinatorics*, Cambridge University Press, Cambridge.

Volenec, V.

[1981] Extension of Toyoda's theorem on entropic groupoids, *Math. Nachr.* **102**, 183–188.

Wechler, W.

[1992] *Universal Algebra for Computer Scientists*, Springer Verlag, Berlin.

Weissglass, J.

[1973] Semigroup rings and semilattice sums of rings, *Proc. Amer. Math. Soc.* **39**, 471–477.

Werner, H.

[1974] Congruences on products of algebras, *Algebra Universalis* **4**, 99–105.

Wigner, E. P.

[1960] The unreasonable effectiveness of mathematics in the natural sciences, *Comm. Pure Appl. Math.* **13** 1–14.

Winker, S.

[1984] *Quandles, Knot Invariants and the n-Fold Branched Cover*, Ph.D. Thesis, University of Illinois, Chicago, Illinois.

Wraith, G. C.

[1970] *Algebraic Theories*, Aarhus University Lecture Notes Series **22**.

Yamada, M.

[1955] On the greatest decomposition of commutative semigroups, *Kodai Math. Sem. Rep.* **7**, 59–62.
[1956] Composition of semigroups, *Kodai Math. Sem. Rep.* **8**, 107–111.

Yamada, M. and Kimura, N.

[1958] Note on idempotent semigroups II, *Proc. Japan Acad.* **34**, 110–112.

Zając, E.

[1988] The lattice of varieties of left normal bandoids, in K. Hałkowska and B. Stawski (eds.), *Universal and Applied Algebra*, World Scientific Publishing, New Jersey, pp. 366–377.

[1990] *On Some Algebras Derived from the Theory of Distributive Bisemilattices*, Ph.D. Thesis, Warsaw University of Technology.

[1991] Constructions of left normal bandoids, *Demonstratio Math.* **24**, 191–206.

[1992] Subdirectly irreducible left normal bandoids, I, *Demonstratio Math.* **25**, 927–946.

[1994] Subdirectly irreducible left normal bandoids, II, *Demonstratio Math.* **27**, 859–877.

Zirilli, F.

[1968] Sopra certi quasigruppi semisimmetrici mediali, *Ricerche Mat.* **17**, 234–253.

INDEX OF SYMBOLS

$\rho \circ \sigma$ [relational product] 2
$A!$ [set of all bijections from A to A] 3
$\widehat{A}$ [diagonal subset of A^2] 4
a^ρ [equivalence class of element a under equivalence relation ρ] 4
$a \parallel b$ [a and b incomparable] 5
$[a, b)$ [interval $\{x \mid a \leq x < b\}$ in poset] 6
$\langle X \rangle$ [subalgebra generated by subset X] 18
X^+ [free semigroup over X] 36
X^* [free monoid over X] 36
$X\Omega$ [absolutely free algebra over X] 36
$\models$ [satisfies] 45
$\widetilde{\underline{K}}$ [regularization of $\underline{K}$] 48
$/$ [right division in (right) quasigroup] 55
$\backslash$ [left division in (left) quasigroup] 55
$\underline{f}$ [binary weighted mean] 61
$\underline{R}$ [class of affine spaces over ring R] 65
$\bot$ [initial object of category] 81
$\leq_s$ [subdirect product] 106
$[a]$ [principal wall generated by a] 131
$X^{*\kappa}$ [free commutative monoid over X] 142
$\underline{V}_1 \times \cdots \times \underline{V}_n$ [products of $\underline{V}_i$-algebras] 150
$\langle X|\sigma \rangle$ [algebra presented] 154
$\underline{R} \circ_K \underline{S}$ [Mal'cev product] 175
$\underline{R} \circ \underline{S}$ [Mal'cev product] 175, 270
$\preceq$ [algebraic quasi-order] 182
(I) [semilattice I as small category] 187
$A \otimes B$ [tensor product of algebras] 238
$'$ [involution $r \mapsto 1 - r$] 290
$(\gamma|\beta)$ [centreing congruence] 313
$\underline{V}_1 \otimes \underline{V}_2$ [tensor product of varieties] 366
$\underline{V}_1 \oslash \underline{V}_2$ [independent join of varieties] 470

A^* [predual of barycentric algebra A] 510
α_f [center of gravity under function f] 521

$\underline{\text{AGp}}$ [category of abelian groups] 79
$A(w)$ [set of addresses of all subwords of w] 441
$A_m(w)$ [set of addresses of length m] 441
Ann [annihilator] 57
arg [argument map] 38
$\underline{\underline{B}}$ [class of real barycentric algebras] 67
$\underline{\underline{B}}$ [class of barycentric algebras over subfield of $\mathbb{R}$] 289
$\underline{\text{BAlg}}$ [category of Boolean algebras] 85
$\underline{\text{BM}}$ [variety of all binary modes] 270
$\underline{\underline{B}}_n$ [quasivariety generated by T_n] 404
$\underline{\underline{\widetilde{B}}}_n$ [quasiregularization of $\underline{\underline{B}}_n$] 404
$\underline{\underline{B}}_\omega$ [quasivariety generated by all T_n] 404
$\underline{\underline{\widetilde{B}}}_\omega$ [quasiregularization of $\underline{\underline{B}}_\omega$] 404
$\underline{\text{BRing}}$ [category of Boolean rings] 85
$\underline{\underline{C}}$ [class of real convex sets] 67
$\underline{\underline{C}}$ [variety of commutative binary modes] 263
$\overline{\mathcal{C}}(a,b)$ [class of $\mathcal{C}$-morphisms from a to b] 78
cgX [congruence generated by subset X] 23
$cg(a,b)$ [principal congruence generated by pair] 23
$Cg(A,\Omega)$ [congruence poset] 23
$cg_K(a,b)$ [$\underline{K}$-congruence generated] 108
$Cg_K(A,\Omega)$ [poset of $\underline{K}$-congruences] 107
$\underline{\underline{C}}_{2k+1}$ [commutative binary modes satisfying $xyw_{2k+1} = x$] 360
$\underline{\underline{Cl}}$ [quasivariety of cancellative binary modes] 493
c_l [class of cancellative algebras] 386
$\text{Clo } A$ [clone of all operations on A] 44
$\underline{\text{CMon}}$ [category of commutative monoids] 79
$\mathcal{C}^{op}$ [opposite of category $\mathcal{C}$] 80
$\underline{\underline{CQ}}_m$ [subvariety of $\underline{\underline{CQM}}$] 357
$\underline{\underline{CQM}}$ [variety of commutative quasigroup modes] 355
$\underline{\text{CRing}}$ [category of commutative rings] 79
$\underline{\text{CSgp}}$ [category of commutative semigroups] 79
$\underline{Cv}$ [class of convex sets over subfield of $\mathbb{R}$] 289
d [differential (operator)] 268, 281
D [decomposition functor] 515
$\mathbb{D}$ [ring of dyadic rationals] 263
$\mathbb{D}_1$ [dyadic unit interval] 263
D_d [codomain for support functions] 556
$\underline{\underline{D}}_{j,j+p}$ [differential groupoids satisfying $xy^j = xy^{j+p}$] 371

SYMBOL INDEX

$_D\underline{\underline{K}}$ [class of directed colimits] 158
$\underline{\underline{Dm}}$ [variety of differential modes] 269
Δ_n [n-dimensional simplex] 66
$\underline{\underline{D}}_{0,p}$ [binary modes satisfying $x = xy^p$] 368
$\mathbb{D}_{p^r}$ [copy of $\mathbb{Z}/p^r\mathbb{Z}$ in $\mathbb{D}_{p^\infty}$] 379
$\mathbb{D}_{p^\infty}$ [$\{(m2^k/p^n) + \mathbb{D} \mid k, m, n \in \mathbb{Z}\}$] 374
$\varepsilon_{\alpha\beta}(w)$ [word obtained by subword interchange] 444
$\underline{\underline{EGr}}$ [variety of entropic groupoids] 445
$e_i(\omega)$ [translation] 415
End [set of endomorphisms] 56
epi f [epigraph of function f] 38
Eq [equational theory] 146
$Eqv\, A$ [lattice of equivalence relations on set A] 26
$/F$ [reduced product by filter F] 167
$Fi(B)$ [filter poset of Boolean algebra B] 165
$\text{Fix}(X, M)$ [set of fixed points of M-set] 29
$\mathfrak{F}_{pr}$ [set of non-trivial groupoids $(\mathbb{Z}_p, \underline{r})$] 455
F_V [minimal cogenerator] 374
$GCD(m, n)$ [greatest common divisor] 20
GF [Galois field] 62
$\underline{\underline{G}}(n, t)$ [binary modes satisfying $x^n y = y$, $x^t y = yx$] 341
$\underline{\underline{Gp}}$ [category of groups] 79
$\underline{\underline{G}}(q)$ [binary modes satisfying $x^{q-1}y = y$, $x^u y = y^w x$] 342
H [entropy function] 519
$H(C, x)$ [support function] 376
$_H\underline{\underline{K}}$ [class of homomorphic images] 49
$\mathbb{I}$ [quasivariety of naturally quasi-ordered modes] 386
I° [open unit interval $(0, 1)$] 17
I^∞ [extension of unit interval I] 406
$\text{Id}(\underline{\underline{K}})$ [set of identities holding in $\underline{\underline{K}}$] 46
$\text{Id}_{lr}(\underline{\underline{K}})$ [set of left regular identities holding in $\underline{\underline{K}}$] 191
$\text{Id}_r(\underline{\underline{K}})$ [set of regular identities holding in $\underline{\underline{K}}$] 191
$\text{Id}_{rr}(\underline{\underline{K}})$ [set of right regular identities holding in $\underline{\underline{K}}$] 191
$_I\underline{\underline{K}}$ [class of isomorphic images] 50
Imp [set of defining implications] 171
inf B[least upper bound of subset B of poset] 15
Inn Q [inner automorphism group of quandle] 498
K [compact subset function] 376
$K(\ )$ [set of compact elements of (complete lattice)] 115
$\underline{\underline{Kei}}$ [variety of kei modes] 257
Ker [group kernel] 57

ker f [kernel relation of function f] 3
L [Lallement sum functor] 514
$L(a), L_a$ [left multiplications] 71, 183
$\mathcal{L}(\ )$ [lattice of subvarieties] 149
$\mathcal{L}_{ir}(\ )$ [lattice of irregular subvarieties] 480
$\mathcal{L}_q(\ )$ [lattice of subquasivarieties] 205
$\mathcal{L}_r(\ )$ [lattice of regular subvarieties] 480
Λ [empty address] 441
$LCD(m,n)$ [least common multiple] 20
$\underline{LCl}$ [quasivariety of left cancellative binary modes] 493
$\varinjlim$ [(directed) colimit] 98
$\varprojlim$ [limit] 101
$\underline{Ln}$ [variety of left normal bands] 200
$\underline{Lz}$ [variety of left zero bands] 152
$\underline{Mod_S}$ [category of S-modules] 79
$\underline{Mon}$ [category of monoids] 79
Mor$\mathcal{C}$ [class of morphisms of category $\mathcal{C}$] 78
$\underline{M\tau}$ [variety of modes of type τ] 367
N [finitely generated non-empty sink function] 400
$N(A_1, \ldots)$ [class of barycentric algebras excluding subalgebras $A_1, \ldots$] 407
nat ρ [natural projection onto quotient by equivalence relation ρ] 4
$\underline{NB}$ [variety of normal bands] 200
$\mathbb{N}_c$ [chain of natural numbers with usual order] 478
$\mathbb{N}_d$ [lattice of natural numbers under divisibility] 478
$N_k(w,x)$ [indicator for x in w at address with k ones] 453
(Ω) [category of all Ω-algebras] 79
$\Omega(A,B)$ [set of all homomorphisms between Ω-algebras] 21
Ob$\mathcal{C}$ [class of objects of category $\mathcal{C}$] 78
^{OP} [opposite groupoid, variety of opposites] 259
P [Mal'cev operation] 61, 312
P [finitely generated non-empty subalgebra function] 242, 554
$\mathfrak{P}$ [ordered countably infinite set of variables] 38
$\mathcal{P}(A)$ [power set of A] 1
$P\underline{K}$ [class of (algebras isomorphic to) products] 50
$P_r\underline{K}$ [class of reduced products] 167
$P_S\underline{K}$ [class of (algebras isomorphic to) subdirect products] 153
$P_u\underline{K}$ [class of (algebras isomorphic to) ultraproducts] 167
$\prod B$ [greatest lower bound of subset B of poset] 15
q [special quasi-identity] 203
$\underline{Q}$ [category of quasigroups] 79
Q_a [subclass of quasi-affines] 372

SYMBOL INDEX

q^β [canonical distribution given by potential β] 515
$Q(\gamma_1,\dots)$ [quasivariety of barycentric algebras satisfying $\gamma_1,\dots$] 407
$\mathrm{Qid}(\underline{K})$ [set of quasi-identities holding in $\underline{K}$] 171
$Q(\underline{K})$ [quasi-equational class generated by $\underline{K}$] 173
$\underline{QM}$ [variety of quasigroup modes] 353
$\overline{Q(\Sigma)}$ [class axiomatized by quasi-identities] 70
R [prevariety generated] 51
$\mathbb{R}^\infty$ [extended reals] 297
$R(a), R_a$ [right multiplications] 70, 183
$\underline{RCl}$ [quasivariety of right cancellative binary modes] 493
$\underline{Re}$ [variety of rectangular bands] 152
$\underline{Ring}$ [category of unital rings] 79
$\underline{R}_{m,n}$ [variety of reductive modes] 469
(R_n) [n-reductive identity] 282
$\underline{R}_n$ [variety of n-reductive modes] 283
$\underline{Rn}$ [variety of right normal bands] 200
$R_{n,m}$ [ring $\mathbb{Z}[X]/\langle X^n - 1, (1 - X)^m - 1\rangle$] 340
$\underline{R}_{n,m}$ [affine spaces over $\mathbb{Z}[X]/\langle X^n - 1, (1 - X)^m - 1\rangle$] 340
$R^*_{n,m}$ [ring $\mathbb{Z}[X]/\langle X^n(1 - X)^m - 1\rangle$] 482
$\underline{R}^*_{n,m}$ [affine spaces over $\mathbb{Z}[X]/\langle X^n(1 - X)^m - 1\rangle$] 482
$R(n,t)$ [ring $\mathbb{Z}[X]/\langle X^n - 1, X^t + X - 1\rangle$] 340
$R(V)$ [ring of affinization of $\underline{V}$] 367
$\underline{Rz}$ [variety of right zero bands] 152
S [non-empty subalgebra function] 242
$\underline{S}$ [variety of symmetric binary modes] 260
$\sum B$ [least upper bound of subset B of poset] 15
$Sb(A, \Omega)$ [subalgebra poset] 17
s_e [class of separable algebras] 384
$\underline{Set}(A, B)$ [set of functions from A to B] 2
s_f [class of (algebras isomorphic to) functorial sums] 190
$\underline{Sgp}$ [category of semigroups] 79
$Sk(A, \Omega)$ [sink poset] 17
$s\underline{K}$ [class of (algebras isomorphic to) subalgebras] 50
(s_{2k+1}) [identity $(\dots(yx\cdot y)x\dots)y = x$] 346
$\underline{S}_{2k+1}$ [symmetric binary modes satisfying (s_{2k+1})] 347
$\underline{Sl}$ [variety of semilattices] 199
s_Ω [class of (algebras isomorphic to) Ω-subreducts] 372
sp [span] 209
s_p [class of (algebras isomorphic to) Płonka sums] 190
sup B [least upper bound of subset B of poset] 15

$\underline{\mathbb{T}}$ [category of barycentric structures] 514
τ' [derived type] 39
$\bar{\tau}$ [closure of type] 39
$\underline{\tau}(A, B)$ [set of all homomorphisms between τ-algebras] 21
$\underline{TF}$ [class of torion-free groups] 138
T_n [relatively subdirectly irreducible barycentric algebra] 403
$\underline{\text{Tr}}$ [variety of trivial bands] 199
Trans Q [transvection group of quandle] 498
$\mathcal{V}$ [variety generated] 51
$\underline{V}^n$ [iterated Mal'cev power of $\underline{V}$] 285
$\underline{V}_{n,m}$ [variety of binary modes satisfying $x^n y = y = y x^m$] 340
$\underline{V}^*_{n,m}$ [variety of binary modes satisfying $x^n y x^m = y$] 482
$\underline{V}_q$ [variety of binary modes generated by $(\mathbb{Z}_{q^\infty}, \underline{q})$] 487
$\underline{\widetilde{V}}_q$ [subquasivariety of $\underline{\widetilde{V}}$ defined by q] 205
$\underline{V}_{SI}$ [class of subdirectly irreducible $\underline{V}$-algebras] 205
W_d [set of standard binary mode words of depth d] 448
$Wl(A, \Omega)$ [wall poset] 17
Z [non-empty sink function] 400
$\zeta(\ )$ [center congruence] 324
$Z(\beta)$ [partition function] 517
$\mathbb{Z}_{p^\infty}$ [quasi-cyclic group] 112
$\mathbb{Z}[p, q, r]$ [ring] 353

INDEX

Abelian 304
abelian 236, 321
abelian group 13
abelian group sum 539
abelian quandle 497
absolutely free algebra 36
absorbing zero 59
absorption law 14
abstract class 50
action (binary) 304
action (of Boolean ring) 437
action (of monoid) 29
act strongly transitively 60
address 441
adequate envelope 220
adjoint 90
adjunction 91
affine geometry 252
affine map 64
affine module 65
affine space 60, 65, 247, 329
affine subspace 63
affine variety 439
affinization 367
Agassiz sum 188
AG-sum 539
algebra 11
algebra, barycentric 67
algebra, Boolean 16
algebra, central 315
algebra, diagonal 246
algebra, diagonally normal 319
algebra, entropic 235
algebra, free 126

algebraically open 132
algebraic closure operator 114
algebraic closure system 115
algebraic interval 301
algebraic lattice 115
algebraic quasi-order 182
algebra, idempotent 17, 238
algebra, indexing 188
algebra, injective 74
algebra, Mal'cev 26
alphabet 35
alternation 236
analytic variety 198
annihilator 57
antisymmetric 2
antitone 6
argument 38
argument map 38
arity (of operator) 10
arity (of operation) 10
arrow 78
associative 12, 246
associativity, partial 257
atom 150
augment 18
automorphism 21
axiomatized (by identities) 45
axioms, set of 45

balanced identity 48
Banaschewski-Herrlich Theorem 137
band 17
band, (left/right) normal 53

band, rectangular 32
barycentric algebra 67
barycentric structure 514
basic operation 11
basic reduct 18
basis (for identities) 45
bi-commutative 236
bijective 2
binar 11
binary action 304
binary identity 192
binary operation 10
binary relation 1
Birkhoff's Variety Theorem 145
bisymmetric 236
block 253
Boolean algebra 16
Boolean space 559
boundary 268
boundary operator 268
bounded lattice 15
branch (of parsing tree) 441
breaks 193

cancellation 71, 293
cancellative 293
cancellative (mode) 385
canonical distribution 517
canonical embedding 227, 389
canonical Lallement sum 216
canonically separable 524
canonical representation 216
cardinal (number) 7
cardinality (of multiset) 7
cardinality (of set) 7
cartesian product 1, 3
cartesian power 3
category 78
category of small categories 83
center of gravity 62, 521
central algebra 315

centralize 313
central variety 321
centreing congruence 313
chain 5
character of semilattice 529
circle composition 299
class, abstract 50
class defined (by identities) 45
class defined (by quasi-identities) 70
class, equational 45
class, implicational 72
class, irregular 48
class, quasi-equational 70
class, regular 48
clone 40, 43
clone operation 40
closed operator domain 39
closed set of operations 40, 43
closed subset 113
closed under directed limits 158
closed under replicas 135
closure (of operator domain) 39
closure (of operation set) 40
closure (of type) 39
closure operator 113
closure operator, algebraic 114
closure system 112
closure system, algebraic 115
closure, transitive 2
closure under (co-)products 96
cobordic 271
cocyclic 271
codomain 78
coequalizer 101
cofinite subset 178
cogenerator 372
coherent Lallement sum 215, 233
combinatorial geometry 252
comma category 91
commutative 236
commutative binary mode 262
commutative diagram 3

commutative (dual monoid) 299
commutative group 13
commutative (multiplication) 12
commutative (operations) 33
commutative (quasigroup) 253
commutative ring 13
commutative semiring 59
commute (operations) 33
compact element 115
compatibility property 22
complemented lattice 15
complete lattice 15
completely meet irreducible 112, 119
Completeness Theorem 149
complete sublattice 28
complex algebra 242
complex product 48, 241
complex set algebra 241
component 87
component homomorphism 514
composite 88
composition, circle 299
composition (of functions) 3
composition (of functors) 83
composition (of morphisms) 78
composition (of operations) 43
concave 512
conclusion 70
congruence 22, 107
congruence class 22
congruence generated 23
congruence modular 71
congruence permutable 25
congruence poset 23
congruence, principal 23
constant 10
continuous 277, 278
contravariant functor 85
converse 1
Convexity Lemma 554
convexity theory 309
convex, logarithmically 513

convex module 309
convexor 308
convex prestructure 308
coproduct 95, 95
counit (of adjunction) 91
covariant functor 83
cover 5
crossbar 511
cycle 268
cyclic groupoid 307

decomposable (algebra) 31
decomposition functor 515
decomposition (operation) 34, 194
Decomposition Theorem 132
decomposition word 150
deductive 549
defined class 45
De Morgan law 299
dense functor 88
depth (of word) 441
derivative 277
derived operation 40
derived operator 39
derived type 39
diagonal algebra 246
diagonally normal algebra 319
diagonally normal variety 321
diagonal mode 246
diagonal operation 54
diagonal relation 4, 23
difference 1
differentiable 277
differential 268
differential group 268, 281
differential groupoid 259
differential mode 259, 282
differential operator 268, 281
directed 5
directed colimit 97, 100
directed limit 101

directed (quasi-order) 6
directed system 96
directly indecomposable 32
directoid 186
direct power 30
direct product 20, 30
disjoint union 1
dissemilattice 202
distribute 202, 553
distributive (dissemilattice) 202
distributive (lattice) 15
distributive (ring) 13
division groupoid 184
division (in quasigroup) 55
domain (of maorphism) 78
domain, integral 13
down-set 6
dual category 80
dual equivalence 557
dual isomorphism 6
duality 557
dualiz(abl)e 559
dual monoid 299
dual numbers 268
dual representation 559
dyadic rational numbers 263
dyadic unit interval 263

edge (of parsing tree) 441
embedding 2, 21
emvedding, subdirect 106
Embedding Theorem 415, 433
empty word 36
endomorphism 21
enrichment 18
entropic 305
entropic algebra 235
entropic identity 53
entropy 518
entropy function 520
entropy (physical) 530

envelope 215
epigraph 38
epimorphism 80
equalizer 102
equation 45
equational basis 45
equational implication 72
equationally complete 150
equational(ly definable) class 45
equational theory 146
equihoop 354
equilibrium 530
equipollent 7
equivalence class 4
equivalence (of categories) 88
equivalence relation 4
equivalent categories 88
equivalent (clonally) 42, 86
equivalent (polynomially) 42
essential extension 463
evaluation 558
expected value 518
Extended Structure Theorem 516
extension 16

factor congruence 31
faithful binary action 304
faithful functor 88
faithful (module) 57
family (of sets) 3
field 13
filter 165
filter generated 165
finitary (type) 10
finite (cardinal) 7
finite equational basis 46
finite part 268
finitely based 46
finitely generated (algebra) 18
finitely generated (prevariety) 51
finitely generated (variety) 51

finitely presented 154
finitely separable:
— algebra in variety 374
— barycentric algebra 510
First Isomorphism Theorem 4, 24
fixed point 29
flag 5
flat 252
flexibility 257
forgetful functor 84
free, absolutely 36
free (algebra) 126
free generator 129
free (monoid) 36
free product 134
free (semigroup) 36
full functor 88
full (quasi-order) 6
full (subcategory) 79
full (word) 441
fully invariant congruence 147
function 2
function, bijective 2
function, identity 2
function, inclusion 2
function, injective 2
function, one-to-one 2
function, onto 2
function, surjective 2
functor 83
functorial sum 187

generated (congruence) 23
generated, finitely 18
generated, freely 129
generated (prevariety) 51
generated (subalgebra) 18
generated (variety) 51
generating function 199
generator 18
geodesic (of parsing tree) 440

graph 511
greatest element 5, 15
group 12
group, abelian 13
group, commutative 13
groupoid 11

Hasse diagram 5
head 78
Helmholtz free energy 530
heterotypical 48
hold 45, 70
homologous 268
homology class 268
homology group 268
homology set 272
homomorphic image 26
homomorphism 21
homotypical 48
hyperidentity 239
hypervariety 239
hypograph 512

idempotent (algebra) 17, 238
idenpotent (element) 17
idempotent law 14
identity 45
identity, balanced 48
identity, binary 192
identity element 12
identity function 2
identity (in dual monoid) 299
identity (in ring) 13
identity (in semiring) 58
identity, irregular 47
identity, left regular 52
identity, linear 48
identity (morphism) 78
identity (of mode) 415
identity, regular 47

identity, right regular 52
identity, symmetric 260
identity transformation 88
image, homomorphic 26
implicational class 72
implication, equational 72
impoverishment 18
improper congruence 23
inclusion 1
inclusion function 2
inclusion functor 83
indecomposable 32
independent join 151
independent varieties 150
indexing algebra 188
inductive 5
infimum 15
infinite (cardinal) 7
infinitesimal 268
infix notation 10
initial object 81
injective algebra 74
injective function 2
injective hull 463
inner automorphism group 498
insertion 95
integral domain 13
interpretation 87
intersection 1
inverse function 2
inverse morphism 81
inverse functor 85
invertible 81
involutary 299
irregular class 48
irregular identity 47
isomorphic (algebras) 21
isomorphic (categories) 85
isomorphic, weakly 153
isomorphism 2, 21, 81
isomorphism, dual 6
isomorphism (of categories) 85

isomorphism (of ordered sets) 6
iterated Mal'cev power 285

join semilattice 14

kei mode 256
kernel 3, 22
Klein-Hilbert part relation 436
Kronecker product of varieties 366
Kuratowski-Zorn Lemma 5

Lallement sum 215
Lallement sum functor 514
large submodule 463
Latin square 56
lattice 14
lattice, bounded 15
lattice, complemented 15
lattice, complete 15
law 45
leaf (of parsing tree) 440
least element 5, 15
left adjoint 90
left branch 441
left cancellation 71
left cancellative 492
left division groupoid 183
left division (in quasigroup) 55
left multiplication 71
left normal 269
left normal band 53
left regular identity 52
left trivial 12
left zero 12
length (of word) 36, 37
limit ordinal 8
Lindenbaum-Tarski duality 559
linear identity 48
linearization 435

INDEX

linear (operation) 40
linear (operator) 39
linear (ordering) 5
LIR-groupoid 308
locally finite 144
locally strongly solvable 541
logarithmically convex 513
log odds function 518
lower bound 5, 15
lower set 6
Lz-Lz sum 280

macrostate 517
M-action (over monoid M) 29
Magari's Theorem 143
magma 11
Mal'cev algebra 26
Mal'cev identities 312
Mal'cev operation 61, 312
Mal'cev power, 285
Mal'cev product 175
Mal'cev's Quasivariety Theorem 171
Mal'cev variety 311
maximal 5
medial 236, 305
meet semilattice 14
membership 1
minimal 5
minimal cogenerator (algebra) 373
minimal cogenerator (module) 372
modal 553
mode 239
modular law 71
module 56
monoid 12
monoid action 29
monolith 109
monomorphism 81
monotone 6
Monotonicity Lemma 553
morphism 78

morphism part 83
M-set (over monoid M) 29
multiplication (left) 71
multiplication (quasigroup) 54
multiplication (right) 70
multiset 7
mutually orthogonal 254

natural duality 558
natural full duality 559
natural projection 4, 24
natural isomorphism 87
naturally quasi-ordered 184
natural transformation 87
neutral element 12
node (of parsing tree) 440
nofix notation 10
non-empty word 35
normal band 53
normal form 47
normal identity 48
nullary operation 10

object 78
object part 83
one-to-one 2
onto 2
open, algebraically 132
operation 9
operation, basic 11
operation, clone 40
operation, derived 40
operation, diagonal 54
operation, polynomial 42
operation symbol 10
operator 10
operator, basic 39
operator, derived 39
operator domain 10
operator, projection 40

opposite category 80
opposite groupoid 259
opposite variety 259
orbit 29, 271
order (relation) 5
ordered set 5
order-preserving 6
order-reversing 6
order type 7
ordinal 7
Ω-ring 544
Ω-semigroup 19
Ω-word 37

pair of factor congruences 31
parallelogram operation 61
parallel pair 101
parsing tree 440
partial associativity 257
partial order 5
partition 4
partition function 517
partition operation 194
permutable 25
physical entropy 530
plain 23
Płonka's Theorem 193
Płonka sum 188
plural type 11
point 78
Polish notation 10
polynomial form of word 453
polynomial operation 42
polynomial ring 13
polynomial word 453
polytope functor 554
Pontryagin duality 529
poset 5
postfix notation 10
potential 517
power, direct 30

power (of a set) 7
power set 1
predual 510
pre-envelope 515
prefix notation 10
premise 70
pre-order 6
presentation 154
preserve (operation) 21
preserve (subalgebra) 215
prevariety 51
prevariety generated 51
prime sink 400
primitive class 45
primitive element 238
principal congruence 23
principal filter 165
principal wall 131
product, cartesian 1, 3
product, complex 48, 241
product, direct 20, 30
product, free 134
product (in category) 94, 95
product, Kronecker (of varieties) 366
product, reduced 167
product, relational 2
product, subdirect 106
product, tensor (of algebras) 238
product, tensor (of varieties) 366
projection 21, 30, 94, 188, 214
projection function 3
projection (natural) 4, 24
projection operator 40
projection subalgebra 246
projective space 551
proper filter 165
proper functorial sum 188
proper sum 216

quandle 497
quasi-affine 372

INDEX

quasi-cyclic group **112**
quasi-equational(ly **definable**) 70
quasigroup 54
quasigroup mode 252
quasi-identity 69
quasi-order 6
Quasi-Płonka's Theorem 205
quasiregularization 203
quasivariety 158
Quasivariety Theorem 171
quaternary operation 10
quotient 4, 22

rational variety 199
rectangular (band) 32
reduced product 167
reduct, basic 18
reduct 40
reduction 269
reductive 283
reductive mode 469
reductive, right 467
reductive variety 439
reflexion 256
reflexion binary mode 256
reflexive 2
regular class 48
regular identity 47
regularization 48
regularized Pontryagin duality 529
regularly conjugate 247
regular quasi-identity 209
relation 1
relation, antisymmetric 2
relation, binary 1
relation, diagonal 4
relation, equivalence 4
relation monoid 25
relation, n-ary 1
relation, reflexive 2
relation, symmetric 2

relation, transitive 2
relation, universal 4
relational product 2
relatively free algebra 126
relatively subdirectly irreducible 108
relative Mal'cev product 175
relative monolith 109
relative subdirect representation 119
replica 122, 125
replica functor 123
respect reflexivity 313
respect symmetry 313
respect transitivity 313
restriction functor 84
reverse Polish notation 10
Riemann zeta function 528
right adjoint 90
right branch 441
right cancellation 71
right cancellative 492
right division groupoid 183
right division (in quasigroup) 55
right M-set (over monoid M) 29
right multiplication 70
right multiplication group 498
right normal band 53
right quasigroup 67
right reductive 467
right regular identity 52
right trivial 12
right zero 12
ring 13
ring, commutative 13
ring, polynomial 13
ring, unital 13
ring with identity 13
root (of parsing tree) 440
$\underline{r}$-sum 286

satisfy (identity) 45
satisfy (quasi-identity) 70

scaling factor 510
schizophrenic 558
Second Isomorphism Theorem 24
self-replicating 550
semiaffine space 542
semiconvex set 308
semigroup 11
semilattice 14
semilattice, join 14
semilattice Lallement sum 514
semilattice, meet 14
semilattice mode 541
semilattice sum 216
semilattice sum of rings 232
semimodule 59
semiring 58
semiring, commutative 59
separable (barycentric algebra) 510
separable (in variety) 381
set of axioms 45
signature (of algebra) 11
signature (of type) 10
simple 23
simplex 66
sink 16
sink poset 17
skeleton 88
small category 80
span 209
standard binary mode word 448
standard form 47
standard form (in binary mode) 450
Steiner quasigroup 253
Steiner triple system 253
Stone space 559
strictly simple 23
strict sum 216
strongly irregular variety 184
strongly transitive action 60
strong semilattice 230
structure, barycentric 514
Structure Theorem 399

Structure Theorem, Extended 516
subalgebra 16
subalgebra generated 18
subalgebra poset 17
subcategory 79
subdirectly irreducible 108
subdirect product 106
Subdirect Representation 110
sublattice, complete 28
submode functor 554
subprevariety 174
subreduct 40
subspace, affine 63
substantial virtual species 535
subtype 386
subvariety 141
subword 440
sum 1
sum, abelian group 539
sum, AG- 539
sum, Agassiz 188
sum fibre 214
sum, functorial 187
sum homomorphism 214
sum, Lallement 215
sum, Lz-Lz 280
sum of algebras 214
sum, Płonka 188
sum, proper 216
sum, semilattice 216
sum, semilattice Lallement 514
sum, strict 216
Sum-Superiority Lemma 554
Sum Theorem 215
superalgebra 16
supremum 15
sum commutative 236
surjective 2
symmetric 2
symmetric binary mode 260
symmetric identity 260
synonymous 45

INDEX

tail 78
tensor product (of algebras) 238
tensor product (of varieties) 366
term 39
term algebra 39
term condition 321
term equivalent 42
terminal object 81
term operation 40
ternary operation 10
Third Isomorphism Theorem 25
totality 555
totally symmetric 253
total (ordering) 5
Transfinite Induction 8
transitive 2
transitive closure 2
translation (of potential) 510
translation (of space) 60
translation (of mode) 415
transvection group 498
trivial algebra 11
trivial congruence 23
trivial homology 272
trivial variety 51, 51
true 45, 70
τ-word 37
type derived 39
type (finitary) 10
type (of algebra) 11
type, plural 11

ultrafilter 165
ultraproduct 167
unary operation 10
underlying set functor 84
union 1
union, disjoint 1
unit (of adjunction) 91
universal congruence 23
universality property:

— (of coequalizer) 102
— (of coproduct) 95
— (of directed colimit) 98
— (of directed limit) 101
— (of free algebras) 126
— (of free monoid) 36
— (of product) 94
— (of replication) 123
— (of word algebra) 37
universal relation 4
upper bound 5, 15
upper set 6
up-set 6

variable 35
variable-uniform 48
variety 51
variety generated 51
variety, trivial 51
vertex (of parsing tree) 440
vertical 511
virtual species 535

wall 16
wall poset 17
weakly isomorphic 153
weight 62
well ordered set 5
word 35, 37
word algebra 36
word algebra functor 84

zero, absorbing 59
zero object 81
zero (of algebra) 16
zero (of dual monoid) 299
zero (of semiring) 59
Zustandssumme 517